SOLUTIONS TO EXERCISES

Organic Chemistry

SOLUTIONS TO EXERCISES
Organic Chemistry

Thomas N. Sorrell
UNIVERSITY OF NORTH CAROLINA AT CHAPEL HILL

University Science Books
Sausalito, California

University Science Books
55D Gate Five Road
Sausalito, CA 94965

Fax: (415) 332-5393
www.uscibooks.com

Production manager: *Susanna Tadlock*
Design consultant: *Robert Ishi*
Printer & binder: *Edwards Brothers, Inc.*

This book is printed on acid-free paper.

Library of Congress Catalog Card Number: 98-061679

ISBN 1-891389-03-3

Printed in the United States of America
10 9 8 7 6 5 4 3 2 1

CONTENTS

Preface   vii

Solutions to Exercises for:

CHAPTER 2. Nomenclature and Functional Groups   1

3. Bonding in Organic Molecules   35

4. Conformations of Organic Molecules   59

5. Stereochemistry of Organic Molecules   87

6. Chemical Reactions and Mechanisms   119

7. Substitution Reactions of Alkyl Halides, Alcohols, and Related Compounds   141

8. Elimination Reactions of Alkyl Halides, Alcohols, and Related Compounds   169

9. Electrophilic Addition Reactions of Alkenes and Alkynes   195

10. Concerted Addition Reactions of Alkenes and Alkynes   221

11. Free Radicals: Substitution and Addition Reactions   245

12. Electrophilic Aromatic Substitution Reactions of Benzene and its Derivatives   269

13. Organometallic Chemistry, the Formation of C-C Bonds, and Chemical Synthesis   301

14. Spectroscopic Identification of Functional Groups   331

15. Proton and Carbon Nuclear Magnetic Resonance Spectroscopy   345

16. Asymmetric Reactions and Synthesis   365

17. Nucleophilic Addition Reactions of Aldehydes and Ketones   391

18. Addition–Elimination Reactions of Aldehydes and Ketones   429

19. Addition–Elimination Reactions of Carboxylic Acids and Derivatives   463

20. Acid–Base Chemistry of Carbonyl Compounds   497

21. Nucleophilic Addition Reactions of Enolate Ions to Carbonyl Groups   523

22. Nucleophilic Addition to α, β-Unsaturated Carbonyl Compounds   553

23. The Chemistry of Amines and Other Nitrogen-Containing Compounds   587

24. The Chemistry of Heterocyclic Compounds   617

25. Amino Acids, Peptides, and Proteins   647

26. Nucleic Acids and Molecular Recognition   669

APPENDIX A. Naming Organic Compounds   695

SOLUTIONS TO EXERCISES

Organic Chemistry

PREFACE

You have to work problems to learn organic chemistry

The exercises interspersed throughout the textbook, *Organic Chemistry*, are intended to reinforce your understanding of the concepts that have just been introduced. Some problems at the ends of chapters have the same format as those within the body of the text; this will test your mastery of the material. However, some chapter-ending exercises require you to consider more than one concept before you formulate an answer, and the point is to encourage you to think more broadly about what you have learned and how it relates to chemistry that you have not seen previously.

This book contains solutions to exercises that appear in the textbook. Instead of simply providing an answer for each exercise, I have included a short discussion of each solution, giving an overview of important material or ideas and presenting a way to approach solving the problem. I have also supplied references to relevant pages in the text where a specific discussion is located, especially if that topic is not covered in the chapter in which the exercise is placed.

Nomenclature exercises appear throughout the text, starting with Chapter 2. In solutions to those exercises, less detail is provided in the later chapters than in the earlier ones. Exercises having to do with mechanisms are also common in the textbook, starting with Chapter 6. Because a knowledge of mechanisms is vital for understanding biological aspects of organic chemistry, I have provided detailed answers to these types of problems throughout the manual. Mechanisms that are very common, such as ester hydrolysis, are given less attention in the answers that appear in later chapters, but references are given to earlier problems that show every step of the solution. Synthesis problems appear starting with Chapter 13, and normally only a single solution is provided in this book. Realize, however, that a synthesis problem may have many solutions, so other routes that you propose may be correct, even if the answer I give does not match yours.

Many students at the University of North Carolina who have studied organic chemistry with me during the past several years have worked problems from my textbook. Two students, however, deserve specific mention because they helped me to compile and check these solutions against the exercises. Julius Lucks, a freshman who was enrolled in my organic chemistry courses during 1997–98, worked nearly every one of the 1000+ exercises. He also proofread the text and found many errors during the final stages of production. Sarah Greer, who was enrolled in the same classes as Julius, proofread the *Solutions Manual* in detail. I thank both of them for their hard work that ensured the success of this project.

The responsibility for any remaining errors rests with me. I have tried to minimize mistakes in this manual, but inevitably, some remain. Please feel free to bring these errors to my attention so that I can correct them in future editions. I also invite any comments that you have about the exercises or their solutions or about the topics that appear in the text. Enjoy your study of organic chemistry!

—*Tom Sorrell*

Chapel Hill, 1999
sorrell@unc.edu

NOMENCLATURE AND FUNCTIONAL GROUPS

2.1. Functional groups in a molecule are easily recognized because they consist of heteroatoms (those atoms other than C or H) or multiple bonds. Identities of the common functional groups are summarized in Table 2.1 (text page 33). Functional groups normally include the carbon atom to which the heteroatom is attached.

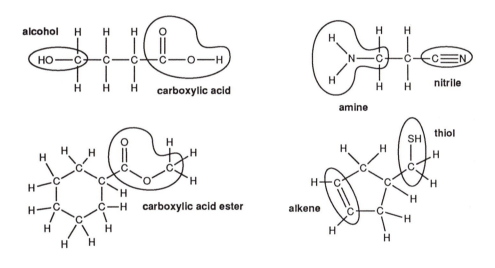

2.2. To generate condensed formulas, express hydrocarbon units by use of the designations CH, CH_2, or CH_3. Functional groups are indicated by the formulas given in Table 2.1 (text page 33).

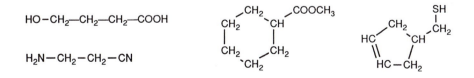

2.3. To generate a full structure from a condensed one, draw a chain of atoms that includes any carbon, nitrogen, oxygen, and sulfur atoms, recognizing the structures of the common functional groups (for example –COOH has a carbon atom double-bonded to one oxygen atom and single-bonded to the other). Ignore hydrogen and halogen atoms in this first step. Then, by including the hydrogen and halogen atoms and any multiple bonds, give four bonds to each carbon atom, three bonds to each nitrogen atom, and two bonds to each oxygen and sulfur atom. For cyclic compounds, draw the ring first, then give the correct number of bonds to each non-hydrogen and non-halogen atom.

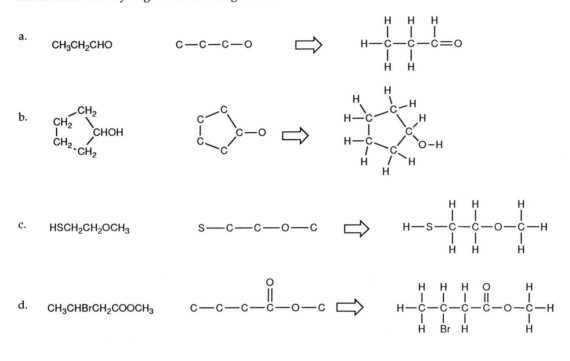

2.4. Follow the same procedure given in the solution to exercise 2.3.

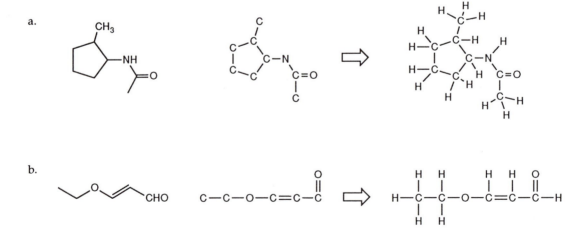

2.5. To convert a full structure to a condensed line structure, represent the carbon skeleton first with use of a polygon for a cyclic molecule or a zigzag line for an acyclic molecule. Then include any double or triple bond as well as heteroatoms attached to any carbon atom. If a heteroatom is attached via a multiple bond to the carbon atom at the end of a chain (for example in a nitrile, aldehyde, carboxylic acid, or ester group), include the carbon atom in the condensed structure, but make certain that the total number of carbon atoms does not change.

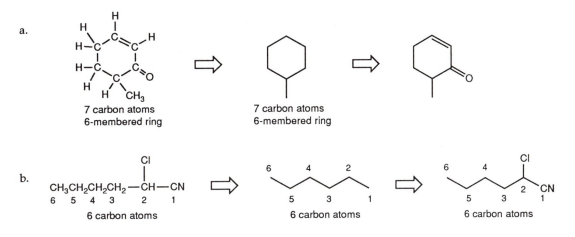

2.6. For the compounds in this exercise, it is only necessary to convert the name of the compound root to a representation that is a chain or ring, then include any double or triple bonds between carbon atoms as specified by the suffix. Use the numeral, if given, to define the first carbon atom in the chain or ring at which the unsaturation starts. If no numeral is given, unsaturation starts at C1 of the ring or chain.

a.	3-hexene	hex ene	6 carbon atoms one double bond that starts at C3	
b.	4-octyne	oct yne	8 carbon atoms one triple bond that starts at C4	
c.	1,3-pentadiene	penta diene	5 carbon atoms two double bonds that start at C1 and C3	
d.	cyclobutene	cyclobut ene	four-membered ring of carbon atoms one double bond that starts at C1, by convention	

2.7. The principal functional group in a compound is denoted by the suffix of the compound's name. The identities and structures of common functional groups are summarized in Table 2.1 (text page 33).

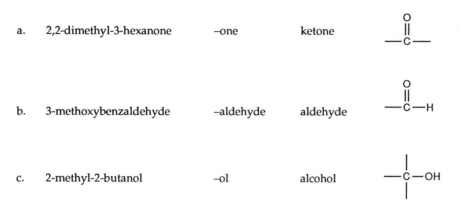

a.   2,2-dimethyl-3-hexanone   –one   ketone

b.   3-methoxybenzaldehyde   –aldehyde   aldehyde

c.   2-methyl-2-butanol   –ol   alcohol

2.8. To simplify a given structure, identify and draw the structure of the principal functional group in each molecule, then attach "R" to the functional group by drawing the bond.

a.   R—C(=O)—H   b.   R—C(=O)—NH₂   c.   R—C≡C—H

2.9. The prefix associated with the substituents must match the quantity of numerals that precede the name. The only exceptions are those in which there is no ambiguity.

a.   2,3,4-Hydroxyhexanal should be 2,3,4-<u>tri</u>hydroxyhexanal because there are three numerals preceding the prefix.

b.   2,2,4,4-Tetrachloropentane is correct because there are four numerals and the prefix tetra– precedes the substituent name, chloro-.

c.   Triiodomethane is correct because there is only one carbon atom to which the three iodine atoms can be attached, so the name is unambiguous.

2.10.   Break each name into its constituent parts, then combine the pieces to form the structure.

a.   4-fluorobutanal

but	4 carbon atoms
an	no double or triple bonds
al	aldehyde functional group at C1, by convention
4-fluoro	fluorine atom at C4

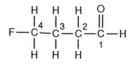

2.10.   (continued)

b.   3-mercapto-2-pentanol

pent	5 carbon atoms
an	no double or triple bonds
ol	OH group at C2
3-mercapto	SH group at C3

$$H-\overset{\overset{\displaystyle H}{|}}{\underset{\underset{\displaystyle H}{|}}{C}}_1 - \overset{\overset{\displaystyle OH}{|}}{\underset{\underset{\displaystyle H}{|}}{C}}_2 - \overset{\overset{\displaystyle SH}{|}}{\underset{\underset{\displaystyle H}{|}}{C}}_3 - \overset{\overset{\displaystyle H}{|}}{\underset{\underset{\displaystyle H}{|}}{C}}_4 - \overset{\overset{\displaystyle H}{|}}{\underset{\underset{\displaystyle H}{|}}{C}}_5 - H$$

c.   trichloroethanenitrile

eth	2 carbon atoms
ane	no double or triple bonds
nitrile	CN group; carbon atom defines C1
trichloro	three Cl atoms must be on C2 because the nitrile carbon atom already has four bonds

$$Cl-\overset{\overset{\displaystyle Cl}{|}}{\underset{\underset{\displaystyle Cl}{|}}{C}}-C\equiv N$$

2.11.   Break each name into its constituent parts, then combine the pieces to form the structure.

a.   3-oxopentanoic acid

pent	5 carbon atoms
an	no double or triple bonds
oic acid	COOH group; carbon atom defines C1
3-oxo	oxygen atom at C3 double-bonded to the carbon atom

$$HO-\overset{\overset{\displaystyle O}{\|}}{\underset{}{C}}_1 - \overset{\overset{\displaystyle H}{|}}{\underset{\underset{\displaystyle H}{|}}{C}}_2 - \overset{\overset{\displaystyle O}{\|}}{\underset{}{C}}_3 - \overset{\overset{\displaystyle H}{|}}{\underset{\underset{\displaystyle H}{|}}{C}}_4 - \overset{\overset{\displaystyle H}{|}}{\underset{\underset{\displaystyle H}{|}}{C}}_5 - H$$

b.   2-nitropentanal

pent	5 carbon atoms
an	no double or triple bonds
al	aldehyde functional group; carbon atom defines C1
2-nitro	nitro group at C2

$$H-\overset{\overset{\displaystyle O}{\|}}{\underset{}{C}}_1 - \overset{\overset{\displaystyle NO_2}{|}}{\underset{\underset{\displaystyle H}{|}}{C}}_2 - \overset{\overset{\displaystyle H}{|}}{\underset{\underset{\displaystyle H}{|}}{C}}_3 - \overset{\overset{\displaystyle H}{|}}{\underset{\underset{\displaystyle H}{|}}{C}}_4 - \overset{\overset{\displaystyle H}{|}}{\underset{\underset{\displaystyle H}{|}}{C}}_5 - H$$

c.   3,3-dicyanohexane

hex	6 carbon atoms
ane	no double or triple bonds
3,3 dicyano	two cyano groups at C3

$$H-\overset{\overset{\displaystyle H}{|}}{\underset{\underset{\displaystyle H}{|}}{C}}_1 - \overset{\overset{\displaystyle H}{|}}{\underset{\underset{\displaystyle H}{|}}{C}}_2 - \overset{\overset{\displaystyle CN}{|}}{\underset{\underset{\displaystyle CN}{|}}{C}}_3 - \overset{\overset{\displaystyle H}{|}}{\underset{\underset{\displaystyle H}{|}}{C}}_4 - \overset{\overset{\displaystyle H}{|}}{\underset{\underset{\displaystyle H}{|}}{C}}_5 - \overset{\overset{\displaystyle H}{|}}{\underset{\underset{\displaystyle H}{|}}{C}}_6 - H$$

2.12.   First, draw each structure in its expanded form. A carbon atom attached to one other carbon atom is primary (1°); to two other carbon atoms is secondary (2°); to three other carbon atoms is tertiary (3°); and to four other carbon atoms is quaternary (4°).   A carbon atom attached to another atom by a double or triple bond is not defined by this system.

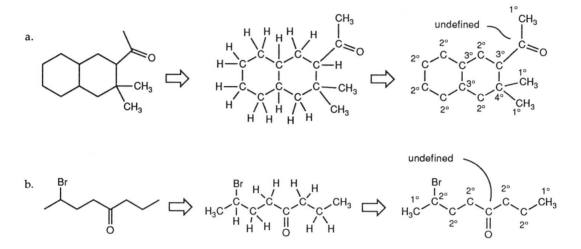

2.13.   Break each name into its constituent parts, then combine the pieces to form the structure.

a.   isobutylbenzene

        benzene   six-membered ring with alternating
                           single and double bonds
        isobutyl   a 4-carbon fragment with a branch,
                           attached via one of its 1° carbon atoms

b.   3-*tert*-butyl-1-hexanol

           hex   6 carbon atoms
           an   no double or triple bonds
           ol   OH group at C1
           3-*tert*-butyl   a 4-carbon branched fragment attached via
                           its 3° carbon atom. This substituent
                           is attached to C3 of the parent chain

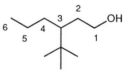

2.14. Break each name into its constituent parts, then combine the pieces to form the structure.

2-(1,1-dimethylpropyl)hexanoic acid

hex	6 carbon atoms
an	no double or triple bonds
oic acid	carboxylic acid, defines C1 of the main chain
2-(•••)	substituent attached at C2

••• = **substituent:**　1,1-dimethylpropyl

prop	3 carbon atoms; by convention this side chain is attached through its C1
yl	a suffix that indicates this group is a substituent
1,1-dimethyl	two methyl groups attached at C1 of the substituent

---

2.15. Break each name into its constituent parts, then combine the pieces to form the structure.

2-(2-chloro-1-propenyl)heptanoic acid

hept	7 carbon atoms
an	no double or triple bonds
oic acid	carboxylic acid, defines C1 of the main chain
3-(•••)	substituent attached at C3

••• = **substituent:**　2-chloro-1-propenyl

prop	3 carbon atoms; by convention this side chain is attached through its C1
en	a double bond, starting at C1 of this group
yl	a suffix that indicates this group is a substituent
2-chloro	a chlorine atom is attached at C2 of the substituent

---

2.16. Break each name into its constituent parts, then combine the pieces to form the structure.

a.　2-bromobenzoic acid

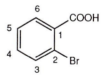

benzoic acid	carboxylic acid derivative of benzene; the acid group defines C1 of the ring
2-bromo	bromine atom attached at C2 of the ring

b.　3-methyl-2-cyclohexenone

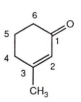

cyclohex	six-membered ring of carbon atoms
en	a double bond, starting at C2
one	ketone functional group; in a ring, the ketone group defines C1
3-methyl	methyl group attached at C3

2.17.   Break each name into its constituent parts using the alternate root word that is derived from the older systems (where needed), then combine the pieces to form the structure. Greek letters indicate substitution at the appropriate positions relative to the functional group. For functional groups that have only a single bond between the heteroatom and a carbon atom, that carbon atom is considered the α-position.

a.   phenylacetone

  acetone the ketone that has 3 carbon atoms

  phenyl   substituent; no Greek letter is needed because the
        phenyl group must be attached to one of the
        non-carbonyl carbon atoms

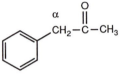

b.   α-bromopropionitrile

  propio 3 carbon atoms

  nitrile nitrile functional group

  α-bromo bromine atom attached to carbon atom
       adjacent to the nitrile functional group

c.   trichloroacetic acid

  acetic acid the carboxylic acid that has two carbon atoms

  trichloro three chlorine substituents; no Greek letters are needed
       because only one position can bear substituents

d.   α,α′–dibromoacetone

  acetone   the ketone that has 3 carbon atoms

  α,α′–dibromo two bromine atoms: as indicated by the use of the
         prime mark, one bromine atoms is attached to a
         carbon atom on each side of the carbonyl group

$$\overset{\alpha}{Br-CH_2}-\underset{\underset{O}{\|}}{C}-\overset{\alpha'}{CH_2-Br}$$

e.   β-mercaptoethanol

  ethanol   the alcohol that has two carbon atoms

  β-mercapto SH group on the carbon atom two
        away from the OH functional group

$$\overset{\beta}{HS-CH_2}-\overset{\alpha}{CH_2}-OH$$

2.18. Greek letters indicate substitution or placement of a double or triple bond at positions relative to the functional group.

a. a β,γ-unsaturated aldehyde

b. an α,α-dichloro nitrile

c. a γ-hydroxy ketone

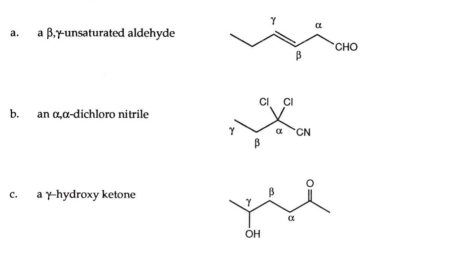

2.19. Break each name into its constituent parts, then combine the pieces to from the structure. The terms *cis* and *trans* specify the geometry of the carbon-carbon double bond.

a. *cis*-4-hexenal

hex	6 carbon atoms
en	a double bond, starting at C4
al	aldehyde at C1, by convention
cis	the hydrogen atoms attached to the carbon-carbon double bond are on the same side of the bond axis

b. 4-fluoro-*trans*-2-pentene

pent	5 carbon atoms
ene	a double bond, starting at C2
trans	the hydrogen atoms attached to the carbon-carbon double bond are on opposite sides of the bond axis
4-fluoro	fluorine atom attached at C4

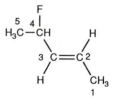

2.20.   To draw all of the possible isomers of a hydrocarbon, generate the carbon skeletons that can exist. Isomers may differ in the number of carbon atoms in the longest chain or by placement of substituents. For this exercise, first draw a straight chain with the requisite number of carbon atoms. Then reproduce that structure with one less carbon atom, but connect the omitted carbon atom at different sites along the chain (except the terminal positions). Then draw a chain with two fewer carbon atoms, and attach the two omitted carbon atoms at unique positions along the chain. Consider putting the omitted carbon atoms together, too (see part b of this exercise, where the two carbon atoms are combined to form an ethyl group instead of two methyl groups). Repeat this process, drawing the chain each time with one less carbon atom in the row. Number the chain from each end to make certain you have not just drawn isomers in a different orientation.

a.   Draw the five isomers of C_6H_{14}.

Following the procedure outlined above:

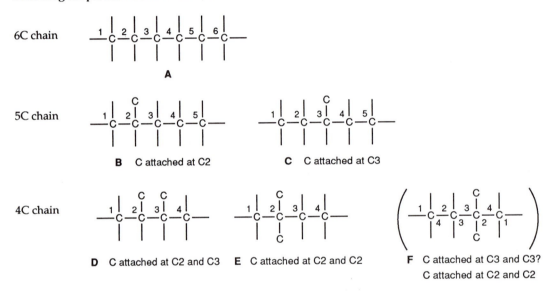

In this problem, notice that connecting two carbon atoms at C3 (**F**) is the same as **E** if you simply number the chain from the right end instead of the left.

Finish by adding hydrogen atoms so that each carbon atom forms four bonds.

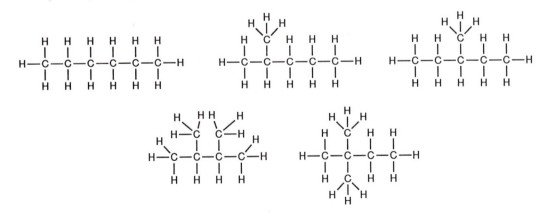

2.20.   (continued)

b.   Draw the nine isomers of C_7H_{16}.

7C chain

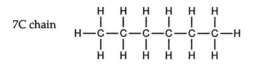

6C chain

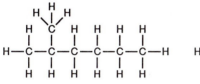

methyl group attached at C2   methyl group attached at C3

5C chain

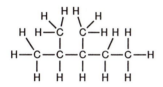

methyl groups attached at C2 and C3   methyl groups attached at C2 and C4

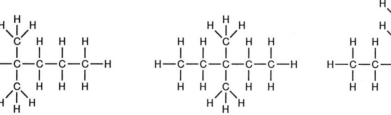

methyl groups attached at C2 and C2   methyl groups attached at C3 and C3   ethyl group attached at C3

4C chain

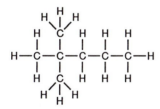

methyl groups attached
at C2, C2, and C3

2.21   To generate condensed formulas, express the hydrocarbon units with use of the groups CH, CH_2, and CH_3.

a.

$CH_3CH_2CH_2CH_2CH_2CH_3$

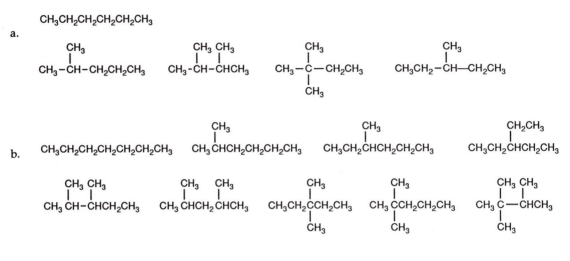

b. $CH_3CH_2CH_2CH_2CH_2CH_2CH_3$

2.22.   Follow the procedure outlined in the solution to exercise 2.3.

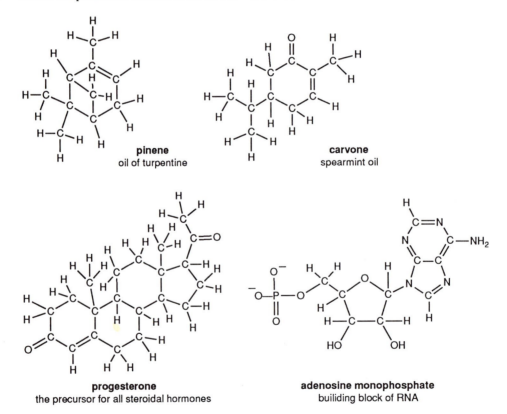

pinene
oil of turpentine

carvone
spearmint oil

progesterone
the precursor for all steroidal hormones

adenosine monophosphate
builiding block of RNA

2.22.   (continued)

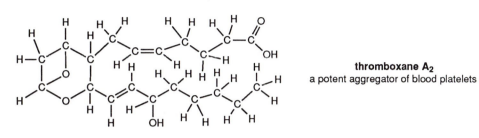

thromboxane A$_2$
a potent aggregator of blood platelets

2.23.   An exercise like this reinforces your knowledge of the identities of different functional groups and compound types. Table 2.1 (text page 33) lists the common functional groups. An aromatic compound is a derivative of benzene. An alicyclic compound has a ring but does not have alternating single and double bonds. There are many possible answers for each type of compound. Two examples are illustrated for each type of compound; there are many other possibilities.

a.   a ketone with the formula C$_5$H$_{10}$O

b.   a chloro ketone with four carbon atoms

c.   an aromatic amine

d.   an aldehyde with six carbon atoms

e.   a hydroxy aldehyde

f.   an alicyclic carboxylic acid

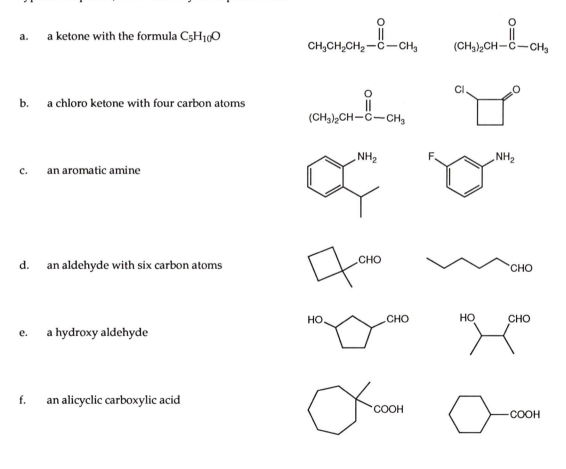

2.24.   Follow the procedure outlined in the solution to exercise 2.1.

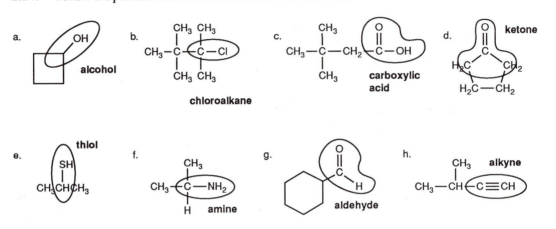

2.25.   Follow the procedure outlined in the solution to exercise 2.12.

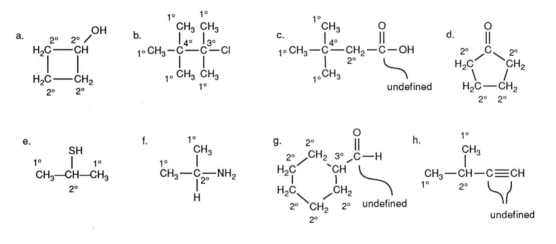

2.26.   Follow the procedure outlined in the solution to exercises 2.1 and 2.3.

Metoprolol

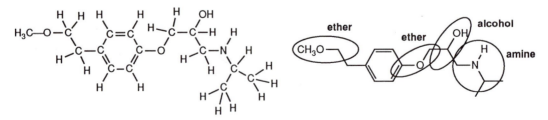

2.26. (continued)

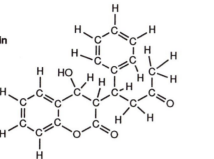

Warfarin

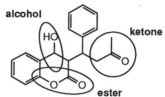

alcohol
ketone
ester

Chloramphenicol

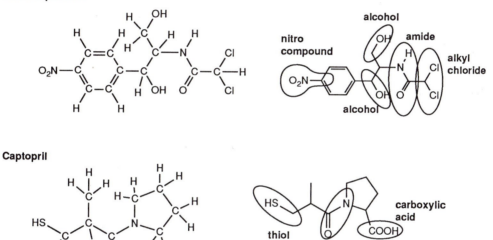

nitro
compound
alcohol
amide
alkyl
chloride
alcohol

Captopril

HS
thiol
amide
carboxylic
acid
COOH

2.27. Look at each structure in its expanded form. If a carbon atom is attached to one carbon atom, it is primary (1°); two carbon atoms, secondary (2°); three carbon atoms, tertiary (3°); and four carbon atoms, quaternary (4°). A carbon atom attached to another atom by a double or triple bond is not defined in this classification system. A methyl group attached to a heteroatom is classified as a methyl carbon atom.

Metoprolol   **Warfarin**

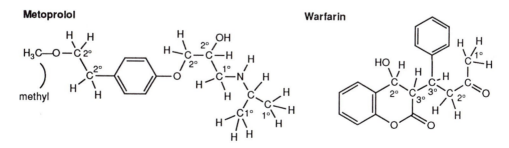

2.27.   (continued)

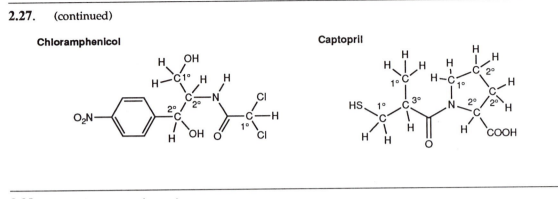

Chloramphenicol

Captopril

2.28.   Possible sources of error in a name are:

1.   The substituent can be located at several positions, so a numeral is required to indicate the point of attachment.
2.   The longest carbon chain containing the principal functional group was not chosen as the root word. In the examples below, the *incorrect* numbering is shown italicized.
3.   The numeric prefix does not match the substituent prefix (di-, tri-, etc.)

a.   **Methylheptane:** the methyl group can be placed at several positions along the 7C chain.

b.   **3-Propylhexane:** the longest chain is seven carbon atoms. This compound should be named 4-ethylheptane.

$$\underset{1}{CH_3}-\underset{2}{CH_2}-\underset{3}{\underset{4}{CH}}-\underset{4}{\underset{5}{CH_2}}-\underset{5}{\underset{6}{CH_2}}-\underset{6}{\underset{7}{CH_3}}$$
$$\underset{3}{CH_2}-\underset{2}{CH_2}-\underset{1}{CH_3}$$

c.   **2,2-Dimethyl-3-ethylbutane:** the longest chain is five carbon atoms. This compound should be named 2,2,3-trimethylpentane.

d.   **2-Dimethylpentane:** there should be two numerals in the prefix. The name should read **2,2-dimethylpentane.**

e.   **2-Isopropyl-1-propanol:** the longest chain that contains the OH group is four carbon atoms. This compound should be named 2,3-dimethyl-1-butanol.

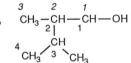

f.   **Dichloroheptane:** the two chlorine atoms can be placed at several positions along the 7C chain, therefore numerals need to appear at the beginning of the name.

2.29. Follow the procedure outlined in the solution to exercise 2.20.

6C chain

$$H—C≡C—CH_2—CH_2\text{-}CH_2\text{-}CH_3 \qquad H_3C—C≡C—CH_2—CH_2—CH_3$$

$$H_3C—CH_2—C≡C—CH_2—CH_3$$

5C chain

$$H—C≡C—CH_2—\underset{\underset{CH_3}{|}}{CH}—CH_3 \qquad H—C≡C—\underset{\underset{CH_3}{|}}{CH}—CH_2\text{-}CH_3$$

$$H_3C—C≡C—\underset{\underset{CH_3}{|}}{CH}—CH_3$$

4C chain

$$H—C≡C—\overset{\overset{CH_3}{|}}{\underset{\underset{CH_3}{|}}{C}}—CH_3$$

2.30.   To assess whether two compounds are identical, confirm that the longest carbon chain (or ring system) is the same. Then make certain that any substituents are attached at the same positions on the chain (or ring). Number the chains from each end when checking substitution patterns. To decide if compounds are isomers, make certain that the molecular formulas are identical. If the formulas are the same, but the compounds are not identical, the two substances are isomers.

a.   **A** and **C** are identical. Each has a chain of four carbon atoms.

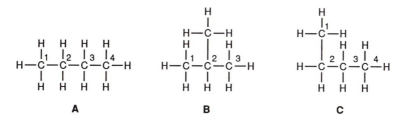

b.   All of these compounds are the same. Each has four carbon atoms with a chlorine atom attached at C2.

2.30.   (continued)

c.   **B** and **C** are identical. Each has a methyl and propyl group attached to an oxygen atom.

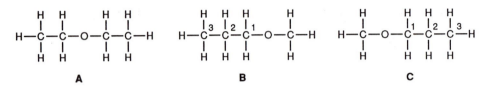

d.   **A**, **B** and **C** are identical. Each has four carbon atoms with an OH group attached at C2.

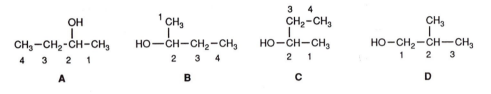

2.31.   Constitutional isomers are those that have different functional groups. Two functional groups listed in Table 2.1 (text page 33) that contain sulfur are thiol, RSH, and sulfide, RSR. Therefore, it is only necessary to replace "R" in the general structures with carbon and hydrogen atoms.

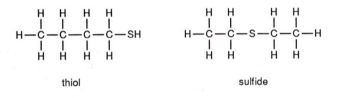

2.32.   Structurally, an alcohol is generated from an alkane simply by inserting an oxygen atom into a carbon-hydrogen bond, so the first step in solving this problem is to identify the possible isomers of C_5H_{12}. This can be done by following the procedure outlined in the solution to exercise 2.20; there are three such alkanes.

$$CH_3-CH_2-CH_2-CH_2-CH_3$$

Inserting an oxygen atom into each unique C–H bond of each alkane generates the possible alcohols.

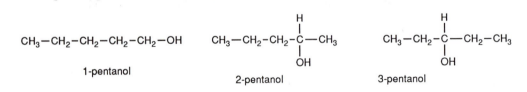

2.32.   (continued)

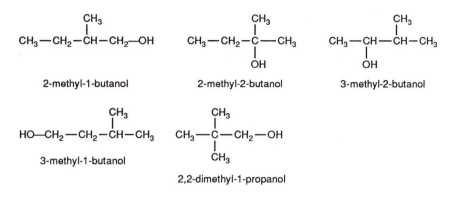

2-methyl-1-butanol   2-methyl-2-butanol   3-methyl-2-butanol

3-methyl-1-butanol

2,2-dimethyl-1-propanol

Assign the type to each carbon atom according to the number of other carbon atoms to which it is attached.

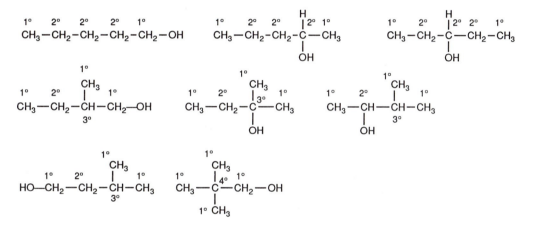

2.33.   Break each name into its constituent parts, then combine the pieces to form the structure.

a.   2-bromobenzyl alcohol   benzyl alcohol   benzene ring with a CH_2OH group attached; the CH_2OH group defines C1 of the ring

2-bromo   bromine atom at C2 of the ring

b.   1,3-dibromo-2-propanol   prop   3 carbon atoms
an   no double or triple bonds
ol   alcohol group at C2
1,3-dibromo   two bromine atoms, one at C1 and one at C3

2.33. (continued)

c.   3-chloro-1-propanol

prop	3 carbon atoms
an	no double or triple bonds
ol	alcohol group at C1
3-chloro	chlorine atom at C3

$$Cl-CH_2-CH_2-CH_2-OH$$
$$321$$

d.   2-methyl-3-buten-2-ol

but	4 carbon atoms
en	double bond starting at C3
ol	alcohol group at C2
2-methyl	methyl group at C2

e.   2,2,2-trifluoroethanol

eth	2 carbon atoms
an	no double or triple bonds
ol	alcohol group at C1, by convention
fluoro	fluorine atom at C2 (three)

f.   2-amino-2-methyl-1-propanol

prop	3 carbon atoms
an	no double or triple bonds
ol	alcohol group at C1
2-amino	NH_2 group at C2
2-methyl	methyl group at C2

2.34. In carbinol nomenclature, substituent groups are attached to the C—OH unit with a bond to the carbon atom that bears the OH group.

a.   triethyl carbinol

b.   diphenyl carbinol

c.   dibutyl carbinol

2.34. (continued)

d.   ethyl dimethyl carbinol

$$CH_3CH_2 - \underset{\underset{CH_3}{|}}{\overset{\overset{CH_3}{|}}{C}} - OH$$

2.35. Break each name into its constituent parts, then combine the pieces to form the structure.

a.   2-aminobenzoic acid

benzoic acid	carboxylic acid derivative of benzene; the point of attachment of the COOH group defines C1
2-amino	amine group at C2

b.   2,3-dibromopropanoic acid

prop	3 carbon atoms
an	no double or triple bonds
oic acid	carboxylic acid functionality, the COOH group defines C1 of the chain
2,2-dibromo	two bromine atoms, one at C2 and one at C3

c.   chloroacetic acid

acetic acid	2 carbon carboxylic acid
chloro	chlorine atom, attached to only position that can bear substituents

$$\overset{2}{Cl} - \overset{}{CH_2} - \overset{1}{COOH}$$

d.   2-chlorocyclohexanecarboxylic acid

cyclohex	6 carbon atoms in a ring
an	no double or triple bonds
carboxylic acid	carboxylic acid functional group; the point of attachment of the COOH group defines C1
2-chloro	chlorine atom at C2

e.   2,2-difluorobutanoic acid

but	4 carbon atoms
an	no double or triple bonds
oic acid	carboxylic acid functional group; the COOH group defines C1
2,2-difluoro	two fluorine atoms at C2

f.   2-phenyl-2-vinylpentanoic acid

pent	5 carbon atoms
an	no double or triple bonds
oic acid	carboxylic acid functionality; the point of attachment of the COOH group defines C1
2-phenyl	phenyl group at C2
2-vinyl	vinyl group at C2

2.36.   Greek letters indicate substitution or placement of a double or triple bond at positions relative to the functional group.

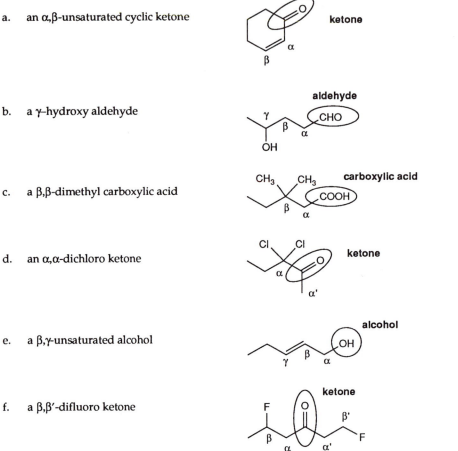

a.   an α,β-unsaturated cyclic ketone

b.   a γ–hydroxy aldehyde

c.   a β,β-dimethyl carboxylic acid

d.   an α,α-dichloro ketone

e.   a β,γ-unsaturated alcohol

f.   a β,β′-difluoro ketone

2.37.   Break each name into its constituent parts, then combine the pieces to form the structure.

a.   *N,N*-dimethylaniline

aniline	amine derivative of benzene the NH$_2$ group defines C1
N,N-dimethyl	two methyl groups attached to the nitrogen atom

b.   1,2-diaminocyclohexane

cyclohex ane	6 carbon atoms in a ring no double or triple bonds
1,2-diamino	two amine groups, one attached to C1 and one to C2

2.37.   (continued)

c.   *N*-methylacetamide   acetamide   2 carbon amide
                            N-methyl   methyl group attached
                                            to the nitrogen atom

d.   1,5-diaminopentane   pent   5 carbon atoms
                           ane   no double or triple bonds
                           1,5-diamino two amine groups, one
                                          attached to C1 and one to C5

e.   4-phenyl-1-butylamine   butyl   4 carbon atoms
                              amine   NH$_2$ group attached to C1
                              4-phenyl   phenyl group at C4

f.   2,4-dimethylaniline   aniline   amine derivative of benzene; the
                                            NH$_2$ group defines C1
                            2,4-dimethyl   two methyl groups, one attached
                                            to C2 and one to C4

2.38.   Break each name into its constituent parts, then combine the pieces to form the structure.

a.   3-methyl-2-butanone   but   4 carbon atoms
                            an   no double or triple bonds
                            one   ketone functional group at C2
                            3-methyl   methyl group at C3

b.   1,3-dichloropropanone
                            prop   3 carbon
                            an   no double or triple bonds
                            one   ketone: the carbonyl group
                                         must be at C2 when the compound
                                         has only three carbon atoms
                            1,3-dichloro   two chlorine atoms, one at C1 and
                                             one at C3

2.38.   (continued)

c.   3,3-dimethoxypentanal

pent	5 carbon atoms
an	no double or triple bonds
al	aldehyde; the carbonyl group defines C1
3,3-dimethoxy	two methoxy groups, both attached at C3

d.   2,4-dimethoxybenzaldehyde

benzaldehyde	the aldehyde derivative of benzene; the CHO group defines C1 of the ring
2,4-dimethoxy	two methoxy groups, one at C2 and one at C4

e.   4-bromobenzaldehyde

benzaldehyde	the aldehyde derivative of benzene; the CHO group defines C1 of the ring
4-bromo	bromine atom attached at C4

f.   1,3-diacetylbenzene

benzene	
1,3-diacetyl	two acetyl groups, one at C1 and one at C3

2.39.   Break each name into its constituent parts, then combine the pieces to from the structure.

a.   methyl phenylacetate

acetate	ester of acetic acid; the separate, unnumbered prefix is the group attached to the oxygen atom—methyl
phenyl	substituent on the root; no numeral is necessary because there is only one carbon atom that can have a substituent

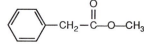

b.   *tert*-butyl propanoate

prop	3 carbon atoms
an	no double or triple bonds
oate	ester functional group; the separate, unnumbered prefix is the group attached to the oxygen atom—*tert*-butyl

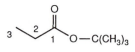

c.   phenyl acetate

acetate	an ester of acetic acid; the separate, unnumbered prefix is the group attached to oxygen—phenyl

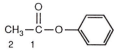

2.39.   (continued)

d.   ethyl 3-chlorocyclohexanecarboxylate

cyclohexane	six-membered ring; no double or triple bonds
carboxylate	ester functional group; the carboxy group is attached to the ring; the separate, unnumbered prefix is the group attached to oxygen—ethyl
3-chloro	substituent attached to the ring at C3

e.   methyl 2,4-dihydroxybenzoate

benzoate	an ester of benzoic acid; the separate, unnumbered prefix is the group attached to oxygen—methyl
2,4-dihydroxy	two hydroxy groups, one at C2 and one at C4 of the benzene ring.

f.   ethyl 4-bromobutanoate

but	4 carbon atom
an	no carbon-carbon double or triple bond
oate	ester functional group; the separate, unnumbered prefix is the group attached to oxygen—ethyl
4-bromo	bromine atom at C4

$$\underset{4}{Br}-\underset{}{CH_2}-\underset{3}{CH_2}-\underset{2}{CH_2}-\underset{1}{\overset{\overset{O}{\|}}{C}}-O-CH_2CH_3$$

2.40. Break each name into its constituent parts, then combine the pieces to from the structure.

a.   1-chloro-2-hexyne

hex	6 carbon atoms
yne	triple bond starts at C2
1-chloro	chlorine atom at C1

$$\underset{6}{CH_3}-\underset{5}{CH_2}-\underset{4}{CH_2}-\underset{3}{C}\equiv\underset{2}{C}-\underset{1}{CH_2}-Cl$$

b.   1,3-dicyanobenzene

benzene	6-membered ring arene
1,3-dicyano	cyanide groups at C1 and C3

c.   2-octanone

oct	8 carbon atoms
an	no double or triple bonds
one	ketone functional group at C2

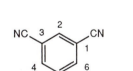

d.   1-iodobutane

but	4 carbon atoms
ane	no double or triple bonds
1-iodo	iodine atom at C1

$$\underset{4}{CH_3}-\underset{3}{CH_2}-\underset{2}{CH_2}-\underset{1}{CH_2}-I$$

2.40. (continued)

e.   1-bromo-2,2-dimethoxypropane

prop	3 carbon atoms
an	no double or triple bonds
1-bromo	bromine atom at C1
2,2-dimethoxy	two methoxy groups at C2

f.   2-chloro-2-methylpropane

prop	3 carbon atoms
ane	no double or triple bonds
2-chloro	chlorine atom at C2
2-methyl	methyl group at C2

g.   4-bromotoluene

toluene	benzene derivative with a methyl group attached to the ring at C1
4-bromo	bromine atom at C4

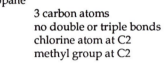

h.   allyl phenyl ether

ether	an oxygen atom attached to two carbon-containing groups
allyl	a three-carbon fragment having a double bond between C2 and C3
phenyl	a six-carbon fragment derived from benzene

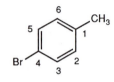

i.   dipropyl ether

ether	an oxygen atom attached to two carbon-containing groups
dipropyl	two three-carbon fragments attached to the oxygen atom

$CH_3CH_2CH_2$ —O— $CH_2CH_2CH_3$

j.   2-mercapto-2-methylpropane

prop	3 carbon atoms
ane	no double or triple bonds
2-mercapto	SH group at C2
2-methyl	methyl group at C2

2.41. Break each name into its constituent parts, then combine the pieces to from the structure.

a. 1,3-Bis(bromomethyl)benzene

benzene   6-membered ring arene

1,3-bis(•••)   two substituents attached at C1 and C3

••• = **substituent:** bromomethyl $BrCH_2-$

$BrCH_2$ 3   2 CH_2Br 1   4   6   5

b. Tetrakis(hydroxymethyl)methane

methane   1 carbon atom

tetrakis(•••)   four substituents attached to the carbon atom

••• = **substituent::** (hydroxymethyl) $HOCH_2-$

CH_2OH
$HOCH_2-C-CH_2OH$
CH_2OH

c. Tris(2-chloroethyl)amine

amine   nitrogen atom attached to three groups

tris(•••)   three substituents attached to the nitrogen atom

••• = **substituent::** (2-chloroethyl)

ethyl   2 carbon chain attached at C1, by convention

2-chloro   chlorine atom attached at C2 of the ethyl groups.

$ClCH_2CH_2-N-CH_2CH_2Cl$
CH_2CH_2Cl

d   Pentakis(1,1-difluoroethyl)phenol

phenol   benzene derivative in which OH defines C1

pentakis(•••)   five substituents attached to benzene
                    ring at the other positions

••• = **substituent::** (1,1-difluoroethyl)

ethyl   2 carbon chain, attached at C1
             by convention

1,1-difluoro   two fluorine atoms attached to each ethyl group

CF_2CH_3 CH_3CF_2 OH CH_3CF_2 CF_2CH_3 CF_2CH_3

2.42. Break each name into its constituent parts, then combine the pieces to from the structure.

a. 1,4-Cyclohexanedione

cyclohex   6 carbon atoms in a ring

an   no double or triple bonds

dione   ketone functional groups at C1 and C4

b. 1,3-Propanedial

prop   3 carbon atoms

an no   double or triple bonds

dial   C1 and C3 are –CHO groups

3   2   1
$OHC-CH_2-CHO$

2.42.   (continued)

c.   2,3,4-Hexanetriol

hex	6 carbon atoms
an	no double or triple bonds
triol	OH groups at C2, C3, and C4

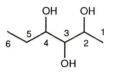

d.   *N,N'*-Dimethyl-2,6-hexanediamine

hex	6 carbon atoms
an	no double or triple bonds
diamine	NH_2 groups at C2 and C6
N,N'-Dimethyl	one methyl group attached to each nitrogen atom

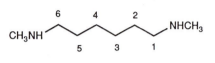

e.   1,4-Benzenedicarboxylic acid

benzene	6-membered ring arene
dicarboxylic acid	COOH groups at C1 and C4

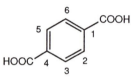

f.   2,5-Dimethyl-2,3,4-hexatriene

hexa	6 carbon atoms
triene	three double bonds starting at C2, C3, and C4
2,5-dimethyl	methyl groups at C2 and C5

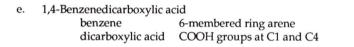

g.   3,3-Dimethyl-1,5-pentanedioic acid

pent	5 carbon atoms
an	no double or triple bonds
dioic acid	C1 and C5 are COOH groups
3,3-dimethyl	two methyl groups at C3

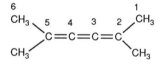

j.   3-Vinyl-1,4-pentanediol

pent	5 carbon atoms
an	no double or triple bonds
diol	OH groups at C1 and C4
3-vinyl	$-CH=CH_2$ group at C3

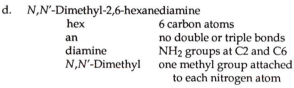

2.43.   Break each name into its constituent parts, then combine the pieces to from the structure.

a.   6-bromohexanoic acid

hex	6 carbon atoms
an	no double or triple bonds
oic acid	carboxylic functional group; defines C1
6-bromo	bromine atom at C6

$Br\diagdown\diagup\diagdown\diagup\diagdown\diagup COOH$

6   4   2

5   3   1

b.   1,2,4,5-benzenetetracarboxylic acid

benzene	six-membered ring arene
tetracarboxylic acid	COOH groups at C1, C2, C4, and C5

HOOC, H 6 COOH
5   1
4   2
HOOC 3 COOH
H

c.   1-nitropropane

prop	3 carbon atoms
aneno	double or triple bonds
1-nitro	NO_2 group at C1

3   2   1
$CH_3{-}CH_2{-}CH_2{-}NO_2$

d.   4-pentyn-1-ol

pent	5 carbon atoms
yne	triple bond starting at C4
ol	alcohol functional group: OH at C1

5   4   3   2   1
$H{-}C{\equiv}C{-}CH_2{-}CH_2{-}CH_2OH$

e.   2-amino-3-methylbenzoic acid

benzoic acid	benzene with a carboxylic acid functional group; the point of attachment of the COOH group defines C1 of the ring
2-amino	NH_2 groups at C2
3-methyl	CH_3 group at C3

6 COOH
5   1
4   2
3 NH_2
CH_3

f.   1,2-Dichlorobenzene

benzene	six-membered ring arene
1,2-dichloro	chlorine atoms attached at C1 and C2

6 Cl
5   1
4   2 Cl
3

g.   2-Chloro-5-methylanisole

anisole	a derivative of benzene having a methoxy group attached to the ring; the CH_3O group defines C1
2-chloro	chlorine atom attached to the ring at C2
5 methyl	methyl group attached to the ring at C5

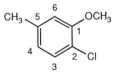

2.43.   (continued)

h.   1-Decanethiol

dec	10 carbon atoms
an	no double or triple bonds
thiol	SH functional group at C1

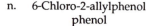

i   Chloroacetonitrile

acetonitrile	2 carbon nitrile
chloro	chlorine atom attached to only carbon atom besides the one that is part of the CN group

$$Cl-CH_2-CN$$
$$\quad\quad 2\quad\quad 1$$

j.   2-Nitrobenzaldehyde

benzaldehyde	benzene with an aldehyde functionality; the point of attachment of the CHO group defines C1 of the ring
2-nitro	NO_2 group at C2

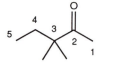

k.   Diphenylacetylene

acetylene	the 2 carbon alkyne, $H-C{\equiv}C-H$
diphenyl	a phenyl group attached to each carbon atom

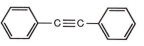

l.   3,3-Dimethyl-2-pentanone

pent	5 carbon atoms
an	no double or triple bonds
one	ketone functional group at C2
3,3-dimethyl	two methyl groups at C3

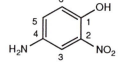

m.   4-Amino-2-nitrophenol

phenol	a derivative of benzene having an –OH group attached to the ring; the point of attachment of the OH group defines C1
4-amino	NH_2 group attached to the ring at C4
2-nitro	NO_2 group attached to the ring at C2

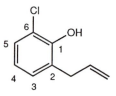

n.   6-Chloro-2-allylphenol

phenol	a derivative of benzene having an –OH group attached to the ring; the point of attachment of the OH group defines C1
6-chloro	chlorine atom attached to the ring at C6
2-allyl	allyl group attached to the ring at C2

2.43. (continued)

o.   **3-Chloro-2-nitrobenzoic acid**

benzoic acid   benzene with a carboxylic acid functional
group; the point of attachment of the
COOH group defines C1
2-nitro NO_2 groups at C2
3-chloro   chlorine atom at C3

p.   **3-Chloro-2-methylpropene**

prop   3 carbon atoms
~~eneone~~   double bond starting at C1, by convention
2-methyl   methyl group at C2
3-chloro   chlorine atom at C3

q.   **2-Chloro-5-nitroaniline**

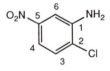

aniline   benzene with an NH_2 group; the point
of attachment of the NH_2 group defines C1
2-chloro   chlorine atom at C2
5-nitro NO_2 group at C5

r.   **2-*tert*-Butylphenol**

phenol   a derivative of benzene having an –OH
group attached to the ring; the point
of attachment of the OH group defines C1
2-*tert*-butyl   *tert*-butyl group attached to the ring at C2

2.44.   Break each name into its constituent parts, then combine the pieces to from the structure.

a.   Methyl 2-(1,1-dimethylethyl)pentanoate

pent   5 carbon atoms
an   no double or triple bonds
oate   ester functional group; the separate,
unnumbered prefix is the group
attached to oxygen—methyl
2-(•••)   substituent at C2

••• = **substituent**:
ethyl   2 carbon atoms
1,1-dimethyl   two methyl groups attached at C1 of the substituent group

2.44. (continued)

b. 3-(2-Chloroethyl)-4-(2-methyl-1-butenyl)-1,8-octanediol

oct	8 carbon atoms
an	no double or triple bonds
diol	two alcohol functional groups; an OH group is attached at C1 and at C8
3-(•••)	substituent #1 attached at C3
4-(•••)	substituent #2 attached at C4

••• = substituent #1:
    ethyl   2 carbon atoms
    2-chloro   chlorine atom attached to C1 of the substituent
••• = substituent #2:
    but   4 carbon atoms
    en   a double bond starting at C1 of the substituent
    2-methyl a methyl group attached at C2 of the substituent

c. 5-Chloro-6-(1-hydroxyethyl)-2-nitrophenol

phenol	a derivative of benzene having an OH group attached to the ring; the point of attachment of the OH group defines C1
2-nitro	NO_2 group attached to the ring at C2
5-chloro	chlorine atom attached to the ring at C5
6-(•••)	substituent attached at C6 of the ring

••• = substituent::
    ethyl   2 carbon atoms
    1-hydroxy OH group attached to C1 of the substituent

d. 3-(4-Nitro-2-vinylphenyl)butanoic acid

but	4 carbon atoms
an	no double or triple bonds
oic acid	carboxylic acid functionality; defines C1
3-(•••)	substituent at C3

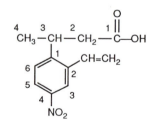

••• = substituent::
    phenyl   benzene derivative; attachment is at C1 of the ring
    2-vinyl   vinyl group attached at C2 of the substituent
    4-nitro NO_2 group attached at C4 of the substituent

2.44.   (continued)

e.   4-(1-Methylethyl)-5-methyl-3-hexenal

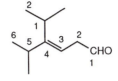

hex	6 carbon atoms
en	double bond starting at C3
al	aldehyde functional group; the point of attachment of the CHO group defines C1
5-methyl	methyl group attached at C5
4-(•••)	substituent at C4

••• = **substituent::**
ethyl   2 carbon atoms
1-methyl   methyl group attached at C1 of the substituent

f.   3,4-Difluoro-5-(3-methyl-1-butyl)benzonitrile

benzonitrile	a benzene ring with a –CN group attached
3,4-difluoro	fluorine atoms attached at C3 and C4
5-(•••)	substituent at C5

••• = **substituent::**
butyl   4 carbon atoms
3-methyl   methyl group attached at C3 of the substituent

BONDING IN ORGANIC MOLECULES

3.1. To draw a Lewis structure for a molecule, first draw its expanded line structure according to the procedure outlined in the solution to exercise 2.3. Calculate the total number of valence electrons, then subtract the number of electrons that constitute the bonds already shown (2 electrons per bond). Distribute the remaining electrons in pairs to any heteroatom present, giving each non-hydrogen atom 8 electrons.

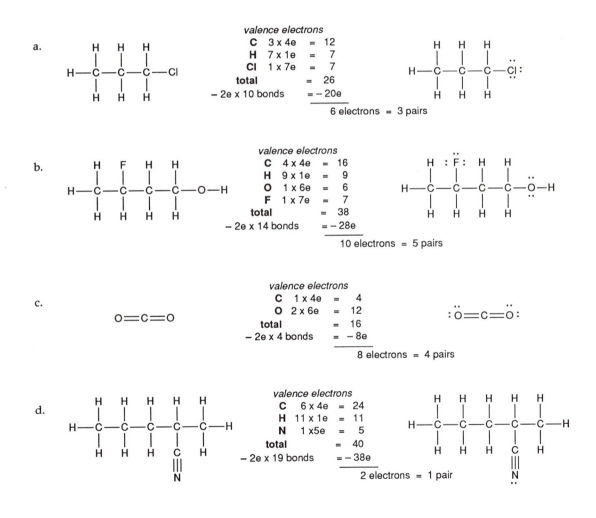

a.

valence electrons
C 3 x 4e = 12
H 7 x 1e = 7
Cl 1 x 7e = 7
total   = 26
– 2e x 10 bonds = – 20e

6 electrons = 3 pairs

b.

valence electrons
C 4 x 4e = 16
H 9 x 1e = 9
O 1 x 6e = 6
F 1 x 7e = 7
total   = 38
– 2e x 14 bonds = – 28e

10 electrons = 5 pairs

c.

valence electrons
C 1 x 4e = 4
O 2 x 6e = 12
total   = 16
– 2e x 4 bonds = – 8e

8 electrons = 4 pairs

d.

valence electrons
C 6 x 4e = 24
H 11 x 1e = 11
N 1 x 5e = 5
total   = 40
– 2e x 19 bonds = – 38e

2 electrons = 1 pair

35

3.1. (continued)

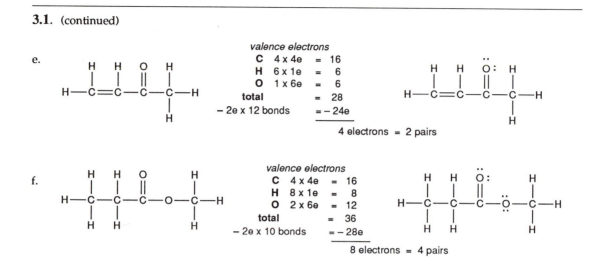

e.

valence electrons
C	4 x 4e	= 16
H	6 x 1e	= 6
O	1 x 6e	= 6
total		= 28
− 2e x 12 bonds		= − 24e

4 electrons = 2 pairs

f.

valence electrons
C	4 x 4e	= 16
H	8 x 1e	= 8
O	2 x 6e	= 12
total		= 36
− 2e x 10 bonds		= − 28e

8 electrons = 4 pairs

3.2. Draw the Lewis structure of each compound by following the procedure outlined in the solution to exercise 3.1. To calculate the formal charge on each atom, apply the following formula:

Formal charge = # valence electrons − (# bonds + # non-bonding electrons)

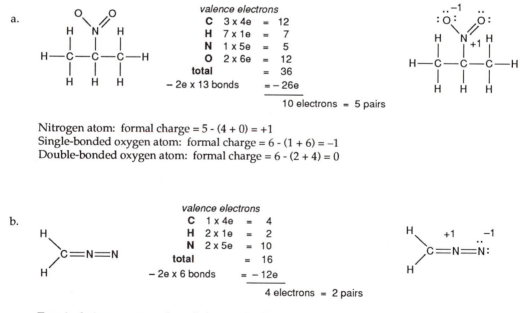

a.

valence electrons
C	3 x 4e	= 12
H	7 x 1e	= 7
N	1 x 5e	= 5
O	2 x 6e	= 12
total		= 36
− 2e x 13 bonds		= − 26e

10 electrons = 5 pairs

Nitrogen atom: formal charge = 5 - (4 + 0) = +1
Single-bonded oxygen atom: formal charge = 6 - (1 + 6) = −1
Double-bonded oxygen atom: formal charge = 6 - (2 + 4) = 0

b.

valence electrons
C	1 x 4e	= 4
H	2 x 1e	= 2
N	2 x 5e	= 10
total		= 16
− 2e x 6 bonds		= − 12e

4 electrons = 2 pairs

Terminal nitrogen atom: formal charge = 5 - (2 + 4) = −1
Other nitrogen atom: formal charge = 5 - (4 + 0) = +1

3.2. (continued)

c.

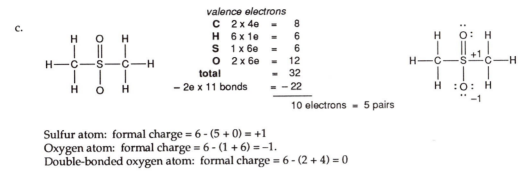

valence electrons

C	2 x 4e	=	8
H	6 x 1e	=	6
S	1 x 6e	=	6
O	2 x 6e	=	12
total		=	32
− 2e x 11 bonds		=	− 22

10 electrons = 5 pairs

Sulfur atom: formal charge = 6 - (5 + 0) = +1
Oxygen atom: formal charge = 6 - (1 + 6) = −1.
Double-bonded oxygen atom: formal charge = 6 - (2 + 4) = 0

3.3. To decide about bond polarity, consider that the element with a higher electronegativity value will be δ⁻; the other element will be δ⁺. The assumption is made that carbon-carbon and carbon-hydrogen bonds are not polarized to a significant extent.

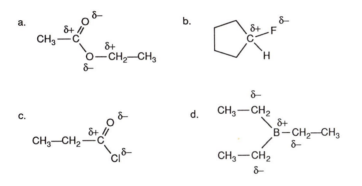

3.4. Follow the procedure outlined in example 3.4 (text page 82).

a. Pyridine is structurally like benzene, so two of its structures should resemble those drawn for benzene. Any other structure that is drawn has only 6 electrons around one of the carbon atoms.

b. The phosphate ion has resonance structures in which a double bond exists between the phosphorus atom and one of the four oxygen atoms. Each single-bonded oxygen atom carries a −1 charge.

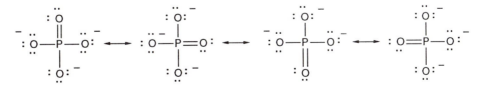

3.4.   (continued)

c.   Methyl azide has two excellent resonance structures. Any other form places either 6 or 10 electrons around one of the nitrogen atoms.

$$H_3C—\overset{..}{N}\!=\!\overset{+}{N}\!=\!\overset{..}{\underset{}{N}}:\!^- \quad\longleftrightarrow\quad H_3C—\overset{}{\underset{..}{N}}\!—\!\overset{+}{N}\!\equiv\!N:$$

3.5. The Lewis structure for water has an octet of electrons around the oxygen atom, and a bond to each of the hydrogen atoms.

$$\begin{array}{c} H \\ | \\ :\underset{..}{O}—H \end{array}$$

b.   An orbital representation using the atomic orbitals for oxygen ($1s^2\,2s^2\,2p^4$) places an electron in each of two p-orbitals and a pair of electrons in the $2s$ and $2p_z$ orbitals. Overlap with the hydrogen $1s$ orbital creates the O–H bonds. The expected ∠H–O–H would be 90° in this depiction because the p_x and p_y -orbitals are orthogonal.

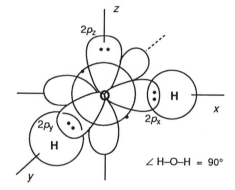

∠H–O–H = 90°

c.   The LCAO representation for water that utilizes hybrid orbitals has four, sp^3 orbitals arranged in a tetrahedral fashion. Overlap with the hydrogen $1s$ orbital creates the O–H bonds. The expected ∠H–O–H would be 109.5° .

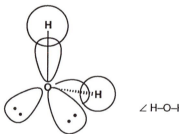

∠H–O–H = 109.5°

d.   The use of hybrid orbitals seems to be justified here because the bond angles are more accurately mimicked. In fact, an angle of slightly less than 109.5 degrees suggests that repulsion of the unshared electrons is greater than the repulsive forces between the two hydrogen atoms, so ∠H–O–H is compressed slightly from the idealized value expected for a tetrahedral arrangement of orbitals.

3.6. To draw an LCAO representation, first generate a Lewis structure to make certain the electrons are located appropriately. Then decide on the hybridization of the carbon atoms according to whether they have four single bonds (sp^3), two single bonds and a double bond(sp^2), one single bond and a triple bond (sp), or two double bonds (sp). Create overlap of hybrid orbitals to form one bond between each pair of atoms, and use overlap between adjacent p-orbitals to create multiple bonds, if they are present. Overlap between a hybrid orbital of carbon with the hydrogen 1s orbital generates the C–H bonds.

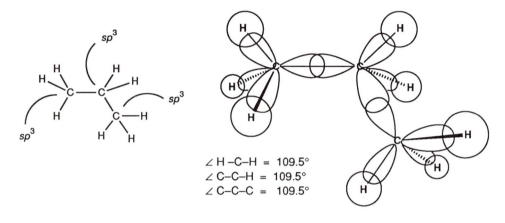

∠ H –C–H = 109.5°
∠ C–C–H = 109.5°
∠ C–C–C = 109.5°

3.7. Follow the procedure outlined in the solution to exercise 3.6.

∠ H –C–H = 120°
∠ C–C–H = 120°
∠ C–C–C = 120°

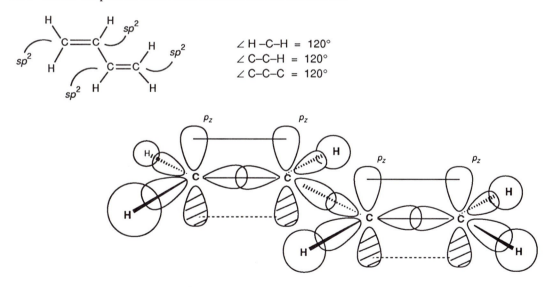

3.8. Follow the procedure outlined in the solution to exercise 3.6.

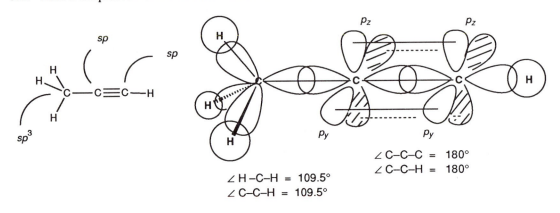

$$\angle\,C\text{–}C\text{–}C \;=\; 180°$$
$$\angle\,C\text{–}C\text{–}H \;=\; 180°$$

$$\angle\,H\text{–}C\text{–}H \;=\; 109.5°$$
$$\angle\,C\text{–}C\text{–}H \;=\; 109.5°$$

3.9. Follow the procedure outlined in the solution to exercise 3.6. Notice that the hydrogen atoms at each end of the molecule lie in the same plane.

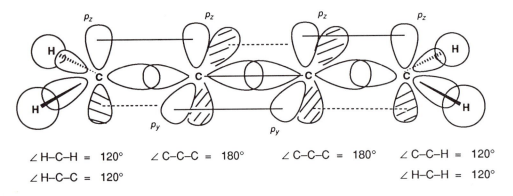

$$\angle\,H\text{–}C\text{–}H \;=\; 120° \qquad \angle\,C\text{–}C\text{–}C \;=\; 180° \qquad \angle\,C\text{–}C\text{–}C \;=\; 180° \qquad \angle\,C\text{–}C\text{–}H \;=\; 120°$$
$$\angle\,H\text{–}C\text{–}C \;=\; 120° \qquad\qquad\qquad\qquad\qquad\qquad\qquad\qquad\qquad \angle\,H\text{–}C\text{–}H \;=\; 120°$$

3.10. Follow the procedure outlined in the solution to exercise 3.6. When a heteroatom is present, give it the same hybridization as the carbon atom to which it is attached. If a heteroatom is attached to carbon atoms with different hybridization, then give it the one with less *p*-orbital character (that is, *sp*² rather than *sp*³.)

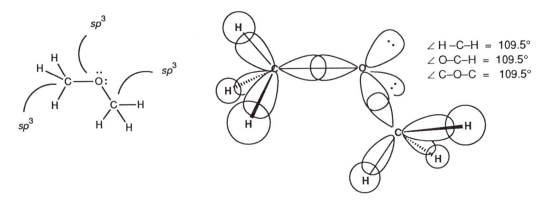

$$\angle\,H\text{–}C\text{–}H \;=\; 109.5°$$
$$\angle\,O\text{–}C\text{–}H \;=\; 109.5°$$
$$\angle\,C\text{–}O\text{–}C \;=\; 109.5°$$

3.11. Follow the procedure outlined in the solution to exercise 3.10.

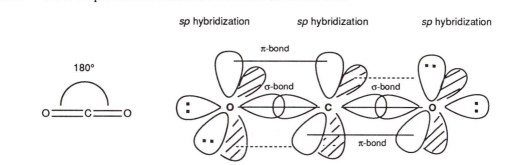

3.12. Follow the procedure outlined in the solution to exercise 3.10.

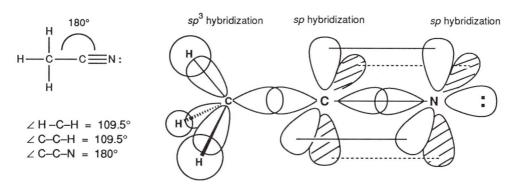

3.13. A bonding molecular orbital is formed by the additive combination of the two orbitals being considered. Because the large portion of a hybrid orbital and the hydrogen 1s orbital have the same sign, they combine to form a unit, as shown. The antibonding orbital is made by subtracting the hydrogen atom 1s orbital from the hybrid orbital contribution. The mathematical signs of the lobes of the hybrid orbital do not change, but their sizes are inverted. The portion of the σ* orbital around hydrogen does change in sign because it has been subtracted (the dark portions in the figures below are portions of the orbital having a negative sign.)

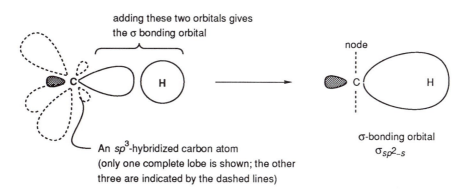

3.13.   (continued)

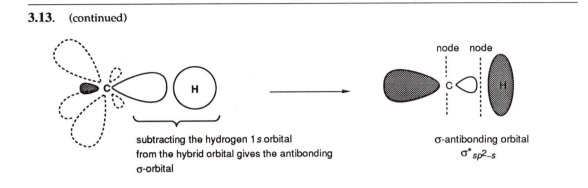

subtracting the hydrogen 1 s orbital
from the hybrid orbital gives the antibonding
σ-orbital

σ-antibonding orbital
$\sigma^*_{sp^2-s}$

3.14.   Resonance forms are drawn according to the procedure outlined in example 3.4 (text page 82). To draw the LCAO representation, follow the instructions given in the solution to exercise 3.10.

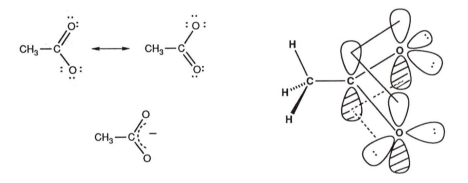

3.15.   Resonance forms are drawn according to the procedure outlined in example 3.4 (text page 82).

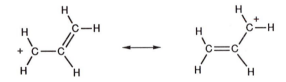

3.16.   Resonance forms are drawn according to the procedure outlined in example 3.4 (text page 82). To draw the LCAO representation, follow the instructions given in the solution to exercise 3.10.

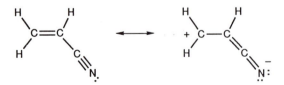

3.16.   (continued)

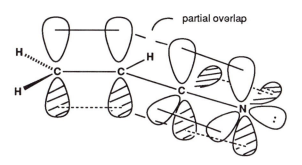

3.17.   For the allyl radical, delocalization takes place to stabilize the electron-deficient carbon atom. Each carbon atom has sp^2 hybridization, so overlap occurs between the orbital with the unshared electron and the π bond of the alkene portion of the molecule.

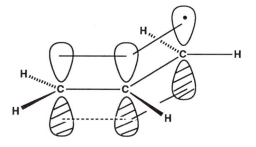

3.18.   The benzyl cation has a structure similar to that of the benzyl radical, shown in Figure 3.31, text page 109, except that only six electrons are distributed among the seven p-orbitals. (The carbon and hydrogen atoms in the back of the following figure have been omitted for clarity.)

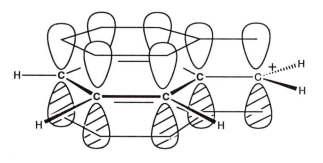

3.19. Hydrogen bonding takes place when a hydrogen atom that is attached to a heteroatom (usually O, N, or S) is proximal to an unshared pair of electrons on a second heteroatom, usually O or N.

a. In ethanol, the proton donor is the alcohol OH group, and the acceptor is the oxygen atom of another molecule of ethanol with its unshared pair of electrons.

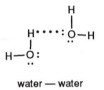

ethanol — ethanol

b. In pure water, the proton donor is the water OH group, and the acceptor is the oxygen atom of another molecule of water with its unshared pair of electrons.

water — water

c. In a mixture of water and ethanol, the proton donor is the OH group of either water or alcohol, and the acceptor is the oxygen atom of either water or alcohol. There are four combinations possible.

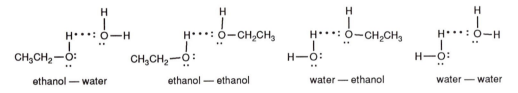

ethanol — water ethanol — ethanol water — ethanol water — water

d. Dimethyl ether cannot form hydrogen-bonds with itself because there is no proton donor group. All of the hydrogen atoms are attached to carbon.

dimethyl ether

e. In a mixture of water and dimethyl ether, the proton donor must be the OH group of water, but the acceptor can be the oxygen atom of either water or the ether, so two combinations are possible.

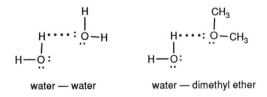

water — water water — dimethyl ether

3.20.   Follow the instructions given in the solution to exercise 3.1

a.   2-methyl-3-pentanone

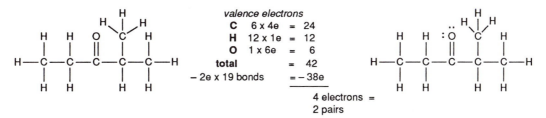

valence electrons
C   6 x 4e   = 24
H   12 x 1e   = 12
O   1 x 6e   =   6
total   = 42
− 2e x 19 bonds   = − 38e
_____
4 electrons =
2 pairs

b.   3,3-dibromopropanenitrile

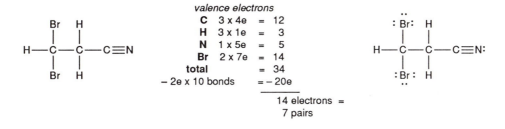

valence electrons
C   3 x 4e   = 12
H   3 x 1e   =   3
N   1 x 5e   =   5
Br   2 x 7e   = 14
total   = 34
− 2e x 10 bonds   = − 20e
_____
14 electrons =
7 pairs

c.   3-hydroxycyclopentanecarboxylic acid

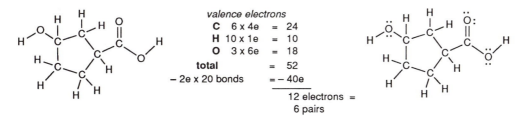

valence electrons
C   6 x 4e   = 24
H   10 x 1e   = 10
O   3 x 6e   = 18
total   = 52
− 2e x 20 bonds   = − 40e
_____
12 electrons =
6 pairs

d.   3-methoxybenzamide

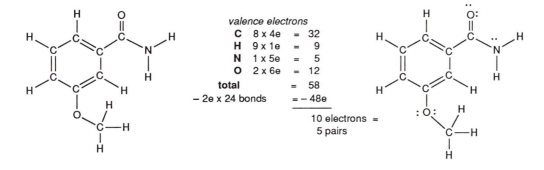

valence electrons
C   8 x 4e   = 32
H   9 x 1e   =   9
N   1 x 5e   =   5
O   2 x 6e   = 12
total   = 58
− 2e x 24 bonds   = − 48e
_____
10 electrons =
5 pairs

3.21.   To assign the hybridization of a carbon atom, look at the number of bonds its forms with neighboring elements. If there are four single bonds, its hybridization is *sp*³. If the carbon atom has a double bond, its hybridization is *sp*². If the carbon atom has a triple bond or two double bonds attached to it, its hybridization is *sp*. A heteroatom is assigned the same hybridization as the type of carbon to which it is attached. When more than one carbon atom is attached to a heteroatom, assign the same hybridization to the heteroatom as that of the attached carbon atom with less *p*-orbital character.

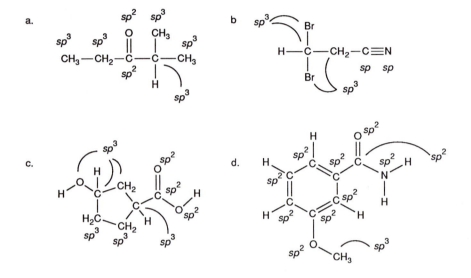

3.22.   One bond between any two atoms is always a sigma bond. If there are additional bonds, for instance when there is a double or triple bond, then the additional ones are π bonds. In the figures below, bold lines are used to indicate π bonds; the others are sigma bonds.

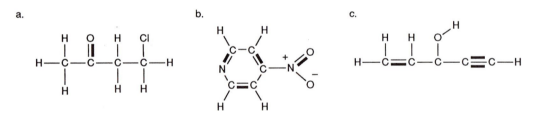

3.23.   Bond length is related to the type of bond: a single bond is longer than a double bond, which in turn is longer than a triple bond. If the same type of bond is being evaluated, then bonds to smaller atoms are shorter. Atom size decreases toward the upper right corner of the periodic table. In the following figures, the shorter bond is indicated by the arrow.

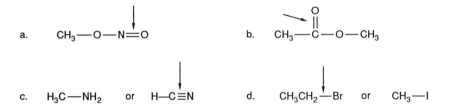

3.24.   Follow the procedure outlined in the solution to exercise 3.2.

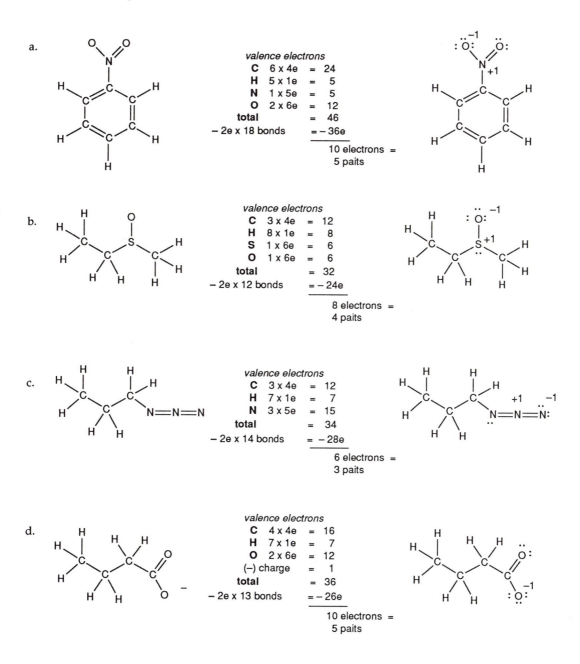

a.

valence electrons

C	6 x 4e	=	24
H	5 x 1e	=	5
N	1 x 5e	=	5
O	2 x 6e	=	12
total			46
− 2e x 18 bonds		=	− 36e

10 electrons =
5 paits

b.

valence electrons

C	3 x 4e	=	12
H	8 x 1e	=	8
S	1 x 6e	=	6
O	1 x 6e	=	6
total		=	32
− 2e x 12 bonds		=	− 24e

8 electrons =
4 paits

c.

valence electrons

C	3 x 4e	=	12
H	7 x 1e	=	7
N	3 x 5e	=	15
total		=	34
− 2e x 14 bonds		=	− 28e

6 electrons =
3 paits

d.

valence electrons

C	4 x 4e	=	16
H	7 x 1e	=	7
O	2 x 6e	=	12
(−) charge		=	1
total		=	36
− 2e x 13 bonds		=	− 26e

10 electrons =
5 paits

3.25. Follow the procedure outlined in the solution to exercise 3.21.

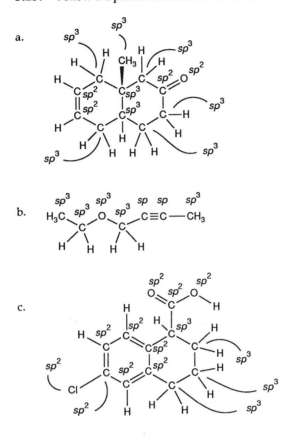

3.26. Follow the procedure outlined in the solution to exercise 3.2. It is not surprising that the adduct between boron trifluoride and diethyl ether is stable because each non-hydrogen atom has an octet of electrons.

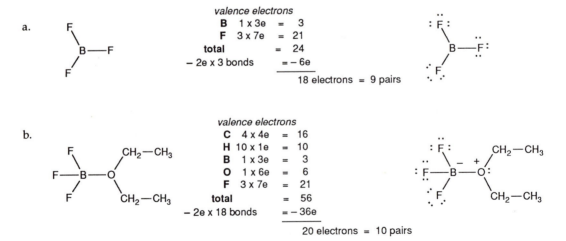

3.27. Follow the instructions given for the solution to exercise 3.21. In the structures below, any unlabeled, non-hydrogen atom has *sp*³ hybridization. There are no atoms with *sp* hybridization in these molecules.

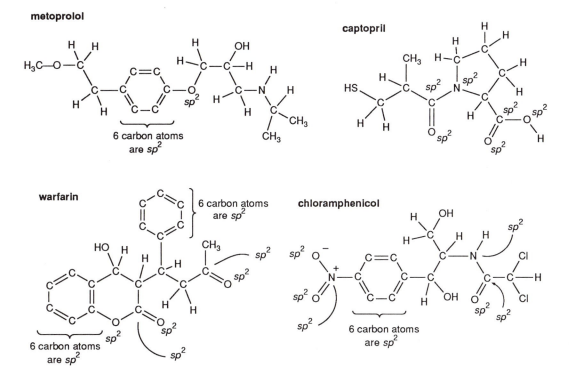

3.28. When drawing resonance structures, nuclei positions must not change, and the number of electrons and atoms must be the same. Pairs of structures in (b), (d), and (f) are legitimate resonance structures.

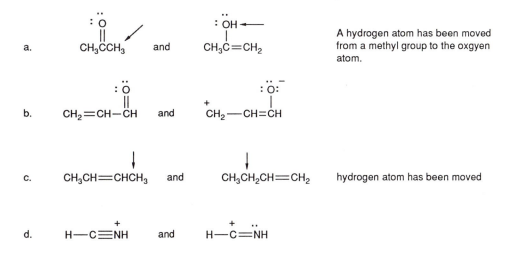

3.28. (continued)

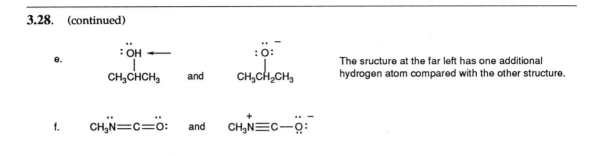

e.   The sructure at the far left has one additional
hydrogen atom compared with the other structure.

f.

3.29.   Follow the procedure outlined in the solution to exercise 3.2.   If we consider only those structures in which each atom has an octet of electrons, we find that three equivalent resonance forms can be drawn for SO_3, whereas a single Lewis structure suffices to represent electron distribution in the sulfite ion if.   Sulfur trioxide should have shorter sulfur–oxygen bonds than the sulfite ion will because each sulfur–oxygen bond has some double bond character.

Sulfur trioxide: *Lewis structure*:

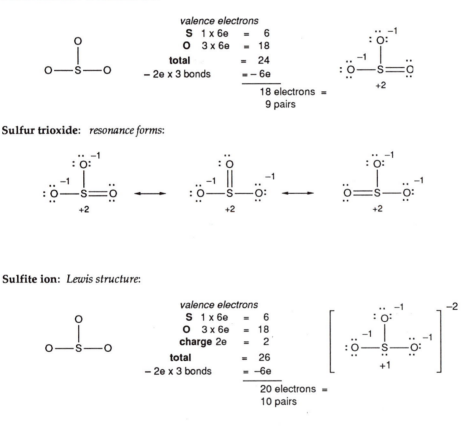

Sulfur trioxide: *resonance forms*:

Sulfite ion: *Lewis structure*:

3.30. The best resonance structures are those in which each non-hydrogen atom has an octet of electrons and no formal charges. The next best structures may have formal charges, but the charges are small (+1 or –1), and a negative charge resides on the more electronegative atom.

a.   Structure **I** is the most important because each atom has an octet of electrons. Structure **II** is also a reasonable contributor because the negative formal charge is on oxygen, a highly electronegative element.

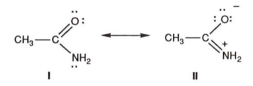

b.   Structure **I** is more important because each atom has an octet of electrons and there are no charges on the atoms. Structure **II** is also reasonable because the negative formal charge is on oxygen, a highly electronegative element.

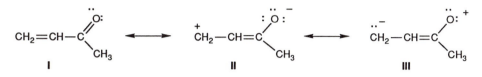

c.   Structure **I** is the most important because each atom has an octet of electrons, and no atom carries a charge. Structure **II** is reasonable because each atom has an octet of electrons. It is less important than **I** because oxygen is a highly electronegative element yet it carries a formal charge of +1.   Structure **III** has a carbon atom with only six electrons, so it is the least important of the resonance forms.

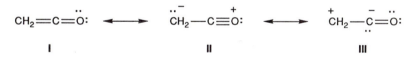

d.   Structures **I** and **II** are equally good and contribute significantly to the resonance hybrid. Structure **III** is much less important because one carbon atom lacks an octet of electrons.

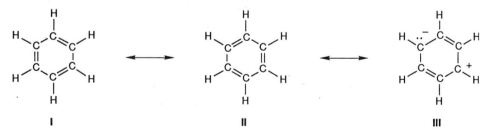

3.31. Follow the procedure outlined in Example 3.4 (text page 82).

a. The anion of 2,4-pentanedione has an overall charge of –1, so each resonance form must have a –1 charge, too. Three good structures can be drawn in which each atom has an octet of electrons.

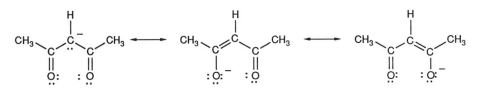

b. Two good structures can be drawn for methyl nitrite in which each non-hydrogen atom has an octet of electrons. Each oxygen atoms in the second structure bears a formal charge.

$$CH_3-\overset{..}{\underset{..}{O}}-N=\overset{..}{\underset{..}{O}}: \longleftrightarrow CH_3-\overset{+1}{\underset{..}{O}}=N-\overset{-1}{\underset{..}{O}}:$$

c. The cation of cycloheptatriene has seven equivalent resonance forms, each one having a positive charge.

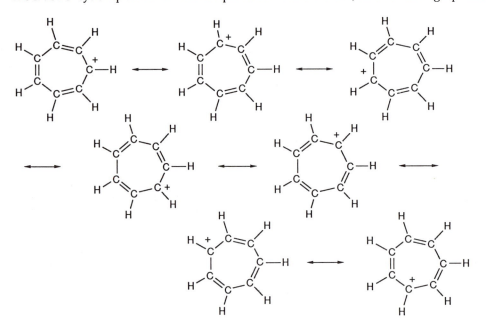

3.32.   Follow the procedure outlined in the solution to exercise 3.10.

a.   This LCAO representation of this compound has already been presented in the solution to exercise 3.16.

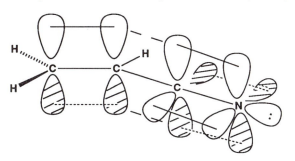

b.   The LCAO representation of acetamide has a double bond between carbon and oxygen because each atom has sp^2 hybridization. The nitrogen atom is also assigned sp^2 hybridization, and its unshared pair of electrons partially overlaps with the carbon-oxygen π bond.

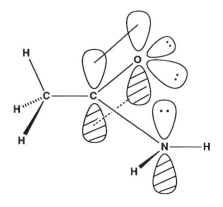

c.   The LCAO representation of methyl vinyl ether has a double bond between the two carbon atoms, both of which have sp^2 hybridization. The oxygen atom is also assigned sp^2 hybridization, and its p-orbital partially overlaps with the carbon-carbon π bond.

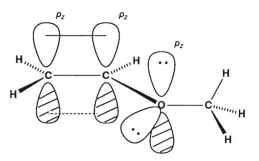

3.32.   (continued)

d.   The LCAO representation of 2-methylpropene has a double bond between two of the carbon atoms that have sp^2 hybridization. The carbon atoms of the two methyl groups have sp^3 hybridization.

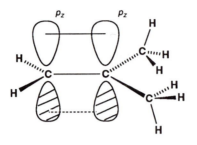

3.33.   Bond angles are determined by the hybridization of the middle atom among the three that defines an angle. If the central atom has sp hybridization, the angle is 180°. If the central atom has sp^2 hybridization, the angle is 120°. If the central atom has sp^3 hybridization, the angle is 109.5°

a.   ∠C2—C1—N   180°   ∠C1—C2—C3 120°   ∠H—C3—H 120°

b.   ∠C—C—N   120°   ∠C—C—O   120°   ∠N—C—O   120°   ∠H—C—H   109.5°

c.   ∠C2—O—C3   120°   ∠C1—C2—O 120°   ∠H—C1—H 120°   ∠H—C3—O 109.5°

d   ∠C4—C2—C1 120°   ∠C4—C2—C3 120°   ∠H—C3—C2 109.5°

3.34.   Follow the procedure outlined in the solution to exercise 3.10. Sigma bonds are shown as lines, and the carbon and hydrogen atoms in the back portion of toluene and pyridine are omitted for clarity.

a.   In toluene, a methyl group takes the place of a hydrogen atom of benzene.

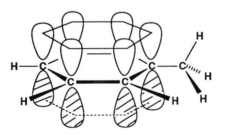

3.34.   (continued)

b.   In pyridine, the unshared pair of electrons on the nitrogen atom take the place of a C–H bond of benzene.

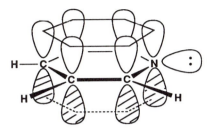

c.   Thiophene has only five atoms in the ring, so one carbon atom of benzene is removed from its representation to produce a drawing for the orbital picture of thiophene.   One unshared pair of electrons on the sulfur atom takes the place of a C–H bond of benzene, and the other unshared pair in its p-orbital overlaps with the p-orbitals of the adjacent carbon atoms.

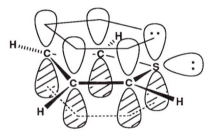

3.35.   Triplet methylene has two electrons with the same spin, and they will repel each other.   If each is in a separate orbital, then repulsive forces are minimized.   Therefore, triplet methylene is likely to have a carbon atom with sp^3 hybridization.

Singlet methylene has two electrons with opposite spins, so they can occupy the same orbital.   According to the VSEPR model, three electron pairs (the two that form the C–H bonds plus the unshared pair) adopt a trigonal arrangement, so the hybridization is likely to be sp^2.

triplet methylene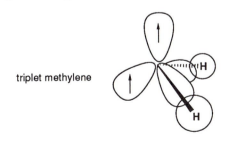

singlet methylene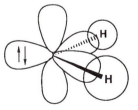

3.36. A hydrogen bond <u>donor</u> is a compound that has at least one O–H, N–H, or S–H bond. A hydrogen bond <u>acceptor</u> has a heteroatom, usually O or N, that also has an unshared pair of electrons (therefore, R_4N^+ compounds are not in this category). Any compound that has a hydrogen bond donor group is also normally a hydrogen bond acceptor because it also has a heteroatom with an unshared pair of electrons.

Compounds that can act as hydrogen bond donors are circled. These will be hydrogen bond donors toward water and toward other like molecules. Those that are <u>only</u> hydrogen bond acceptors are enclosed in a box. These will form a hydrogen bond with water in which the O–H bond of water is the donor. They will not, however, be hydrogen bond donors toward like molecules. The alkene (e) is neither a hydrogen bond donor nor acceptor.

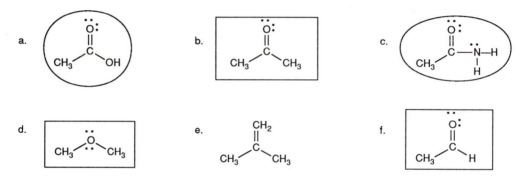

3.37. A compound that forms hydrogen bonds with like molecules will have strong attractive forces that have to be overcome to reach boiling. More heat , therefore, has to be applied to these substances to vaporize them, and their boiling points will be high.

The ketone (a) and the ether (c) do not have an O–H or N–H bond, so these molecules cannot undergo hydrogen bonding with like molecules. The amide has both an N–H bond and a heteroatom with unshared electron pairs (the oxygen atom), so intermolecular hydrogen bonds are formed, as illustrated at the right.

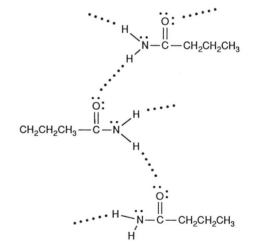

boiling points:

2-pentanone	≈ 105°
butanamide	≈ 220°
ethyl propyl ether	≈ 72°

3.38.   A compound that forms hydrogen bonds with water will be more soluble than those that do not. Even though the ketone and ether can act as a hydrogen bond acceptor toward water, butanamide can be both a hydrogen bond donor and acceptor, so it likely will be more soluble.

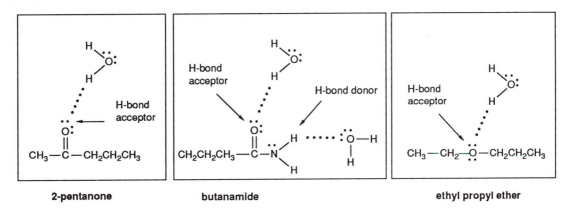

2-pentanone   butanamide   ethyl propyl ether

3.39.   Hydrogen bonds form when an N–H or O–H group comes within about 3Å of a heteroatom that has an unshared electron pair. In the reverse turn of a protein, the amide N–H bonds form hydrogen bonds with the unshared electron pairs on the carbonyl group of another amide bond 3 or 4 amino acids away. It is also possible that the NH group of residue #3 can hydrogen bond with the carbonyl oxygen atom of residue #1.

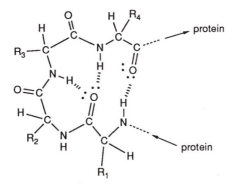

3.40.   Hydrogen bonds form when an N–H or O–H group comes within about 3Å of a heteroatom that has an unshared electron pair. In the compounds shown, the amide N–H bonds form hydrogen bonds with an electron pair on the oxygen atoms of the carbonyl groups of the other compound.

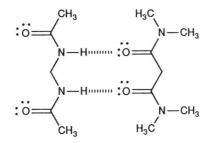

CONFORMATIONS OF ORGANIC MOLECULES

4.1. To construct the sawhorse representation of a molecule, draw an angled line that corresponds to the central carbon-carbon bond. Attach the two carbon atoms that will be methyl groups. In the syn-periplanar conformation, these carbon atoms are eclipsed. Finally, attach hydrogen atoms to each carbon atom.

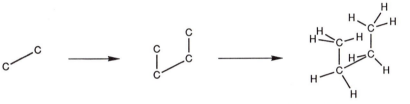

The methyl groups are eclipsed.

Notice that the hydrogen atoms on the two central carbon atoms are eclipsed, but one hydrogen atom on each of the terminal carbon atoms is staggered with respect to the central carbon-carbon bond. These staggered relationships between carbon-hydrogen and carbon-carbon bonds are emphasized in the figure below with use of bold lines.

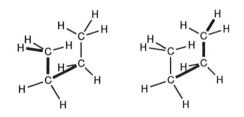

4.2. The most stable conformation of an alkane exists when the carbon-carbon bonds are staggered with respect to neighboring bonds. Other atoms are also staggered, and any atom bigger than hydrogen will tend to be as far as possible from any large atom attached to the adjacent carbon atom. To draw the sawhorse projection, follow the steps shown in the solution to exercise 4.1. To convert that representation to a Newman projection, sight along the bond that you drew originally, and add the substituents as they appear in the sawhorse projection.

4.2.   (continued)

a.   **Propane (C1-C2 bond).** This compound is not unlike butane (exercise 4.1) except that it has one less methyl group. The staggered conformation is the most stable, so the methyl group should be as far as possible from the hydrogen atoms on C1.

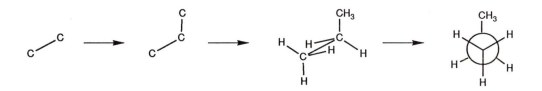

b.   **2-Methylpropane (C1-C2 bond).** This compound looks like propane itself except that one hydrogen atom at C2 is replaced by a methyl group.

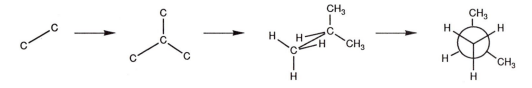

c.   **Hexane (all atoms).** This alkane adopts a conformation in which every bond is staggered with respect to the bonds on adjacent carbon atoms.

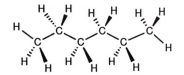

d.   **2,2-Dichlorobutane (C2-C3 bond).** This compound looks like butane itself except that two of the hydrogen atoms at C2 are replaced by chlorine atoms. A chlorine atom is smaller than a methyl group, so the two methyl groups (C1 and C4) of the butane adopt an *anti* orientation.

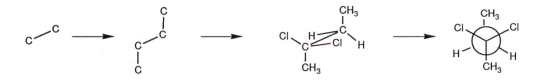

4.2. (continued)

The energy diagrams for propane and for 2-methylpropane reflect the limiting conformations that exist. For each compound, only two are unique: staggered and eclipsed, so a plot of energy vs. dihedral angle looks like the diagram that we construct for the conformations of ethane.

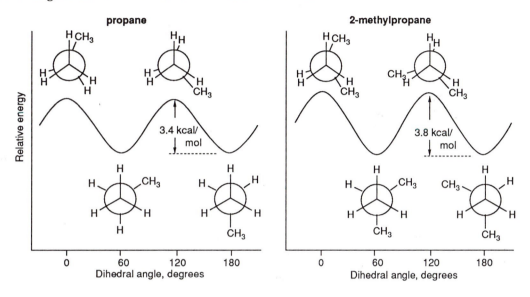

The only difference is the magnitude of the energy difference for the limiting conformations. Each H•••H eclipsing interaction is about 1 kcal/mol and each H•••CH$_3$ eclipsing interaction is about 1.4 kcal/mol. Therefore, the differences in energy between eclipsed and staggered for propane has a value of 1 + 1 + 1.4 = 3.4 kcal/mol, and that for 2-methylpropane has a value of 1+ 1.4 + 1.4 = 3.8 kcal/mol.

4.3. Draw a Newman projection by sighting along each of the carbon–carbon bonds in turn. Place the substituents in the appropriate orientations. Sighting along the C1-C2 and the C1-C6 bonds is illustrated in the text on page 134.. The views that you see looking along the other bonds are illustrated below.

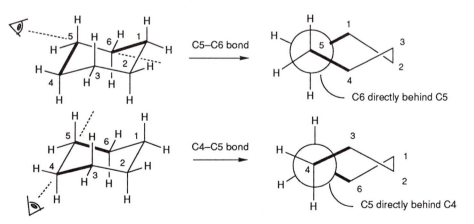

4.3.   (continued)

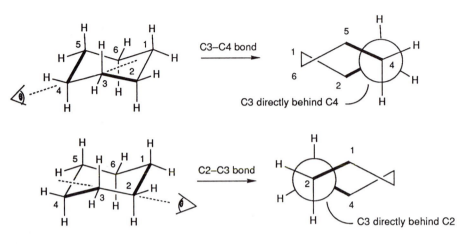

4.4.   Two of the eclipsing interactions in the half chair and boat conformations of cyclohexane are shown in the text on page 137. There are four such interactions in each of the two forms because adjacent carbon atoms are in the same plane. This coplanarity orients some hydrogen atoms so that they point in the same direction, which makes them eclipsing. All sets of eclipsing interactions are indicated below.

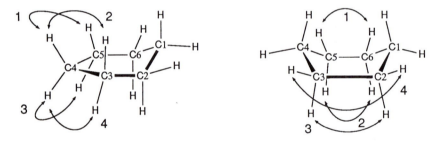

4.5.   The Newman projection of the C1-C2 bond in the boat conformation of cyclohexane is generated by sighting along the C1–C2 bond and orienting the substituents as they appear. The C2–C3 bond has a pseudo-*gauche* orientation with respect to C1–C6.

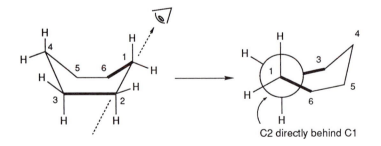

4.6. Cyclopentane itself has two *gauche* and two pseudo-*gauche* interactions within the cyclic framework. When an axial methyl group is included, two additional *gauche* interactions are created (shown below in bold lines). When the methyl group is equatorial, its relationship to the carbon-carbon bonds within the ring are *anti*. Therefore, the conformation having an equatorial methyl group is expected to be more stable.

Cyclopentane

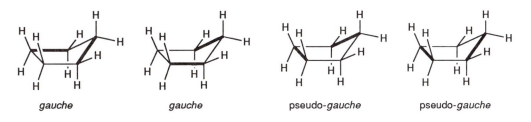

gauche   gauche   pseudo-*gauche*   pseudo-*gauche*

Methylcyclopentane (axial)

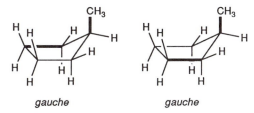

gauche   gauche

Methylcyclopentane (equatorial)

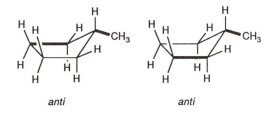

anti   anti

4.7. Use equations 4.2 and 4.3 (text page 142) to calculate K_{eq} and the percentage of B.

a. $\Delta G^\circ = +5.0 \text{ kcal/mol}$ $K_{eq} = 1.99 \times 10^{-4}$ %B = 1.99%

b. $\Delta G^\circ = -0.5 \text{ kcal/mol}$ $K_{eq} = 5.03 \times 10^{3}$ %B = 99.98%

c. $\Delta G^\circ = -5.0 \text{ kcal/mol}$ $K_{eq} = 2.35$ %B = 70.0%

4.8.   For the substituted cyclohexane derivatives, use equations 4.2 and 4.3 (text page 142) to calculate K_{eq} and the percentage of equatorial isomer (= "B").

Substituent	$\Delta G°$	K_{eq}	% equatorial isomer
—H	0.0	1.00	50
—F, —CN	−0.2	1.41	59
—Cl, —Br, —C≡CH	−0.5	2.34	70
—OCH$_3$	−0.6	2.78	74
—OH	−1.0	5.50	85
—COOH	−1.4	10.8	92
—CH$_3$, —CH=CH$_2$	−1.7	18.1	95
—CH$_2$CH$_3$	−1.8	21.5	96
—CH(CH$_3$)$_2$	−2.1	35.9	97
—C$_6$H$_5$	−2.9	140	99
—C(CH$_3$)$_3$	−5.4	9.9×10^4	≈ 100

4.9. The unique *gauche* interactions in a substituted cyclohexane derivative exist between the carbon-carbon bonds in the side chain and those in the ring that are separated by a dihedral angle of 60°. It is easier to see these *gauche* interactions by looking at Newman projections along the bond from the alkyl group to C1 of the cyclohexane ring. These are illustrated below each chair form of the specific cyclohexane derivative. In all of the structures, carbon-carbon bonds that have a *gauche* relationship are indicated by bold lines.

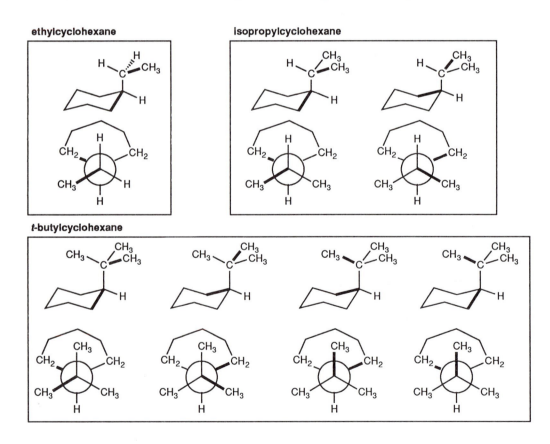

4.10.   Besides the usual *gauche* interactions that exist within the ring of a cyclohexane ring, *gauche* interactions also exist between a methyl group and any carbon-carbon bonds in the ring that is separated by a dihedral angle of 60° or between the two methyl groups themselves if they are 60° apart.   In all of the structures shown below, carbon-carbon bonds that have a *gauche* relationship are indicated by bold lines.

Newman projections are generated by looking along the bond in the direction indicated by the arrow. The curved line in the Newman projections is being used to indicate the remaining portion of the cyclohexane ring.   Notice that the *cis* isomer has three additional *gauche* interactions besides those in the six-membered ring.   The *trans* isomer has only one additional *gauche* interaction when both methyl groups are equatorial. The *trans* isomer has four additional *gauche* interactions when both methyl groups are axial.   This result illustrates why larger groups are normally equatorial.

***cis*-1,2-dimethylcyclohexane:**

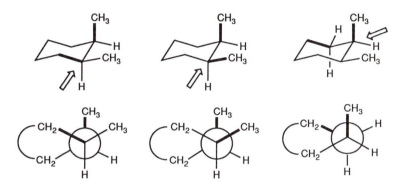

***trans*-1,2-dimethylcyclohexane:** (methyl groups equatorial):

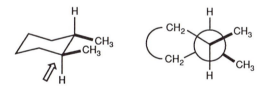

***trans*-1,2-dimethylcyclohexane** (methyl groups axial):

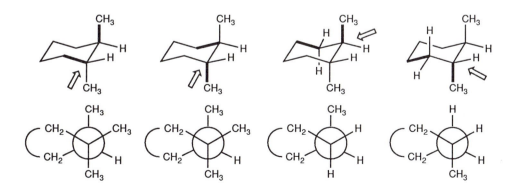

4.11.   To draw the structures of the conformers of the 1,4-dimethylcyclohexane isomers, first draw the chair form and place a methyl group at C1 in the axial position. Then attach the methyl group at C4, also in the axial position. This is clearly the *trans* isomer. A ring flip keeps the configuration at each carbon atom the same, but generates the more stable conformer, which has both methyl groups in equatorial positions.

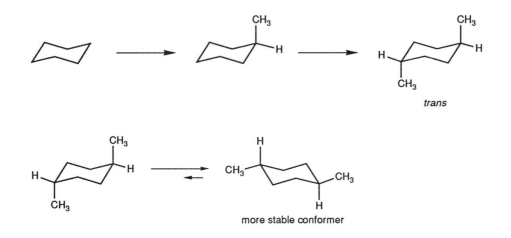

Switching the configuration at C4 generates the *cis*-isomer, the conformers of which have equal energies.

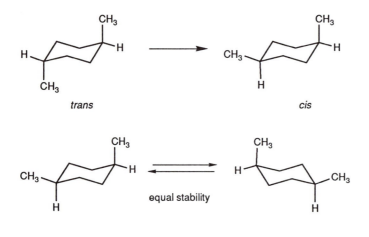

4.12.   To draw the more stable chair conformation of a disubstituted cyclohexane derivative, place one substituent in the axial position at C1. Then attach the other group in the axial position at the appropriate position of the ring. Next, ask yourself, "is this the correct isomer?" If not, then switch the position of the second substituent so that it is equatorial instead of axial. Finally, perform a ring flip and decide which conformer is more stable based on having the larger group (Table 4.2, text page 143) in the equatorial positions.

4.12.   (continued)

a.   ***trans*-1-Bromo-3-fluorocyclohexane**

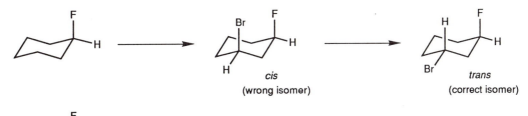

cis
(wrong isomer)

trans
(correct isomer)

more stable conformer

b.   ***cis*-2-Ethylcyclohexanecarboxylic acid**

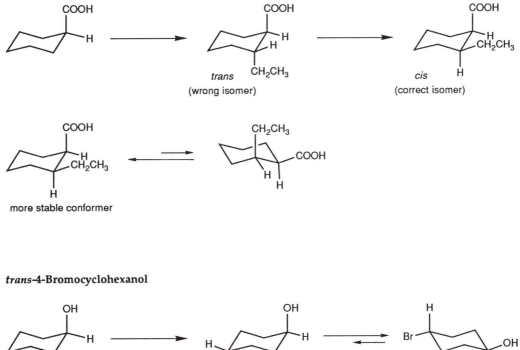

trans
(wrong isomer)

cis
(correct isomer)

more stable conformer

c.   ***trans*-4-Bromocyclohexanol***

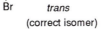

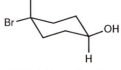

trans
(correct isomer)

more stable conformer

4.12. (continued)

d. *cis*--1-*tert*-Butyl -2-methyl cyclohexane

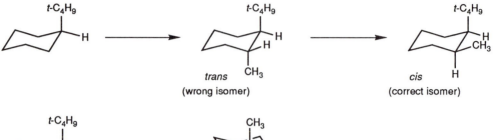

 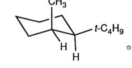

more stable conformer

e. *cis*-1-Isopropyl-4-phenylcyclohexane

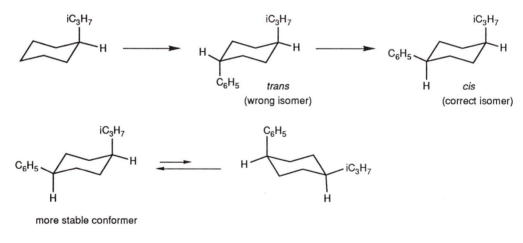

f. *trans*-1,2-Dimethoxycyclohexane

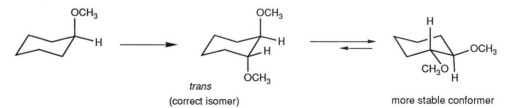

4.13. To construct the conformers of decalin, first draw the ring system for the two isomers, *trans-* and *cis-*decalin. Next attach a methyl group at C2 in the axial and equatorial positions of each isomer. For *cis-*decalin, perform a ring flip and decide which conformer is more stable. In the structures of the *cis-*isomers drawn below, the more stable conformation is the one with fewer alkyl substituents in axial positions. These axial alkyl groups are indicated with use of bold lines.

2-Methyl-*trans*-decalin

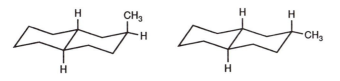

2-Methyl-*cis*-decalin

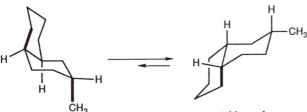

more stable conformer

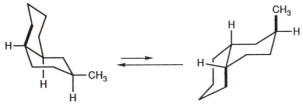

more stable conformer

4.14. Use equations 4.2 and 4.3 (text page 142) to calculate K_{eq} and percentage of the equatorial conformer of 4-methylcyclohexene:

$\Delta G° = -1.0 \text{ kcal/mol}$ $K_{eq} = 5.50$ %conformer having an equatorial methyl group = 85%

4.15. A ring flip of the illustrated conformer of D-glucose puts the larger substituent at each carbon atom in an axial position.

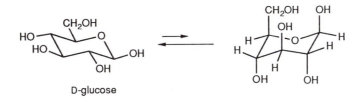

D-glucose

4.16.   The boat conformation of cycloheptane is generated by flipping one end so that it points in the same direction (up or down) that the other end points. There are six eclipsed interactions that can be identified in the boat conformation.

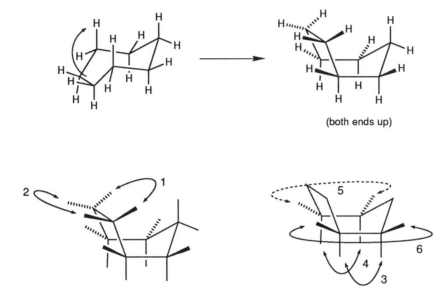

(both ends up)

There are two *gauche* and four pseudo-*gauche* interactions, shown with bold lines on the figures below.

gauche

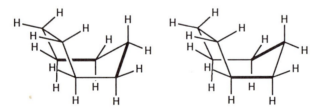

pseudo-*gauche*

4.17. A spiro compound is a bicyclic molecule with two rings that share a common atom, and the numbers in brackets indicate how many atoms are in the links that form each ring. Bridged bicyclic compounds have two carbon atoms in common, and the numbers in brackets indicate the numbers of atoms that link these two bridgehead positions. When one of the numbers in the brackets is zero, a bridged bicyclic compound has fused rings. The total number of carbon atoms in all bicyclic compounds is given by the root name.

a.   spiro[5.3]nonane
         nonane: 9 carbon atoms
         spiro: shared atom, bridged by 3 and 5 carbon atoms.

b.   bicyclo[3.2.1]octane
         octane: 8 carbon atoms
         bicyclo: bridgehead carbon atoms are linked
                    by 1, 2, and 3 carbon atoms.

c.   bicyclo[3.3.0]octane
         octane: 8 carbon atoms
         bicyclo: bridgehead carbon atoms are linked
                    by 0, 3, and 3 carbon atoms.

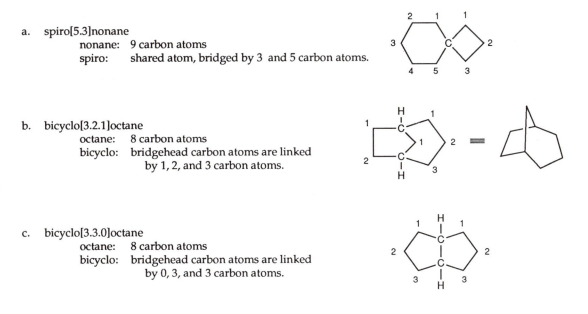

4.18. Follow the procedure outlined in the solution to exercise 4.1.

a.   **1-Chloropropane (C1–C2 bond).** This compound is like butane (exercise 4.1) except that a chlorine atom replaces one of the terminal methyl groups. The *anti* conformation orients the C3 methyl group and chlorine atom as far as possible from each other.

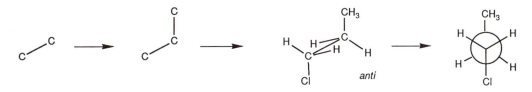

The *gauche* conformation orients the C3 methyl group and chlorine atom 60° apart as you look down the C1–C2 bond.

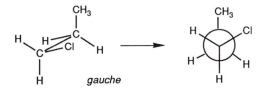

4.18.   (continued)

There are two eclipsed conformations. One has the two large groups pointing in the same direction. The other has the chlorine atom in front of a hydrogen atom.

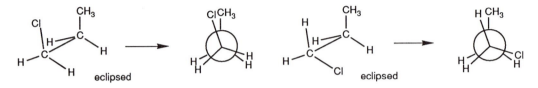

b.   **2-Iodobutane (C2–C3 bond).** The *anti* conformation is that in which the methyl groups attached to C2 and C3 are as far from each other as possible. The iodine atom is *gauche* to the methyl group attached to C3.

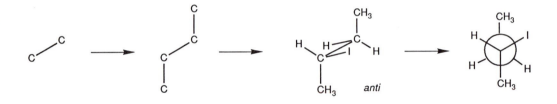

There are two eclipsed conformations, one in which a methyl group is in front of the other, and one in which the iodine atom is in front of the methyl group attached at C3.

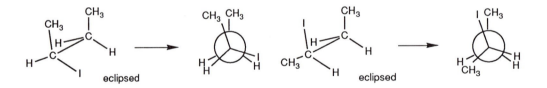

The conformation in which the methyl groups are *gauche* puts the iodine atom either *gauche* or *anti* to the methyl group. The conformation with the iodine atom *anti* to a methyl group is more stable because it puts the two largest groups farthest apart.

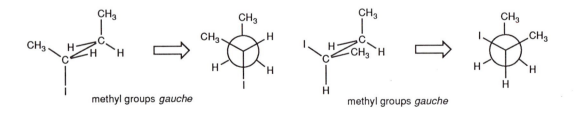

4.19. To calculate ΔG° for an equilibrium situation, determine the value of K_{eq}, which for a simple equilibrium between two species, is equal to the concentration of the product over the concentration of the reactant.

$$K_{eq} = [gauche]/[anti] = 11/89 = 0.1236 \qquad \Delta G^\circ = -2.303RT\log K_{eq} = -2.303(1.986)(298)\log(0.1236)$$
$$= 1238 \text{ cal/mol}$$
$$= 1.24 \text{ kcal/mol}$$

Compare this value of 1.24 kcal/mol with those calculated in exercise 4.8.

4.20.   The rotational barrier in ethane is about 3 kcal/mole, and C1 has three hydrogen atoms attached to it. The rotational barrier in methanol is 1 kcal/mol, and the oxygen atom has only 1 hydrogen atom attached to it. Therefore, an unshared pair has essentially no effect on the rotational energy barrier. Methylamine, with two hydrogen atoms attached to nitrogen, probably has a barrier of about 2 kcal/mol.

The conformation in oxygen-containing heterocycles is consistent with the notion that an unshared electron pair does not affect conformations (Figure 4.25, text page 155). If the repulsive force of an unshared pair is inconsequential, then a methyl group in the axial position of a six-membered ring oxygen heterocycle experiences no repulsion from an unshared electron pair of an oxygen atom at position 3.

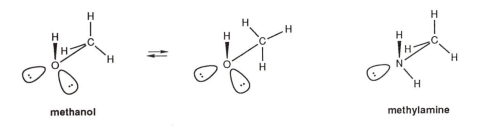

methanol   methylamine

4.21.   Follow the procedure outlined in the solution to exercise 4.2.

a. **Methyl propanoate (C2–C3 bond).** This compound is like propane (exercise 4.2) except that an ester group replaces the methyl group. The staggered conformation is the more stable and the eclipsed conformation is the less stable.

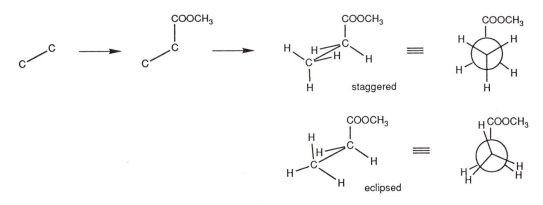

4.21.   (continued)

b.   **1-Butyne (C3–C4 bond).** This compound has the same conformations as methyl propanoate. The staggered conformation is the more stable and the eclipsed conformation is the less stable.

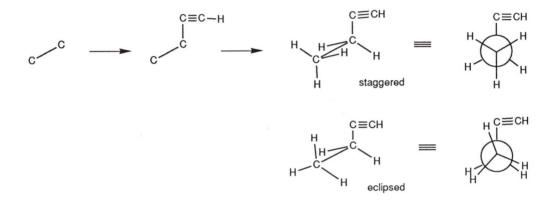

4.22.   Break each name into its constituent parts, then combine the pieces to form the structure.

a.   *cis*-1-bromo-2-methylcyclopentane

cyclopent	five-membered ring
ane	no double or triple bonds
1-bromo	bromine atom at C1 of the ring
2-methyl	methyl group at C2 of the ring
cis-	substituents are on the same side of the ring (both up)

b.   2,2,-difluorocyclohexanone

cyclohex	six-membered ring
an	no double or triple bonds
one	ketone functional group (defines C1)
2,2-difluoro	two fluorine atoms at C2 of the ring

c.   2-cyclohexen-1-ol

cyclohex	six-membered ring
en	a double bond starting at C2
1-ol	alcohol functional group; OH at C1

4.22.   (continued)

d.   *trans*-4-*tert*-butylcyclohexanecarboxylic acid

cyclohex	six-membered ring
an	no double or triple bonds
carboxylic acid	carboxylic acid functional group; –COOH at C1
4-*tert*-butyl	a *tertiary*-butyl group at C4 of the ring
trans-	the *tert*-butyl group and the –COOH group are on opposite sides of the ring (one up, one down)

e.   *cis*-3-bromocyclobutanol

cyclobut	four-membered ring
an	no double or triple bonds
ol	alcohol functional group; the point of attachment of the OH group defines C1
3-bromo	bromine atom at C3 of the ring
cis-	Br and OH groups are on the same side of the ring

4.23.   To evaluate the stereochemical and isomeric relationships between compounds, consider the position of attachment for each substituent, determine the stereochemistry (*cis* or *trans*), then evaluate conformations (axial, equatorial, staggered, *gauche*, eclipsed, *anti*, *syn*-periplanar). If the point of attachment of substituents is different, the compounds are structural isomers. If the stereochemical relationships of substituents differ, the compounds are geometric isomers. And if conformational relationships are different, the substances are conformers.

a.   The attachment points are the same, and the substituents in each compound are *cis*. Only the conformation of **B**. is different   Therefore, **A** and **C** are identical; **A** and **B** are conformers; and **B** and **C** are conformers.

A	B	C
1-chloro	1-chloro	1-chloro
2-methyl	2-methyl	2-methyl
cis	*cis*	*cis*
axial; equatorial	equatorial; axial	axial; equatorial

4.23.   (continued)

b.   The attachment points are the same, and there is no stereochemistry to consider. Only the conformation of **C** is different. Therefore, **A** and **B** are identical; **A** and **C** are conformers; and **B** and **C** are conformers.

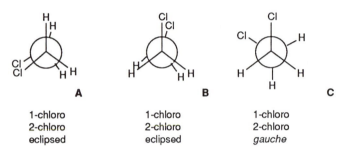

A	**B**	**C**
1-chloro	1-chloro	1-chloro
2-chloro	2-chloro	2-chloro
eclipsed	eclipsed	*gauche*

c.   The attachment points are the same for **A** and **C**, but **B** is different (1,4–). **A** and **C** differ in the stereochemical relationships of the substituents (*cis* versus *trans*.) Therefore, **A** and **B** are structural isomers; **A** and **C** are geometric isomers; and **B** and **C** are geometric isomers. If compounds are structural or geometric isomers, then we do not have to consider conformational differences.

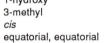

A	**B**	**C**
1-hydroxy	1-hydroxy	1-hydroxy
3-methyl	4-methyl	3-methyl
trans	*cis*	*cis*
axial; equatorial	equatorial; axial	equatorial, equatorial

4.24.   Follow the procedure outlined in the solution to exercise 4.12.

a.   *cis*-**1-Bromo-2-ethylcyclohexane**

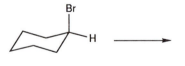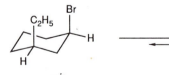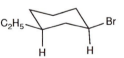

cis
(correct isomer)

more stable conformer

4.24.   (continued)

b.   *cis*-**4-Hydroxycyclohexanecarboxylic acid**

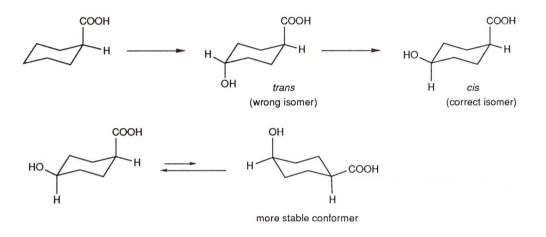

c.   *trans*-**2-Isopropylcyclohexanol**

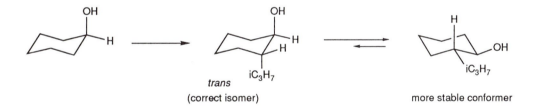

d.   **4-Bromocyclohexanone.** This compound cannot exist as geometric isomers. Place the carbonyl group at C1, then attach the bromine atom at C4. The compound with the bromine atom in the equatorial position is more stable.

4.25. Follow the procedure outlined in Appendix A. First, identify the highest priority functional group. The ring attached to the functional group is chosen as the root word. Numbering of the ring is done to give the principal functional group and substituents the lowest possible numbers, and the necessary suffix for the principal functional group (Table 2.1, text page 33) is added to the root word. Then the substituents are included in the front field of the name, along with stereochemical descriptors.

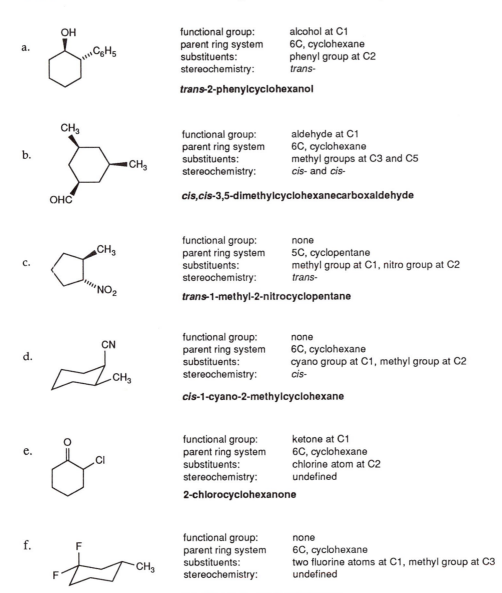

a.

functional group:	alcohol at C1
parent ring system	6C, cyclohexane
substituents:	phenyl group at C2
stereochemistry:	*trans-*

trans-2-phenylcyclohexanol

b.

functional group:	aldehyde at C1
parent ring system	6C, cyclohexane
substituents:	methyl groups at C3 and C5
stereochemistry:	*cis-* and *cis-*

cis,cis-3,5-dimethylcyclohexanecarboxaldehyde

c.

functional group:	none
parent ring system	5C, cyclopentane
substituents:	methyl group at C1, nitro group at C2
stereochemistry:	*trans-*

trans-1-methyl-2-nitrocyclopentane

d.

functional group:	none
parent ring system	6C, cyclohexane
substituents:	cyano group at C1, methyl group at C2
stereochemistry:	*cis-*

cis-1-cyano-2-methylcyclohexane

e.

functional group:	ketone at C1
parent ring system	6C, cyclohexane
substituents:	chlorine atom at C2
stereochemistry:	undefined

2-chlorocyclohexanone

f.

functional group:	none
parent ring system	6C, cyclohexane
substituents:	two fluorine atoms at C1, methyl group at C3
stereochemistry:	undefined

1,1-difluoro-3-methylcyclohexane

4.26. For a cyclohexane ring with three substituents, follow the same procedure that was outlined in the solution to exercise 4.24.

a.   *cis*-4-Ethyl-*trans*-3-methylcyclohexanol

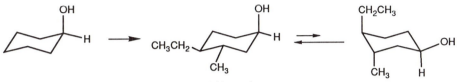

more stable conformer

b.   *cis*-2-Bromo-*cis*-3-methylcyclohexanecarboxylic acid

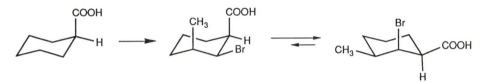

more stable conformer

c.   *trans*-3-Bromo-*trans*-4-chloro-1-methylcyclohexane

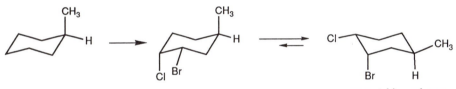

more stable conformer

d.   *trans*-4-Methyl-*cis*-1,3-cyclohexanediol

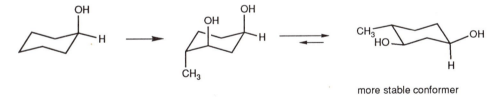

more stable conformer

4.27. A nitrile group is linear, so it has small steric requirements. Steric interactions are therefore slight when a CN group is in the axial position. A methyl group occupies a larger volume, which leads to greater steric interactions with axial hydrogen atoms at C3 and C5.

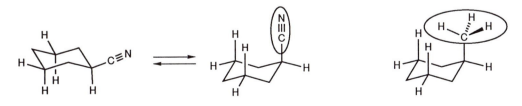

4.28.   Isoprene is 2-methyl-1,3-butadiene.

2-Methyl-1,3-butadiene

but	four carbon atoms
diene	two double bonds, one starting at C1 and one starting at C3
2-methyl	methyl group at C2

For a bridged bicyclic hydrocarbon, the root name reflects the total number of carbon atoms in the parent structure (erase the methyl and isopropyl substituents first). Next, add the prefix bicyclo-. Within the brackets, list in decreasing numerical value the number of carbon atoms needed to link the two bridgehead carbon atoms (indicated as "c" in the structures shown below).

pinane	**bornane**	**carane**	**thujane**
bicyclo[3.1.1]heptane	bicyclo[2.2.1]heptane	bicyclo[4.1.0]heptane	bicyclo[3.1.0]hexane

---

4.29.   The IUPAC name of menthol is derived by specifying the individual substituents attached to the six-membered ring, as indicated below:

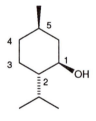

cyclohexane: six-membered ring;
alcohol (OH group) defines C1: suffix -ol
isopropyl group at C2, *trans* to OH group
methyl group at C5, *cis* to OH group

trans-2-isopropyl-*cis*-5-methyl-1-cyclohexanol

From the structure given, draw its conformation by placing the methyl group at C1 in the axial position, then add the OH group (also up) and the isopropyl group (down). After a ring flip, the CH₃ and OH groups are still on the top side of the ring, and the isopropyl group remains down. The more stable conformation has the larger groups in equatorial positions.

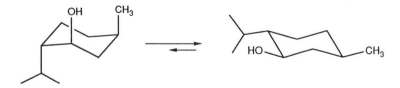

4.30.   In the boat conformation of cyclohexane, there are two types of carbon atoms: the four in the plane and the two above the plane. Four unique conformations therefore exist based on placing a substituent in the axial or equatorial position of each of the two unique carbon atom environments.

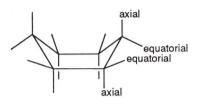

The four different conformers are generated by putting the methyl group at each possible point of attachment. Unfavorable interactions are indicated below each structure and the problematic bonds are shown in bold. Only one conformer lacks unfavorable interactions among the methyl group and bonds within the ring.

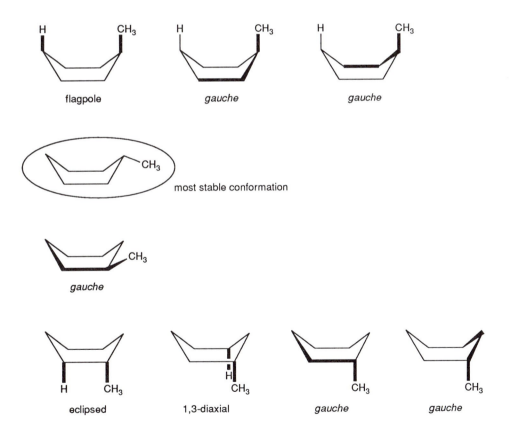

4.31.   Follow the procedure outlined in the solution to exercise 4.17. To carry out conformational analysis of bicyclic compounds, consider the typical conformations that exist for the different rings. A six-membered ring will be in the chair form with atoms in the other ring in equatorial positions, if possible. A five-membered ring will be in the envelope form, and a four-membered ring has a butterfly shape.

a.   **spiro[3.5]nonane**
        nonane   nine carbon atoms
        links: three and five carbon atoms

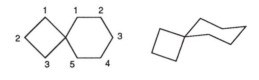

b.   ***trans*-bicyclo[3.3.0]octane**
        octane   eight carbon atoms
        links: three, three, and zero carbon
                        atoms
        ring junction *trans*

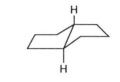

c.   **spiro[2.4]heptane**
        heptane   seven carbon atoms
        links: two and four carbon atoms

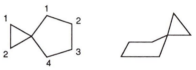

d.   ***trans*-2,2-dibromobicyclo[4.4.0]decane**
        decane   10 carbon atoms
        links: four, four, and zero carbon atoms
        ring junction *trans*
        substituents   two bromine atoms at C2

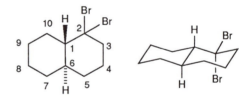

e.   **2-methylspiro[3.3]heptane**
        heptane   seven carbon atoms
        links: three and three carbon atoms
        substituent   methyl group at C2

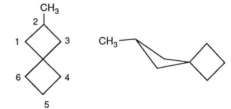

f.   **bicyclo[4.3.1]decane**
        decane   10 carbon atoms
        links: four, three, and one carbon atom

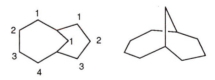

4.32. The process for drawing isomers and conformers of a four-membered ring is not much different than that used for the case of cyclohexane. Attach the substituents at the appropriate carbon atoms so that each combination of *cis* and *trans* geometries are formed. Then perform a ring-flip. Conformers with a greater number of equatorial substituents are expected to be more stable.

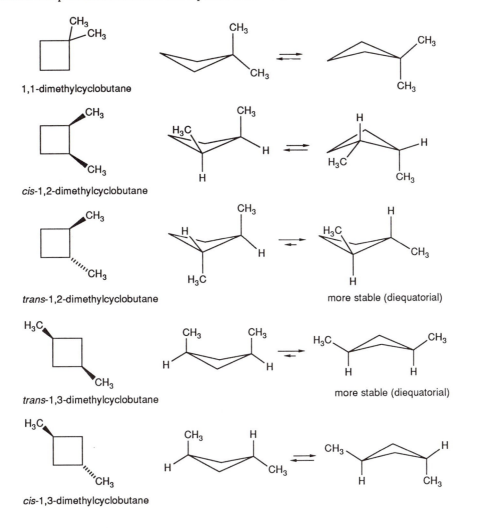

1,1-dimethylcyclobutane

cis-1,2-dimethylcyclobutane

trans-1,2-dimethylcyclobutane   more stable (diequatorial)

trans-1,3-dimethylcyclobutane

more stable (diequatorial)

cis-1,3-dimethylcyclobutane

4.33. Like decalin, the *trans* isomer of hydrindane has only a single conformation. The *cis* isomer undergoes a ring-flip, and the two conformations have equal energies because the five-membered ring is attached in both axial and equatorial positions to the six-membered ring.

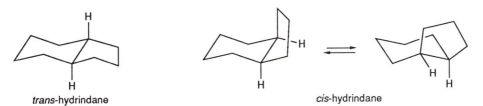

trans-hydrindane   *cis*-hydrindane

4.34.   If *cis*-1,4-di-*tert*-butylcyclohexane exists in the chair conformation, one of the *tert*-butyl groups would have to be in an axial position.

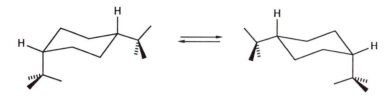

By adopting a twist-boat conformation, the two large groups occupy pseudo-equatorial or pseudo-axial positions. The conformation with the two pseudo-equatorial groups is more stable because there are fewer *gauche* interactions.

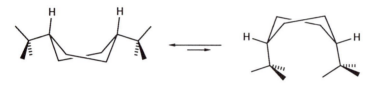

4.35.   In exercise 4.20 you learned that the principal factor that influences conformations in acyclic compounds is the number of gauche interactions between substituents attached to adjacent atoms. An unshared electron pair has essentially no influence on the energy of any conformation; we therefore infer that steric effects that involve an unshared electron pair are minimal. Because 1,3-diaxial effects are fundamentally steric effects, fewer such interactions exist when an electron pair is axial. This lack of steric influence of an electron pair means that any substituent attached to a nitrogen atom tends to be in the equatorial position.

4.36. *trans*-Decalin has no *gauche* interactions besides those already present within the six-membered rings. *cis*-Decalin, in contrast, has three *gauche* interactions between bonds in separate rings. Two 1,3-diaxial interactions also exist between a carbon-carbon bond in one ring and two hydrogen atoms attached to the other ring. These steric interactions in the *cis* isomer make it less stable than *trans*-decalin.

| *gauche* | *gauche* | *gauche* | 1,3-diaxial | 1,3-diaxial |

4.37. The most stable conformation of a cyclohexane derivative is the chair form with the "larger" groups in equatorial positions. Because a methyl group is larger than a carboxylic acid group, as indicated by the greater value for $\Delta G\,^{\circ}$ in Table 4.2 (text page 143), methyl groups occupy the equatorial positions in Kemp's triacid.

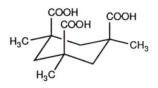

4.38. In methylcyclohexane, the principle destabilizing factor, when the methyl group is axial, comprises two 1,3-diaxial interactions, indicated by the curved arrows in the structure directly below:

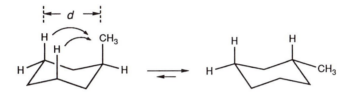

In 4-methylcyclohexene, only one such diaxial interaction exists. The distance, d between the methyl group and the hydrogen atom is longer in the unsaturated compound than it is in methylcyclohexane.

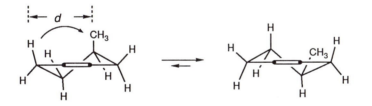

4.39. Hydrogen bonding between OH groups can provide stabilizing interactions in six-membered rings, but the most stable conformation of a substituted cyclohexane is that in which any group larger than hydrogen is equatorial. The conformation expected to be most stable therefore has the six OH groups equatorial.

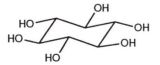

STEREOCHEMISTRY OF ORGANIC MOLECULES

5.1. The stereochemical classification of an alkene double bond is made as follows: for the pair of substituents at each end of the double bond, one is assigned a higher priority ranking based on atomic number of the atom (or atoms) that constitute the substituent. In the answers given below, the group that has the higher priority or the portion that gives the substituent its higher priority is circled. Functional groups that have multiple bonds are first converted to their single bond equivalents.

Thus, if the higher priority groups at the ends of the double bond are on the same side of the bond axis (shown as a dashed line in the structures below), the prefix (Z) is appropriate; if the higher priority groups at the ends of the double bond are on opposite sides of the bond axis, then (E) is the designation.

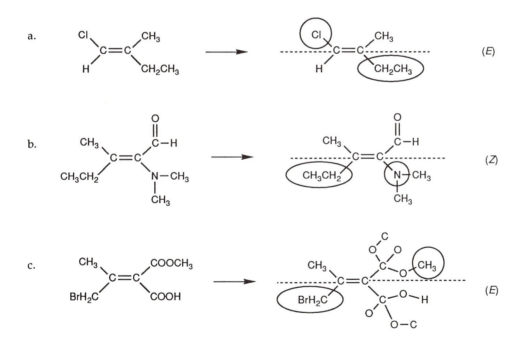

5.2.   To interpret a compound's name when stereochemical descriptors are present, first draw the structure by the normal procedures that were outlined in Chapter 2. A prefix of *cis* or *trans* indicates that the double bond is disubstituted: *cis* means that the hydrogen atoms are on the same side of the bond axis and *trans* means they are on opposite sides of the axis. A prefix of (*E*) or (*Z*) denotes that the Cahn-Prelog-Ingold system is being used to specify the double bond geometry according to which side of the bond axis the "higher priority" groups are attached—follow the procedure outlined in the solution to exercise 5.1.

a.   **cis-3-octene**

oct	8 carbon atoms
ene	a carbon-carbon double bond starting at C3
cis	the hydrogen atoms are on the same side of the bond axis

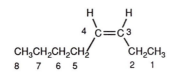

b.   **(Z)-3-methoxy-2-octenal**

oct	8 carbon atoms
en	a carbon-carbon double bond starting at C2
al	an aldehyde functional group, at C1 by convention
3-methoxy	a methoxy group at C3
(Z)	the higher priority groups at the ends of the double bond are on the same side of the bond axis

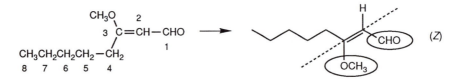

c.   **(E)-1,3-dichloro-2-methyl-2-hexene**

hex	6 carbon atoms
ene	a carbon-carbon double bond, starting at C2
1,3-dichoro	two chlorine atoms, one at C1 and one at C3
2-methyl	a methyl group at C2
(E)	the higher priority groups at the ends of the double bond are on opposite sides of the bond axis

5.2.   (continued)

d.   **(E)-3,4-dibromo-3-heptene**

hept	7 carbon atoms
ene	a carbon-carbon double bond, starting at C3
3,4-dibromo	two bromine atoms, one at C3 and one at C4
(E)	the higher priority groups at the ends of the double bond are on opposite sides of the bond axis

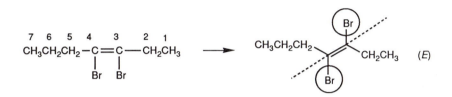

5.3.   The most straightforward way to evaluate whether compounds are either identical or stereoisomers is to make a model of the first compound, then see if it superimposes on a model of the second, in which case the molecules are identical.   Only the molecules in (c) are stereoisomers. The other pairs are identical.

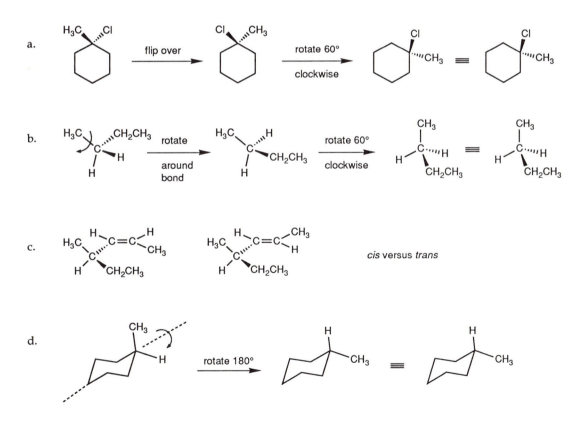

5.4.   To draw the enantiomer of a compound, first make certain that it is chiral (not superimposable on its mirror image). If there is only one stereogenic center, switch any two of the substituents attached to the chiral center. If there is more than one stereogenic center, or if there are no asymmetric carbon atoms, draw the mirror image of the compound.

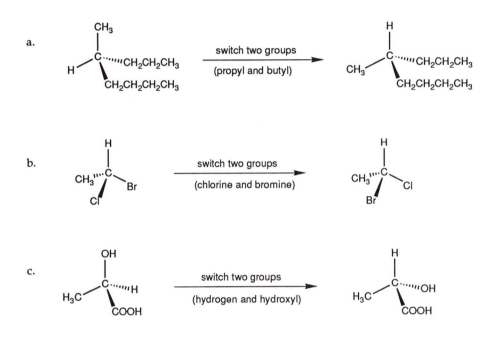

a.

b.

c.

5.5.   To convert the structure of a D compound to its L isomer, interchange any two groups on each asymmetric carbon atom. In glyceraldehyde there is only one stereogenic center, so switching two groups (H and OH, below) produces the enantiomeric structure.

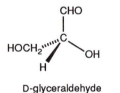

D-glyceraldehyde   L-glyceraldehyde

5.6. The stereochemical classification of a stereogenic center is made as follows: a priority is assigned to each substituent based on its atomic number. Functional groups that have multiple bond may first have to be converted to their single bond equivalents. Reorient the molecule so that the lowest priority group is pointing away from you, then follow the instructions given on page 178 of the text.

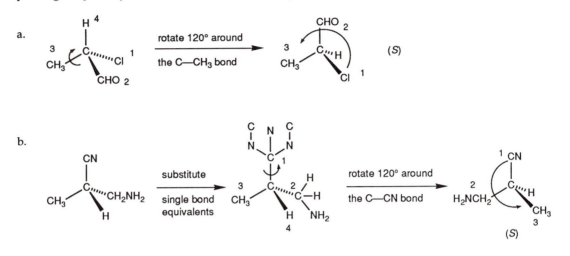

a.

b.

5.7.   To interpret a compound's name when stereochemical descriptors are present, first draw a "flat" structure by the normal procedures that were outlined in Chapter 2. Then use the Cahn-Prelog-Ingold system to designate a priority for each substituent attached to any carbon atom that has four different groups. If the carbon atom has an attached hydrogen atom, introduce it so that it points away from you, then assign the configuration, R or S. If that configuration is incorrect, simply exchange any two groups, which inverts the configuration.

a.   **(R)-2-bromobutanoic acid**

but	4 carbon atoms
an	no double or triple bonds
oic acid	carboxylic acid functional group, defines C1
2-bromo	bromine atom at C2

priority numbering

add H in back;
assign configuration

(carbon atom numbering)

(R)
(correct)

Final structure with
correct stereochemistry

5.7.   (continued)

b.   **(*S*)-2-pentanol**
pent	5 carbon atoms
an	no double or triple bonds
ol	alcohol functional group at C2

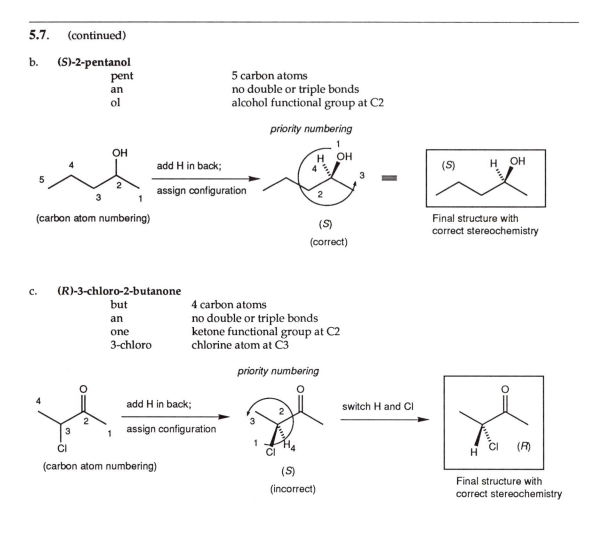

c.   **(*R*)-3-chloro-2-butanone**
but	4 carbon atoms
an	no double or triple bonds
one	ketone functional group at C2
3-chloro	chlorine atom at C3

5.8. To assign the absolute configuration of a stereogenic carbon atom in a ring, break the ring at the bond or atom directly across from the stereogenic center, then assign priorities to each substituent of the chiral carbon atom. Functional groups that have multiple bonds may first have to be converted to their single bond equivalents.   Reorient the molecule so that the lowest priority group is pointing away from you, then follow the instructions given on page 178 of the text.

a.

5.8.   (continued)

b.

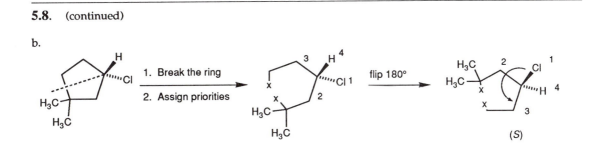

5.9.   The total number of stereoisomers that exist for a compound without symmetry is 2^n, where n is the total number of stereogenic centers. The compound in this exercise has two potentially stereogenic centers, so there exist a maximum of 4 stereoisomers. All four stereoisomers are unique because there is no symmetry.

First, draw all of the isomers. Then, compare the configurations of the stereogenic centers in two isomers [you can make the assignments of (R) and (S) if necessary]. Pairs of compounds in which every chiral center is inverted relative to the other are enantiomers . If only one configuration has changed, then the pair of compounds represents diastereomers.

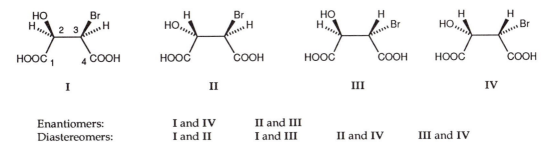

Enantiomers:	I and IV	II and III		
Diastereomers:	I and II	I and III	II and IV	III and IV

5.10.   To decide if a compound can have *meso* forms, you first have to interpret the name and draw all possible stereoisomers. If the molecule has a mirror plane (shown by a dashed line, below), it is a *meso* compound. Compounds (a) and (b) exist only in three stereoisomeric forms (the isomers labeled as *meso* are actually identical substances).

a.   **2,4-dichloropentane**

pent	5 carbon atoms
ane	no carbon-carbon double or triple bonds
2,4-dichloro	chlorine atoms at C2 and C4

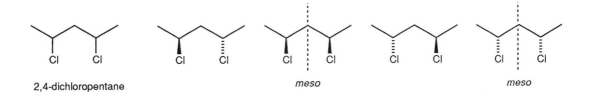

2,4-dichloropentane   *meso*   *meso*

5.10.   (continued)

b.   **1,3-dichlorocyclohexane**

cyclohex	six-membered ring of carbon atoms
ane	no carbon-carbon double or triple bonds
1,3-dichloro	chlorine atoms at C1 and C3

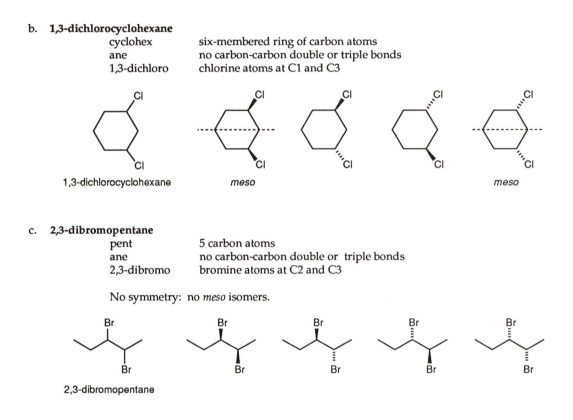

1,3-dichlorocyclohexane	*meso*			*meso*

c.   **2,3-dibromopentane**

pent	5 carbon atoms
ane	no carbon-carbon double or triple bonds
2,3-dibromo	bromine atoms at C2 and C3

No symmetry: no *meso* isomers.

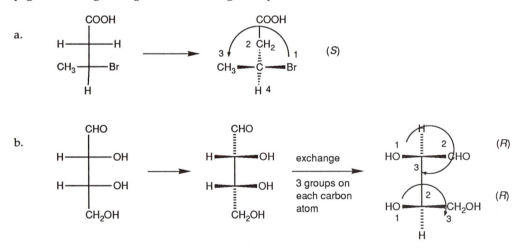

2,3-dibromopentane

5.11.   To assign absolute configurations in a Fischer projection, first convert the Fischer form to a perspective representation by drawing wedges to indicate that the groups attached to the horizontal bonds are in front of the plane of the page and that the substituents at the top and bottom are behind the plane of the page. Then assign configurations according to the procedure outlined in the solution to exercise 5.6.

a.

b.

5.12. Assign configurations of chiral centers according to the procedure outlined in the solution to exercise 5.6.

a.

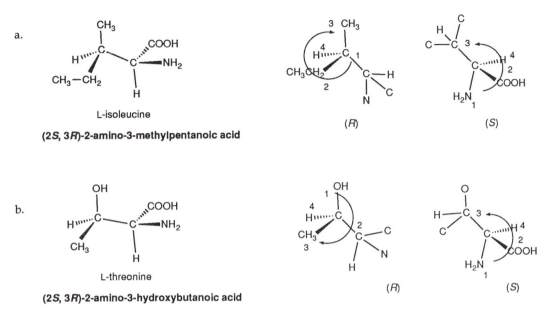

L-isoleucine

(2S, 3R)-2-amino-3-methylpentanoic acid

(R)

(S)

b.

L-threonine

(2S, 3R)-2-amino-3-hydroxybutanoic acid

(R)

(S)

5.13. To convert a D compound to its L isomer, interchange any two groups attached to every asymmetric carbon atom.

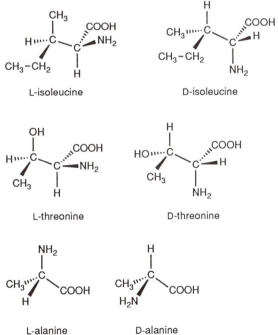

L-isoleucine

D-isoleucine

L-threonine

D-threonine

L-alanine

D-alanine

5.14. To draw a Fischer projection in its correct orientation, first reorient the perspective drawing so that the carboxylic acid group is at the top and is pointed behind the plane of the page. Make certain that the configuration of the other stereogenic carbon atom is correct. The orientation of the second chiral center has the longest carbon chain aligned vertically.

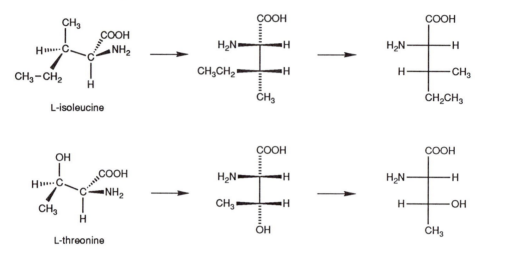

5.15. Follow the procedure outlined in the solution to exercise 5.6.

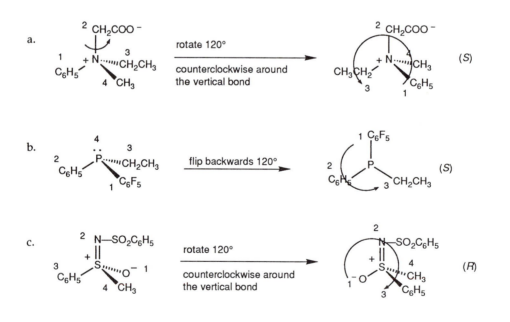

5.16.   First, generate a structure according to the procedure outlined in Chapter 2, but ignore stereo-chemistry. Then consider how the stereochemistry will be indicated. If the prefix that appears is different from the one needed, then consider how the name is to be modified.

a.   ***trans*-2-chloro-3-methyl-2-pentene**

pent	5 carbon atoms
ene	a carbon-carbon double bond, starting at C2
2-chloro	a chlorine atom at C2
3-methyl	a methyl group at C3

A tetrasubstituted double bond needs to have (*E*) or (*Z*) as the stereochemical descriptor. The structure shown should have the name **(*E*)-2-chloro-3-methyl-2-pentene** because the chlorine atom and the ethyl group are on opposite sides of the double bond axis.

b.   **3,4-dimethylcyclohexene**

cyclohex	six-membered ring of carbon atoms
ene	a carbon-carbon double bond, starting at C1
3,4-dimethyl	two methyl groups, one at C3 and one at C4

The substituents may be on the same or different sides of the ring, but without a stereochemical designation, we cannot tell. If we draw each on the same side, the structure shown would have the name ***cis*-3,4-dimethylcyclohexene**. If the methyl groups are on opposite sides of the ring, the compound is ***trans*-3,4-dimethylcyclohexene**.

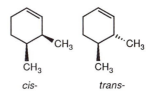

cis-   *trans-*

c.   **3-methylcyclopentene**

pent	5 carbon atoms
ene	a carbon-carbon double bond, starting at C1
3-methyl	a methyl group at C3

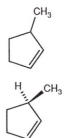

The methyl substituent may be up or down, but without a stereochemical designation, we cannot tell. If we draw the structure so that the methyl group is above the plane of the ring, the structure shown would have the name **(*R*)-3-methylcyclopentene**.

d.   **3-hexene**

hex	6 carbon atoms
ene	a carbon-carbon double bond, starting at C3

A disubstituted double bond needs to have *cis*, *trans*, (*E*), or (*Z*) as a stereochemical descriptor. The structures shown below have the names ***trans*-3-hexene** and ***cis*-3-hexene**, respectively.

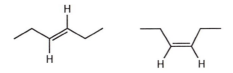

5.17.   First, draw a structure according to the procedure outlined in the solution to exercise 5.16. Consider which geometric isomers can be generated by switching substituents at each end of the double bond(s). Then use the Cahn-Prelog-Ingold system to designate the stereochemistry of any double bond for which a stereochemical descriptor is valid.

a.   **1,3-pentadiene**
   penta   5 carbon atoms
   diene   two carbon-carbon double bonds, starting at C1 and C3

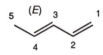

   (E)-1,3-pentadiene   **(Z)-1,3-pentadiene**

b.   **3-methyl-2,4-hexadiene**
   hexa   six carbon atoms
   diene   two carbon-carbon double bonds, starting at C2 and C4
   3-methyl   a methyl group at C3

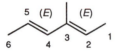

   (2E, 4E)–3-methyl-2,4-hexadiene   **(2E4Z)–3-methyl-2,4-hexadiene**

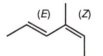

   (2Z, 4E)–3-methyl-2,4-hexadiene   **(2Z,4Z)-3-methyl-2,4-hexadiene**

c.   **2,3-dimethyl-2-butene**
   but   four carbon atoms
   ene   a carbon-carbon double bond, starting at C2
   2,3-dimethyl   two methyl groups, one at C2 and one at C3

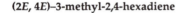   No isomers are possible; the name is correct as **2,3-dimethyl-2-butene**.

5.18. Follow the procedure outlined in the solution to exercise 5.2.

a.   **(Z)-3-bromo-2-hexene**

hex	6 carbon atoms
ene	a carbon-carbon double bond starting at C2
3-bromo	a bromine atom at C3
(Z)	the higher priority groups at the ends of the double bond are on the same side of the bond axis

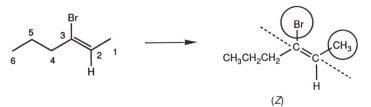

(Z)

b.   **(3Z, 6E)-1,3,6-octatriene**

octa	8 carbon atoms
triene	three carbon-carbon double bonds, starting at C1, C3, and C6
(3Z)	the higher priority groups at the ends of the double bond are on the same side of the bond axis
(6E)	the higher priority groups at the ends of the double bond are on opposite sides of the bond axis

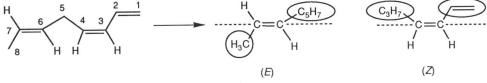

(E)   (Z)

c.   **(Z)-1-methylcyclononene**

cyclonon	nine-membered ring of carbon atoms
ene	a carbon-carbon double bond starting at C1
1-methyl	a methyl group at C1
(Z)	the higher priority groups at the ends of the double bond (shown circled) are on the same side of the bond axis

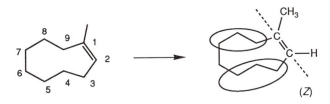

(Z)

5.19. Follow the procedure outlined in the solution to exercise 5.1.   If two groups at either end of the double bond are the same, then no stereochemistry is designated.

a.   (E)

b.   neither

c.   (E)

d.   (Z)

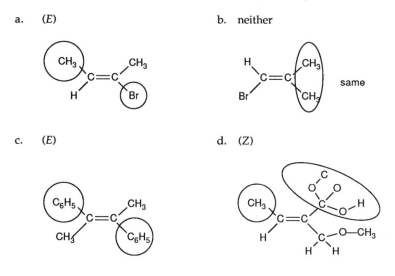

5.20. The priority of a group in the Cahn-Ingold-Prelog system is related to the atomic number of the atom attached to the bond that is shown. If that atom is the same for two different groups, atomic numbers of the elements in the next shell are compared. A higher atomic number for any one of those atoms gives the group a higher priority. Functional groups that have multiple bonds are first converted to their the single bond equivalents before comparing shells of atoms beyond the initial point of attachment.

a. $-\overset{\overset{\text{H}}{|}}{\underset{\underset{\text{H}}{|}}{\text{C}}}-\text{CH}_3$ < $-\overset{\overset{\text{N}}{|}}{\underset{\underset{\text{N—C}}{|}}{\text{C}}}-\text{N—C}$ < $-\overset{\overset{\text{O}}{|}}{\underset{\underset{\text{O—H}}{|}}{\text{C}}}-\text{O—C}$ < $-\text{O—CH}_3$

b. $-\text{H}$ < $-\overset{\overset{\text{H}}{|}}{\underset{\underset{\text{H}}{|}}{\text{N}}}$ < $-\text{O—CH}_3$ < $-\text{Cl}$

c. $-\overset{\overset{\text{H}}{|}}{\underset{\underset{\text{H}}{|}}{\text{C}}}-\text{CH}_3$ < $-\overset{\overset{\text{O}}{|}}{\underset{\underset{\text{H}}{|}}{\text{C}}}-\text{O—C}$ < $-\text{O—H}$ < $-\text{Br}$

d. $-\overset{\overset{\text{H}}{|}}{\underset{\underset{\text{H}}{|}}{\text{C}}}-\text{H}$ < $-\overset{\overset{\text{C—C}}{|}}{\underset{\underset{\text{C—H}}{|}}{\text{C}}}-\text{C—C}$ < $-\overset{\overset{\text{N}}{|}}{\underset{\underset{\text{N—C}}{|}}{\text{C}}}-\text{N—C}$ < $-\overset{\overset{\text{O}}{|}}{\underset{\underset{\text{O—CH}_3}{|}}{\text{C}}}-\text{O—C}$

5.21. Follow the procedures outlined in solutions to exercises 5.1 and 5.19.

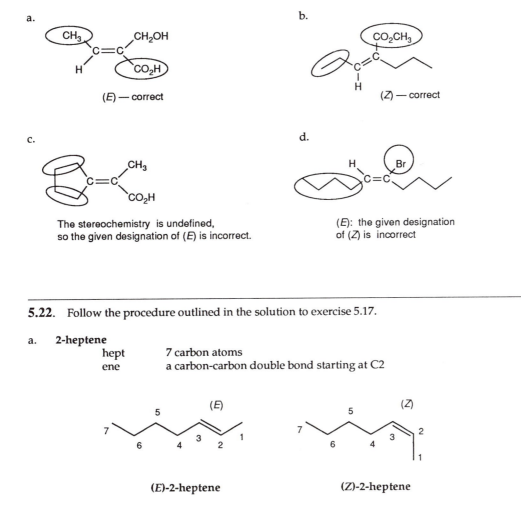

a.

(E) — correct

b.

(Z) — correct

c.

The stereochemistry is undefined,
so the given designation of (E) is incorrect.

d.

(E): the given designation
of (Z) is incorrect

5.22. Follow the procedure outlined in the solution to exercise 5.17.

a. **2-heptene**

| hept | 7 carbon atoms |
| ene | a carbon-carbon double bond starting at C2 |

(E)-2-heptene   (Z)-2-heptene

b. **1-chlorocyclobutene**

cyclobut	four-membered ring of carbon atoms
ene	a carbon-carbon double bond starting at C1
1-chloro	a chlorine atom at C1

No isomers of the double bond are possible because the ring size is
too small to accommodate a *trans* or (E) double bond.

5.22.   (continued)

c. **1,3-butadiene**

buta   4 carbon atoms
diene   two carbon-carbon double bonds, starting at C1 and C3

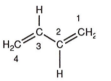

No isomers of the double bonds are possible because one end of each double bond has two like substituents (hydrogen atoms).

d. **3,4,-dimethyl-3-hexene**

hex   6 carbon atoms
ene   a carbon-carbon double bond starting at C3
3,4,-dimethyl   methyl groups at C3 and C4

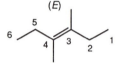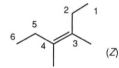

(*E*)-3,4,-dimethyl-3-hexene   (*Z*)-3,4,-dimethyl-3-hexene

e. **3-phenyl-2-pentene**

pent   5 carbon atoms
ene   a carbon-carbon double bond starting at C2
3-phenyl   a phenyl group at C3

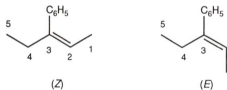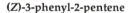

(*Z*)-3-phenyl-2-pentene   (*E*)-3-phenyl-2-pentene

5.23.   The prefix *"trans"* is not valid here because the double bond is trisubstituted. The name should be (*E*)-2-methyl-2-butenal.

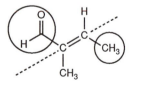

The higher priority group at each end of the carbon-carbon double bond are on opposite sides of the bond axis: (*E*)

5.24. The prefix *"cis"* is not valid here because the double bond is trisubstituted. The name should be **(Z)-aconitate.**

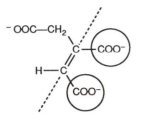

The higher priority group at each end of the carbon-carbon double bond are on the same side of the bond axis: (*Z*)

5.25. A thing is chiral if it is not superimposable on its mirror image. Often, an item that cannot be used the same by right- and left-handed persons is chiral.

a.	a golf club	chiral
b.	a pair of scissors	chiral
c.	a baseball glove	chiral
d.	a corkscrew	chiral
e.	a telephone	chiral
f.	a pencil	achiral
g.	a basketball	achiral
h.	a hammer	achiral
i.	a spiral staircase	chiral

5.26. Follow the procedure outlined in the solution to exercise 5.6.

a.

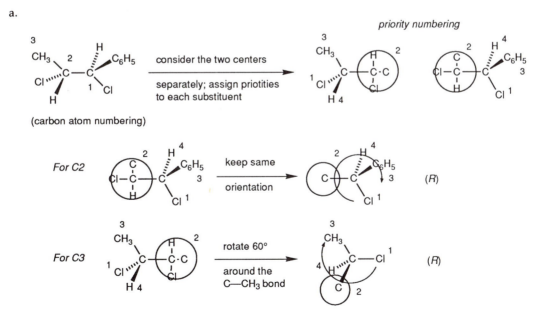

5.26. (continued)

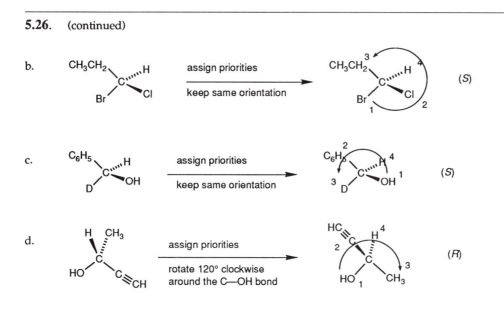

b.

c.

d.

5.27. If a molecule has no symmetry, then the total number of stereoisomers is 2^n, where n is the total number of stereogenic centers and double bonds that can exist as (E) and (Z) isomers. It is best to expand the structure so that the carbon atoms with four different groups can be identified. If the compound has a mirror plane, then it may be necessary to draw each stereoisomer separately so that the identical ones can be identified.

a.

$$C_6H_5-\underset{\underset{H}{|}}{\overset{\overset{OH}{|}}{C}}-\underset{\underset{H}{|}}{\overset{\overset{OH}{|}}{C}}-CH_3$$

There are 2 stereogenic centers in this compound and there is no symmetry. Therefore, 4 stereoisomers of this compound exist: ($1R, 2S$), ($1S, 2S$), ($1R, 2R$), and ($1S, 2R$).

b.

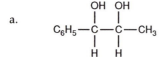

There are 2 stereogenic centers in this compound (indicated by *) and there is no symmetry. Therefore, 4 stereoisomers of this compound exist: ($2R, 3S$), ($2S, 3S$), ($2R, 3R$), and ($2S, 3R$).

c.

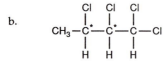

There are 2 stereogenic centers in this compound, but there is a mirror plane of symmetry (indicated by the dashed line.) Therefore, a maximum of 4 stereoisomers exist for this compound, but two of them are the same and *meso*. The unique isomers are ($1R, 2R$), ($1S, 2S$), and *meso*.

d.

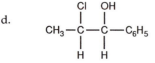

There are 2 stereogenic centers in this compound and there is no symmetry. Therefore, 4 stereoisomers of this compound exist: ($1R, 2S$), ($1S, 2S$), ($1R, 2R$), and ($1S, 2R$).

5.28.   Follow the procedure outlined in the solution to exercise 5.8.

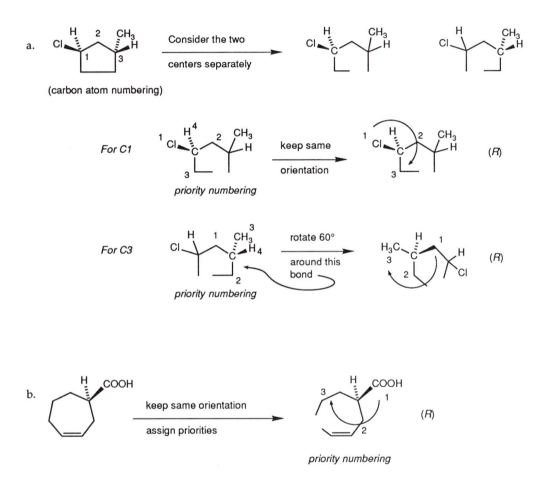

5.28.   (continued)

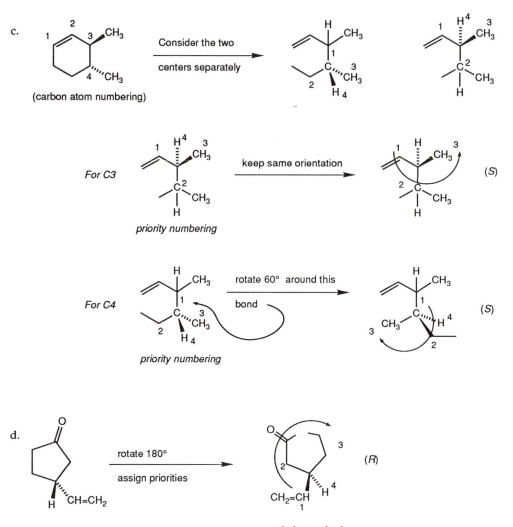

c.

(carbon atom numbering)

Consider the two
centers separately

For C3

priority numbering

keep same orientation

(S)

For C4

priority numbering

rotate 60° around this
bond

(S)

d.

rotate 180°

assign priorities

(R)

CH₂=CH

priority numbering

5.29.   Follow the procedure outlined in the solution to exercise 5.7.

a.   **(3S, 4S)-4-methyl-3-hexanol**

hex	6 carbon atoms
an	no double or triple bonds
ol	alcohol functional group; OH group at C3
4-methyl	methyl group at C4

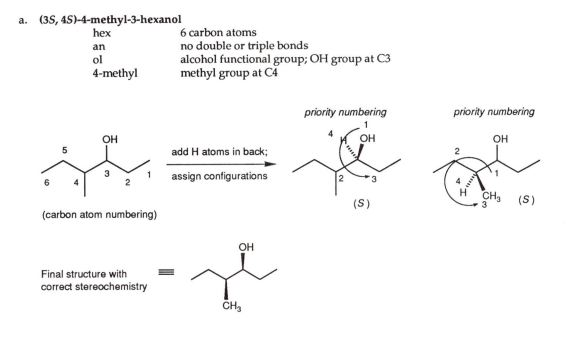

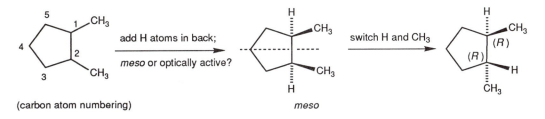

b.   an optically active isomer of **1,2-dimethylcyclopentane**

cyclopent	five-membered ring of carbon atoms
ane	no double or triple bonds
1,2-dimethyl	two methyl groups, one at C1 and one at C2

If we put the hydrogen atoms in going away from us, then the compound has a mirror plane and is optically inactive. Therefore, we have to exchange the hydrogen atom and methyl group attached to one of the carbon atoms.

5.29.   (continued)

c.   **(1R, 3S)-3-methylcyclohexanol**

cyclohex	six-membered ring of carbon atoms
an	no double or triple bonds
ol	alcohol functional group; OH group is at C1
3-methyl	methyl group at C3

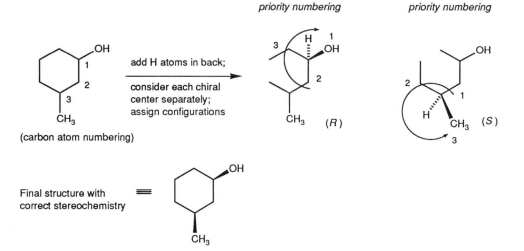

Final structure with correct stereochemistry   ≡

d.   the *meso* isomer of **1,3-dimethylcyclopentane**

cyclopent	five-membered ring of carbon atoms
ane	no double or triple bonds
1,2-dimethyl	two methyl groups, one at C1 and one at C2

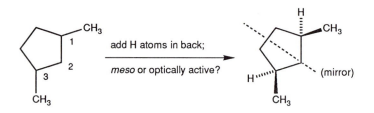

meso–1,3-dimethylcyclopentane

If we draw the hydrogen atoms receding behind the plane of the page, then the compound has a mirror plane and is *meso*.

5.30.   If a molecule has no symmetry, then the total number of stereoisomers is 2^n, where n is the total number of stereogenic centers. The compound under consideration has three potentially stereogenic centers, so a maximum of 8 stereoisomers exists. The symmetry of the compound reduces that number, however, so only four stereoisomers exist, two of which are *meso*.

Draw each of the eight isomers, then determine which are identical (making models may be helpful). The *meso* compounds have a mirror plane of symmetry that passes through the diagonal of the square and includes the carbon atom attached to the carboxylic acid group.

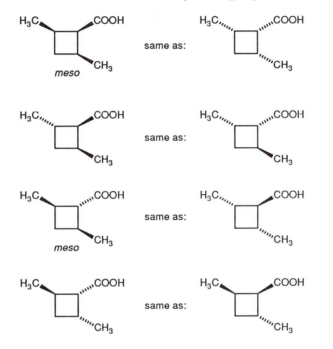

5.31.   Of the isomeric alcohols being considered, three have a carbon atom with four different groups attached. We assign the configurations as described in the solution to exercise 5.26.

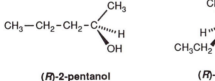 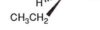 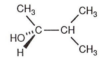 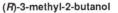

(*R*)-2-pentanol   **(*R*)-2-methyl-1-butanol**   **(*R*)-3-methyl-2-butanol**

To generate the other stereoisomer of each alcohols, we switch the position of two groups attached to the stereogenic carbon atom.

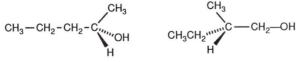 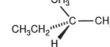 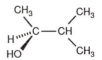

(*S*)-2-pentanol   **(*S*)-2-methyl-1-butanol**   **(*S*)-3-methyl-2-butanol**

5.32. To convert a dimensional structure to a Fischer projection, orient the molecule so that the groups at the top and bottom of the asymmetric carbon atom recede behind the plane of the paper.

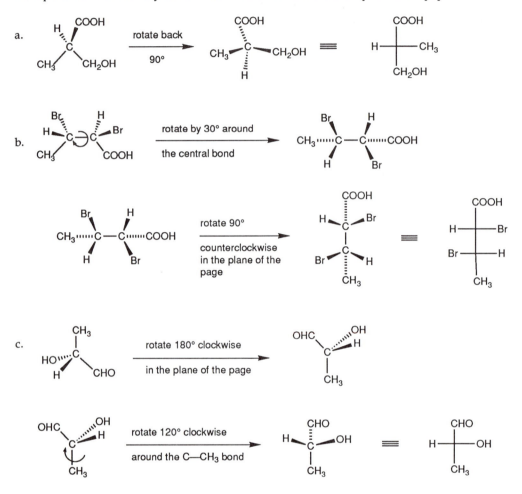

5.33. Follow the procedure outlined in the solution to exercise 5.26. To convert from a Fischer projection to a perspective formula, simply show the horizontal groups to be in front of the plane of the page, and the two substituents at the top and bottom to be behind the plane of the page. A *meso* compound is recognized by the presence of an internal mirror plane in the molecule.

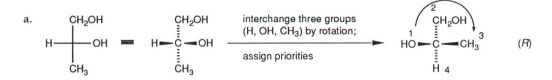

5.33.   (continued)

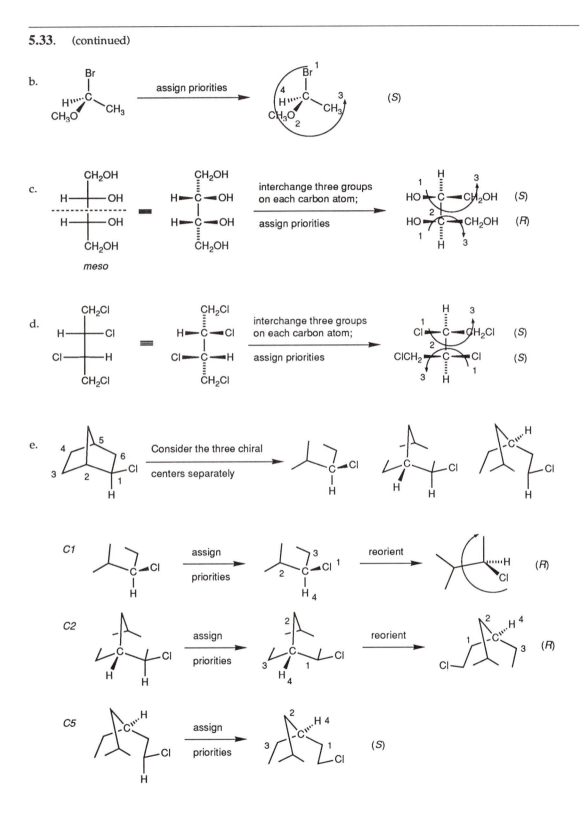

5.33. (continued)

f.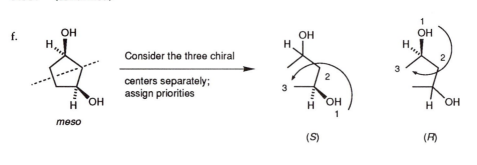

(S)   (R)

5.34. A *meso* form is readily identified in a Fischer projection, so we draw it first (the internal mirror plane is shown in the structures below as a dashed line). To convert from a Fischer projection to a perspective formula, simply show the horizontal groups to be in front of the plane of the page, and the two substituents at the top and bottom to be behind the plane of the page. These can be reoriented to sawhorse projections if desired, but that conversion is not required.

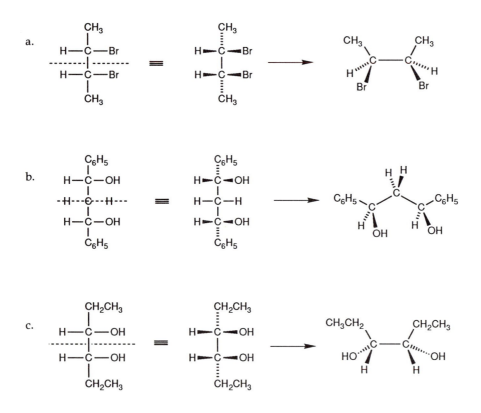

5.35. If a molecule has no symmetry, then the total number of stereoisomers is 2^n, where n is the total number of stereogenic centers. The compound under consideration has three potentially stereogenic centers, so a maximum of 8 stereoisomers exists. The symmetry of the compound reduces that number, however, so only four stereoisomers exist, two of which are *meso*.

Draw each of the eight isomers, then determine which are identical (making models may be helpful). The *meso* compounds have a mirror plane of symmetry that passes through the central carbon atom (shown as a dashed line in the structures below).

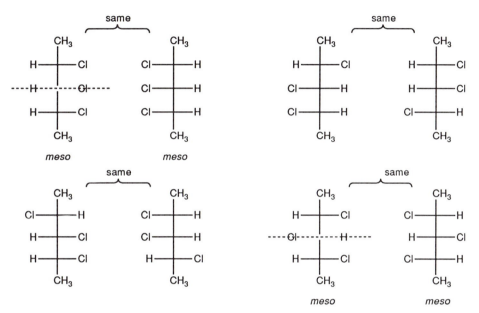

To assign absolute configurations, it is helpful to convert each structure first to a perspective formula. Then follow the procedure outlined in the solution to exercise 5.33d.

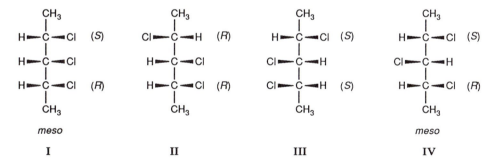

When a stereogenic center in a *meso* compound sits on a mirror plane, it is chiral if there are four different groups attached to it. For instance in structure **I** just above, the four groups are H, Cl, (*S*)–CHClCH₃ and (*R*)–CHClCH₃. When two structurally "identical" groups differ only because they have a different configuration, the priority rules state that the (*R*) group has a higher priority than the one that is (*S*). Therefore, C3 (the central carbon atom) in **I** has the (*S*) configuration, and C3 in structure **IV** has the (*R*) configuration.

For the other two structures, however, C3 is achiral because it has two like groups attached to it, namely (*R*)–CHClCH₃ and (*R*)–CHClCH₃ [structure **II**] or (*S*)–CHClCH₃ and (*S*)–CHClCH₃ [structure **III**].

5.36.   Follow the procedure outlined in the solution to exercise 5.34.

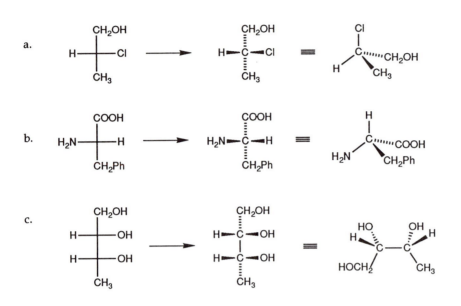

a.

b.

c.

5.37.   Follow the procedure outlined in the solution to exercise 5.29 to generate a perspective drawing. Then convert that structure to a Fischer projection according to the procedure outlined in the solution to exercise 5.32.

a.   **(R)-2-hydroxybutanal**

but	4 carbon atoms
an	no double or triple bonds
al	aldehyde functional group, the CHO group is C1
2-hydroxy	OH group at C2

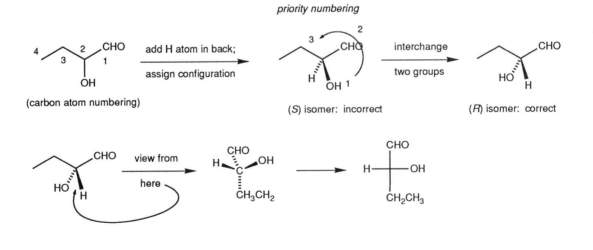

5.37.   (continued)

b.   **(3S, 4R)-1,3,4-trihydroxy-2-pentanone**

pent	5 carbon atoms
an	no double or triple bonds
one	ketone functional group at C2
1,3,4-trihydroxy	three OH groups, one each at C1, C3, and C4

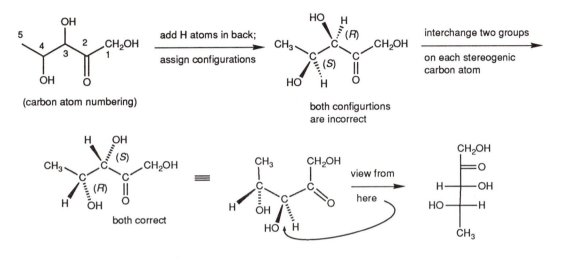

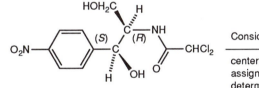

5.38.   Follow the procedure outlined in the solution to exercise 5.26.

chloramphenicol

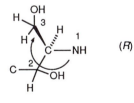

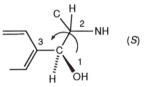

5.38. (continued)

captopril

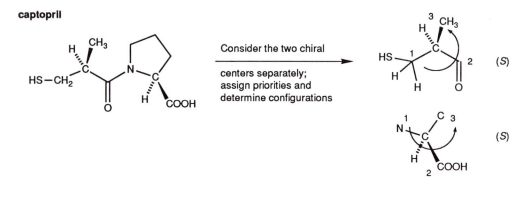

5.39. To draw the enantiomer of a compound that has only one stereogenic center, switch two of the substituents attached to the chiral carbon atom. If more than one stereogenic center is present, or if there are no asymmetric carbon atoms, draw the mirror image of the compound.

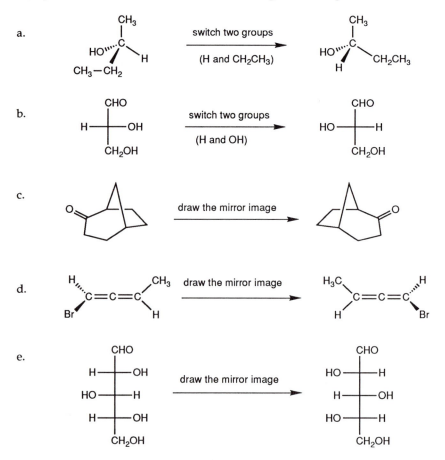

5.40.   Any nitrogen atom with three groups attached undergoes inversion via a pyramidal intermediate in which the hybridization of the nitrogen atom changes from sp^3 to sp^2. This same phenomenon accounts for racemization of a potentially stereogenic nitrogen atom of an amine.

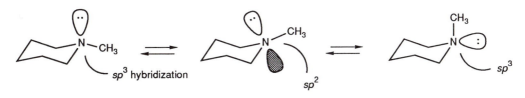

5.41.   Draw the stereoisomers according to the procedure outlined in the solution to exercise 5.35. If every stereogenic center has the opposite configuration [you can make the assignments of (R) and (S) if necessary], then the compounds are enantiomers. If more than one chiral center is present in the molecule and if not every configuration is changed, then the compounds are diastereomers.

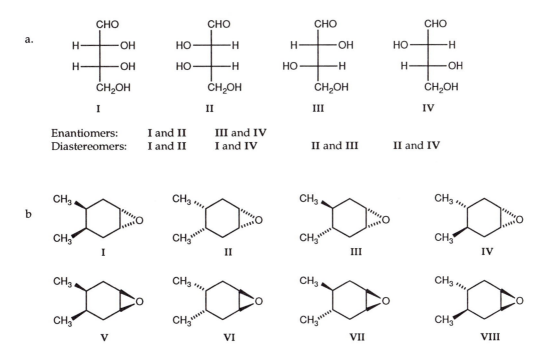

a.

	I and II	III and IV		
Enantiomers:	I and II	III and IV		
Diastereomers:	I and II	I and IV	II and III	II and IV

b

The epoxide ring must be *cis*, but it can be either up or down relative to the plane of the six-membered ring. The methyl groups are *cis* or *trans* to each other, so there are eight possible isomers. Several are identical because of symmetry; only **I** through **IV** are unique.

Identical:	I and VI	II and V	III and VII	IV and VIII	
meso compounds	I and II (these are not optically active)				
Enantiomer:s	III and IV				
Diastereomers:	I and III	I and II	I and IV	II and III	II and IV

5.42. The D-amino acids are enantiomers of the L-amino acids, so we need only change the configuration of every stereogenic center in the molecule. Inverting the configuration is readily done by interchanging any two groups attached to the stereogenic carbon atom.

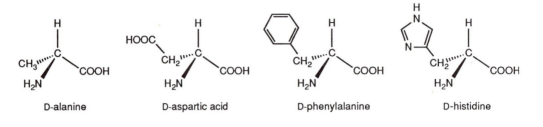

D-alanine   D-aspartic acid   D-phenylalanine   D-histidine

CHEMICAL REACTIONS
AND MECHANISMS

6.1.   A reactant that contains atoms that do not appear in the product is most likely acting as a catalyst.

6.2   To classify a reaction by one of the six given types, use the definitions given in the text on pages 214-216.   Drawing the full structure of reactants to show the functional groups that are involved in a transformation will reveal when multiple bonds are made or broken.

a.   **addition**: methyllithium undergoes addition to a carbon-oxygen double bond to form a product having only single bonds.

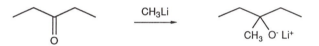

b.   **substitution**: the OCH_3 group of the ester is replaced by an NH_2 group to form an amide.   A proton is also transferred from N to O during this transformation.

c.   **elimination**: water is lost as a carbon-carbon double bond is created.   Sulfuric acid is a catalyst for this reaction.

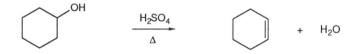

d.   **oxidation**: an alcohol is converted to a carboxylic acid, a functional group that has fewer hydrogen atoms on the carbon atom undergoing reaction.   This process is also considered a combination of substitution and elimination because a hydrogen atom is replaced by an oxygen atom, and a new π bond (C=O) is created.

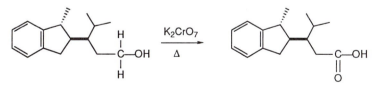

6.3   A nucleophile is a species that has at least one "accessible" pair of electrons. These electrons can be either an unshared pair or a π bond.   An electrophile has a positively-charged atom.

a.   : CN: ⁻ has a negatively-charged carbon atom with an unshared electron pair, so it is a nucleophile.

b.   A proton has only one positively-charged atom, so it is an electrophile.

c.   Trimethylamine has a nitrogen atom that bears an unshared electron pair, so it is a nucleophile.

d.   A carbocation has a positively-charged carbon atom, so it is an electrophile.

e.   Methylthiolate ion has a negatively-charged sulfur atom with unshared electrons, so it is a nucleophile.

f.   Ethylene has a carbon-carbon double bond. This π bond, because it is not along the internuclear axis between the carbon atoms, acts as a nucleophile.

6.4   Use curved arrows to show how electrons move according to the formalisms outlined on text page 221.

a.   If charged species are involved in a reaction, electron movement proceeds from a nucleophilic center toward an electrophilic center.   Remember that negatively-charged species are nucleophiles and positively-charged centers are electrophiles. Electrons move in pairs in such steps.

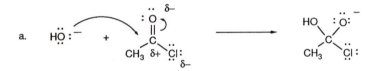

b.   If carbon-carbon double and triple bonds are involved in a reaction with no obvious charged species, then chances are good that a pericyclic mechanism is involved.   Also, if a ring is made or broken, a pericyclic reaction is likely. Electrons move in pairs in such steps.

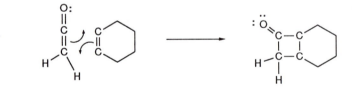

c.   Radical reactions are recognized by the presence of species with an unpaired electron.   Reaction with a radical often involves another species without unpaired electrons, but electron movement involves single electrons, so two arrows are shown for each bond that is made or broken.

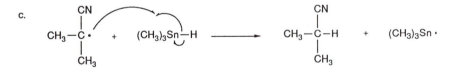

6.5   The conjugate base of a protic acid is made conceptually by removing the most acidic proton and leaving a negatively-charged atom. If the starting species has a positive charge, then the conjugate base is neutral. A proton attached to a heteroatom is usually the most acidic one in a molecule or ion.

a. CH_3CH_2COOH → $CH_3CH_2COO^-$

b. HN_3 → N_3^-

c. $(CH_3)_3CSH$ → $(CH_3)_3CS^-$

d. C_6H_5OH → $C_6H_5O^-$

e. NH_4^+ → NH_3

6.6   Acid-base reactions occur to produce a weaker acid from a stronger one. Identify the acid on **each** side of the equation, then determine which is the weaker (higher pK_a). The equilibrium lies in that direction.

a. CH_3COOH ($pK_a = 5$) vs. HCl (($pK_a < 0$); equilibrium lies to the **left**

b. CH_3OH ($pK_a = 15$) vs. HCN ($pK_a = 9$); equilibrium lies to the **left**

c. C_6H_5OH ($pK_a = 10$) vs. CH_3OH ($pK_a = 15$); equilibrium lies to the **right**

d. NH_4^+ ($pK_a = 9$) vs. CH_3CH_2SH ($pK_a = 11$); equilibrium lies to the **right**

6.7   Use the criteria described in the text on pages 227-229 to predict relative acidities of compounds. For protons attached to atoms in the same *group* of the periodic table, acidity increases with increasing atomic number because of polarizability. For protons attached to atoms in the same *period* of the periodic table, acidity increases with increasing atomic number because of the effect of increasing electronegativity. If the proton is attached to the same type of atom, resonance effects may constitute the primary influence—look for unsaturation next to the atom to which the proton is attached.

a.   **polarizability** (S > O)

b.   **resonance** (adjacent unsaturation)

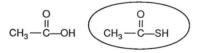

c.   **electronegativity** (O > N)

d.   **resonance** (adjacent unsaturation)

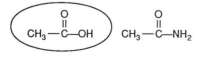

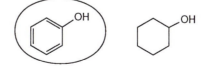

6.8   A Lewis acid is an electron pair acceptor, which includes compounds that have an atom in group 13 of the periodic table, species that carry a positive charge, and elements in period 4 and higher that have vacant, low-energy *d*-orbitals and no unshared electron pairs [Sn(IV), Ti(IV), and Fe(III) are common examples of the latter]. A Lewis base is discerned because one of its atoms has an unshared pair of electrons.

a.   Aluminum chloride has a group 13 element (Al)—it is a Lewis acid.

b.   Ammonia has an atom (N) that bears an unshared electron pair —it is a Lewis base.

c.   Dimethyl ether has an atom (O) that bears two unshared electron pairs —it is a Lewis base.

d.   A proton has no electron pairs and it carries a positive charge —it is a Lewis acid.

e.   Trimethylphosphine has an atom (P) with an unshared electron pair —it is a Lewis base.

f.   The tin atom in tin(IV) tetrachloride has vacant, low energy *d*-orbitals —it is a Lewis acid.

6.9   When given a reaction in which it may not be clear which species is the acid and which is the base, use pK_a values from Table 6.3 (text page 224) or Table 6.4 (text page 227) to determine which species on the left side of an equation is the stronger acid. The other substance in the reaction functions as the base. The products are the protonated form of the base and the deprotonated form of the acid.

a. NH_3 + CH_3COOH $\rightarrow$ NH_4^+ + CH_3COO^-

    base   acid   conj acid   conj base

b. NH_3 + C_6H_5Li $\rightarrow$ $NH_2^-\ Li^+$ + C_6H_6

    acid   base   conj base   conj acid

6.10   A nucleophile is a species that has at least one "accessible" pair of electrons. These electrons can be either an unshared pair or a π bond. An electrophile has a positively-charged atom. The first step of a reaction between an nucleophile and an electrophile occurs by movement of the electron pair of the nucleophile toward the positively-charged atom of the electrophile.

a.   A proton is the most common electrophile, so if a protic acid is present as one of the reactants, it usually functions as the electrophile in a transformation. Any heteroatom or π bond in the other reactant plays the role of the nucleophile. The first step in a reaction in which a protic acid is present takes place when the nucleophile reacts with the acidic proton.

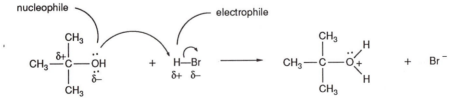

6.10   (continued)

b.   If a reactant carries a negative charge, it usually functions as the nucleophile in a transformation. Cyanide is a good nucleophile, the carbon atom of which bears a formal –1 charge.   A proton source or a carbon atom attached to a heteroatom in the other reactant plays the role of the electrophile because electronegativity differences make the carbon atom partially positive.   The first step is the reaction of the nucleophile with the electrophilic carbon atom.   The π electrons between C and O move so that carbon maintains a maximum of four bonds.

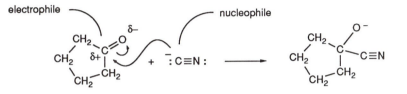

6.11   A stereospecific reaction is one in which the products have a single, predictable stereochemistry.   A stereoselective reaction produces an excess of a product with one stereochemistry versus other possibilities. In this transformation, the starting alkene has a *cis* double bond, and the products have *cis* substitution of the rings, even though there are two of them.   Therefore, this is a stereospecific reaction.

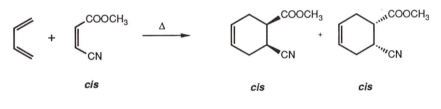

6.12   In a non-productive collision between two reactants, the starting materials and products have identical energies because the same species exist before and after collision.   The energy curve is symmetric about the transition state, which occupies the point of highest energy.

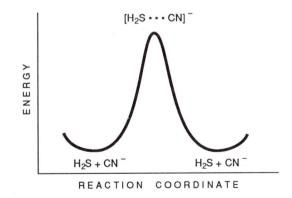

6.13   A slow reaction has a high free energy of activation whereas a fast reaction has a low free energy of activation. An endergonic process is one in which the free energy of the products is higher than the free energy of the reactants.

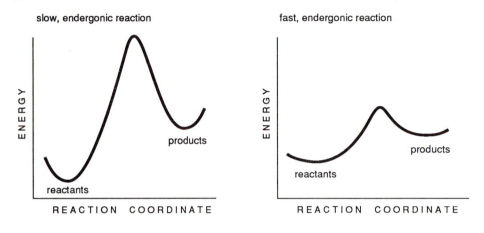

6.14   Equation 6.13 on text age 241 relates the value for the standard free energy, $\Delta G°$, with the value of the equilibrium constant, K_{eq}. Substitute the given values of $\Delta G°$ into the equation and solve it to obtain values for K_{eq}.

$\Delta G° = -2.303(RT \log K_{eq})$ where R = 1.986 cal • mol^{-1} • K^{-1} and T is the temperature in K

a. $\Delta G° = +3.00$ kcal/mol K_{eq} = 0.00629

b. $\Delta G° = -3.00$ kcal/mol K_{eq} = 1.59 x 10^2

c. $\Delta G° = -6.00$ kcal/mol K_{eq} = 2.53 x 10^4

6.15   Follow the procedure outlined in the solution to exercise 6.14 to obtain numerical values for K_{eq}. Substitute those values into the expression given in this exercise to calculate the percentage of B.

a. $\Delta G° = -1.00$ kcal/mol K_{eq} = 5.42   % conversion = 84.4

b. $\Delta G° = -2.00$ kcal/mol K_{eq} = 1.59 x 10^2 % conversion = 99.4

c. $\Delta G° = -3.00$ kcal/mol K_{eq} = 2.53 x 10^4 % conversion = 100

6.16   Like reactions that proceed via a carbocation intermediate, a reaction that occurs with involvement of a radical intermediate comprises reactants, products, and the intermediate itself. The energy diagram, therefore, looks much like that in Figure 6.4 (text page 245), in which an activation barrier exists between reactants and intermediate, and between intermediate and products.

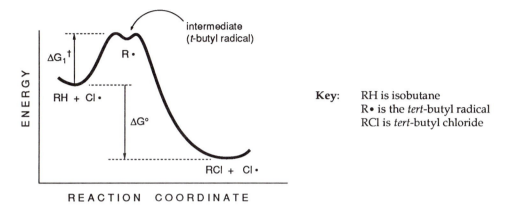

Key: RH is isobutane
        R• is the *tert*-butyl radical
        RCl is *tert*-butyl chloride

6.17   An oligopeptide has amide functional groups that link adjacent amino acids—the carbonyl group from one amino acid is bonded to the amino group of the next amino acid to create the peptide bond. The side chain (Table 6.5, text page 249) then projects from the carbon atom *alpha* to the carbonyl group. The oligopeptide is drawn so that the amino end (the *N*-terminus) is to the left and the carboxy end (the *C*-terminus) is to the right.

a.   Asp–Glu

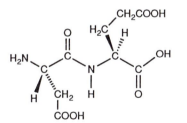

b.   His–Glu–Lys

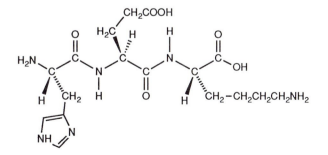

6.17. (continued)

c. Val–Phe–Cys–Tyr

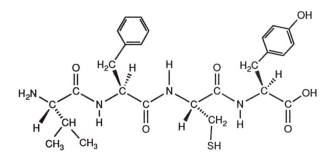

6.18 Follow the instructions given in the solution to exercise 6.2.

a. **elimination**—the elements of water are removed to form a new carbon-carbon double bond.

b. **addition**—the elements of hydrogen add to a carbon-oxygen double bond, converting a carbonyl group to an alcohol functional group that has only single bonds at the carbinol carbon atom. This transformation is also considered a reduction reaction because hydrogen atoms add to the π bond.

6.19 The first step in writing an equation for a reaction described by words is to interpret each of the compound names for reactants and products. Then place each on the appropriate end of the arrow (reactants at the left, products at the right). Include other information over the arrow, numbering separate steps as needed. To classify each reaction, follow the procedure outlined in the solution to exercise 6.2.

a. This is a **reduction** process: two hydrogen atoms add to a carbon-oxygen double bond to form a product that has only single bonds to the carbinol carbon atom.

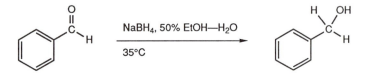

6.19   (continued)

b.   This is a **reduction** process: two hydrogen atoms add to a carbon-oxygen double bond to form a product that has only single bonds to the carbinol carbon atom.

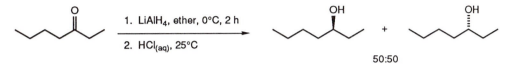

50:50

c.   This is a **substitution** reaction: the sulfur containing group replaces the iodine atom of 1-iodopropane. Triethylamine serves as a base to react with the hydrogen iodide produced.

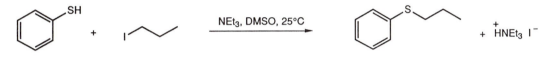

d.   This is an **oxidation** process: a hydrogen atom is replaced by an oxygen atom, and a carbon-oxygen double bond is formed. The creation of a new π bond requires the loss of hydrogen atoms.

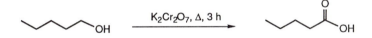

6.20   Follow the procedure outlined in the solution to exercise 6.2.

a.   **Reduction**: two hydrogen atoms replace the oxygen atom of the carbonyl group, and the benzylic carbon atom has only single bonds in the product.

b.   **Addition**: a hydrogen atom and a cyanide ion are attached at the ends of a carbon-oxygen double bond to form a product that has only single bonds at the original carbonyl carbon atom.

c.   **Rearrangement**: the product has the same atoms as the reactant, but the atoms are connected to each other differently.

d.   **Addition**: a hydrogen atom and a bromine atom are attached at the ends of a carbon-carbon double bond to form a product that has only single bonds at the original alkene carbon atoms.

6.21 Follow the procedure outlined in Appendix A. First, identify the highest priority functional group. Next, identify the longest carbon atom chain *that also contains the principal functional group*; the corresponding root word becomes the root word for the name. If the compound is a cyclic hydrocarbon, then the ring attached to the functional group is chosen as the root word. Numbering the chain or ring is done to give the principal functional group the lowest possible number, and the necessary suffix for the principal functional group (Table 2.1, text page 33) is added to the root word. Once the numbering is defined, any double and triple bonds can be specified by inserting the appropriate suffix between the root word and the functional group suffix; use a number to identify the carbon atom at which the double or triple bond begins. Finally, add the substituents to the appropriate position(s) and include stereochemical descriptors.

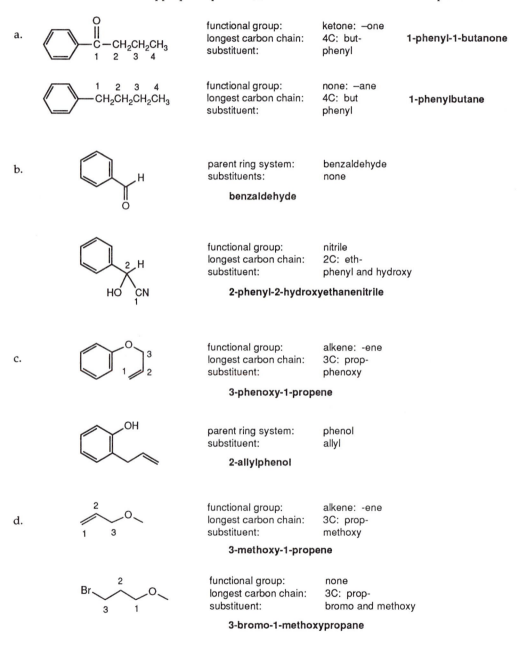

a.

functional group:	ketone: –one
longest carbon chain:	4C: but-
substituent:	phenyl

1-phenyl-1-butanone

functional group:	none: –ane
longest carbon chain:	4C: but
substituent:	phenyl

1-phenylbutane

b.

| parent ring system: | benzaldehyde |
| substituents: | none |

benzaldehyde

functional group:	nitrile
longest carbon chain:	2C: eth-
substituent:	phenyl and hydroxy

2-phenyl-2-hydroxyethanenitrile

c.

functional group:	alkene: -ene
longest carbon chain:	3C: prop-
substituent:	phenoxy

3-phenoxy-1-propene

| parent ring system: | phenol |
| substituent: | allyl |

2-allylphenol

d.

functional group:	alkene: -ene
longest carbon chain:	3C: prop-
substituent:	methoxy

3-methoxy-1-propene

functional group:	none
longest carbon chain:	3C: prop-
substituent:	bromo and methoxy

3-bromo-1-methoxypropane

6.22   The stoichiometry of a reaction is defined by the ratio of reactant molecules. These ratios are most often 1:1, but 2:1 is also not uncommon. A catalyst is identifiable because its atoms do not appear in the product.

a.   The stoichiometry is 1:1 for the reaction between 1-phenyl-1-butanone and molecular hydrogen. Pd is a catalyst for this reaction.

b.   The stoichiometry is 1:1 for the reaction between benzaldehyde and hydrogen cyanide. Cyanide ion is a catalyst for this reaction.

c.   There is only one reactant, so stoichiometry has no meaning. Likewise, there is no catalyst.

d.   The stoichiometry is 1:1 for the reaction between 3-methoxy-1-propene and hydrogen bromide. Molecular oxygen is a catalyst for this reaction.

6.23   The yield of a chemical reaction is the ratio, expressed as a percent, of the number of moles of product obtained divided by the number of moles product expected. The amount of expected product is based on the number of moles of starting material and the stoichiometry of reactant to product.

a.   The reaction begins with $(8.33 \div 90.19) = 0.0924$ mol of thiol starting material, and the molar ratio of reactant to product (1:1) indicates that we expect to obtain 0.0924 mol of product. Instead, $(15.9 \div 180.3) = 0.0882$ mol are obtained. Therefore the yield is $(0.0882 \div 0.0924) \times 100 = 95.4\%$.

b.   The reaction begins with $(3.38 \div 84.12) = 0.0402$ mol of ether starting material, and the molar ratio of reactant to product (1:1) indicates that we expect to obtain 0.0402 mol of product. Instead, $(3.16 \div 84.12) = 0.0376$ mol are obtained. Therefore the yield is $(0.0376 \div 0.0402) \times 100 = 93.5\%$.

6.24   A nucleophile is a species that has at least one "accessible" pair of electrons. These electrons can be either an unshared pair or a π bond. A species that bears a negative charge is a nucleophile. An electrophile has a positively-charged atom.

a. Cl^+ is positively-charged, so it is an electrophile.

b. OH^- has a negatively-charged oxygen atom, so it is a nucleophile.

c. $(C_6H_5)_3P$: has a phosphorus atom that bears an unshared electron pair , so it is a nucleophile.

d. $Br\bullet$ has no charge, so it is neither. It is a radical.

e. $(CH_3)_3Al$ has an aluminum ion that has only 6 valence electrons, so it is an electrophile.

f. I^- is a negatively-charged ion with four unshared pairs of electrons, so it is a nucleophile.

g. $(CH_3)_2S$ has a neutral sulfur atom that bears two unshared electron pairs, so it is a nucleophile.

h. CH_3^+ has a carbon atom with only 6 valence electrons, so it is an electrophile.

I. CH_3^- has a negatively-charged carbon atom with one unshared pair of electrons, so it is a nucleophile.

j.   Li metal carries no charge, so it is neither a nucleophile nor an electrophile.

6.25   When you are asked to write an acid-base reaction between some compound ("X") and a base (in this instance, sodium hydroxide), then compound **X** is reacting as an acid. First identify which proton in the acid has the lowest pK_a value. This proton is usually attached to a heteroatom (O, N, or S are most common). The equation that you write will have the conjugate base of the acid and the conjugate acid of the base (in this case, water) on the right side of the equation arrow.

If the pK_a value for the proton that is removed by the base is less than that of water (pK_a = 15.7), then the equilibrium lies to the right. If the pK_a value for the proton is greater than that for water, then the equilibrium lies to the left. pK_a values are summarized in Table 6.3 (text page 224).

a.   The acid is a carboxylic acid, $pK_a \approx 5$. Its conjugate base is the carboxylate ion. This equilibrium lies to the right.

$$CH_3COOH \quad + \quad NaOH \quad \rightleftharpoons \quad CH_3COO^- \ Na^+ \quad + \quad H_2O$$

b.   The acid is a phenol, $pK_a \approx 10$. Its conjugate base is the phenolate ion. This equilibrium lies to the right.

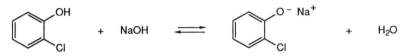

c.   The acid is a thiol, $pK_a \approx 10$. Its conjugate base is the thiolate ion. This equilibrium lies to the right.

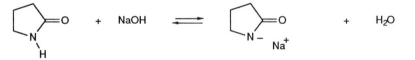

d.   The acid is amide, $pK_a \approx 16$. Its conjugate base is the amide ion. This equilibrium lies to the left.

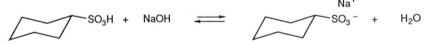

e.   The acid is a sulfonic acid, $pK_a < 0$. Its conjugate base is a sulfonate ion. This equilibrium lies to the right.

f.   The acid is an alcohol, $pK_a \approx 18$. Its conjugate base is the alkoxide ion. This equilibrium lies to the left.

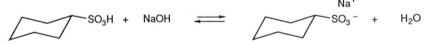

6.26 When you are asked to write an acid-base reaction between some compound ("Y") and an acid (in this instance, hydrochloric acid), then compound Y is acting as a base. First identify which atom in the base is the proton acceptor. This is usually a heteroatom, and N is more basic than O, if they bear the same charge. The equation that you write will have the conjugate base of the acid (chloride ion in this case) and the conjugate acid of the base on the right side of the equation arrow.

   If the pK_a value of the conjugate base that you have drawn is greater than the pK_a value of HCl (pK_a = –7), then the equilibrium lies to the right. Otherwise, the equilibrium lies to the left. pK_a values are summarized in Table 6.3 (text page 224).

a.   The conjugate acid is an ammonium ion derivative, $pK_a \approx 9$. This equilibrium lies to the right.

$$CH_3CH_2NH_2 \quad + \quad HCl \quad \rightleftharpoons \quad CH_3CH_2\overset{+}{N}H_3 \quad + \quad Cl^-$$

b.   The conjugate acid is a protonated ester carbonyl group, $pK_a \approx –6$. This equilibrium lies to the right.

c.   The conjugate acid is a sulfonic acid, $pK_a \approx –6.5$. This equilibrium lies to the right.

d.   The conjugate acid is a protonated alcohol, $pK_a \approx –2$. This equilibrium lies to the right.

6.27 Follow the same procedure outlined in the solution to exercise 6.26. For these reactions, however, if the pK_a value for the conjugate base that you have drawn is greater than the pK_a value for HOAc ($pK_a \approx 5$), then the equilibrium lies to the right. Otherwise, it lies toward the left.

a.   The conjugate acid is an ammonium ion derivative, $pK_a \approx 9$. This equilibrium lies to the right.

$$CH_3CH_2NH_2 \quad + \quad HOAc \quad \rightleftharpoons \quad CH_3CH_2\overset{+}{N}H_3 \quad + \quad AcO^-$$

b.   The conjugate acid is a protonated ester carbonyl group, $pK_a \approx –6$. This equilibrium lies to the left.

6.27 (continued)

c. The conjugate acid is a sulfonic acid, $pK_a \approx -6.5$. This equilibrium lies to the left.

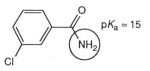

d. The conjugate acid is a protonated alcohol, $pK_a \approx -2$. This equilibrium lies to the left.

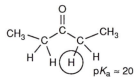

6.28 First, interpret the name, following the procedure outlined in Chapter 2: break the name into the constituent parts, then combine the pieces to form the structure.

The most acidic proton is likely attached to a heteroatom. If only C–H bonds are present, those adjacent to unsaturation or strongly electron-withdrawing groups are usually more acidic. pK_a values are summarized in Table 6.3 (text page 224).

a. **3-Pentanone**

pent	5 carbon atoms
an	no double or triple bonds
one	ketone functional group at C3

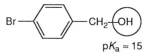

b. **2-Bromobenzyl alcohol**

benzyl alcohol	benzene ring with a CH_2OH group attached; the point of attachment of the CH_2OH group defines C1
4-bromo	bromine atom at C4 of the ring

c. **1,1,1-Trifluoroethane**

eth	2 carbon atoms
ane	no unsaturation or functional groups
1,1,1-trifluoro	three fluorine atoms attached at C1

d. **3-Chlorobenzamide**

benzamide	benzene ring with a $CONH_2$ group attached; the point of attachment of the $CONH_2$ group defines C1
3-chloro	chlorine atom at C3

6.28   (continued)

Arrange the compounds in order of increasing acidity by using the approximate pK_a values. A primary alcohol is usually slightly more acidic than an amide group, but the specific acidity depends upon solubility and other factors.

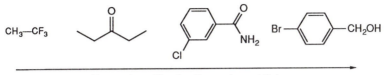

Decreasing pK_a value (increasing acidity)

6.29   Movement of electrons proceeds from a nucleophilic center toward an electrophilic center if charged species are involved. Remember that negatively-charged species are nucleophiles and positively-charged centers are electrophiles. In radical reactions, electrons move as individual entities, one from the original unpaired electron toward a bond. The two electrons that constitute a bond move separately in a radical process, one toward the incoming radical and the other toward one of the atoms (usually carbon).

a.   A radical reaction is readily identified by the presence of a species with an unpaired electron. All four bonds are equivalent in methane, so any can react with the bromine radical. A hydrogen atom is normally abstracted by a halogen radical, generating a carbon-centered radical intermediate.

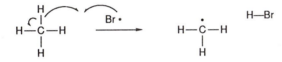

b.   This is a polar reaction, recognizable because the nitronium ion reactant has a positive charge so is electrophilic. The nucleophile is the π system of the benzene ring. The ether oxygen atom is also nucleophilic, but normally reaction takes place at carbon atoms if possible. The first step is reaction of the electrons of the benzene ring with the electrophile.

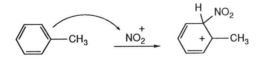

c.   This is a polar reaction, recognizable because the reactant methoxide ion has a negative charge so is nucleophilic. If a reaction occurs, therefore, methoxide ion must react with an electrophile. The carbonyl group is polarized so that its carbon atom has a partial positive charge, so its carbon atom is electrophilic. The methoxide ion reacts with the electrophilic carbon atom, and a pair of electrons must move so that carbon does not have to bear 10 valence electrons. The double bond becomes a single bond, and a pair of electrons moves onto the O atom.

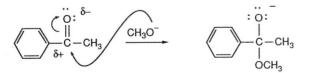

6.29 (continued)

d.   A radical reaction is readily identified by the presence of a species with an unpaired electron. Either a hydrogen atom or the iodine atom (as illustrated below) is abstracted by reaction with the tin radical, generating a carbon-centered radical intermediate.

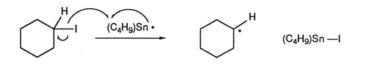

e.   This is a polar reaction, recognizable because the proton reactant has a positive charge so is electrophilic. If a reaction occurs, a nucleophile must be present to interact with the proton. The nucleophile is the π bond of the alkene. The first step is reaction between the electrons of the double bond and the proton. As a new bond is created, a positively-charged intermediate, a carbocation, is generated.

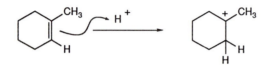

f.   This is a polar reaction, recognizable because the proton reactant has a positive charge so is electrophilic. The nucleophile is the oxygen atom of the alcohol functional group.

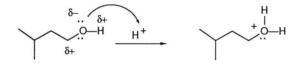

6.30   When assessing pericyclic reactions, expand the structures of reactant and product molecules to make certain that each carbon atom retains four bonds as the electrons move. The movement of electrons occurs in a cyclic fashion starting (usually) with a π bond. If movement of a pair of electrons from a π bond to the adjacent position puts 10 electrons around any C, N, or O atom, then move another pair of electrons from that atom (even if it involves the electrons in a σ bond) until each non-hydrogen atom has only eight electrons.

a.

b.

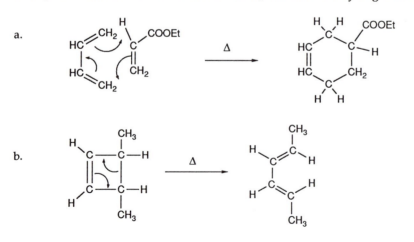

6.30   (continued)

The energy diagram for a pericyclic reaction does not have a discrete intermediate at the transition state. The fact that these reactions need heat to proceed at a reasonable rate indicates that the transition state energy is greater than that of the reactant(s). The fact that the reaction proceeds means that the overall process is exergonic.

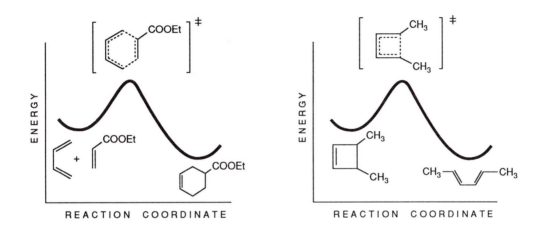

6.31   For a one-step endergonic process, the energy profile has only a transition state maximum (no intermediate). The energy of the products is higher than that of the reactants.

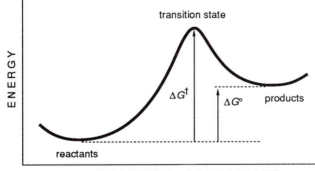

6.32 For a two-step exergonic process, the energy profile has a local minimum near the energy maximum that corresponds to formation of an intermediate. As long as the first barrier is higher than the second one, the first step will be slower than the second. The energy of the products is lower than that of the reactants.

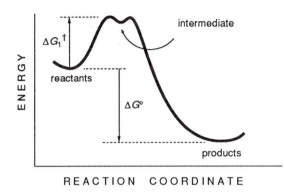

6.33 When $K_{eq} = 1$, the solution of equation 6.13 (text page 241) reveals that $\Delta G = 0$. This means that the reactants and products have the same energy.

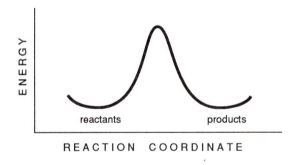

6.34 The functional group in the side chain of aspartic and glutamic acids is a carboxylic acid group, the conjugate base of which is the carboxylate ion. The structures of the side chains in their conjugate base forms are as follows:

$$asp\text{—}CH_2\text{—}COO^- \qquad\qquad glu\text{—}CH_2\text{—}CH_2\text{—}COO^-$$

6.35 The enzyme *triose phosphate isomerase* catalyzes a unimolecular equilibrium, so the equations given in Chapter 6 that relate thermodynamic quantities to simple equilibria are applicable.

a. Calculate K_{eq} using equation 6.13 (text page 241): $\Delta G^{o\prime} = -2.303(RT)\log K_{eq}$.

$K_{eq} = 10^{(1800)/-2.303(1.986)(298)} = 0.048$

With this value for the equilibrium constant, the concentration of glyceraldehyde-3-phosphate ("B") to dihydroxacetone phosphate ("A") is slightly less than 95:5.

b. The transformation of glyceraldehyde-3-phosphate to dihydroxacetone phosphate is a rearrangement reaction because the compounds are isomers.

c. A qualitative energy diagram looks like that pictured in the solution to exercise 6.31 because it is a one-step endergonic process.

d. D-glyceraldehyde is a three carbon aldehyde with hydroxy groups at C2 and C3. Therefore, its IUPAC name is (R)-2,3-dihydroxypropanal. Dihydroxacetone is a three carbon ketone with hydroxy groups at C1 and C3 and the carbonyl group at C2. Its IUPAC name is 1,3-dihydroxy-2-propanone.

6.36 Follow the procedure outlined in the solution to exercise 6.35.

a. Calculate K_{eq} using equation 6.13 (text page 241): $\Delta G^{o\prime} = -2.303(RT)\log K_{eq}$.

$K_{eq} = 10^{(400)/-2.303(1.986)(298)} = 0.5$

With this value for the equilibrium constant, the concentration of D-fructose-6-phosphate ("B") to D-glucose-6-phosphate ("A") is 1:2

b. A qualitative energy diagram looks like that pictured in the solution to exercise 6.31 because it is a one-step endergonic process.

6.37 Lysine has an amino group in its side chain, so the conjugate acid form is an ammonium salt. Its pK_a value is approximately 9 according to the values in Table 6.3 (text page 224). Histidine has an imidazole ring in its side chain. The conjugate acid form has a protonated nitrogen atom.

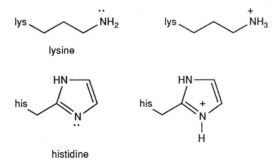

6.38 Each zwitterion of an amino acid has a carboxylate group and an ammonium ion. To generate a zwitterion structure, move the proton from the carboxylic acid to the amino group of the amino acid structures illustrated in Table 6.5 (text page 249). The full structure of each amino acid is formed by attaching the indicated side chain in place of the "R" group in the general structure at the top of that table.

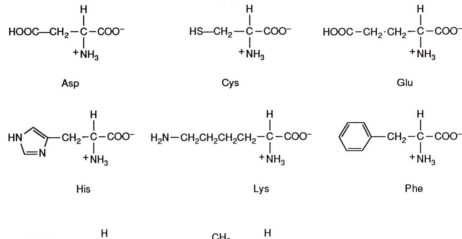

6.39 Follow the procedure outlined in the solution to exercise 6.3.

Nucleophilic side chains

    Cys (the sulfur atom has an electron pair)
    Lys and **His** (the nitrogen atom has an electron pair).

Electrophilic side chain

    Asp and **Glu** (the carboxylic acid group is a proton donor)
    Tyr (the phenol group is a proton donor)

        All of these side chains are nucleophilic in their conjugate base forms because the oxygen atoms that are present have unshared pairs of electrons.

Neither
    Phe and **Val** have only hydrocarbon groups in their side chains.

6.40   Follow the procedure outlined in the solution to exercise 6.2.

a.   **Elimination.** Two hydrogen atoms are removed from the proline ring to form a carbon-nitrogen double bond. This is also an oxidation process because hydrogen atoms have been removed.

b.   **Substitution.** A hydrogen atom on the benzene ring has been substituted by a hydroxy group. This is also an oxidation because an oxygen atom has been introduced into the structure.

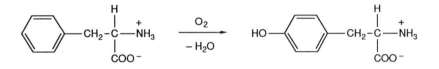

c.   **Addition.** The elements of water have added to a carbon-carbon π bond generating new sigma bonds.

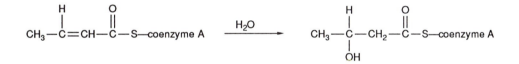

SUBSTITUTION REACTIONS OF ALKYL HALIDES, ALCOHOLS AND RELATED COMPOUNDS

7.1. As described in Chapter 6, a nucleophile is the species in a reaction that carries a negative charge or has at least one unshared pair of electrons. To identify the nucleophile in the reactions collected in Table 7.1 (text page 264), look for the portion of the product that has changed relative to the structure of the starting material. If a protic acid is one of the reactants, the nucleophile is likely its conjugate base.

a. CH_3S^-

b. Br^- (the conjugate base of the acid HBr)

c. Br^-. This transformation is a bit tricky because you are not yet familiar with how phosphorus reagents react. A bromine atom in the product has replaced the OH group in the starting material, however, so bromide ion is a likely choice for the identity of the nucleophile.

d. $(CH_3)_3N$: The nitrogen atom of an amine is nucleophilic because it has an unshared pair of electrons.

e. I^- (the conjugate base of the acid HI)

7.2. In assessing nucleophilicity, remember that for a given atom type, nucleophilicity increases with basicity. Within a group of atoms in the periodic table (a vertical column), nucleophilicity increases as the atoms becomes more polarizable ($S > O$; $I > Br > Cl$, etc.). The effect of polarizability is usually the more important factor.

a. $H_2O < OH^- < SH^-$ The sulfur-containing species is the most nucleophilic ($S > O$). Hydroxide ion is more nucleophilic than water because its basicity is greater.

b. $NH_3 < NH_2^- < (CH_3)_3P$ The phosphorus-containing species is the most nucleophilic ($P > N$), and the amide ion is more nucleophilic than ammonia because it is charged, hence more basic.

7.3. The reactant in the transformation given in this exercise is a tertiary alkyl halide, so an S_N1 reaction is likely. The first step is dissociation of chloride ion with forming of a carbocation. Then the nucleophile, methanol, reacts with the electrophilic carbon atom. Finally, the solvent acts as a base to deprotonate the oxonium ion intermediate, forming the ether.

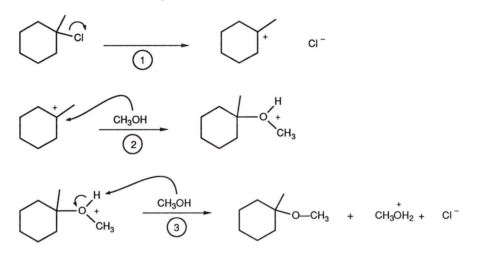

7.4. Hyperconjugative structures place a positive charge on a proton attached to a carbon atom adjacent to the one that bears the positive charge initially. This delocalization also produces a carbon-carbon double bond. Only the carbon-hydrogen bond that is aligned with the *p*-orbital on the cationic carbon atom is able to participate, which explains why each methyl group contributes only one resonance form (the choice of the hydrogen atom on each methyl group in the following structures is arbitrary).

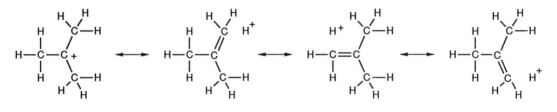

7.5. The phenyl cation has a π system that is the same as benzene's. The only structural difference is that a phenyl cation has one less hydrogen atom. Removal of the hydrogen atom along with the electrons that constitute the carbon-hydrogen bond leaves a positive charge on carbon. This "empty" hybrid orbital is perpendicular to the conjugated π bonds. If a nucleophile approaches a phenyl cation, it would react with the positively charged carbon atom within the plane of the ring. Notice that no overlap exists between the hybrid orbital and the π system, so no resonance stabilization is possible for the positive charge in this species.

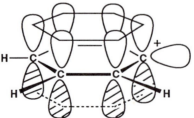

7.6.   The order of substrate reactivity for a reaction that occurs by an S_N1 pathway is benzylic, allylic > 3° > 2° >> 1°, methyl, phenyl, and vinyl. The latter types of organohalides do not react via an S_N1 pathway.

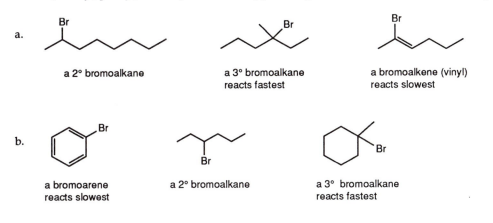

a.

a 2° bromoalkane

a 3° bromoalkane
reacts fastest

a bromoalkene (vinyl)
reacts slowest

b.

a bromoarene
reacts slowest

a 2° bromoalkane

a 3° bromoalkane
reacts fastest

7.7.   As explained in this exercise, the given reaction occurs by an S_N1 pathway, which is recognizable anyway because the solvent that is used is aqueous–alcohol, a reagent combination that comprises only weak nucleophiles. A carbocation is formed in an S_N1 reaction, so besides direct reaction between a nucleophile and the first-formed carbocation, rearrangements are also possible.

direct substitution:

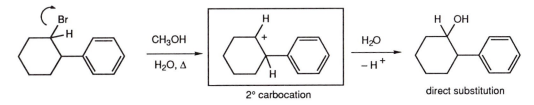

2° carbocation   direct substitution

To predict likely rearrangement products, look at the structure of the initially generated carbocation and see if it is possible to generate a more stable carbocation simply by moving a hydrogen atom or alkyl group from adjacent atom(s).

If a more stable carbocation *can* be generated, possible substitution products will result from reaction of the nucleophile with the more stable (rearranged) carbocation.

rearrangement:

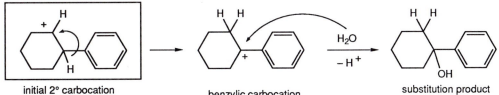

initial 2° carbocation
from dissociation of Br⁻

benzylic carbocation

substitution product
after rearrangement

7.8. This reaction occurs by an S_N1 pathway, recognizable because the solvent that is used is aqueous–alcohol, a reagent combination that comprises only weak nucleophiles. In addition, the substrate is a benzylic sulfonate ester, which is prone to form a stable carbocation upon dissociation of the leaving group. When a carbocation forms, the stereochemical course of a reaction is racemization at that center. The products are enantiomers.

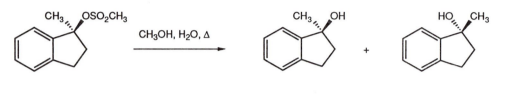

7.9. These reactions proceed via an S_N2 pathway, inferred because an aprotic solvent and a strong nucleophile [hydroxide ion in part a. and iodide ion in part b.] are present in each. The stereochemical course of an S_N2 reaction is inversion of configuration. The assignment of configurations is made according to the procedure described in Chapter 5.

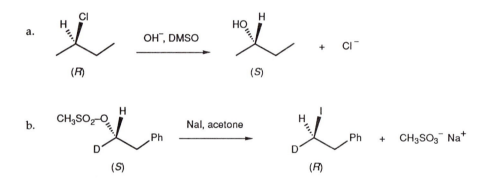

7.10. The reaction between a primary amine and a primary alkyl halide normally gives both mono- and di-substitution products, in which the nitrogen atom of the amine replaces the leaving group, in this case bromide ion. The initial substitution product subsequently reacts with a second portion of alkyl halide, so the second product of this transformation has two alkyl groups from the original alkyl halide attached to the nitrogen atom of the original amine component. The structures of the two products are shown below.

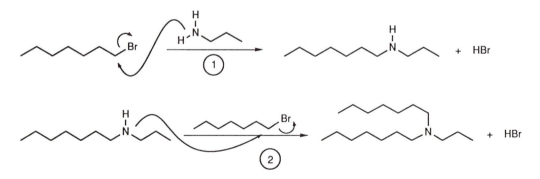

7.11. The starting material of the reaction shown in this exercise has a neopentyl structure, which means that it has a primary carbon atom with leaving group that is adjacent to a tertiary carbon atom. The expected substitution products result from direct replacement of the leaving group by the nucleophile (water in this example), as well from reaction of the nucleophile with rearranged intermediates. These intermediates are tertiary carbocations produced by migration of each of the two possible alkyl groups to the primary center as the leaving group is departing. The nucleophile forms a bond with the tertiary carbon atom.

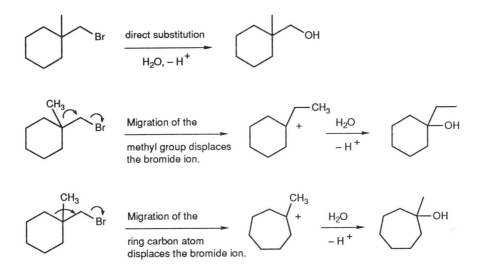

7.12. The conversion of a 3° alcohol to a 3° alkyl bromide by reaction with HCl occurs by an S_N1 pathway. The alcohol OH group is protonated in step (1) to produce a good leaving group, which departs in step (2) to form the carbocation intermediate. Chloride ion subsequently traps the carbocation to form the product.

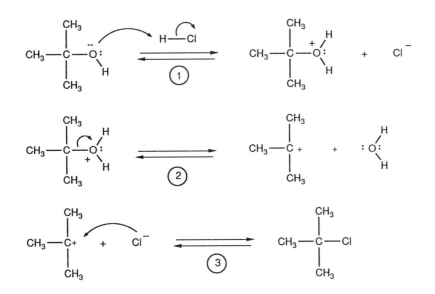

7.13. The OH group of a 1° or 2° alcohol is replaced by conversion to its sulfonate ester derivative, in which the alkylsulfonate ion can act as a good leaving group. If the desired transformation involves inversion of configuration at the original carbinol carbon atom, S_N2 conditions (strong nucleophile) are called for. For the transformation shown in this exercise, the sulfonate ester derivative is treated with cyanide ion to form the desired product, the nitrile. Remember that sulfonate ester formation occurs with **retention** of configuration at the carbinol carbon atom; displacement by the nucleophile occurs with **inversion** of configuration at the carbon atom to which the leaving group is attached.

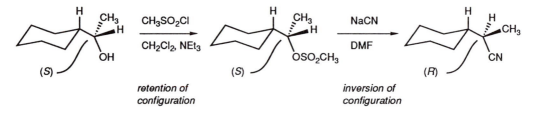

7.14. The mechanism for the last step is the same as the mechanism for the step that immediately precedes it. The oxygen atom that is double-bonded to phosphorus is protonated by reaction with the HBr. Then bromide ion reacts at the alkyl group with backside displacement, generating the alkyl bromide and H_3PO_3.

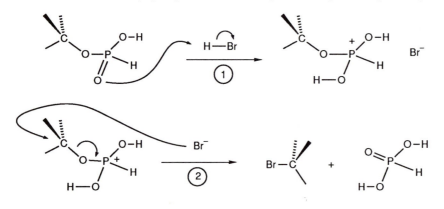

7.15. The transformation shown in the exercise occurs by an S_N2 pathway in which bromide ion functions as the nucleophile. The alcohol reacts with PBr_3 with retention of stereochemistry of the carbinol carbon atom because the carbon-oxygen bond remains intact as the P–O bond is formed. When that intermediate reacts with bromide ion in the second step, the carbon-oxygen bond is broken, leading to inversion of configuration.

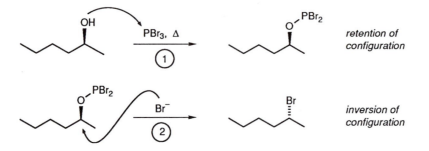

7.16.   The transformation shown in the exercise occurs by an S_N2 pathway in which chloride ion is the nucleophile. The alcohol reacts with $SOCl_2$ without affecting the stereochemistry of the carbinol carbon atom because the carbon-oxygen bond remains intact as the S–O bond is formed. When that intermediate reacts with chloride ion in the second step, the carbon-oxygen bond is broken, leading to inversion of configuration.

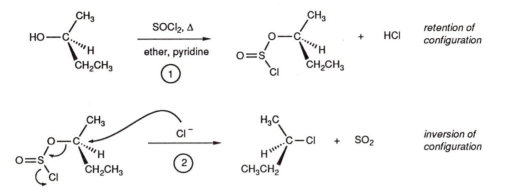

7.17.   The alcohol reacts with thionyl chloride in the first step to produce the chlorosulfinyl ester. Then the chlorine atom of HCl, which carries a partially negative charge, functions as a nucleophile to displace the leaving group. This is shown below as a concerted process (all bonds made and broken in a synchronous process), but it may take place in separate steps. The product is 2-chloroethyl methyl sulfide.

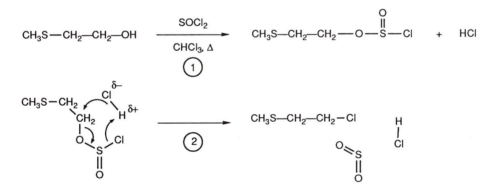

7.18.   In the Mitsunobu reaction, the first step creates the triphenylphosphonium salt. The counterion is azide ion, which is generated from HN_3.

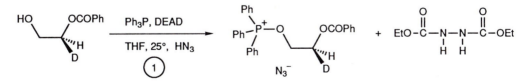

7.18. (continued)

In the second step, azide ion reacts at the primary carbon atom, displacing the leaving group, which is converted to triphenylphosphine oxide. Notice that the configuration of the carbon atom *adjacent* to the carbon atom with the leaving group is not affected by the substitution process.

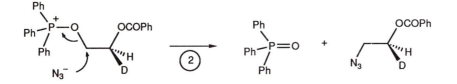

7.19. The Mitsunobu reaction occurs by an S_N2 pathway, which means that the configuration of a stereogenic center bearing a leaving group undergoes inversion. In a Mitsunobu reaction, the nucleophile that replaces the leaving group is the conjugate base of the acidic reagent. For this reaction, the nucleophile is benzenethiolate ion, PhS^-.

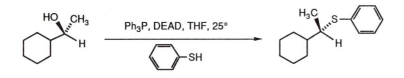

7.20. Acidic cleavage of an ether is a two-step process. First, the ether oxygen atom is protonated by reaction with the acid. Then the nucleophile reacts at the carbon atom attached to the positively-charged oxygen atom, displacing the phenol as the leaving group. The aryl C–O bond of an aryl ether or phenol is not cleaved under these conditions because the carbon atom has sp^2 hybridization. Nucleophilic substitution reactions occur only with substrates in which the leaving group is attached to a carbon atom that has sp^3 hybridization.

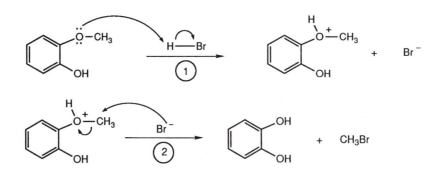

7.21.   Cleavage of a dialkyl ether with HBr proceeds in four stages.   First, the ether oxygen atom is protonated by reaction with hydrogen bromide. The nucleophile reacts at the carbon atom attached to the positively-charged oxygen atom, displacing an alcohol as the leaving group. This alcohol subsequently reacts in the same way that the original ether did: the alcohol oxygen atom is protonated by reaction with hydrogen bromide. The nucleophile reacts at the carbon atom attached to the positively-charged oxygen atom, displacing water as the leaving group. A cyclic ether, as this exercise illustrates, reacts with HBr to form an α,ω-dihaloalkane as the product.

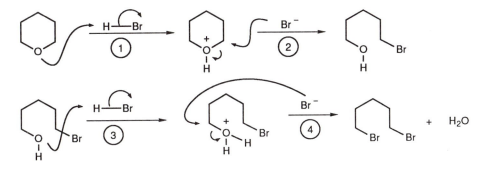

7.22.   Cleavage of an alkyl aryl ether by trimethylsilyl iodide occurs by reaction of the ether oxygen atom with the silicon atom by a pathway that is analogous to an S_N2 reaction. The iodide ion then reacts in the second step at the alkyl carbon atom attached to the positively-charged oxygen atom. The aryl C–O bond of an aryl ether or phenol is not cleaved under these conditions because the carbon atom has sp^2 hybridization. Nucleophilic substitution reactions occur only with substrates in which the leaving group is attached to a carbon atom that has sp^3 hybridization.

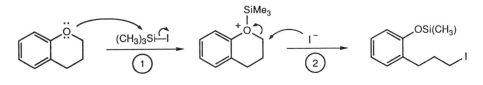

7.23.   1,2-Epoxycyclopentane is a *meso* compound, so it is achiral, even though each carbon atom is a stereogenic center. Azide ion reacts at either carbon atom of the epoxide ring, leading to opening of the ring. Because there is an equally likely chance that the nucleophile will react at either carbon atom, enantiomers are produced in a ratio of 1:1, which defines what a racemic mixture is.

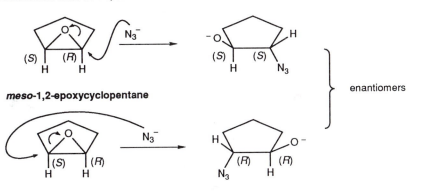

7.24. An epoxide, like any ether, is protonated at oxygen in step (1) by reaction with the electrophile H^+.

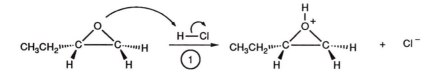

The nucleophilic chloride ion subsequently reacts at each of the ring carbon atoms to form the products.

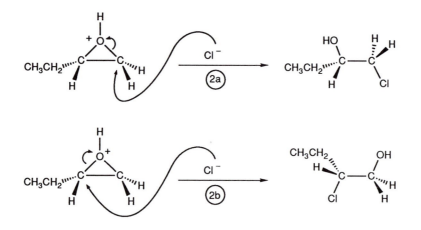

7.25. An epoxide undergoes reaction with a negatively-charged nucleophile at its less highly-substituted carbon atom. In this case, the nucleophile is formally $H:^-$.

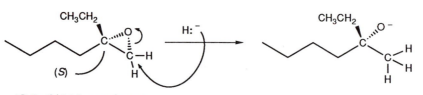

(*S*)-2-ethyl-1,2-epoxyhexane

In the second step, the alkoxide ion is protonated by reaction with water in the hydrolysis step, forming the product. The configuration of the stereogenic carbon atom in the product is the same as that in the starting epoxide.

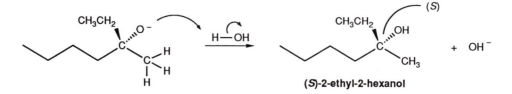

(*S*)-2-ethyl-2-hexanol

7.26. The substitution reaction described in this exercise is a biologically important reaction that follows the same course as any nucleophilic substitution reaction that occurs in the laboratory: the nucleophilic oxygen atom of a carboxylate group reacts with the electrophilic carbon atom of the methyl group of *S*-adenosylmethionine. The product is an ester derivative of the protein. *S*-Adenosylhomocysteine functions as the leaving group.

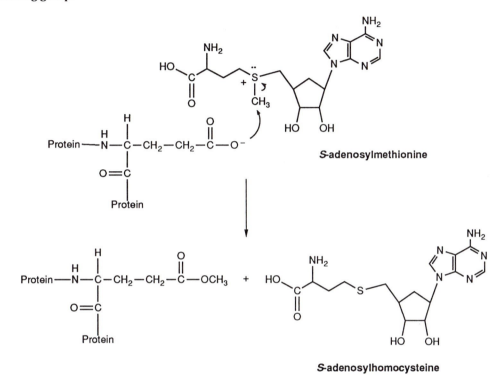

S-adenosylmethionine

S-adenosylhomocysteine

7.27. To interpret the name, follow the procedures outlined in Chapter 2:   break the name into its constituent parts and form the structure.

a. **2-Methyl-2-hexanol**

hex	6 carbon atoms
an	no double or triple bonds
ol	alcohol functional group at C2
2-methyl	methyl group attached at C2

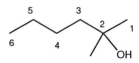

b. **3,3-Dichloro-1-pentanol**

pent	5 carbon atoms
an	no double or triple bonds
ol	alcohol functional group at C1
3,3-dichloro	two chlorine atoms attached at C3

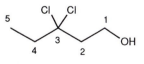

7.27. (continued)

c. ***trans*-2-Phenyl-1-cyclohexanol**

cyclohex	six-membered ring of carbon atoms
an	no double or triple bonds
ol	alcohol functional group at C1
2-phenyl	phenyl group attached at C2
trans	phenyl and OH groups are on opposite sides of the ring

To draw the most stable conformation, convert the flat structure to the chair form, then perform a ring flip. The more stable form has the larger groups in equatorial positions.

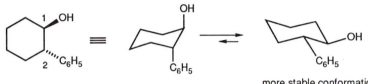

more stable conformation

d. **Cyclobutyl methanesulfonate**

methanesulfonate	CH_3SO_3 —
cyclobutyl	four-membered ring of carbon atoms attached to the CH_3SO_3 — group through one of the oxygen atoms

---

7.28. Follow the procedure outlined in Appendix A. First, identify the highest priority functional group. Next, identify the longest carbon atom chain *that also contains the principal functional group*; the corresponding root word becomes the root word for the name. If the compound is a cyclic hydrocarbon, then the ring attached to the functional group is chosen as the root word. Numbering the chain or ring is done to give the principal functional group the lowest possible number, and the necessary suffix for the principal functional group (Table 2.1, text page 33) is added to the root word. Once the numbering is defined, any double and triple bonds can be specified by inserting the appropriate suffix between the root word and the functional group suffix; use a number to identify the carbon atom at which the double or triple bond begins. Finally, add the substituents to the appropriate position(s) and include stereochemical descriptors.

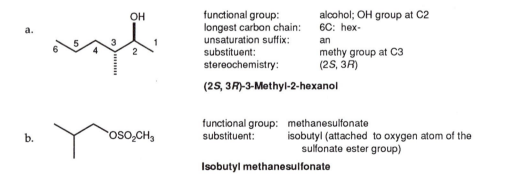

a.

functional group:	alcohol; OH group at C2
longest carbon chain:	6C: hex-
unsaturation suffix:	an
substituent:	methy group at C3
stereochemistry:	(2*S*, 3*R*)

(2*S*, 3*R*)-3-Methyl-2-hexanol

b.

functional group:	methanesulfonate
substituent:	isobutyl (attached to oxygen atom of the sulfonate ester group)

Isobutyl methanesulfonate

7.28.   (continued)

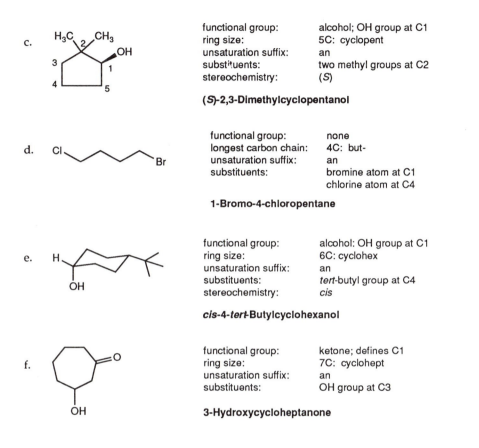

c.

functional group:	alcohol; OH group at C1
ring size:	5C: cyclopent
unsaturation suffix:	an
substituents:	two methyl groups at C2
stereochemistry:	(S)

(S)-2,3-Dimethylcyclopentanol

d.

functional group:	none
longest carbon chain:	4C: but-
unsaturation suffix:	an
substituents:	bromine atom at C1
	chlorine atom at C4

1-Bromo-4-chloropentane

e.

functional group:	alcohol: OH group at C1
ring size:	6C: cyclohex
unsaturation suffix:	an
substituents:	*tert*-butyl group at C4
stereochemistry:	*cis*

***cis*-4-*tert*-Butylcyclohexanol**

f.

functional group:	ketone; defines C1
ring size:	7C: cyclohept
unsaturation suffix:	an
substituents:	OH group at C3

3-Hydroxycycloheptanone

7.29.   For a substitution reaction to be viable, two features are required: a good leaving group attached to an sp^3-hybridized carbon atom, and a nucleophile. Each of the following reactions has both, so we can be certain that a reaction will occur.

If the leaving group is attached to a 1° or 2° carbon atom, then the mechanism is likely S_N2, and the configuration of any chiral center to which the leaving group is attached will be inverted.

a.   The nucleophile is cyanide ion, which reacts to replace the bromide ion with inversion of configuration at the 2° carbon atom.

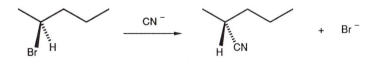

7.29.   (continued)

b.   The nucleophile is methoxide ion, which replaces the mesylate ion with inversion of configuration at the 2° carbon atom.

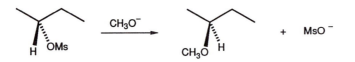

c.   The nucleophile is trimethylamine, an uncharged nucleophile. It replaces the bromide ion, but the product is ionic because both reactants are neutral, and there is no proton that can dissociate. Stereochemistry is not an issue here because the substrate is a primary alkyl bromide.

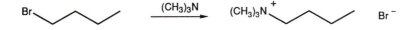

d.   The nucleophile is methanethiolate ion, which replaces the bromide ion with inversion of configuration at the 2° carbon atom. Even though the carbon atom that bears the leaving group is not chiral, its relative position is defined by another substituent in the ring, so a geometric isomer is formed.

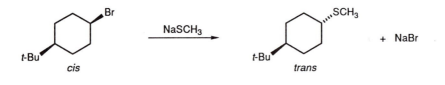

7.30.   The rate of a substitution reaction depends mainly on the strength of the nucleophile (better nucleophiles react faster) and the structure of organic substrate undergoing substitution. For an S_N1 reaction (usually a solvolysis), the rate decreases in the order 3° > 2° >> 1°. For an S_N2 reaction, the rate decreases in the order methyl > 1° > 2° >> 3°, neopentyl.

a.   The substrate in each reaction is the same, so we consider the relative strengths of the nucleophiles. For a given atom type with the same charge, nucleophilicity parallels base strength. Hydroxide ion is a stronger base than acetate ion (in both, the oxygen atom is the nucleophilic atom), so the first reaction goes faster. Methyl substrates react via an S_N2 pathway.

$$CH_3I \quad + \quad OH^- \quad \xrightarrow{\ H_2O\ } \quad CH_3OH \quad + \quad I^-$$

$$CH_3I \quad + \quad OAc^- \quad \xrightarrow{\ H_2O\ } \quad CH_3OAc \quad + \quad I^-$$

7.30. (continued)

b.   The nucleophile in each reaction is the same, so we consider substrate structures. The nucleophile is the solvent, which is water, so each transformation is a solvolysis reaction and probably occurs via an S_N1 pathway. Therefore, the order of reactivity is $3° > 2°$, so the first reaction occurs faster.

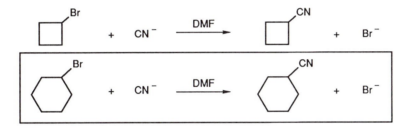

c.   The nucleophile in each reaction is the same, so we consider substrate structures. The nucleophile is a charged species, and the substrates are both 2° alkyl bromides, so these transformations probably occur by an S_N2 pathway. In an S_N2 reaction, the transition state is five-coordinate, and the carbon–carbon bond angles within the substrate approach 120°. The cyclobutyl compound has more angle strain, so going from 109° in the reactant to 120° in the transition state is more difficult for the substrate with the four-membered ring. Therefore, bromocyclohexane reacts faster.

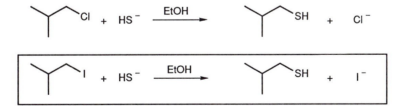

d.   The nucleophile in each reaction is the same, so we consider substrate structures. The nucleophile is a charged species, and the substrates are both 1° alkyl halides, so these transformations probably occur by an S_N2 pathway. Iodide ion is a better leaving group than is chloride ion, so the second reaction occurs faster.

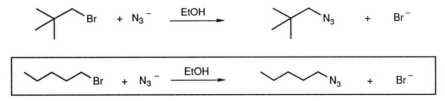

e.   The nucleophile in each reaction is the same, so we consider substrate structures. The nucleophile is a charged species, and the substrates are both 1° alkyl bromides, so these transformations likely occur by an S_N2 pathway. Neopentyl substrates react extremely slowly by an S_N2 mechanism, so the second reaction proceeds more rapidly.

7.31. In a solvolysis reaction, a carbocation is formed as an intermediate, so rearrangement reactions are possible. Consider, therefore, if an atom or group can readily migrate within the intermediate to rationalize formation of a rearranged product.

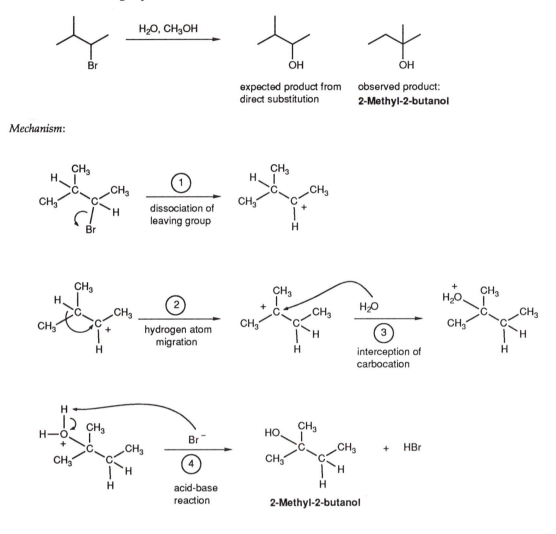

Mechanism:

7.32. For a substitution reaction to be viable, two features are required: a good leaving group attached to an sp^3-hybridized carbon atom, and a nucleophile. An S$_N$2 process occurs with 1° and many 2° substrates. A chiral center in a secondary alkyl halide undergoes inversion of configuration. An S$_N$1 reaction occurs most readily with 3° substrates (and 2° substrates in solvolysis processes). The stereochemical outcome of an S$_N$1 reaction is racemization.

a.   The OH group is not a good leaving group, so no reaction occurs.

7.32. (continued)

b.   The mesylate group is a good leaving group, and cyanide ion is a good nucleophile. The carbon atom to which the leaving group is attached is 2°, so inversion of configuration occurs.

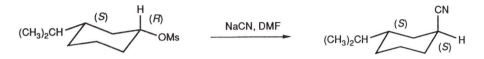

c.   The alcohol OH group is converted to a good leaving group (tosylate) in the first step, which proceeds with retention of configuration. Methanethiolate ion is a good nucleophile. The carbon atom to which the leaving group is attached is 2°, so inversion of configuration occurs.

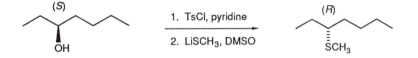

d.   Bromide ion is a good leaving group, and phenylmercaptide ion is a good nucleophile. The carbon atom to which the leaving group is attached is 2°, so inversion of configuration will occur.

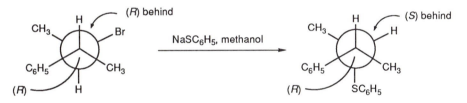

e.   Bromide ion is attached to a carbon atom that has sp^2 hybridization.   Aryl and vinyl halide do not undergo S_N1 or S_N2 reactions.

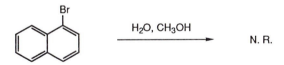

f.   Ethers are cleaved only under conditions in which a good leaving group can be formed. HI and HBr are two common reagents that promote cleavage of carbon-oxygen bonds in ethers. No stereogenic carbon atoms are present in this transformation, so stereochemistry is not an issue.

7.33. To prepare an ether via an S_N2 reaction (the Williamson ether synthesis), the alkyl halide reactant must be methyl, 1°, or 2°. Therefore, to prepare an ether with a 3° carbon atom attached to oxygen requires the use of an alkoxide ion derived from the 3° alcohol together with a 1° or 2° alkyl halide. To decide which compounds to use, break one of the oxygen–carbon bonds; designate the portion removed as the alkyl halide (shown below in the box). The other portion corresponds to the alcohol starting material. Normally, the alcohol is converted to the alkoxide ion with sodium hydride in a polar, aprotic solvent such as DMF. The alkoxide salt reacts with the alkyl halide by an S_N2 pathway.

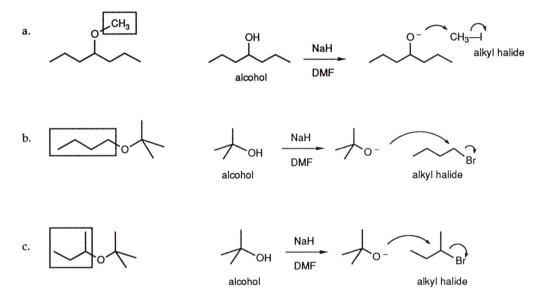

7.34. For a nucleophilic substitution reaction to be facile, the substrate must have a good leaving group, and a good nucleophile must be present. The reaction shown in part b. will not take place because a strong base has to function as the leaving group. The reaction shown in part e. will also not take place because a carbon-carbon bond is rarely broken in a transformation that involves nucleophilic substitution.

a. CH_3CH_2Cl + I^- ⟶ CH_3CH_2I + Cl^-

b. CH_3CH_2OH + Br^- ⟶ CH_3CH_2Br + OH^-

c. CH_3CH_2Br + CN^- ⟶ CH_3CH_2CN + Br^-

d. $CH_3CH_2OSO_2CH_3$ + N_3^- ⟶ $CH_3CH_2N_3$ + $CH_3SO_3^-$

e. CH_3CH_2CN + SCH_3^- ⟶ $CH_3CH_2SCH_3$ + CN^-

7.35. An S$_N$1 reaction proceeds via formation of a carbocation intermediate, so consider first how the carbocation is formed. When an alcohol is treated with a strong mineral acid, the oxygen atom of the alcohol group is protonated, and water dissociates to produce the cationic intermediate. The order of reactivity is benzylic > 3° > > 2°. Primary alcohols do not normally react via formation of a carbocation intermediate. There are two alcohol functional groups in the starting material. One is 1° and benzylic and the other is 3° and benzylic. The latter is the more likely to react and produce a carbocation. A carbocation is susceptible to reaction with a nucleophile, which in this case is the oxygen atom of the other alcohol functional group.

protonation of the OH group to create a good leaving group

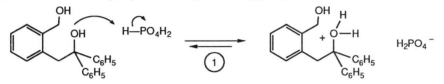

dissociation of the leaving group (water) to form the carbocation intermediate

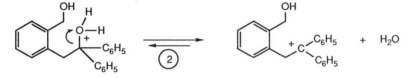

reaction of the carbocation intermediate with the nucleophilic oxygen atom

deprotonation to form the neutral product

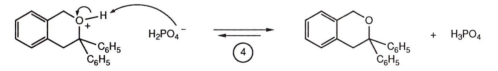

Important points to remember about writing mechanisms:

- charges on both sides of the equation must balance
- electrons move from nucleophiles toward electrophiles
- heteroatoms often have unshared pairs; these are the nucleophilic centers

7.36. In substitution reactions that occur in biological systems, the phosphate or diphosphate ion often functions as a leaving group. The fact that diphosphate (also called "pyrophosphate") is replaced by the nitrogen atom of the amine tells us that the nitrogen atom acts as the nucleophile in this transformation. Therefore, we write a mechanism as follows. After substitution occurs, an acid-base reaction [step (2)] leads to formation of the neutral amino group.

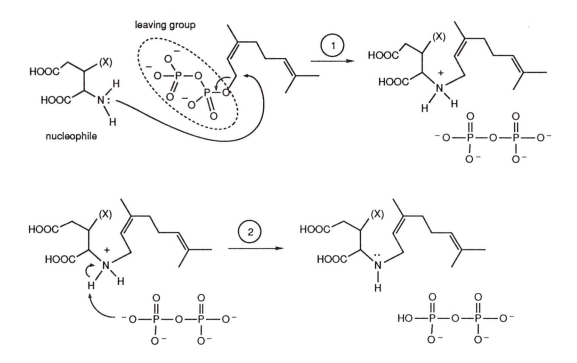

Assign the configuration of each stereogenic centers in the final product by following the procedures outlined in Chapter 5. Each stereogenic carbon atom has the (*S*) configuration.

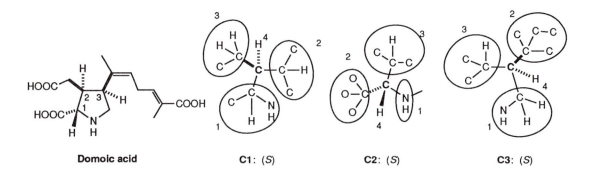

Domoic acid　　　　**C1**: (*S*)　　　　**C2**: (*S*)　　　　**C3**: (*S*)

7.37. The course of this transformation is analogous to that for the reaction between an amine and an alkyl halide. The first product, the thiol, undergoes an acid-base reaction with the nucleophile to form an alkylthiolate ion, which itself is a good nucleophile.

$$CH_3CH_2CH_2CH_2\text{—}Br \quad + \quad HS^- \quad \xrightarrow{S_N2} \quad CH_3CH_2CH_2CH_2\text{—}SH \quad + \quad Br^-$$

$$CH_3CH_2CH_2CH_2\text{—}S\text{—}H \quad + \quad HS^- \quad \rightleftharpoons \quad CH_3CH_2CH_2CH_2\text{—}S^- \quad + \quad H_2S$$
$$pK_a = 11 \qquad\qquad\qquad\qquad\qquad\qquad\qquad pK_a = 7$$

The alkylthiolate ion then reacts with the original alkyl halide to form the dialkylsulfide.

$$CH_3CH_2CH_2CH_2\text{—}S^- \quad + \quad CH_3CH_2CH_2CH_2\text{—}Br \quad \longrightarrow \quad (CH_3CH_2CH_2CH_2)_2S \quad + \quad Br^-$$

The overall process is similar to what happens in reactions between an alkyl halide and ammonia or amines, as described on text pages 291-293.

7.38. There are three steps that take place in the conversion of an alkyl halide to a thiol: substitution, hydrolysis, and an acid-base reaction. (Don't be concerned if you did not predict that potassium acetate is a product of the second step; we will look at the details of this reaction in Chapter 19).

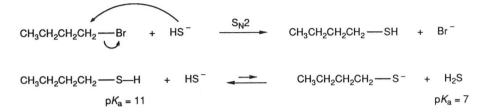

$$CH_3CH_2CH_2CH_2\text{—}Br \quad + \quad K^+ {}^-S\text{—}COCH_3 \quad \xrightarrow{\;\;①\;\;} \quad CH_3CH_2CH_2CH_2\text{—}S\text{—}COCH_3 \quad + \quad KBr$$

$$CH_3CH_2CH_2CH_2\text{—}S\text{—}COCH_3 \quad + \quad KOH \quad \xrightarrow[②]{H_2O} \quad CH_3CH_2CH_2CH_2\text{—}S^- \quad + \quad CH_3COO^-\,K^+$$

$$CH_3CH_2CH_2CH_2\text{—}S^- \quad + \quad H_2SO_4 \quad \xrightarrow[③a]{H_2O} \quad CH_3CH_2CH_2CH_2\text{—}SH \quad + \quad HSO_4^-$$

$$CH_3COO^-\,K^+ \quad + \quad H_2SO_4 \quad \xrightarrow[③b]{H_2O} \quad CH_3COOH \quad + \quad K^+\,HSO_4^-$$

7.39. The reaction described in this exercise is straightforward, comprising substitution of bromide ion by iodide ion with concomitant racemization.

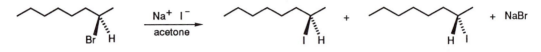

The fact that the reaction rate is dependent on both the concentration of substrate and nucleophile tells us that the mechanism is S_N2. The normal stereochemical result of an S_N2 reaction, however, is inversion of configuration which is in contrast with the experimental results.

Consider the structure of the *initial* S_N2 reaction, a reaction that is a typical S_N2 reaction:

The initial product has a good leaving group (iodide ion) attached to an aliphatic carbon atom, so it also undergoes the S_N2 reaction. The product of this second substitution reaction has the same bonding structure as the starting material, but it has the inverted stereochemistry. A racemic product is formed, therefore, after some time has passed.

Racemization of a carbon atom with a good leaving group will occur any time that the nucleophile and the leaving group are the same. This process normally only occurs when the nucleophile is a halide ion.

7.40. The reaction of an epoxide with acid occurs via protonation of the oxygen atom. In the case of phosphoric acid, dihydrogen phosphate ion is a poor nucleophile, and the solvent, methanol, is a weak nucleophile, too.

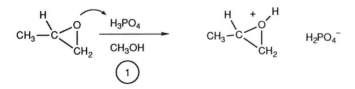

With no strong nucleophile present, the protonated epoxide ring has time to undergo ring-opening, and the 2° carbocation that forms reacts with methanol, leading to the 2-methoxy substituted product as the major one.

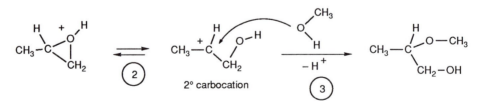

7.40. (continued)

When HCl is added to the same epoxide, the first step is the same, namely protonation of the oxygen atom. Chloride ion is potent enough as a nucleophile, however, to react directly with the protonated ring. Because the CH₂ center is less hindered, chloride ion reacts there preferentially, and the secondary alcohol is the major product.

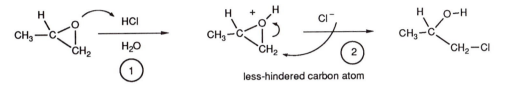

less-hindered carbon atom

7.41. When an alkyl chloride is treated with a silver salt, silver(I) ion reacts with the chlorine atom, which facilitates dissociation of chloride ion and formation of a 2° carbocation.

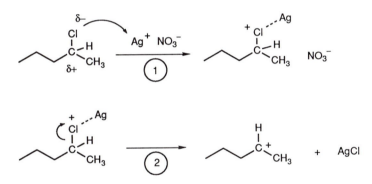

The carbocation is intercepted by reaction with methanol, and this step generates the methoxy compound after an acid-base reaction with nitrate ion.

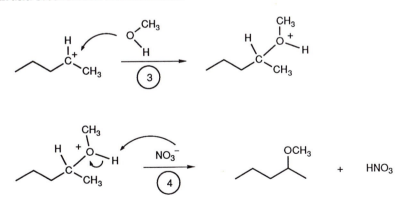

7.42. Draw Lewis structures according to procedures outlined in Chapter 3, for example in the solution to exercise 3.1. In each ion, resonance stabilization is important, and electrons are delocalized among several atoms, which diminishes the nucleophilicity of any one atom and makes the ion only a weak nucleophile.

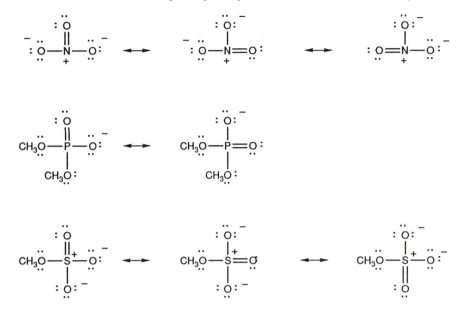

7.43. The Mitsunobu reaction proceeds by an S_N2 pathway, so the configuration of a stereogenic carbon atom that undergoes substitution is inverted. The nucleophile in the Mitsunobu reaction is the conjugate base of the acidic reactant (RSH , ArCOOH, and RCOOH, respectively, in the reactions of this exercise). The configuration of each stereogenic carbon atom is assigned as described in Chapter 5, and the most stable configuration of the *trans*-decalin system in part (c) is the one illustrated in Chapter 4 on text page 151. *trans*-Decalin is a conformationally locked system, so the product has the substituent in the axial position.

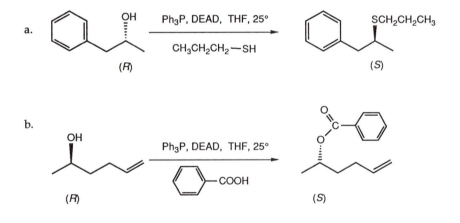

7.43.   (continued)

c.

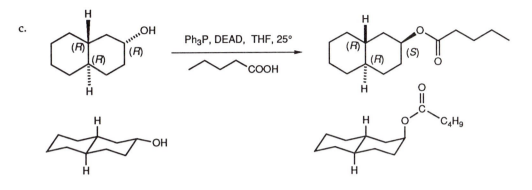

7.44.   The solution to this exercise is like that described in the solution to exercise 7.33. The organohalide that is needed is derived from the portion of the molecule that bears the carboxylic acid group. The oxygen-containing nucleophile in each case is the ion derived from deprotonating a phenol. Remember that an aryl halide cannot be used because the leaving group is attached to a carbon atom with sp^2 hybridization.

a.   To prepare phenoxyacetic acid, we would first treat phenol with sodium hydroxide to form the phenolate ion, which is the required nucleophile.

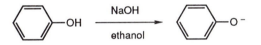

Treating the phenolate ion with chloroacetic acid will lead only to an acid-base reaction. (It is always necessary to consider side reactions like this when thinking about how to make compounds.)

To circumvent the acid-base reaction, we first treat chloroacetic acid with sodium hydroxide to form the sodium salt.

$$Cl-CH_2-COOH \xrightarrow[\text{ethanol}]{\text{NaOH}} Cl-CH_2-COO^-$$

Now the desired substitution reaction can be done; acid workup yields the final product.

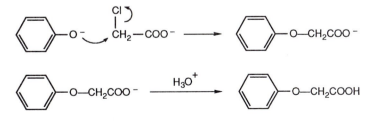

7.44. (continued)

b.

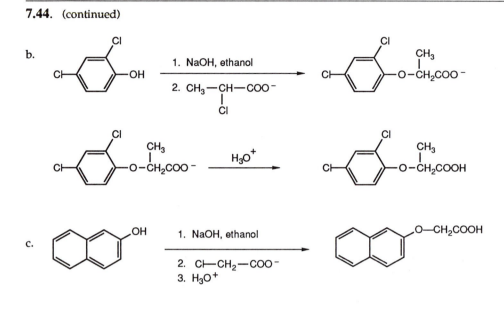

c.

7.45. Boron tribromide is a good Lewis acid because the boron atom has only six electrons. It reacts with an ether, a Lewis base, forming an adduct:

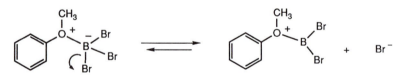

This compound dissociates at the boron-oxygen bond to regenerate starting materials, or a bromide ion dissociates to form a different Lewis acid that has a positively charged oxygen atom.

The oxygen atom carries a positive charge, so along with its substituents it is a good leaving group. Bromide ion reacts as a nucleophile at the methyl carbon atom by an S_N2 pathway.

This ether cleavage reaction works best with aryl methyl ethers because the methyl carbon atom is accessible to reaction with a nucleophile (it is not sterically hindered compared with other types of carbon atoms).

7.46. In the reaction of an allylic halide, the S_N2 pathway is straightforward: the nucleophile reacts backside at the carbon atom that bears the leaving group, displacing bromide ion.

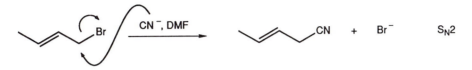

An S_N2' process works in the same way except that the π electrons of the carbon-carbon double bond are involved, too. The π electrons react backside at the carbon atom bearing the leaving group, and that movement leads to displacement of the bromide ion.

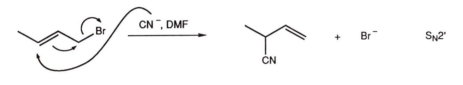

7.47. In the S_N2' pathway, as illustrated in the solution to exercise 7.46, the π electrons of the carbon-carbon double bond are involved as an "auxiliary" nucleophile, reacting backside at the carbon atom that bears the leaving group.

In the S_N1 pathway, a carbocation is formed, and resonance stabilization creates two reactive centers:

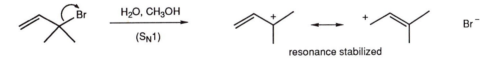

The nucleophile reacts at either resonance form of the carbocation. In this example, water is the nucleophile, so the products are isomeric alcohols.

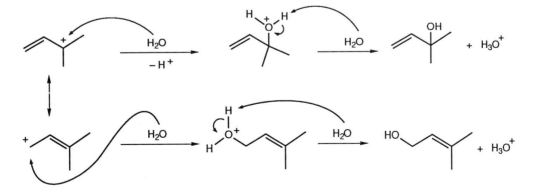

ELIMINATION REACTIONS OF ALKYL HALIDES, ALCOHOLS, AND RELATED COMPOUNDS

8.1. When an alcohol is heated with a strong acid such as H_2SO_4 or H_3PO_4, dehydration takes place to produce the most stable alkene. Alkene stability follows these general trends:

- more highly substituted > less highly substituted
- *endo* > *exo*
- *trans* > *cis*
- (*E*) > (*Z*)

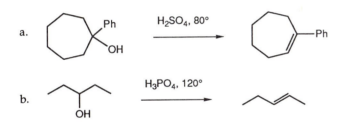

8.2. As noted in the solution to exercise 8.1, water is eliminated when an alcohol is heated with strong acid. Any hydrogen atoms attached to a carbon atom adjacent to one that bears the OH group can be removed to form the double bond. The major product from alcohol dehydration is the one that has the most highly-substituted double bond.

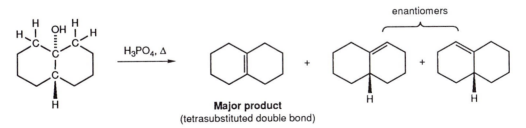

8.3.   The amount of heat evolved during hydrogenation of an alkene double bond is inversely proportional to the stability of the double bond undergoing addition. A tetrasubstituted alkene, therefore, gives off less heat than a trisubstituted one, for example. Use the scheme at the bottom of text page 331 to evaluate the relative stabilities of the given alkenes.

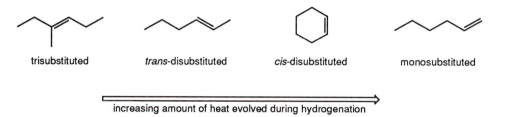

| trisubstituted | *trans*-disubstituted | *cis*-disubstituted | monosubstituted |

increasing amount of heat evolved during hydrogenation

8.4.   For compounds that have an exocyclic double bond the substituent is named with the "normal" root word, but its suffix is –ylidene (except when the group has one carbon atom, in which case it is "methylene"). The ylidene group is added to the name of a compound in the same way that any substituent is.

a.

functional group:	none
ring size:	7C: cyclohept
unsaturation suffix:	ane
substituents:	chlorine atom at C2
	methylene group at C1
stereochemistry:	C2: (*S*)

(*S*)-2-chloro-1-methylenecycloheptane

b.

functional group:	none
ring size:	5C: cyclopent
unsaturation suffix:	ane
substituent:	ethylidene group

ethylidenecyclopentane

c.

functional group:	alcohol (defines C1); suffix = -ol
ring size:	6C: cyclohex
unsaturation suffix:	an
substituent::	benzylidene group at C3

3-benzylidenecyclohexanol

8.5. The starting material in this reaction is a tertiary alkyl halide, and no strong nucleophile is present. This reaction, therefore, most likely proceeds by an E1 pathway, which means that a carbocation is formed by dissociation of the halide ion. Elimination occurs with removal of a proton from any carbon atom adjacent to the carbocation center, so two products are possible. An endocyclic double bond is normally more stable than an exocyclic one, so 1-methylcyclohexene is expected to be the major product.

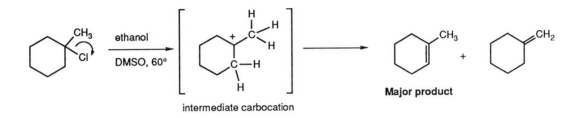

intermediate carbocation

Major product

8.6. The dehydrohalogenation reaction of the 2° alkyl halide shown in this exercise proceeds via an E2 pathway—the substrate is an alkyl halide that has a proton attached to a carbon atom adjacent to the one with the leaving group, and the reaction conditions include strong base and heat. An E2 reaction requires that the proton to be removed is *anti* to the leaving group. Rotation about the central carbon-carbon bond occurs so that the proton and halogen atom achieve the correct orientation.

Rotation about the carbon-carbon bond generates the possible staggered conformations:

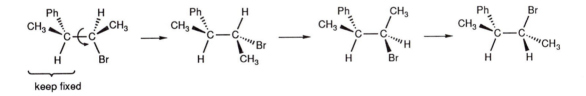

keep fixed

When the proton and leaving are *anti*, the base removes the proton, causing movement of the electrons to displace the leaving group. The product is the (Z) alkene.

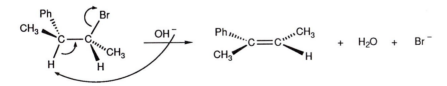

8.7. The reaction given in this exercise proceeds by an E2 pathway—the substrate is an alkyl halide that has a proton attached to an adjacent carbon atom, and the reaction conditions include strong base and heat. An E2 reaction of a bromocyclohexane derivative requires that the proton and leaving group be *trans* to one another as well as diaxial. Therefore, the given substrate must first undergo a ring flip so that the reactive groups assume the proper orientation.

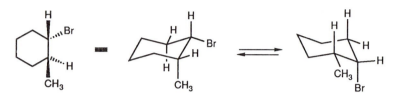

After the correct orientation is achieved, the base reacts with either neighboring proton, causing the movement of electrons that displaces the leaving group. The major product is the one with the more highly-substituted double bond.

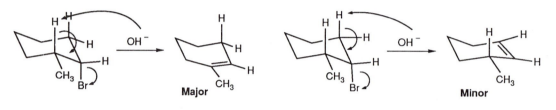

8.8. As shown in the solution to exercise 8.7, you first have to look at possible conformations that orient the H and Br *trans* and diaxial. For the *cis* isomer, the most stable conformation already has the bromine atom in the axial position. Recall that the *tert*-butyl group is always equatorial (text page 144). Elimination is therefore facile.

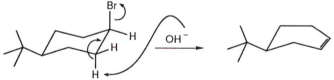

cis-4-Bromo-1-*t*-butylcyclohexane (prefered conformation)

For the *trans* isomer, a ring flip has to occur first to place the bromine atom in the axial position. The *tert*-butyl group prevents this, however, so a less stable boat conformation is the one from which the reaction must proceed. Energy is required to create the boat conformation, so this transformation proceeds more slowly.

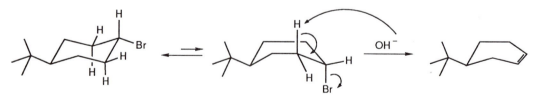

trans-4-Bromo-1-*t*-butylcyclohexane
(chair conformation; Br atom is equatorial)

Adopting the boat conformation keeps the *tert*-butyl group equatorial but allows the bromine atom to be axial so that elimination can take place.

8.9.   The course of dehydration is sometimes different under acidic and basic conditions, as illustrated by the examples in this exercise. If an alcohol is first converted to its sulfonate ester derivative, base can be used for the elimination process, which provides more control of stereochemistry (and sometimes regiochemistry) because the proton and leaving group must be *anti*. This fact accounts for formation of the less stable alkene isomer under E2 conditions in some instances. When dehydration is carried out under acid conditions, a carbocation is formed, so besides elimination, rearrangements of the carbon framework can occur.

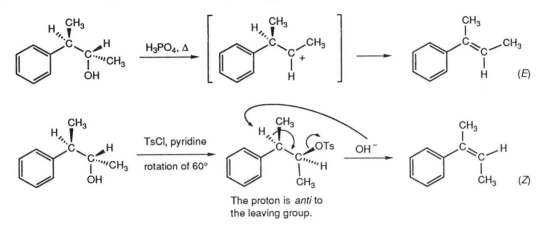

8.10.   To complete the reactions shown in Table 8.1 (text page 343), remove the indicated proton and leaving group from each reactant.

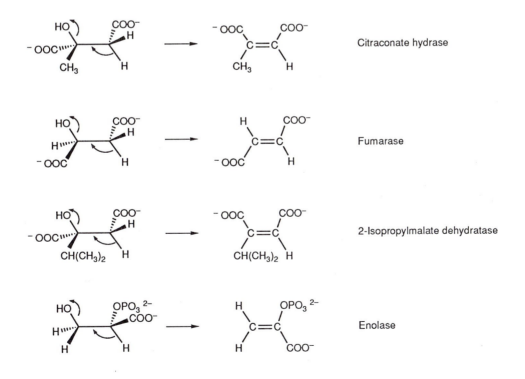

8.10. (continued)

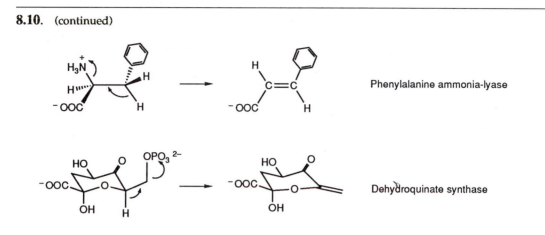

Phenylalanine ammonia-lyase

Dehydroquinate synthase

8.11. When a vicinal dihalide is treated with strong base, an alkyne forms if there are also two hydrogen atoms attached to the same carbon atoms. The base in the transformation shown in this exercise also deprotonates the carboxylic acid groups. Adding sulfuric acid in the second step protonates the carboxylate groups.

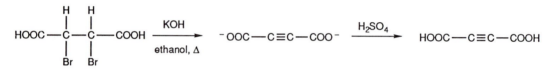

8.12. As described in the paragraph at the bottom of text page 346, a cyclic allene is easier to assemble because only one *sp*-hybridized carbon atom is present in the ring. Working with models, I find that the nine-membered ring allene seems to be the smallest that is stable.

8.13. In writing mechanisms for reactions, first decide what overall process takes place. Each reaction at the bottom of text page 349 involves substitution. Based on the reaction conditions given, as well as the nature of the substrate (all are methyl derivatives), we conclude that S_N2 mechanisms are likely in every instance. The S_N2 mechanism requires that the reactant has a good leaving group and that the nucleophile reacts backside to that leaving group.

The reaction between methanol and HBr begins with protonation of the oxygen atom to form a good leaving group. Bromide ion is the nucleophile and displaces a molecule of water.

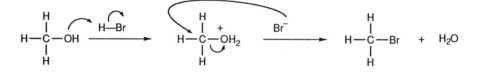

8.13.   (continued)

Bromomethane reacts with methylthiolate ion by nucleophilic displacement of bromide ion.

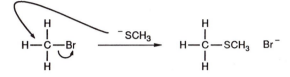

In the given Mitsunobu reaction, triphenylphosphine, DEAD, and hydrazidic acid react to form a phosphonium ion intermediate with azide ion as the counterion. Azide ion reacts in the second step as a nucleophile, displacing the oxyphosphonium group as triphenylphosphine oxide.

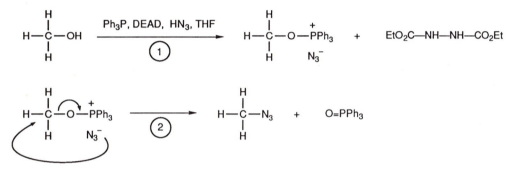

8.14.   The Swern reaction converts an alcohol to the corresponding carbonyl compound. A secondary alcohol, therefore, is converted to a ketone. The mechanism involves a three-step process. In the first step, the activated reagent is formed. In step 2, the alcohol derivative with a sulfur-oxygen bond is formed. In the final step, elimination occurs, producing the carbonyl group.

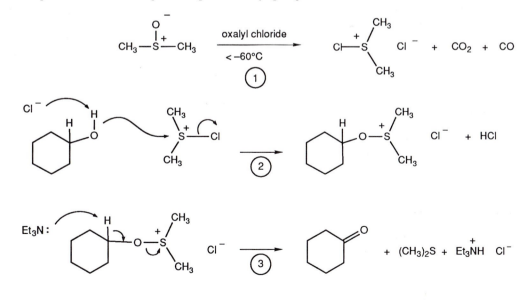

8.15. Oxidation of a 1° alcohol to an aldehyde requires non-aqueous conditions, normally with use of Collins or Corey's reagent. Jones reagent or aqueous permanganate converts a 1° alcohol to a carboxylic acid. All of these reagents convert a 2° alcohol to a ketone. Tertiary alcohols are inert toward these reagents. Manganese(IV) oxide converts benzylic and allylic alcohols to the corresponding carbonyl compound. A primary benzylic or allylic alcohol yields an aldehyde, and a secondary benzylic or allylic alcohol yields a ketone.

a.   The transformation shown is the conversion of a secondary alcohol to a ketone:
      use CrO_3–pyridine, PCC, CrO_3 in aqueous acid, or $KMnO_4$ in aqueous base.

b.   The reaction shown is the conversion of a primary benzylic alcohol to an aldehyde:
      use CrO_3–pyridine, PCC, or MnO_2.

c.   The reaction shown is the conversion of a primary alcohol to an aldehyde:
      use CrO_3–pyridine or PCC.

d.   The reaction shown is the conversion of a primary alcohol to a carboxylic acid:
      use CrO_3 in aqueous acid or $KMnO_4$ in aqueous base followed by workup with aqueous acid.

8.16.   Interpret the name according to the procedures outlined in Chapter 2. The root word, unsaturation index, and principal functional group are listed below. Each substituent is attached to the framework according to the associated numeral. Double bond stereochemistry is interpreted as described in Chapter 4 and the absolute configuration of any stereogenic center is drawn according to procedures outlined in Chapter 5.

a.   **3,3-Dimethyl-1-hexene**
      hex   6 carbon atoms
      ene   carbon-carbon double bond starting at C1

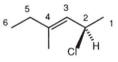

b.   **(R)-2-Chloro-(E)-4-methyl-3-hexene**
      hex   6 carbon atoms
      ene   carbon-carbon double bond starting at C3

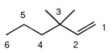

c.   **(Z)-2-Phenyl-1,3-pentadiene**
      penta   5 carbon atoms
      diene   two carbon-carbon double bonds,
               one starting at C1 and one at C3

d.   **(R)-1-Phenyl-1-butanol**
      but   4 carbon atoms
      an   no double or triple bonds
      ol   alcohol functional group; OH group at C1

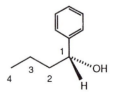

8.17. Follow the procedure outlined in the solution to exercise 7.28.

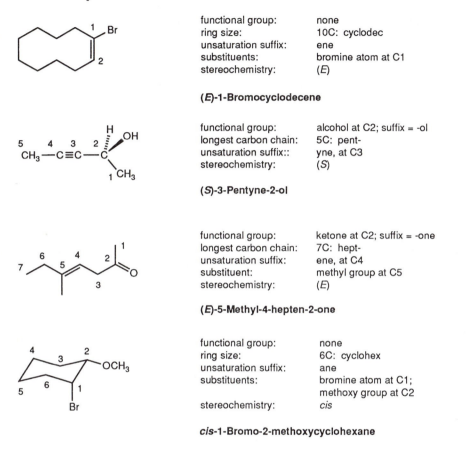

a.

functional group:	none
ring size:	10C: cyclodec
unsaturation suffix:	ene
substituents:	bromine atom at C1
stereochemistry:	(*E*)

(*E*)-1-Bromocyclodecene

b.

functional group:	alcohol at C2; suffix = -ol
longest carbon chain:	5C: pent-
unsaturation suffix::	yne, at C3
stereochemistry:	(*S*)

(*S*)-3-Pentyne-2-ol

c.

functional group:	ketone at C2; suffix = -one
longest carbon chain:	7C: hept-
unsaturation suffix:	ene, at C4
substituent:	methyl group at C5
stereochemistry:	(*E*)

(*E*)-5-Methyl-4-hepten-2-one

d.

functional group:	none
ring size:	6C: cyclohex
unsaturation suffix:	ane
substituents:	bromine atom at C1; methoxy group at C2
stereochemistry:	cis

cis-1-Bromo-2-methoxycyclohexane

8.18. Follow the instructions given in the solution to exercise 2.29.

1-Pentene

trans-2-Pentene

cis-2-Pentene

2-Methyl-1-butene

3-Methyl-1-butene

2-Methyl-2-butene

8.19. To predict what elimination products are formed by dehydrohalogenation, draw the structure of the starting material with the portion around the leaving group in its expanded form. The possible products are formed by removing the halogen atom plus any one of the hydrogen atoms attached to a carbon atom adjacent to the one bearing the halogen atom. A double bond forms between these two carbon atoms, and the major product is the most stable one unless there are conformational effects that only permit certain hydrogen atoms to be *anti* to the halide ion. For a cycloalkyl halide, the hydrogen atom that is removed is one that is *trans* to the halide ion when the halogen atom is axial.

a.   **1-Bromohexane:** the two hydrogen atoms on the adjacent methylene group are equivalent, so only one product is formed.

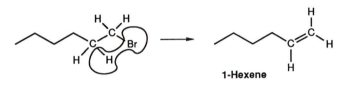

1-Hexene

b.   **1-Bromo-1-methylcyclohexane:** The formation of two products is possible because there are two types of hydrogen atoms: those attached to the adjacent methylene groups in the ring and those attached to the exocyclic methyl group. The product with the endocyclic double bond is more stable, so 1-methyl-cyclohexene is the major product.

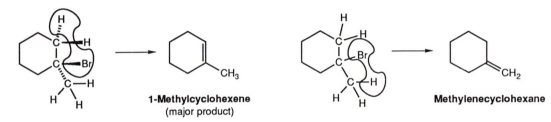

1-Methylcyclohexene
(major product)

Methylenecyclohexane

c.   *cis*-**1-Bromo-2-methylcyclohexane:** The formation of two products is possible because there are two types of hydrogen atoms: those attached to the adjacent methylene group within the ring and the one attached to the adjacent methine group. The compound that has the trisubstituted double bond is more stable.

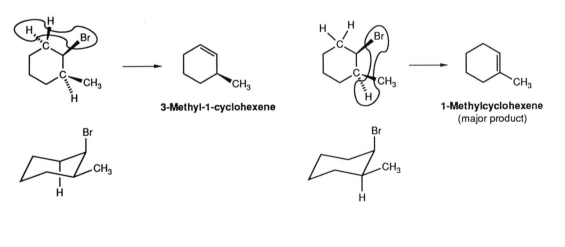

3-Methyl-1-cyclohexene

1-Methylcyclohexene
(major product)

8.19   (continued)

d. (S)-2-Chlorobutane: The formation of three products is possible because there are three types of hydrogen atoms: those attached to the adjacent methylene group (two conformations are possible) and those attached the adjacent methyl group. The compound that has the *trans* disubstituted double bond is the most stable, so *trans*-2-butene is the major product.

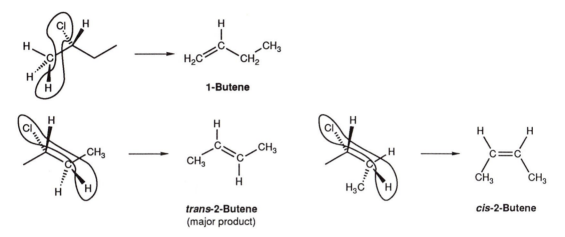

1-Butene

***trans*-2-Butene**
(major product)

***cis*-2-Butene**

e.   4-Iodoheptane: The formation of two products is possible because even though there is only one type of hydrogen atom attached to the adjacent methylene group, different conformations exist because of rotation about the single bonds. The compound that has the *trans* double bond is more stable, so *trans*-3-heptene is the major product.

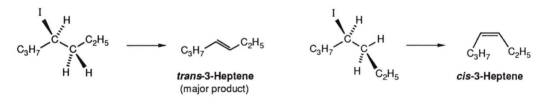

***trans*-3-Heptene**
(major product)

***cis*-3-Heptene**

8.20.   To predict what elimination products are formed by dehydration of an alcohol, follow the same procedure outlined in the solution to exercise 8.19. The difference is that the proton removed from an alcohol under acidic conditions does not have to be *anti* or *trans* to the hydroxyl group, so there is no stereochemical bias for the reaction.

a.   4-heptanol: The compound that has the *trans* double bond is more stable, so *trans*-3-heptene is the major product.

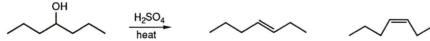

***trans*-3-Heptene**
(major product)

***cis*-3-Heptene**

8.20.   (continued)

b.   **2-Methyl-2-hexanol:** The compound with the more highly-substituted double bond is more stable, so 2-methyl-2-hexene is the major product.

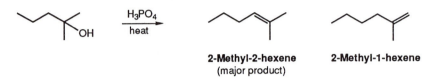

2-Methyl-2-hexene
(major product)

2-Methyl-1-hexene

c.   **1-Hydroxydecalin:** Only one compound is formed by direct elimination of water.

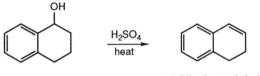

1,2-Dihydronaphthalene

8.21.   Follow the instructions given for the solution to exercise 8.17.

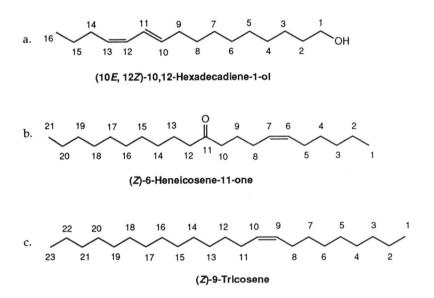

8.22. Follow the procedure outlined in the solution to exercise 8.3.

a.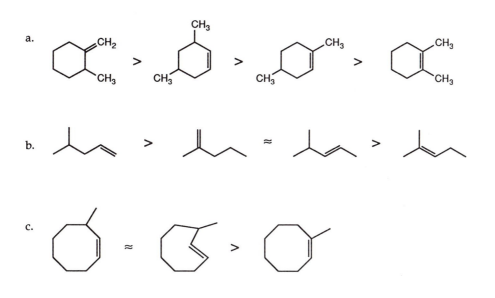

b.

c.

8.23. To decide what role a reagent plays in a given reaction, classify the transformation according to reaction type. If substitution occurs, the reagent other than the organohalide is considered to be a nucleophile. A reagent acts as a base in elimination reactions or in those transformations that involve a protic acid as the other reactant.

a. This is an E2 reaction, so methoxide ion is considered to function as a base.

b. This is a substitution reaction, so methoxide ion is considered to be a nucleophile.

c. This is an acid-base reaction in which phenol is the acid. Methoxide ion functions as a base.

8.24. Acid-catalyzed dehydration reactions occur by formation of a carbocation intermediate after protonation of the hydroxyl group and dissociation of a water molecule. Recall that a carbocation is prone to rearrange, however, which is what occurs in this transformation. Elimination of a proton from the carbon atom adjacent to this more stable carbocation produces the endocyclic double bond.

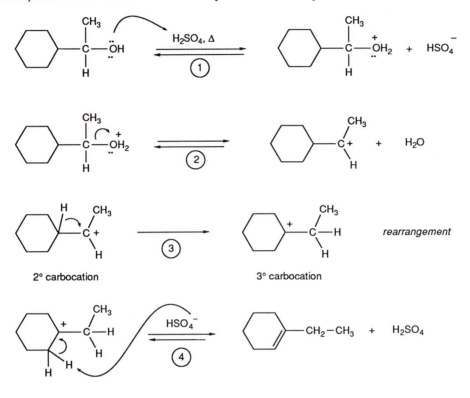

8.25. An elimination reaction that occurs under basic conditions is not normally complicated by rearrangements. Elimination under these conditions requires the presence of a good leaving group and strong base. To convert an OH group to a good leaving group, prepare a sulfonate ester, then use base to promote elimination via an E2 pathway.

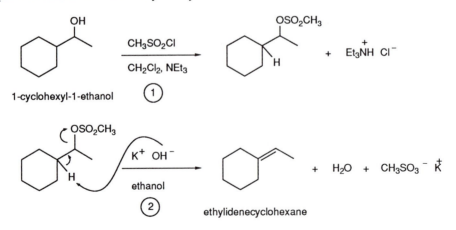

8.26.   Questions about rates of competing reactions require that you consider the possible conformations in which a molecule exists.   An elimination reaction carried out under basic conditions requires an *anti* relationship between the hydrogen atom and leaving group that are removed. The most stable conformation of (1*R*, 2*R*)-1-bromo-1,2-diphenylpropane has the H and Br atoms *gauche* to one another.   To undergo elimination, this isomer undergoes rotation to adopt a less stable conformation. The (1*S*,2*R*) isomer in its most stable conformation has the bromine and hydrogen atoms in the *anti* orientation already, which is ideal for an E2 reaction.

(1*S*, 2*R*)-1-Bromo-1,2-diphenylpropane

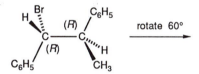   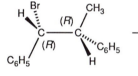

The largest groups (2 phenyl groups) are   (*E*)-1,2-Diphenyl-1-propene
anti, and the Br and H atoms are *anti*

(1*R*, 2*R*)-1-Bromo-1,2-diphenylpropane

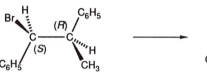

rotate 60°

The largest groups (2 phenyl groups) are   The Br and H atoms   (*Z*)-1,2-Diphenyl-1-propene
anti, and the Br and H atoms are *gauche*   are now *anti*.

8.27.   The mechanism for ketene formation from an acid chloride follows the same course as any E2 process. The base, triethylamine, reacts with the hydrogen atom attached to the carbon atom adjacent to the one bonded to chlorine. Displacement of the leaving group generates the double bond.

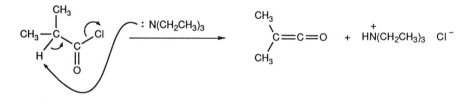

8.28.   The mechanisms for allene and ketene formation are the same, but making a ketene by an elimination reaction requires use of a weaker base than allene formation does. The reason for this will be discussed in Chapter 20, but it has to do with the relative acidity of the proton attached to the carbon atom adjacent to the carbonyl group.

8.29. The mechanism for formation of a nitrile from an oxime acetate is the same as that of any E2 reaction. The acetoxy group attached to the nitrogen atom is a good leaving group for this type of reaction.

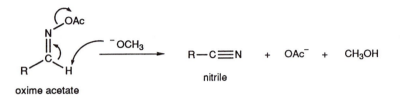

oxime acetate

nitrile

8.30. If an alkene is made by elimination from an organohalide, then possible starting materials are conceptualized by adding a hydrogen atom to one end of the double bond, and a bromine atom to the other end in each of the two possible orientations.

Once you have identified possible starting materials (make certain that you consider all stereoisomers as well), consider what elimination products are formed and which isomer is expected to be the major one. Then follow the procedure outlined in the solution to exercise 8.19.

a.   **1-Methylcyclohexene**

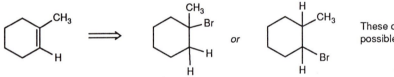

These compounds represent the possible starting materials.

Expected elimination reactions from the possible starting materials:

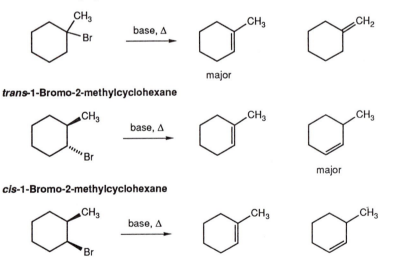

1-Bromo-1-methylcyclohexane

major

***trans*-1-Bromo-2-methylcyclohexane**

major

***cis*-1-Bromo-2-methylcyclohexane**

major

Conclusion: The best way to prepare 1-methylcyclohexene via dehydrohalogenation uses 1-bromo-1-methylcyclohexane or *cis*-1-bromo-2-methylcyclohexane as the starting material.

8.30.   (continued)

b.   ***trans*-5-Methoxy-2-pentene**

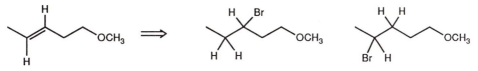

These compounds represent the possible starting materials.

Expected elimination reactions from the possible starting materials:

4-Bromo-1-methoxypentane

these will be produced in approximately equal amounts

5-Bromo-1-methoxypentane

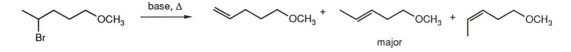

major

Conclusion: The best way to prepare *trans*-5-methoxy-2-pentene via dehydrohalogenation
uses 5-bromo-1-methoxypentane as the starting material.

c.   **(Z)-3-Methyl-3-heptene**

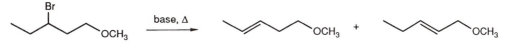

These compounds represent the possible starting materials.

Expected elimination reactions from the possible starting materials:

4-Bromo-3-methylheptane

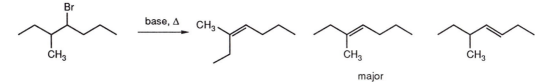

major

8.30.   (continued)

3-Bromo-3-methylheptane

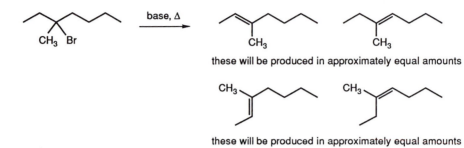

these will be produced in approximately equal amounts

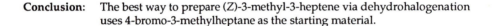

these will be produced in approximately equal amounts

Conclusion: The best way to prepare (Z)-3-methyl-3-heptene via dehydrohalogenation
uses 4-bromo-3-methylheptane as the starting material.

8.31.   Use the information given in Table 8.2 (text page 348) to decide which process—substitution or
elimination—is expected to predominate and which mechanisms are most likely.

a.   This reaction is a solvolysis process, so a unimolecular mechanism (E1 and S_N1) is likely. Heating favors
elimination, and the formation of a carbocation means that the most stable alkene will form. The *trans* alkene
is expected to be the major product.

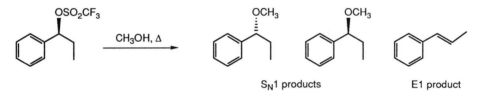

b.   The reagent is a very strong base, so a bimolecular mechanism (E2 and S_N2) is likely. Heating favors
elimination as does the use of such a strong base.

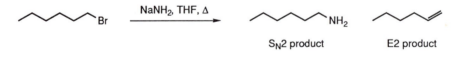

c.   The reaction of a 2° alcohol with a strong acid is normally unimolecular (E1 or S_N1). Heating favors
elimination, and formation of a carbocation means that the most stable alkene is formed.

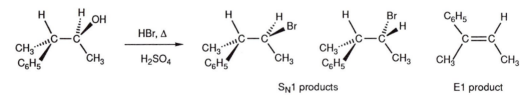

8.31. (continued)

d.   The reagent is a strong nucleophile, so a bimolecular mechanism (E2 and S$_N$2) is likely. Elimination processes require that the H and Br atoms be *trans* and diaxial, which can occur for this substrate after a ring flip. Elimination may predominate here because the reaction is heated, but the results are not as definitive as in the other instances shown in this exercise because cyanide ion is also a good nucleophile.

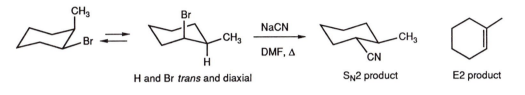

H and Br *trans* and diaxial   S$_N$2 product   E2 product

8.32. The energy diagram for dehydration of *tert*-butyl alcohol has the general form of the curves shown in Figure 8.2 (text page 329). The free energy of activation is equal to the height of the first maximum in the curve, just before the minimum that corresponds to the energy of the carbocation intermediate.

There is a smaller free energy of activation between the intermediate and the product, but the first one is what controls the rate of reaction. The free energy of the reaction, $\Delta G°$ is the difference between the energy of the reactants and that of the products.

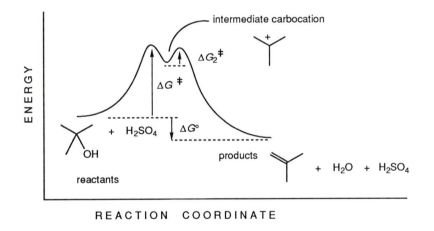

8.33. Two molecules of an alcohol can dimerize to form a molecule of an ether. The presence strong acid in the reaction mixture leads to protonation of the alcohol OH group in step (1), which create a good leaving group.

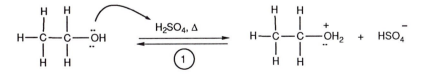

8.33. (continued)

Then, as in all such cases, there is a competition between substitution and elimination. In this case, substitution occurs because another molecule of ethanol functions as a nucleophile.

elimination:

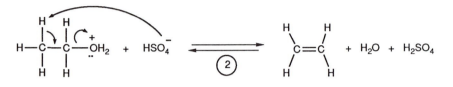

substitution:

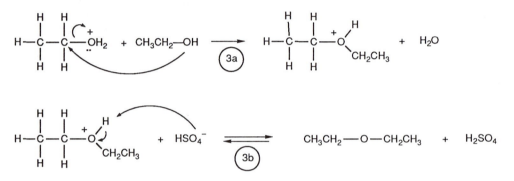

8.34. In elimination reactions, two features are required: a good leaving group attached to an sp^3-hybridized carbon atom (with a very strong base, a halide ion attached to an alkene carbon atom can also eliminate—see part e.), and a base (or conjugate base, if a proton is required to form a good leaving group).

An E2 process occurs under strongly basic conditions, and the proton to be removed must be *anti* to the leaving group. Under solvolysis conditions, an E1 reaction occurs if the mixture is heated. Alcohols undergo elimination of water when heated with strong acid. The most stable alkene is formed under these conditions.

Substitution reactions compete with elimination. Use the data in Table 8.2 (text page 348) to make a decision about which pathway predominates.

a. The OH group is not a good leaving group, but the first step in this transformation converts the alcohol to its sulfonate ester, which is a good leaving group. The strongly basic conditions favor an E2 pathway, and only the proton attached to C6 is *anti* to the leaving group.

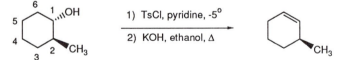

8.34. (continued)

b.   Bromide ion is a good leaving group, and the strongly basic conditions favor an E2 pathway. No stereochemical constraints are imposed by the structure of the substrate, so the most stable double bond is formed.

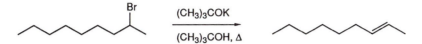

c.   The OH group is not a good leaving group, but the first step in this transformation converts the alcohol to its sulfonate ester, which is a good leaving group. The basic conditions favor an E2 pathway, but the proton at the ring junction cannot be removed because it is *cis* to the leaving group.

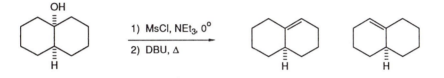

d.   Bromide ion is a good leaving group, and the strongly basic conditions favor an E2 pathway. No stereochemical constraints are imposed by the structure of the substrate, so the most stable double bond is formed.

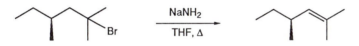

e.   A bromoalkene produces an alkyne under very strongly basic conditions. Acid workup is required to protonate the acetylide ion that is produced in the presence of the strong base.

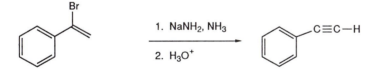

8.35. To assign oxidation level to a specific carbon atom, consider how many heteroatoms are attached to it. The one with more heteroatoms is more oxidized. Before making an assessment, expand double or triple bonds between carbon and a heteroatom to show two or three of that heteroatom, respectively.

8.35. (continued)

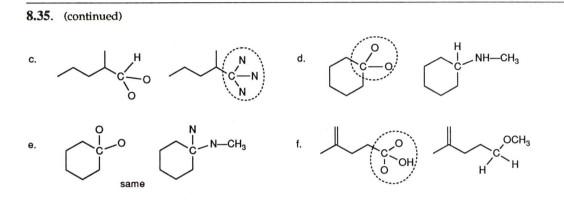

c.

d.

e.

same

f.

8.36. Follow the procedure outlined in the solution to exercise 8.15. In addition, consider use of the Swern oxidation, which is used to convert a 1° alcohol to an aldehyde, and a 2° alcohol to a ketone.

a. The reaction shown is the conversion of a primary alcohol to an aldehyde:
    use Swern, CrO_3–pyridine, or PCC.

b. The reaction shown is the conversion of a primary alcohol to a carboxylic acid:
    use CrO_3 in aqueous acid or $KMnO_4$ in aqueous base followed by acid workup to protonate the carboxylate salt.

c. The reaction shown is the conversion of a secondary alcohol to a ketone:
    use Swern, CrO_3–pyridine, PCC, CrO_3 in aqueous acid, or $KMnO_4$ in aqueous base.

d. The reaction shown is the conversion n of a 2° benzylic alcohol to a ketone:
    use Swern, CrO_3–pyridine, PCC, CrO_3 in aqueous acid, $KMnO_4$ in aqueous base, or MnO_2.

e. The reaction shown is the conversion n of a 2° benzylic alcohol to a ketone and the conversion of a primary alcohol to an aldehyde:
    use Swern, CrO_3–pyridine, or PCC.

8.37. The alcohol in the starting material is 3°, so oxidation does not normally occur. Under acidic conditions, however, the alcohol OH group is protonated, forming a good leaving group. The presence of a double bond renders the leaving group labile, and dissociation occurs (recall that allylic substrates often react by E1 and S_N1 pathways). The intermediate carbocation is intercepted by water in steps (3) and (4) to form a 2° alcohol, and the alcohol is oxidized subsequently to form the ketone.

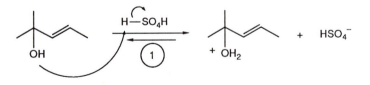

8.37. (continued)

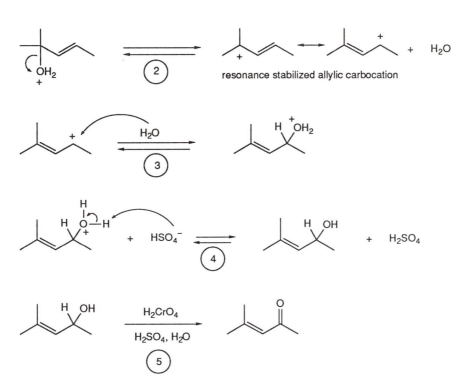

resonance stabilized allylic carbocation

8.38. The involvement of DMSO as an oxidizing agent suggests that a species may be formed that looks like the intermediate in the Swern oxidation. Because DMSO has a negatively-charged oxygen atom, which is certainly nucleophilic, substitution seems likely given that 1° alkyl halides undergo either substitution or elimination reactions as a rule. DMSO is not a strong base, so elimination is unlikely.

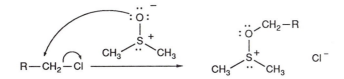

After formation of the sulfoxonium salt, bicarbonate ion subsequently removes a proton from the adjacent carbon atom to form the carbon–oxygen double bond.

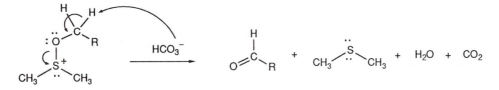

8.39. The fact that the deuterium atom becomes attached to the "front" face of the pyridine ring suggests that the substrate's position must be such that the deuterium can approach from only one direction. The new stereogenic center in the dihydropyridine is (R).

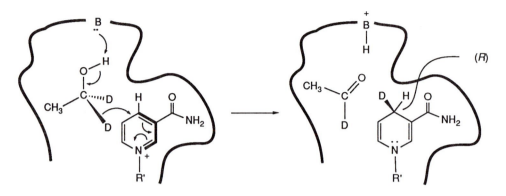

8.40. In an exercise such as this, the first step is to summarize the given information by constructing a flow chart.

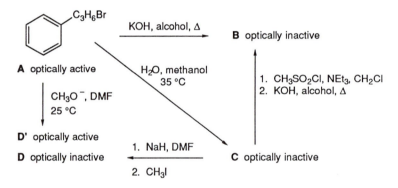

Consider what structures are possible for any compound in the scheme for which you have structural information (if there are too many possibilities, you may have to ignore this step). In this exercise, we can draw five structures. 1-Bromo-3-phenylpropane, **I**, and 2-bromo-2-phenylpropane, **IV**, cannot be optically active, which **A** is supposed to be, so we eliminate those from consideration.

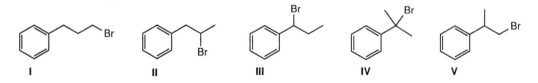

Next, consider the reactions according to the conditions that are specified: **A** to **B** is an E2 reaction, **A** to **C** is an S_N1 reaction, **C** to **B** is an E2 reaction, and **A** to **D** is an S_N2 reaction. The conversion of **A** to **C** would be unlikely for 1-bromo-2-phenylpropane because that compound is a 1° alkyl bromide, which will not undergo an S_N1 reaction. We remove compound **V** from further consideration.

8.40. (continued)

If we now consider the structures of the compounds obtained from the two remaining candidates, **II** and **III**, we see that the structures match the given properties for **A** through **D'**. Compound **III** is probably a better choice because the solvolysis reaction (**A** to **C**) occurs under relatively mild conditions, and a benzylic substrate is more likely to react at 35 °C. (The squiggly bond in the structures shown below denotes that the stereochemistry is unspecified, and the compound is racemic.)

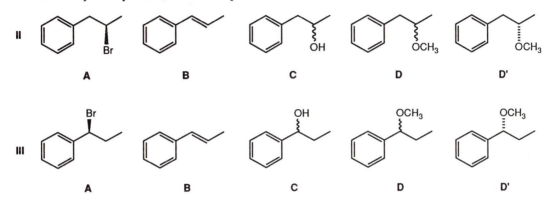

8.41. *Alcohol dehydrogenase* converts a 1° or 2° alcohol to the corresponding carbonyl compound. Formaldehyde is therefore the product of methanol oxidation.

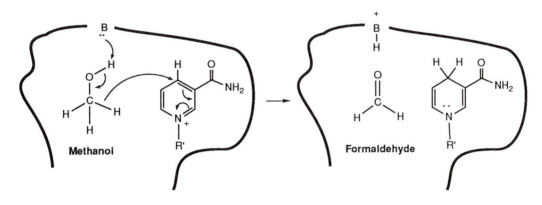

ELECTROPHILIC ADDITION REACTIONS OF ALKENES AND ALKYNES

9.1. For a given molecule, calculate the number of sites of unsaturation that is has by applying the formula on text page 370. Whatever functional groups may be present in a molecule have to be consistent with the number and types of atoms as well as the number of π bonds and rings that the compound has. In this exercise, we are told that the compounds have no rings, so the functional groups shown in the answers must have the calculated number of π bonds.

a. $C_5H_9NO_2$ $\dfrac{2(5) + 2 - (9 - 1)}{2}$ = 2 sites of unsaturation = 2 π bonds

two oxygen atoms		nitrogen atom	carbon–carbon multiple bonds
C—O	alcohol or ether (2)	C—N amine	C=C alkene (2)
C—O	alcohol or ether (2)	C—N amine	C≡C alkyne
C—O	alcohol or ether (2)	C≡N nitrile	none
C—O	alcohol or ether (and amide)	—C—N⟨ amide ‖ O	C=C alkene
C=O	aldehyde or ketone (and amide)	—C—N⟨ amide ‖ O	none
C=O	aldehyde or ketone	C—N amine	C=C alkene
C—O ‖ O	carboxylic acid or ester	C—N amine	C=C alkene
C=O	aldehyde or ketone (2)	C—N amine	none

9.1. (continued)

9.1. $C_8H_{12}Br_2$ $\dfrac{2(8) + 2 - (12 + 2)}{2}$ = 2 sites of unsaturation = 2 π bonds

two bromine atoms	carbon–carbon multiple bonds
C—Br organobromide (2)	C=C alkene (2)
C—Br organobromide (2)	C≡C alkyne

c. $C_9H_{17}ClO$ $\dfrac{2(9) + 2 - (17 + 1)}{2}$ = 1 site of unsaturation = 1 π bond

one oxygen atom	chlorine atom	carbon–carbon multiple bonds
C—O alcohol or ether	C—Cl organochloride	C=C alkene
C=O aldehyde or ketone	C—Cl organochloride	none
(acid chloride)	—C—Cl acid chloride ‖ O	none

9.2. Looking at the product of each reaction in Figure 9.1, we see that a halide ion or hydroxy group appears in each. Knowing from previous chapters that halide and hydroxide ions are nucleophiles, we can deduce from the structure of the reagent which is the electrophilic portion. For reactions carried out in the presence of strong acid, the electrophile is usually the proton.

Reagent combination	electrophile	nucleophile
HX (X = Cl, Br, I)	H^+	X^-
X_2 (X = Cl, Br)	X^+	X^-
X_2, H_2O (X = Cl, Br, I)	X^+	OH^-
H_2O, H_2SO_4	H^+	OH^-
$Hg(OAc)_2$, H_2O	$(AcO)Hg^+$	OH^-

9.3.   Acyclic alkenes react in the same way that cyclohexene does: the proton adds to the double bond to form a carbocation, and the halide ion reacts with that carbocation to form the product.

9.4.   Electrophilic addition of water to a carbon-carbon double bond takes place in three steps. First, the alkene π bond is protonated in step (1) by the mineral acid, producing a carbocation. Then, the carbocation is intercepted by a molecule of water in step (2). Finally, deprotonation yields the alcohol product.

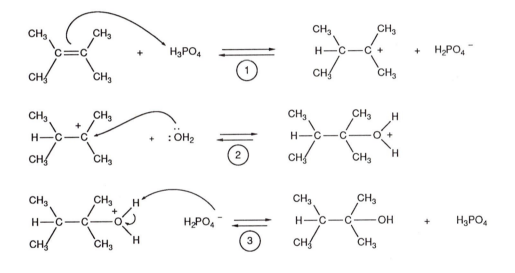

9.5.   Bromine and chlorine add to the double bond of an alkene, forming a product in which one halogen atom is bonded to each carbon atom of the original double bond. The stereochemistry of addition is *trans*, but the product of addition to an acyclic alkene undergoes rotation about the carbon-carbon single bond, so the stereochemical outcome is not apparent as it is for reactions of cyclic alkenes.

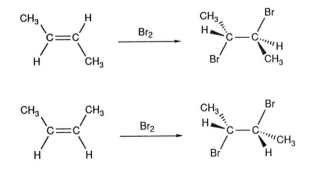

9.6. In the addition reaction of HOBr to cyclohexene, a bridged bromonium ion is formed by the initial reaction between the π bond of the alkene and Br⁺. Because the bridged ion is symmetric, a molecule of water intercepts the cation on the top side of the ring *at either carbon atom*. The products are enantiomers.

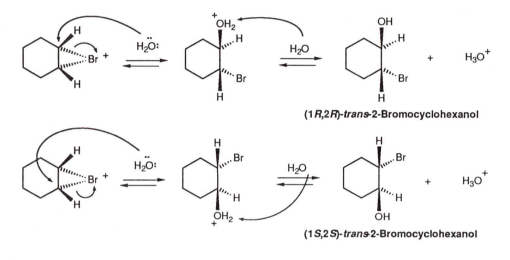

(1*R*,2*R*)-*trans*-2-Bromocyclohexanol

(1*S*,2*S*)-*trans*-2-Bromocyclohexanol

9.7. After formation of the iodinium intermediate in step (1), the oxygen atom of the carboxylic acid group reacts to open the three-membered ring and form a new carbon–oxygen bond. In the first step, the iodine atom can be either above or below the plane of the ring. Reaction from the top works better (2t) because the carboxylic acid group is already above the plane of the six-membered ring, and it has to approach opposite where the iodinium ion is bridged. The final stereochemistry in the major product is such that the 6/5 ring junction is *cis*, and the iodine atom is *trans* to the five-membered ring. If the starting material is optically active, the product is optically active, too.

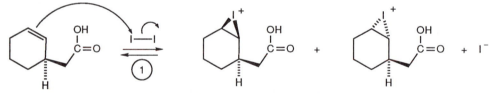

Reaction at the top face of the π bond.   Reaction at the bottom face of the π bond.

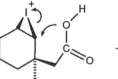

9.8. The mechanism for oxymercuration in an alcohol solvent is the same as the corresponding reaction that takes place in water except that the nucleophile is the oxygen atom of the alcohol molecule.

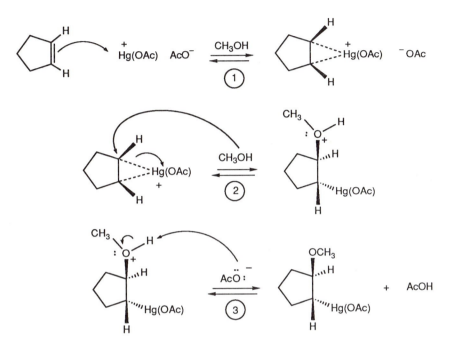

9.9. As you saw in the solution to exercise 9.6, reaction of a nucleophile with each end of a cyclic bromonium ion intermediate leads to formation of enantiomeric products.

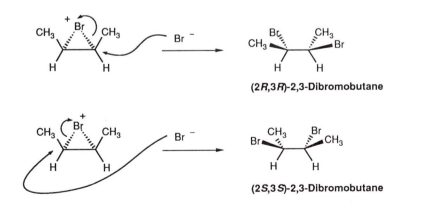

9.10. If we solve this exercise in the same way that we did exercise 9.9, we generate structures of the "two" products. By numbering the carbon chain in the opposite direction, however, we discover that the compounds are the same. The product is therefore *meso*.

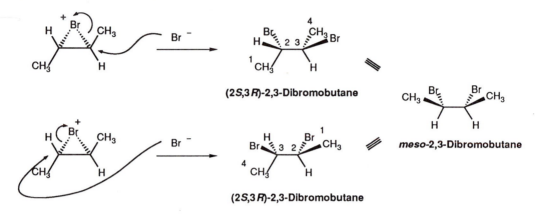

(2S,3R)-2,3-Dibromobutane

(2S,3R)-2,3-Dibromobutane

meso-2,3-Dibromobutane

9.11. To classify a molecule as *erythro* or *threo*, rotate about the central carbon-carbon bond so that the molecule exists in an eclipsed conformation, then determine if one or three groups mirror the others. The mirrored groups appear encircled, below. If one atom or group is mirrored, the molecule is *threo*; if three groups are mirrored, the molecule is *erythro*.

a.   Only the bromine atoms are mirrored; *threo*

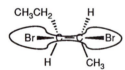

b.   All three groups are mirrored; *erythro*

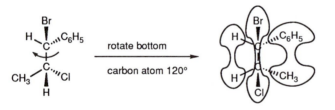

rotate bottom carbon atom 120°

c.   All three groups are mirrored; *erythro*

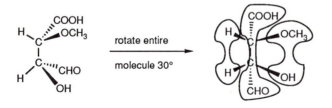

rotate entire molecule 30°

9.12.   The steps involved in hydration of an alkene and dehydration of an alcohol are the reverse of each other. Differences include:

- protonation of the double bond versus the OH group in step (1)
- reaction with versus loss of a water molecule in step (2)
- deprotonation of the OH_2^+ group versus deprotonation adjacent to the carbocation in step (3).

addition of water to a double bond (hydration)

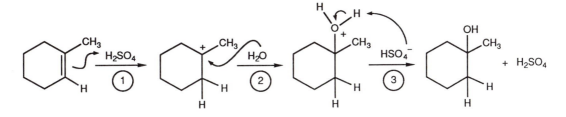

elimination of water from an alcohol (dehydration)

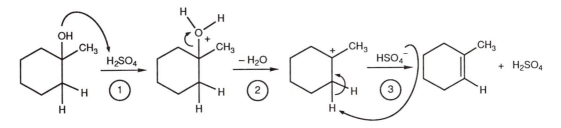

9.13.   Hydration of 1-hexene follows the same course as addition of water to 1-methylcyclohexene. The double bond reacts with a proton in step (1) to form the 2° carbocation. That intermediate in turn is trapped in step (2) by reaction with a molecule of water. Deprotonation of the OH_2^+ group in step (3) yields the 2° alcohol.

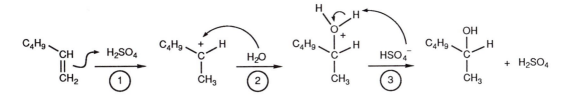

9.14. This transformation follows the same pathway as the reaction shown in the solution to exercise 9.13 except that the electrophile here is Cl⁺. Remember that a halogen cation bridges between carbon atoms to form a three-membered ring, and the bridge forms on either face of the double bond (top or bottom.) The cationic intermediate is subsequently trapped by reaction with a molecule of water, which reacts at the more highly-substituted carbon atom on the side opposite the bridging chlorine atom. Deprotonation of the OH_2^+ group yields a 2° alcohol. The molecule of water reacts at either face of the double bond, so racemic products are formed.

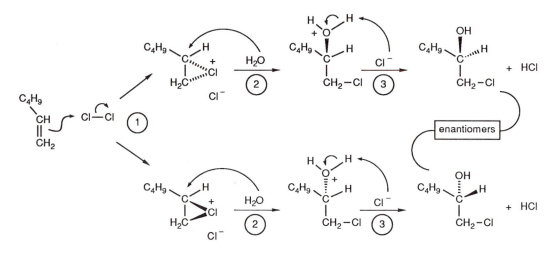

9.15. This transformation follows the same pathway shown in the solution to exercise 9.14. The mercury(II) ion bridges the carbon atoms to form a three-membered ring, and the bridge can form on either face of the double bond (top or bottom). The cationic intermediate is subsequently trapped by reaction with water, which bonds to the more highly-substituted carbon atom on the side opposite the mercury atom. Deprotonation of the OH_2^+ group yields the alcohol. Demercuration with sodium borohydride generates racemic products.

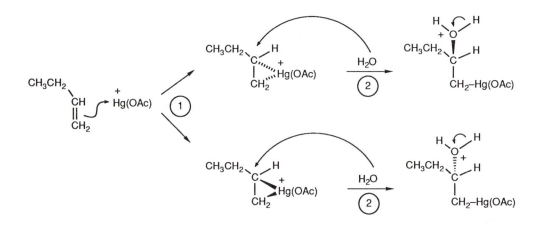

9.15.   (continued)

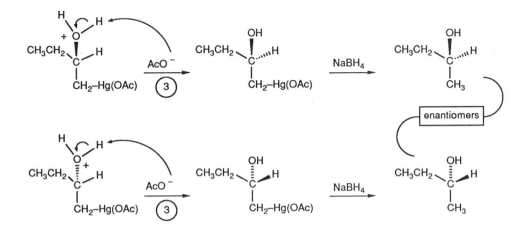

9.16.   An alkyne often undergoes addition with two equivalents of a reagent. When the first equivalent of HBr adds to a triple bond, the first-formed vinyl carbocation is the more highly-substituted one because the positive charge can be stabilized by hyperconjugation, even though stabilization is not as good as it is for an alkyl carbocation. Step (2) produces the vinyl bromide product.

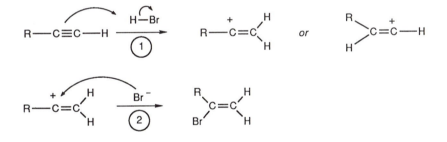

In addition of a second equivalent of HBr, the double bond is protonated in step (3). Protonation generates two possible carbocations. The 2° carbocation is stabilized by hyperconjugation with the R group, and it is also stabilized by resonance delocalization with electron pairs on the bromine atom. In step (4), the more stable carbocation is trapped by bromide ion to form the geminal dibromide.

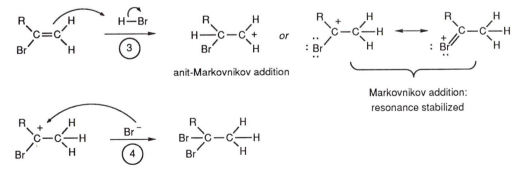

9.17. Just as in the addition of HBr to an alkyne, addition of HCl to a triple bond produces the *gem*-dichloride by Markovnikov addition.

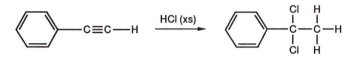

9.18. Markovnikov addition of water to a triple bond produces the corresponding vinyl alcohol, which tautomerizes in step (2) to form the ketone.

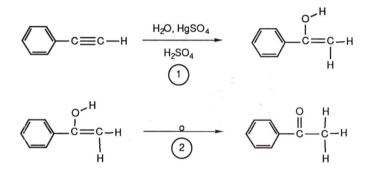

9.19. Addition of an unsymmetric reagent to an internal alkyne produces two regioisomers because the initially formed vinyl cation has the same stability no matter which carbon atom forms a bond to the electrophile. A symmetric reagent does not suffer from this limitation.

a.   The addition of HCl on a silica surface is easy to stop after one equivalent is adds.

$$CH_3CH_2-C\equiv C-CH_3 \xrightarrow{SiO_2 \cdots \{HCl\}} CH_3CH_2-\underset{Cl}{\overset{H}{C}}=C-CH_3 \ + \ CH_3CH_2-C=\underset{Cl}{\overset{H}{C}}-CH_3$$

b.   Addition of bromine to an alkyne triple bond normally gives the tetrabromide.

$$CH_3CH_2-C\equiv C-CH_3 \xrightarrow{Br_2,\ CCl_4} CH_3CH_2-\underset{Br}{\overset{Br}{C}}-\underset{Br}{\overset{Br}{C}}-CH_3$$

c.   Hydration of a triple bond produces a vinyl alcohol, which tautomerizes to the ketone.

$$CH_3CH_2-C\equiv C-CH_3 \xrightarrow{\overset{H_2O,\ HgSO_4}{H_2SO_4}} CH_3CH_2-\overset{O}{\overset{\|}{C}}-CH_2CH_3 \ + \ CH_3CH_2-CH_2-\overset{O}{\overset{\|}{C}}-CH_3$$

9.20.   The kinetic product of electrophilic addition to a 1,3-diene is normally the 1,2-adduct, while the thermodynamic product is formed by 1,4-addition. Addition to a conjugated diene is not stereospecific because the intermediate carbocation is allylic, so a bridged bromonium ion is not formed. If a halogen atom does not bridge between two carbon atoms, stereospecificity is nonexistent because the incoming nucleophile does not have to approach from the opposite side of the ring.

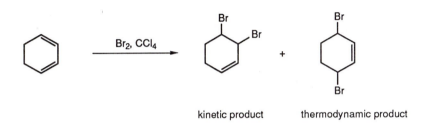

kinetic product   thermodynamic product

9.21.   The elimination reaction that terminates isobutylene dimerization occurs when hydrogen sulfate ion removes a proton from a carbon atom adjacent to the carbocation center, creating the double bond. The major product has the more stable, hence more highly substituted, double bond. Recall that E1 reactions produce the Saytseff product (text page 328).

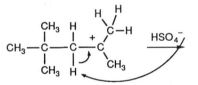
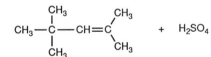

2,4,4-Trimethyl-2-pentene
major product: more highly-substituted double bond

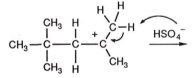
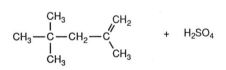

2,4,4-Trimethyl-1-pentene

9.22. The conversion of geranyl to farnesyl pyrophosphate follows the same mechanism as that described for the biosynthesis of geranyl pyrophosphate on text pages 401-402. Isopentenyl pyrophosphate isomerizes to dimethylallyl pyrophosphate, which ionizes to form the resonance-stabilized carbocation. Geranyl pyrophosphate intercepts the carbocation in step (2), and the resulting carbocation loses a proton to form the product. The reaction occurs within the active site of an enzyme, so the carbocation is stabilized by residues in the active site, and a base is available to remove a proton from the carbocation intermediate.

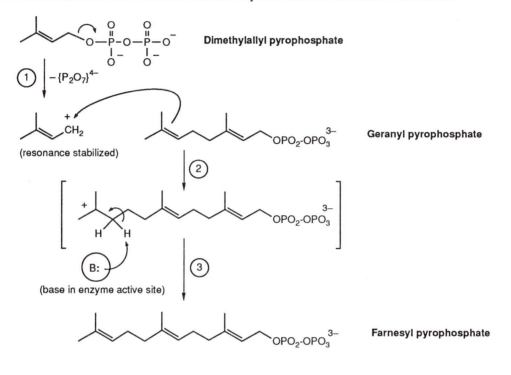

9.23. Copolymerization of isobutylene and isoprene begins with polymerization of isobutylene, as shown on text page 403.

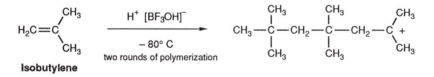

When isoprene is incorporated into the growing polymer chain, one of its double bonds intercepts the carbocation. Isoprene has two types of carbon-carbon double bonds, and the more highly-substituted one reacts with the carbocation to form a 3° carbocation. This cation is also adjacent to a double bond, so it is stabilized by resonance as well.

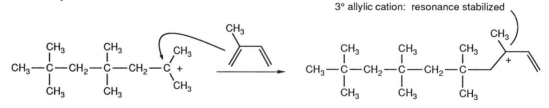

9.24.   In cationic cyclization reactions, an allylic alcohol reacts to form a stabilized carbocation via protonation (step 1) followed by dissociation of a molecule of water (step 2). In this transformation, the triple bond intercepts the carbocation in step (3), producing a vinyl cation *exo* to a five-membered ring. The nucleophilic conjugate base of the original acid traps the cation in step (4) to form the cyclized product.

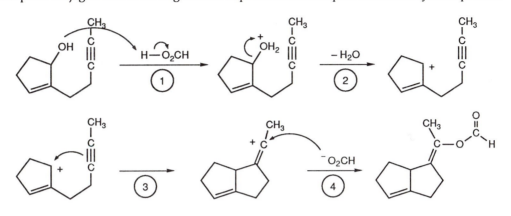

9.25.   Reaction of an allylic alcohol with a Lewis acid such as $SnCl_4$ produces a carbocation in step (1) that is intercepted by each double bond in turn.

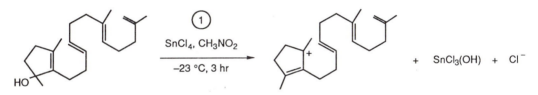

The molecule folds such that the chain looks "chair-like" in each region. The final 3° carbocation is deprotonated by chloride ion that was formed in step (1), which also produced the allylic carbocation.

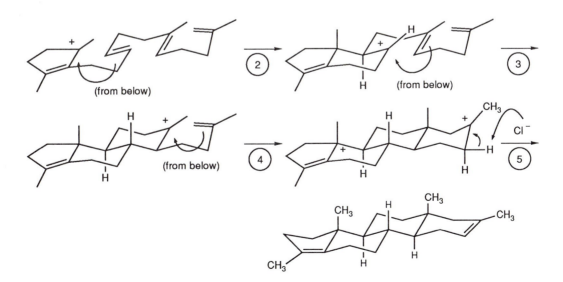

9.26. Follow the procedure outlined in the solution to exercise 7.28.

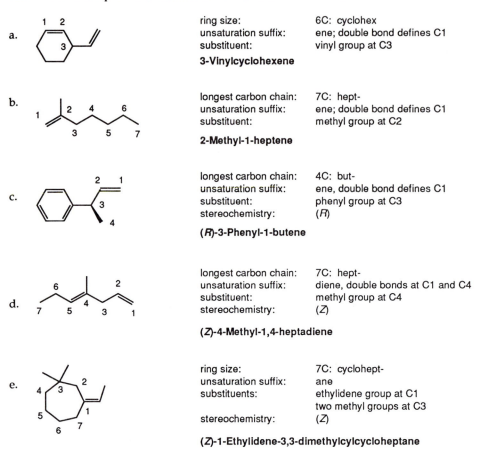

a.

ring size:	6C: cyclohex
unsaturation suffix:	ene; double bond defines C1
substituent:	vinyl group at C3

3-Vinylcyclohexene

b.

longest carbon chain:	7C: hept-
unsaturation suffix:	ene; double bond defines C1
substituent:	methyl group at C2

2-Methyl-1-heptene

c.

longest carbon chain:	4C: but-
unsaturation suffix:	ene, double bond defines C1
substituent:	phenyl group at C3
stereochemistry:	(*R*)

(*R*)-3-Phenyl-1-butene

d.

longest carbon chain:	7C: hept-
unsaturation suffix:	diene, double bonds at C1 and C4
substituent:	methyl group at C4
stereochemistry:	(*Z*)

(*Z*)-4-Methyl-1,4-heptadiene

e.

ring size:	7C: cyclohept-
unsaturation suffix:	ane
substituents:	ethylidene group at C1
	two methyl groups at C3
stereochemistry:	(*Z*)

(*Z*)-1-Ethylidene-3,3-dimethylcylcycloheptane

9.27. To interpret a compound name, follow the procedures outlined in Chapter 2: break the name into its constituent parts, then combine the pieces to form

a. **2-Methyl-(*E*)-3-hexenal**

hex	6 carbon atoms
ene	carbon-carbon double bond starting at C3
al	aldehyde functional group (defines C1)

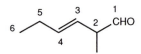

b. **2,2-Difluoro-3-cyclohexene-1-ol**

cyclohex	six-membered ring of carbon atoms
ene	carbon-carbon double bond starting at C3
ol	alcohol functional group; OH group at C1

9.27.   (continued)

c.   **(R)-2-Methylenecyclopentanol**

cyclopent	five-membered ring of carbon atoms
an	no double or triple bond
ol	alcohol functional group; OH group at C1

d.   ***threo*-2,3-Dibromopentane**

pent	5 carbon atoms
ane	no double or triple bond
threo	only one group is mirrored when the compound is viewed in the eclipsed conformation

IUPAC name: **(2R, 3R)-2,3-Dibromopentane**

9.28.   Follow the procedure outlined in the solution to exercise 9.1.

a. $C_6H_{11}NO$ $\dfrac{2(6) + 2 - (11 - 1)}{2}$ = 2 sites of unsaturation = 2 π bonds

one oxygen atom	one nitrogen atom	carbon–carbon multiple bonds
C—O alcohol or ether	C—N amine	C=C alkene (2)
C—O alcohol or ether	C—N amine	C≡C alkyne
C—O alcohol or ether	C≡N nitrile	none
C=O aldehyde or ketone	C—N amine	C=C alkene
(amide)	—C—N< amide ‖ O	C=C alkene

b.   There are no sites of unsaturation in this molecule. The only type of molecule that is consistent with the absence of π bonds and all O, N, and S atoms is a dihaloalkane.

$C_4H_8Br_2$ $\dfrac{2(4) + 2 - (8 + 2)}{2}$ = 0 sites of unsaturation

9.28.   (continued)

c. $C_6H_{10}O_2$ $\dfrac{2(6) + 2 - (10)}{2}$ = 2 sites of unsaturation = $2\,\pi$ bonds

two oxygen atoms		carbon–carbon multiple bonds
C—O	alcohol or ether (2)	C=C alkene (2)
C—O	alcohol or ether (2)	C≡C alkyne
C—O C=O	alcohol or ether aldehyde or ketone	C=C alkene
C=O	aldehyde or ketone (2)	none
C—O with ‖O	carboxylic acid or ester	C=C alkene

9.29.   To predict the major product of an electrophilic addition reaction, attach the components of the reagent (electrophile and nucleophile) to the carbon atoms at each end of the double bond.   The regiochemistry is usually "Markovnikov", so the electrophilic portion of the reagent is attached to the less highly-substituted carbon atom of the original π bond.   The stereochemistry is often *trans*, and an achiral alkene produces a racemic mixture of products.

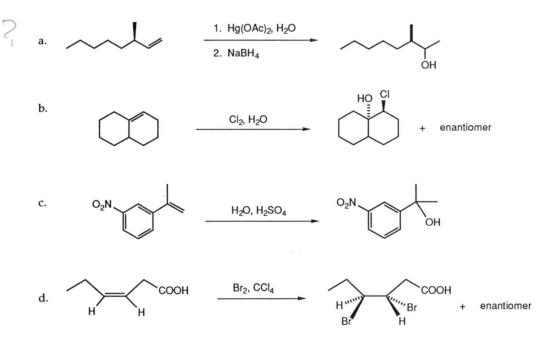

9.29. (continued)

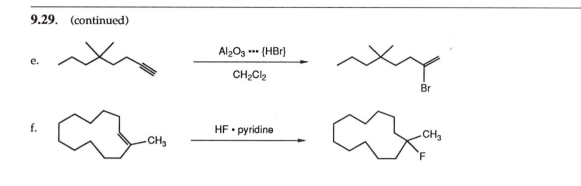

e.

f.

9.30. Electrophilic addition of bromine to a double bond in the presence of an alcohol takes place in the same way that addition of halogen takes place in the presence of water. When an alcohol is the solvent, a vicinal bromo ether is produced instead of a vicinal bromo alcohol. The products are enantiomers.

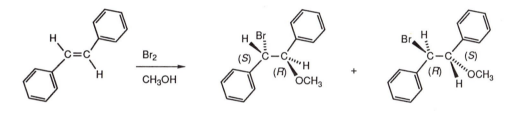

To assess whether the products are *threo* or *erythro*, hold one end fixed and rotate about the central bond to form the eclipsed conformations. If only one set of groups mirror each other when the molecule is in an eclipsed conformation, the compound is the *threo* isomer. If all three sets of substituents mirror each other when the molecule is in an eclipsed conformation, the compound is the *erythro* isomer. The product of this transformation is *erythro*.

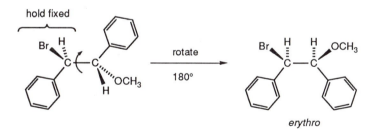

9.31. a. In the reaction of an alkene with mercury(II) ion, the π bond plays the role of nucleophile. A mercury bridged carbocation is formed in the first step.

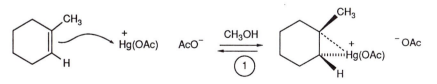

9.31. (continued)

The carbocation intermediate is trapped by reaction with methanol in step (2), and a proton if removed by acetate ion to form the product in step (3).

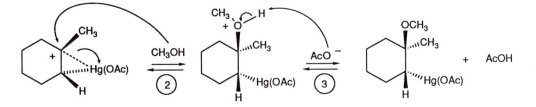

b.   If the product is subsequently treated with sodium borohydride, the mercury-containing group is replaced by a hydrogen atom, and the product is the methyl ether.

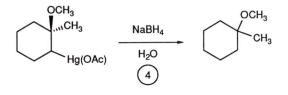

9.32.   Reaction of bromine with an alkene is a stereospecific *trans* addition, so we reproduce the alkene structure and attach one bromine atom to each carbon atom, one above the plane of the π bond, and one below. Rotation about the carbon-carbon sigma bond to form the eclipsed conformation allows us to determine if one or three groups mirror the others.

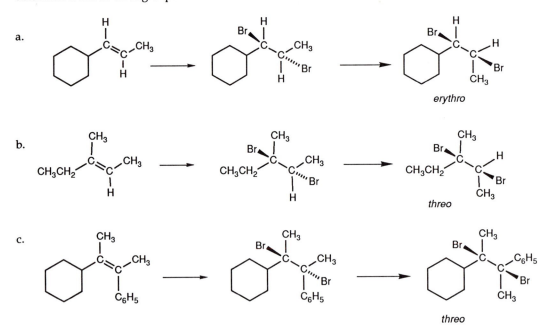

9.33. The energy profile for the reaction between methylenecyclohexane and aqueous acid looks much like that illustrated in Figure 9.3 (text page 386). The 3° carbocation is the more stable of the two possible intermediates that can form, so according to the Hammond postulate, the free energy of activation for its formation will be lower.

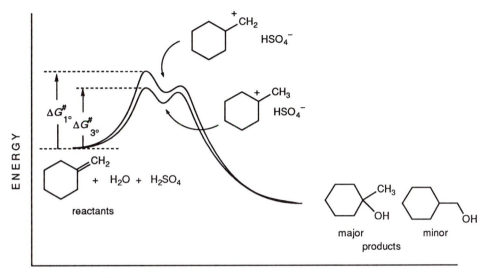

The mechanism for this addition reaction follows the typical pathway: protonation of the double bond in step (1), reaction of the carbocation intermediate with a molecule of water in step (2), and deprotonation of that intermediate to form the alcohol in step (3).

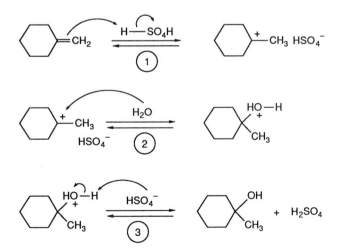

9.34. A vinyl ether reacts like any alkene does with an electrophile. Protonation of the double bond in step (1) generates a carbocation intermediate that is stabilized by resonance.

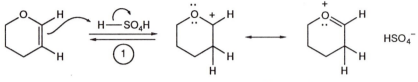

resonance stabilized carbocation

The carbocation undergoes reaction in step (2) with a nucleophile, in this case methanol. Deprotonation in step (3) yields an acetal as the product. This transformation will be discussed in detail in Chapter 17.

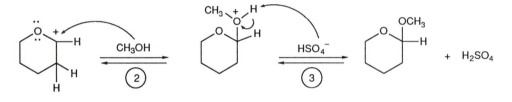

9.35. In the cyclization reaction shown in this exercise, one of the alkene double bonds is protonated by its reaction with the mineral acid. The double bonds are equivalent, so it does not matter which one reacts first.

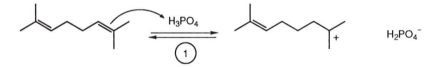

The carbocation is intercepted by the π electrons in the other double bond, which creates the ring in step (2). A rearrangement subsequently takes place in step (3), putting the positive charge on a carbon atom in the ring. If elimination of a proton were to occur instead of rearrangement, an exocyclic double bond would form.

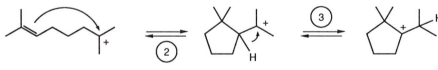

3° *exocyclic* carbocation   3° *endocyclic* carbocation

After rearrangement, the compound is deprotonated to form a trisubstituted, endocyclic double bond, which is more stable than any other double bond that could form.

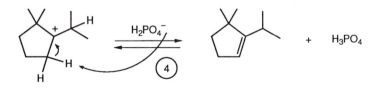

9.36.   Reaction of 5-methyl-5-hexene-1-ol with mercury(II) acetate generates a carbocation intermediate that can be trapped <u>intra</u>molecularly by the nucleophilic oxygen atom of the alcohol functional group. Step (2) is the ring-forming step, and deprotonation by acetate ion in step (3) generates the neutral product.

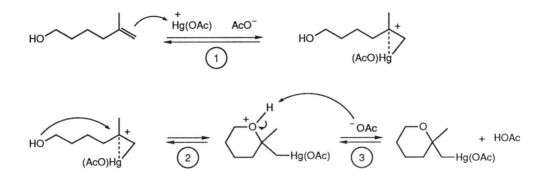

Removal of the acetato(mercury) group with use of sodium borohydride produces the cyclic ether.

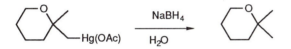

Reaction of the same alkene with bromine generates a bromonium ion intermediate, and the same type of cyclization process occurs. The cyclization product exists in the protonated form, which is deprotonated by reaction with Br⁻.

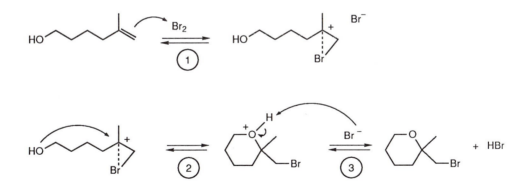

9.37. To predict the product of electrophilic addition for a reagent that you have not encountered before, decide which portion of the reagent is the electrophile and which part is the nucleophile. A formula is normally written so that the electrophilic part appears first. For the reagents in this exercise, therefore, Br^+ and RS^+ are the electrophiles. They become attached to the less highly substituted end of the alkene double bond so as to create a carbocation intermediate at the more highly substituted carbon atom.

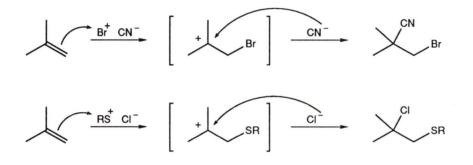

9.38. Styrene and methyl vinyl ether are alkenes that can be polymerized by cation initiation. The mechanism for cationic polymerization follows the steps outlined for polymerization of isobutylene (text pages 403-404). Substrates that undergo cationic polymerization are those that are able to stabilize the carbocation intermediate formed at each stage. The stability of the carbocation accounts for the regiochemistry of each addition step, too.

Styrene

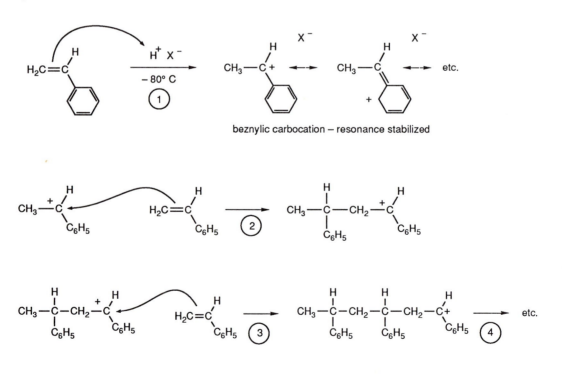

9.38. (continued)

Methyl vinyl ether

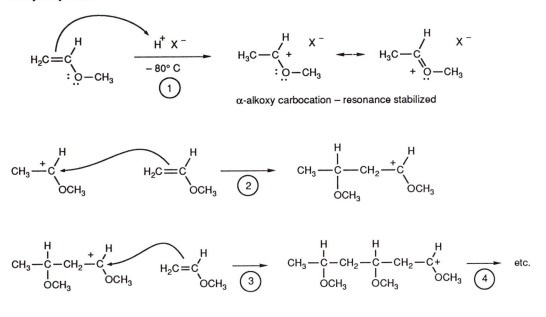

9.39. Bombykol is a conjugated diene, so addition occurs to each of the double bonds separately or to the diene system in a 1,4 fashion. If 1,4-addition occurs, the new double bond can be either *cis* or *trans*. Ignoring possible stereoisomers (enantiomers and diastereomers), the four possible products that are formed are as follows:

(10*E*, 12*Z*)-10,12-Hexadecadiene-1-ol

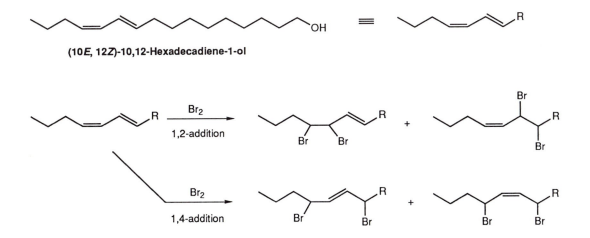

9.40. The mechanism of the cyclization reaction illustrated in this exercise starts with formation of a carbocation by protonation of the alcohol OH group in step (1), followed by dissociation of a molecule of water in step (2). The carbocation intermediate is intercepted by the π bond in step (3), and the new carbocation is trapped by water in step (4). Deprotonation in the last step generates the neutral product.

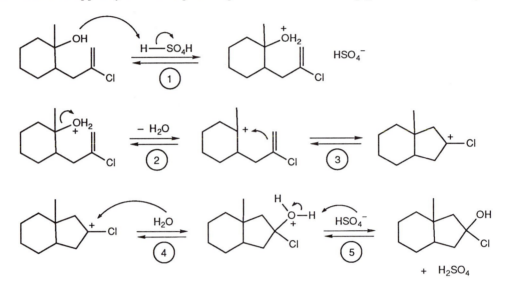

9.41. First, summarize the known information by writing out a flow chart.

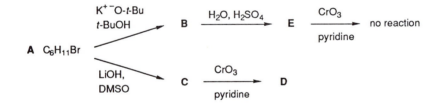

Now consider what the data mean. Compound **C** is converted to an aldehyde upon treatment with CrO₃ in pyridine, which means that **C** is a 1° alcohol. Its preparation from **A** by reaction with a basic nucleophile tells us that **A** is likely a 1° alkyl bromide. Compound **A** has one site of unsaturation, so the bromine atom has to be attached to a methyl group that is a substituent on a ring. Reaction of **B** with water in the presence of acid produces an alcohol (**E**) by addition to a double bond. The fact that compound **E** is not oxidized by chromium oxide indicates that it is a 3° alcohol. The structures are therefore as follows:

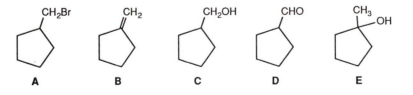

9.42. The interconversion of dimethylallyl pyrophosphate and isopentenyl pyrophosphate consists of two proton transfer steps. Therefore, the acid and base sites in the enzyme must be oriented on opposite sides of the substrate molecule. In addition, there should be a portion of the enzyme that creates a site at which the substrates bind. The negatively-charged phosphate groups of the substrates can be anchored within a binding site by interaction with positively-charged amino acid side chains, denoted by the plus signs below.

Cartoon of the active site for an enzyme that catalyzes the interconversion of dimethylallyl pyrophosphate and isopentenyl pyrophosphate:

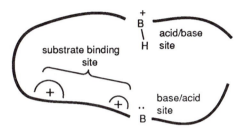

Conversion of dimethylallyl pyrophosphate to isopentenyl pyrophosphate:

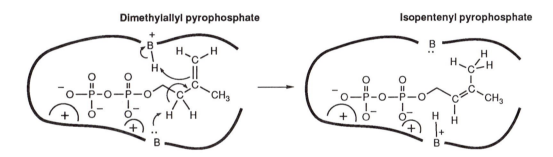

Conversion of dimethylallyl pyrophosphate to isopentenyl pyrophosphate:

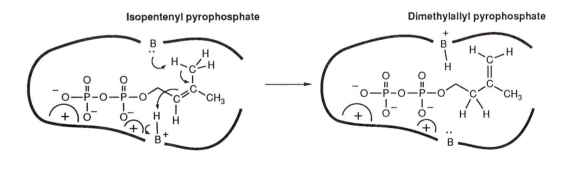

CONCERTED ADDITION REACTIONS OF ALKENES AND ALKYNES

10.1. Hydrogen, in the presence of a palladium catalyst, reacts with carbon-carbon π bonds, adding a hydrogen atom to each carbon atom of a double bond (or two atoms of hydrogen to each carbon atom of a triple bond). Addition of hydrogen to a π bond produces a carbon-carbon single bond at that site. A benzene ring does not react normally under these conditions.

   The stereochemistry of addition of hydrogen to a double bond is *syn*, which means that the hydrogen atoms that add will be *cis* to each other if the stereochemistry of the product can be defined, for example in part a of this exercise.

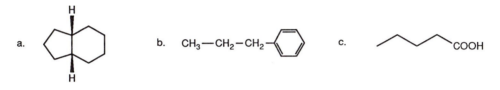

10.2. In a hydroboration reaction, the boron atom becomes attached to the less highly-substituted end of the carbon-carbon double bond. The double bond of 2-methyl-2-butene is sufficiently hindered that only two equivalents of the alkene react with borane, forming disiamylborane.

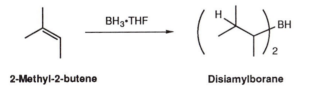

10.3. In a hydroboration reaction, the boron atom becomes attached to the less highly-substituted end of the carbon-carbon double bond. The hydrogen and boron atoms from borane appear *cis* to each other in the product if other groups allow the stereochemistry to be differentiated. The stoichiometry depends on steric hindrance around the ends of the double bond.

a. This alkene is a terminal one, so three equivalents of it react with borane. The boron atom is attached to the unsubstituted end of the double bond.

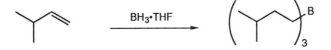

10.3. (continued)

b. This alkene is disubstituted, so three equivalents of it react with borane. There are no electronic effects to differentiate the ends of the double bond, but the left end (as drawn below) is more hindered by methyl groups attached to the adjacent carbon atom. The boron atom therefore becomes attached to the right end of the double bond.

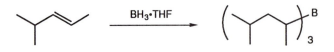

c. This alkene is trisubstituted, so only two equivalents of it are likely to react with borane (the substitution of the double bond resembles that of 2-methyl-2-butene shown in exercise 10.2.). The boron is attached to the less highly-substituted end of the double bond.

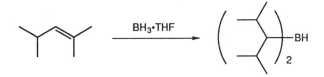

10.4. Borane itself reacts quite selectively with a terminal alkene to form a primary alkyl borane. If the other borane reagents that are mentioned in this exercise are even more selective in their reactions with alkenes, then the product in which B is attached to the 1° carbon atom will be formed in even higher proportions.

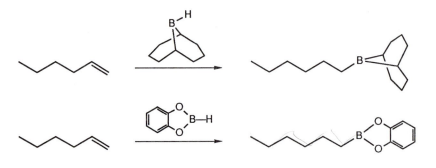

10.5. An alkyne undergoes hydroboration with one equivalent of an active B—H bond (such as the one in 9-BBN–H) to produce a *cis*-substituted alkene. The starting alkyne in this example is symmetric, so regiochemistry of addition is not an issue.

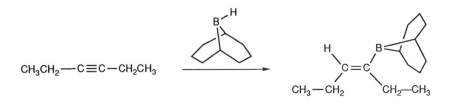

10.6.   The conversion of an alkylborane to the corresponding alkoxyborane follows the mechanism outlined on text page 425. One equivalent of hydroperoxide ion binds to the boron atom, then an alkyl group migrates to the oxygen atom of the OOH group, displacing hydroxide ion as a leaving group. The same steps occur as long as there are B—R bonds. For oxidation of ethoxydiethylborane, two equivalents of hydroperoxide ion react with the boron atom.

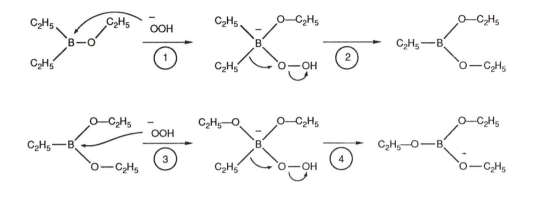

10.7.   In a hydroboration reaction, the boron atom becomes attached to the less highly-substituted end of a carbon-carbon double bond with which it reacts. The oxidation/hydrolysis step replaces the boron atom with a hydroxyl group, so the overall transformation results in the addition of H and OH to the ends of the π bond. The OH group is attached in the product to the less-highly substituted end of the original double bond.

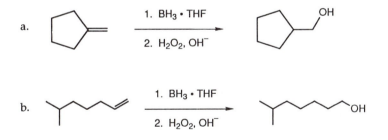

10.8.   Reaction of dimethylborane with a trisubstituted alkene places the boron atom at the less highly-substituted end of the double bond, and the hydrogen and boron atoms are *cis*.

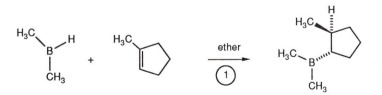

10.8. (continued)

Next, hydroxylamine reacts with the boron atom in step (2), and the cyclohexyl group migrates to the nitrogen atom in step (3). Migration occurs with retention of configuration at the carbon atom of the cyclohexyl group. The amino and methyl groups in the final product are therefore *trans*.

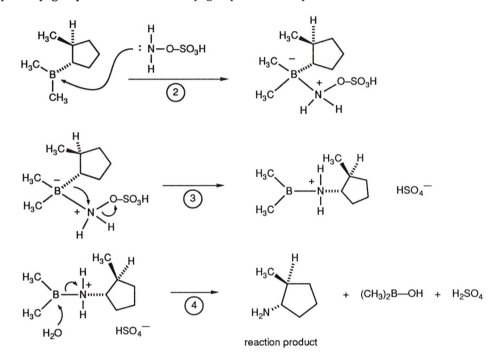

reaction product

10.9. When the elements of dihydrogen, H_2, are added to a π bond with use of a borane reagent, one hydrogen atom comes from the borane reagent and the other comes from a molecule of carboxylic acid that reacts with the organoborane in step 2. Deuterium is incorporated into the product with use of RCOOD.

The two hydrogen atoms are *cis* to each other in the product (if the stereochemistry can be discerned) because the stereochemistry of the hydroboration step is *syn*, and the boron atom is replaced in the second step with retention of configuration.

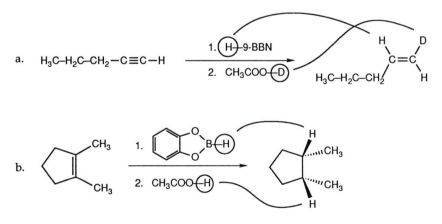

10.10. The Diels-Alder reaction is discussed on text pages 431-439. The product is a cyclohexene derivative in which the substituents have the same relative stereochemistry as substituents in the dienophile or diene. If the product is a bicyclo compound, which is formed when the starting diene is cyclic, the electron-withdrawing group in the dienophile is *syn* with respect to the bridging group that contains the double bond. For the reactions shown in this exercise, there are no stereochemical constraints on the product structure.

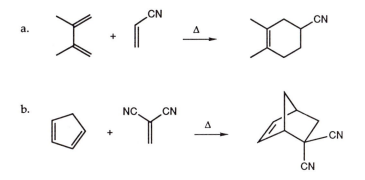

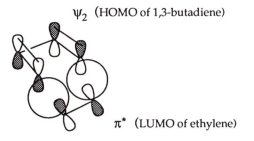

10.11. The LUMO of ethylene is its π^* orbital, and the HOMO of 1,3-butadiene is Ψ_2, which is illustrated on text page 434. When these two orbitals approach each other in a face-to-face fashion, their ends have the appropriate symmetry to overlap productively.

ψ_2 (HOMO of 1,3-butadiene)

π^* (LUMO of ethylene)

10.12. The Diels-Alder reaction is discussed on text pages 431-439. The product is a cyclohexene derivative in which the substituents have the same relative stereochemistry as substituents in the dienophile or diene. If the product is a bicyclo compound, which is formed when the starting diene is cyclic, the electron-withdrawing group in the dienophile is *syn* with respect to the bridging group that contains the double bond.

10.12. (continued)

a. The methoxy group in the diene and ester group in the dienophile will be *cis* to each other in the product because the ester group is under the 1,3-diene π system during the reaction. The hydrogen atoms point in the same direction in the transition state, which means they are *cis* in the product.

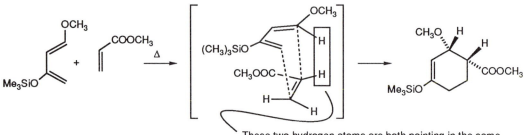

These two hydrogen atoms are both pointing in the same direction here, so they are *cis* to each other in the product.

b. The *endo* product is the major isomer formed in this reaction because the cyano group lies under the 1,3-diene π system in the transition state.

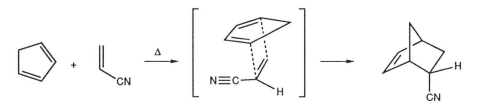

10.13. To predict the products that are formed from ozonolysis of an alkene, first break the double bond. In the second step, attach an oxygen atom to the carbon atoms that were originally double-bonded to each other. Finally, draw the carbon-oxygen double bond and adjust the oxidation state of the carbon atom that bears the oxygen atom, if necessary:

- If the original alkene carbon atom has a hydrogen atom attached to it, that carbon atom becomes an aldehyde under reductive workup conditions and a carboxylic acid under oxidative workup.

- If the end of the double bond lacks a hydrogen atom, then that end appears as a ketone under either set of workup conditions.

a. In this reaction, both alkene carbon atoms in the substrate bear a hydrogen atom, so the product mixture will either comprise two aldehydes or two carboxylic acids.

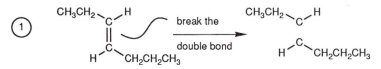

10.13. (continued)

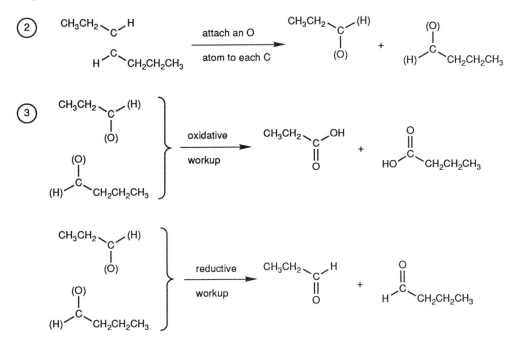

b.   In this reaction, one alkene carbon atom in the substrate bears a hydrogen atom, but the other does not. The alkene is cyclic, too, so the newly oxidized carbon atoms will be part of a single molecule. Under reductive conditions, the product will have one aldehyde and one ketone functional group, and under oxidative conditions, the product will have one carboxylic acid and one ketone functional group.

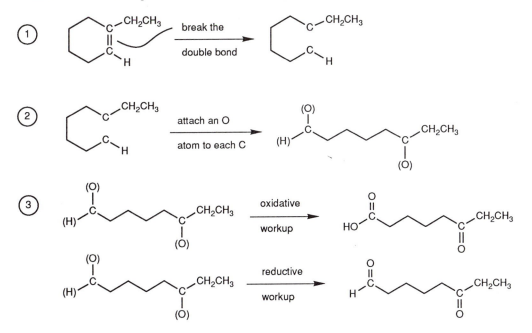

10.14. The following table summarizes the cleavage products of alkenes by different reagents.

Products after reaction with:

Alkene type	O_3 with oxidative workup	$KMnO_4$ (hot and concd)	O_3 with reductive workup	OsO_4 and IO_4^-
H₂C=CHR (H, H / R, H)	RCOOH + HCOOH	RCOOH + CO_2	RCHO + HCHO	
(R', H / R, H)	RCOR' + HCOOH	RCOR' + CO_2	RCOR' + HCHO	
(H, H / R, R')	RCOOH + R'COOH		RCHO + R'CHO	
(R', H / R, R")	RCOR' + R"COOH		RCOR' + R"CHO	
(R', R''' / R, R")	RCOR' + R"COR'''			

The solution to this exercise follows the same procedure that is outlined in the solution to exercise 10.13, making use of the data in the Table directly above.

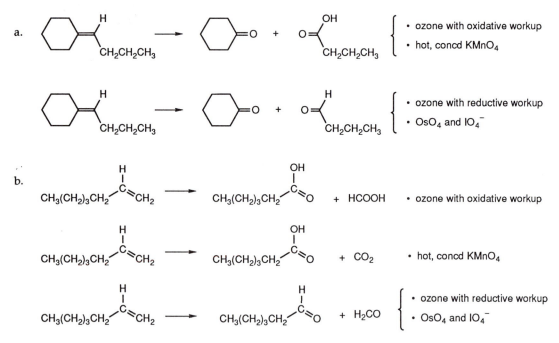

10.15. Epoxidation of an alkene that makes use of a peracid occurs at the more highly substituted double bond, if there is a choice. The stereochemistry of the epoxide is the same as the starting alkene double bond (i.e., *trans*-double bone = *trans* epoxide). If the starting material is achiral racemic products are obtained.

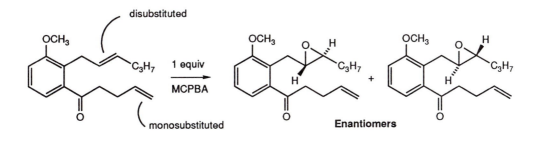

10.16. The vanadium reagent (iPrO)$_3$V=O catalyzes epoxidation reactions only of a double bond adjacent to an alcohol OH group. The stereochemistry at the alcohol carbon atom orients the vanadium ion so that the oxygen atom that forms the epoxide ring attaches at the face of the double bond that is oriented toward the OH group. If the double bond is not in a ring, however, rotation about the sigma bonds exposes both faces of the double bond to the incoming oxygen atom, creating diastereomers.

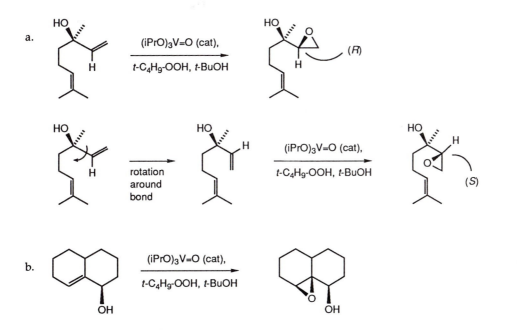

10.17.   Directions for drawing resonance structure were given in Chapter 3 on text page 79.

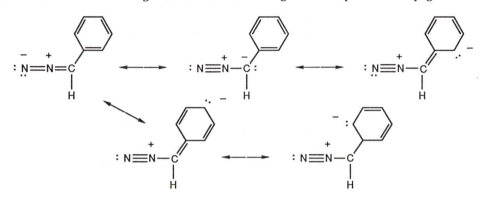

Phenyldiazomethane is more stable than diazomethane itself because of resonance stabilization among the π bonds in the benzene ring.

10.18.   In the reaction of butyl vinyl ether with phenyldiazomethane, the carbene generated from the reaction between phenyldiazomethane and the rhodium reagent adds to either face of the double bond, producing *cis* and *trans* isomers.

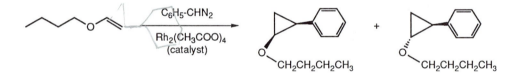

10.19.   Follow the procedure outlined in the solution to exercise 7.28.

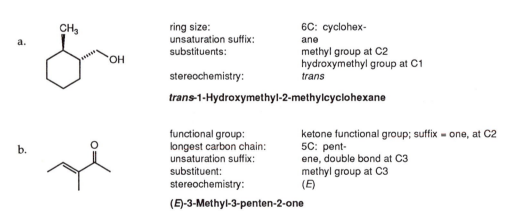

a.

ring size:	6C: cyclohex-
unsaturation suffix:	ane
substituents:	methyl group at C2
	hydroxymethyl group at C1
stereochemistry:	*trans*

trans-1-Hydroxymethyl-2-methylcyclohexane

b.

functional group:	ketone functional group; suffix = one, at C2
longest carbon chain:	5C: pent-
unsaturation suffix:	ene, double bond at C3
substituent:	methyl group at C3
stereochemistry:	(E)

(E)-3-Methyl-3-penten-2-one

10.19.   (continued)

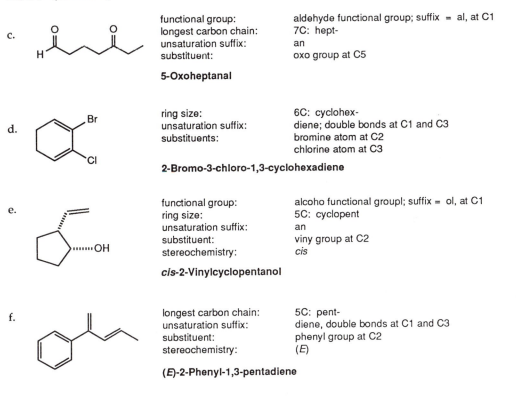

c.

functional group:	aldehyde functional group; suffix = al, at C1
longest carbon chain:	7C: hept-
unsaturation suffix:	an
substituent:	oxo group at C5

5-Oxoheptanal

d.

ring size:	6C: cyclohex-
unsaturation suffix:	diene; double bonds at C1 and C3
substituents:	bromine atom at C2
	chlorine atom at C3

2-Bromo-3-chloro-1,3-cyclohexadiene

e.

functional group:	alcoho functional groupl; suffix = ol, at C1
ring size:	5C: cyclopent
unsaturation suffix:	an
substituent:	viny group at C2
stereochemistry:	*cis*

***cis*-2-Vinylcyclopentanol**

f.

longest carbon chain:	5C: pent-
unsaturation suffix:	diene, double bonds at C1 and C3
substituent:	phenyl group at C2
stereochemistry:	(*E*)

(*E*)-2-Phenyl-1,3-pentadiene

10.20.   Follow the procedure outlined in the solution to exercise 7.27.   Only the root word, unsaturation index, and major functional group are noted.

a.   **3-Methyl-1,3-cyclohexadiene**

| cyclohex | six-membered ring of carbon atoms |
| diene | carbon-carbon double bonds starting at C1 and C3 |

b.   **(*E*)1-Nitro-1-propene**

| prop | 3 carbon atoms |
| ene | carbon-carbon double bond starting at C1 |

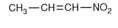

c.   **(*Z*)-2-Bromo-2-butenal**

but	4 carbon atoms
en	carbon-carbon double bond starting at C2
al	aldehyde functional group, defines C1

10.20. (continued)

d. **Tris(*trans*-2-methyl-1-cyclohexyl)borane**

  borane   B atom with three H or R substituents
  R   six-membered ring with a methyl group at C2
        trans to the boron atom

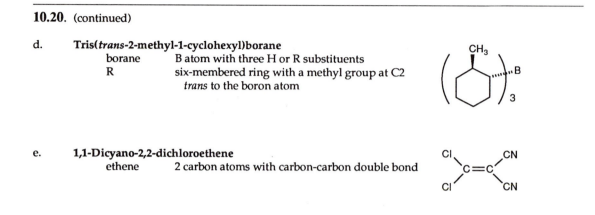

e. **1,1-Dicyano-2,2-dichloroethene**

  ethene   2 carbon atoms with carbon-carbon double bond

10.21. To predict the product of the reaction, you have to recognize the reactivity properties of each reagent as well as the stereochemistry associated with its reactions. Reagents that are polar (electrophile and nucleophile) generally add to a double bond in a stepwise fashion, and addition occurs so that the electrophile becomes attached to the less highly-substituted carbon atom of the double bond. See Chapter 9 for details about the reagents in parts a, b, and h. The other types of reagents add to a double bond in a concerted manner, and the stereochemistry of those addition reactions is normally *syn*.

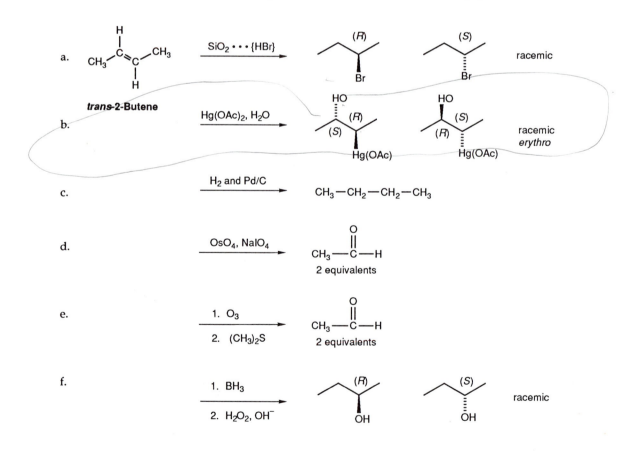

10.21. (continued)

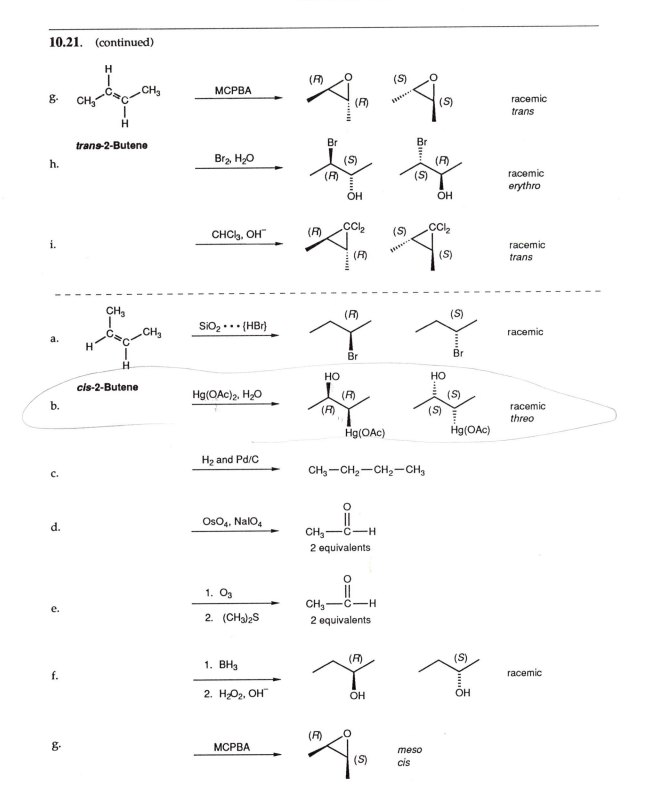

g.

trans-2-Butene

MCPBA → (R) ... (R)   (S) ... (S)   racemic *trans*

h.   Br₂, H₂O → Br (S) (R)   Br (R) (S)   OH   OH   racemic *erythro*

i.   CHCl₃, OH⁻ → (R) CCl₂ (R)   (S) CCl₂ (S)   racemic *trans*

cis-2-Butene

a.   SiO₂ ••• {HBr} → (R)   (S)   racemic
Br   Br

b.   Hg(OAc)₂, H₂O → HO (R) (R)   HO (S) (S)   racemic *threo*
Hg(OAc)   Hg(OAc)

c.   H₂ and Pd/C → CH₃—CH₂—CH₂—CH₃

d.   OsO₄, NaIO₄ → CH₃—C(=O)—H   2 equivalents

e.   1. O₃   2. (CH₃)₂S → CH₃—C(=O)—H   2 equivalents

f.   1. BH₃   2. H₂O₂, OH⁻ → (R) OH   (S) OH   racemic

g.   MCPBA → (R) O (S)   *meso cis*

10.21.   (continued)

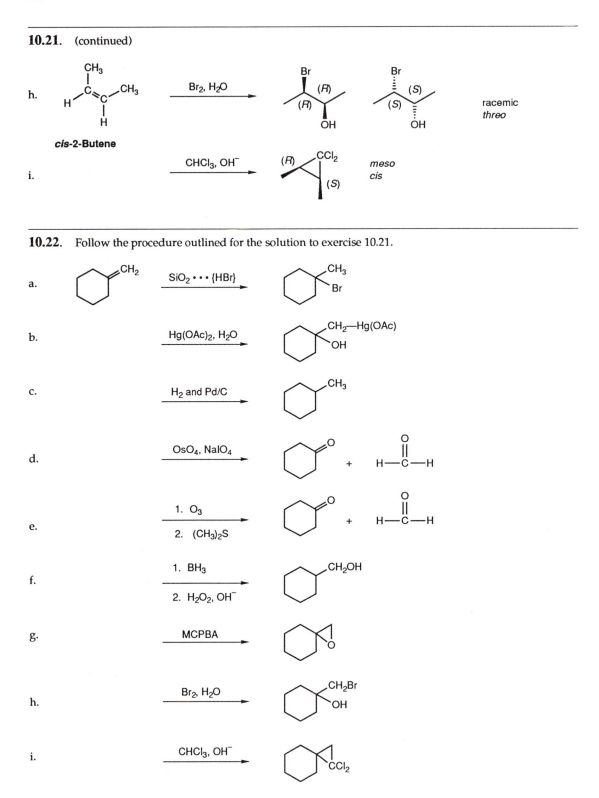

h.

cis-2-Butene

Br₂, H₂O

(R)
(R)
OH

(S)
(S)
OH

racemic
threo

i.

CHCl₃, OH⁻

(R)   CCl₂
(S)

meso
cis

10.22.   Follow the procedure outlined for the solution to exercise 10.21.

a.

SiO₂ • • • {HBr}

b.

Hg(OAc)₂, H₂O

c.

H₂ and Pd/C

d.

OsO₄, NaIO₄

e.

1. O₃

2. (CH₃)₂S

f.

1. BH₃

2. H₂O₂, OH⁻

g.

MCPBA

h.

Br₂, H₂O

i.

CHCl₃, OH⁻

10.23.   a. Hydroboration occurs to put the boron atom at the less highly-substituted end of the double bond. Oxidation converts the organoborane intermediate to the corresponding alcohol. A racemic mixture is formed.

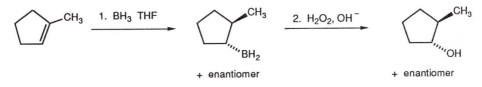

b. A peracid reacts with an alkene to form an epoxide. Ring-opening occurs with water to generate a *trans*-1,2-diol. The products are racemic.

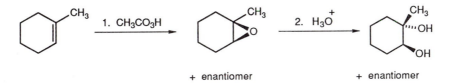

c. Ozone adds to a double bond to form a five-membered ring intermediate. Cleavage under reductive conditions (dimethyl sulfide) converts the intermediate to carbonyl compounds—aldehydes and ketones.

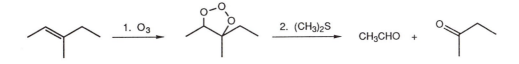

d. Hot, concentrated permanganate ion cleaves a double bond to produce carboxylic acids and ketones.

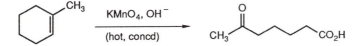

e. A peracid reacts with an alkene to produce an epoxide. A racemic mixture is formed.

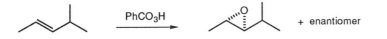

f. Ozone adds to a double bond to form a five-membered ring intermediate. Cleavage under oxidative conditions (hydrogen peroxide) converts the intermediate to the carbonyl compounds—carboxylic acids and ketones.

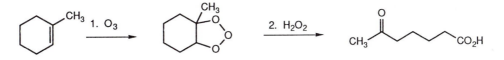

10.23. (continued)

g. Dilute, cold permanganate ion reacts with a double bond to produce a *cis* diol as a racemic mixture.

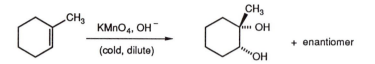

h. Hydrogen atoms add in a *cis* fashion to a double bond to form the corresponding alkane.

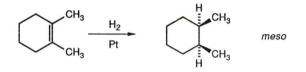

10.24. a. Hydroboration occurs to put the boron atom at the less highly-substituted end of a double bond. Oxidation converts the organoborane intermediate to the corresponding alcohol. A racemic mixture is produced.

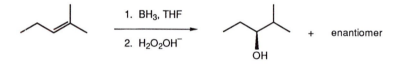

b. Ozonolysis followed by workup under reductive conditions (zinc and water) cleaves the double bond to produce aldehydes and/or ketones, depending on the substitution pattern.

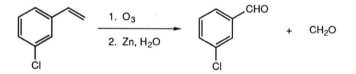

c. Osmium tetroxide together with a peroxo compound reacts with an alkene to form a vicinal diol. When only one new chiral center is produced, a racemic mixture is obtained.

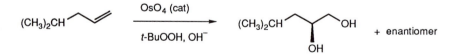

d. Hot, concentrated permanganate ion cleaves a double bond to produce carboxylic acids and/or ketones.

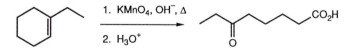

10.25. The first step in each equation of this exercise is an elimination reaction. Review Chapter 8 if necessary to recall how the conditions affect each of these transformations. To decide which reagents are required in the second reaction, consider how the product differs structurally from its starting material, then decide which transformation described in this chapter is appropriate.

a.   The alkyl bromide undergoes elimination with strong base to form **A**. To convert a terminal alkene to a 1°alcohol, hydroboration/oxidation is required.

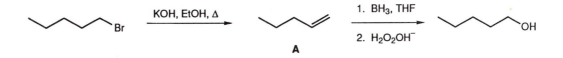

b. An alcohol is converted to its methanesulfonate derivative with mesyl chloride, and that compound undergoes E2 elimination when heated with strong base. To convert a disubstituted alkene to a ketone requires ozonolysis or treatment with hot, concentrated permanganate ion.

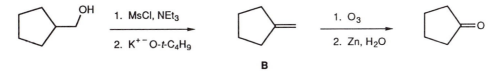

c. Treating a vinyl bromide with LDA produces an alkyne. To convert an alkyne to a *cis* alkene, partial catalytic hydrogenation is required.

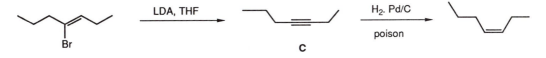

d. An alkyl bromide undergoes elimination with base to form a double bond; a vicinal dibromide produces a 1,3-diene (or a 1,2-diene if the compound is acyclic). The bicyclic structure of the product suggests that a Diels-Alder reaction is involved.

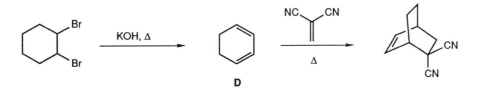

e. Heating a 3°alcohol with strong acid produces an alkene. To convert an alkene to an epoxide, a peracid is required.

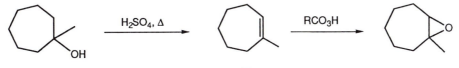

10.25.   (continued)

f. A 2° alcohol is oxidized to a ketone. To prepare a six-membered ring, a Diels-Alder reaction is called for. The unsaturated ketone is a good dienophile, so a 1,3-diene is the other reactant needed.

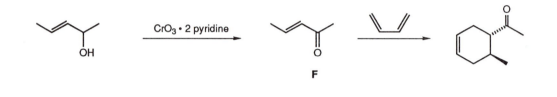

10.26.   Alkenes and alkynes are converted to the corresponding alkane upon reaction with molecular hydrogen and a palladium catalyst. Use of a poisoned catalyst allows one to convert an alkyne to a *cis*-alkene.

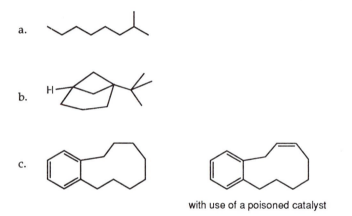

with use of a poisoned catalyst

10.27.   The hydroboration/oxidation sequence that is used to make an alcohol from an alkene puts the OH group on the carbon atom at the less highly-substituted end of the original double bond. A double bond that is substituted equivalently at each carbon atom is not useful for making an alcohol unless the alkene is completely symmetric. The OH group and the new H atom that are added during hydroboration/oxidation will be oriented *cis* to one another.

   When you look at the structure of the alcohol product, consider what alkenes can be used as starting materials. Then consider whether the substitution pattern leads to a regioselective reaction, and look to see if the OH group and H atom in the product have a *cis* relationship.

a.   This alcohol can be prepared from methylenecyclohexane—the less highly-substituted end is the one to which the OH group is attached.

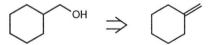

10.27. (continued)

b.   This alcohol can be prepared from 1-methylcyclohexene; the less highly-substituted end of the double bond is the one to which the OH group is attached, and the OH and H are on the same side of the ring.

c.   This alcohol cannot be prepared from either alkene shown. A 3° alcohol is difficult to prepare by hydroboration.

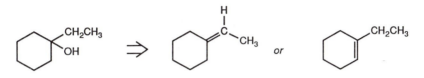

d.   This alcohol cannot be prepared as the major product from either alkene shown. Each terminus of the double bond has a single alkyl substituent, so there is no way to influence the orientation of the incoming boron atom.

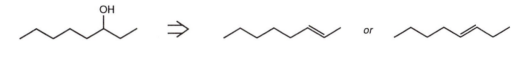

10.28.   The starting materials needed to prepare the alcohols shown in exercise 10.27 are those on the right side of the arrows in parts a and b.

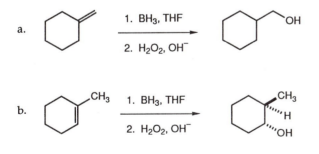

10.29.   To decide what reagents are required to carry out a particular transformation requiring addition to a double or triple bond, consider how the product differs structurally from the starting material, then decide which reagents provide those needed structural elements. When a substituent in the product is attached to the carbon atom that is at the more highly-substituted end of a double bond in the starting material, then electrophilic addition is probably needed to create a carbon-heteroatom bond at that position. When a substituent in the product is attached to the carbon atom that is at the less highly-substituted end of a double bond in the starting material, then hydroboration is called for.

10.29. (continued)

a. The amino group is attached to the less highly-substituted end of the double bond, so hydroboration is appropriate.

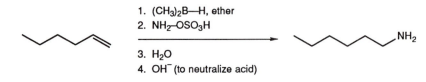

1. (CH₃)₂B—H, ether
2. NH₂-OSO₃H

3. H₂O
4. OH⁻ (to neutralize acid)

b. Ways to introduce an amino group into an organic molecule that you have learned include hydroboration, followed by reaction with hydroxylamine, or nucleophilic substitution with ammonia or an amine (text page 241). For the transformation in this exercise, hydroboration is not applicable because the amino group in the product is attached to the carbon atom that is the more highly-substituted end of the double bond in the starting alkene. A leaving group must therefore be introduced by electrophilic addition, and HBr adsorbed on silica is a good reagent for this transformation. Substitution of the bromide ion by ammonia produces the amine.

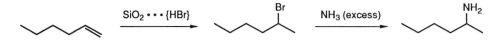

SiO₂ • • • {HBr}   NH₃ (excess)

c. The methoxy group in the product is attached to the carbon atom that is at the less highly-substituted end of the double bond in the starting material, so hydroboration is used to attach an OH group at that position. Conversion of the alcohol to the ether makes use of a substitution reaction between the alkoxide ion and an alkyl halide (text page 305).

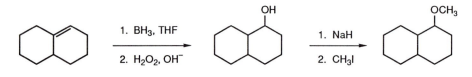

1. BH₃, THF   1. NaH

2. H₂O₂, OH⁻   2. CH₃I

10.30. The Diels-Alder reaction is discussed on text pages 431-439. The stereochemistry of the starting dienophile and diene is retained in the product. If the product is a bicyclo compound, the electron-withdrawing group in the dienophile appears on the same side of a bicyclic system as the bridge that contains the double bond.

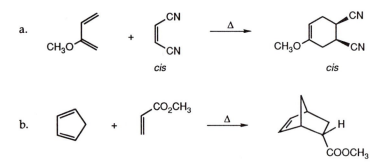

a.

b.

10.30. (continued)

c.

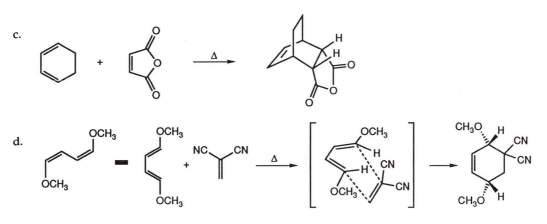

d.

10.31. Nitroethylene and vinyltriphenylphosphonium chloride have a positively-charge atom attached to one of the alkene carbon atoms. The positive charge is an electron-withdrawing group that makes the double bond electron deficient, a characteristic of a good dienophile. The vinylborane has an electron-deficient boron atom (six valence electrons) that withdraws electron density from the π bond.

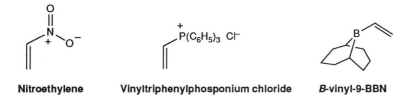

Nitroethylene   **Vinyltriphenylphosponium chloride**   **B-vinyl-9-BBN**

10.32. The dienophiles shown in exercise 10.31 react with a diene in the same manner that an unsaturated carbonyl compound does.

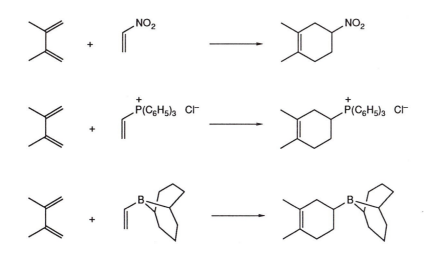

10.33. The organoborane *B*-vinyl-9-BBN reacts with cyclopentadiene to form the *endo* bicyclo compound, which is an alkyl borane. Oxidation and hydrolysis of this organoborane produces the corresponding alcohol with retention of configuration at the carbon atom to which boron is attached.

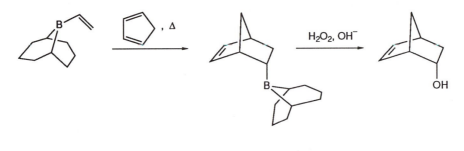

10.34. Among the molecular orbitals (MOs) of a compound, the MO having the lowest energy is the one without nodes between carbon atoms. Each additional node that an MO has raises its energy.

For the molecule ozone, four electrons are placed in the three orbitals, each of which can hold two electrons, starting with the orbital at lowest energy. The middle set is the HOMO because it is the orbital of highest energy that has electrons. For ozone, the orbital at the highest energy level is the LUMO.

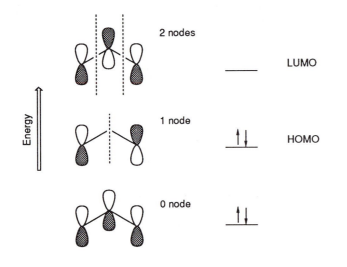

10.35. The reaction between ozone and an alkene is facile because the orbital symmetry correlates for the two combinations that need to occur between HOMO and LUMO, as emphasized by the circled regions in the figures shown below.

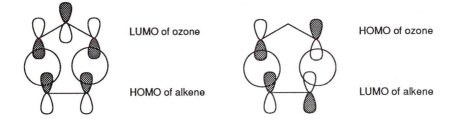

10.36. Given the structures of products formed from ozonolysis of an alkene, you deduce the structure of the starting alkene by drawing a double bond between the two carbon atoms that constitute the carbonyl groups. Conceptually, this is illustrated below, along with the structure drawn in the conventional way at the far right.

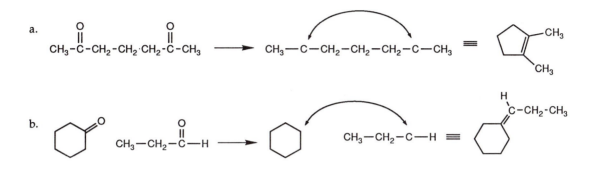

10.37. Given the structure of a cyclopropane, remove the "unique" carbon atom as a carbene and draw a double bond between the other two carbon atoms. A carbonyl-containing carbene normally derives from a diazo compound, and a dihalo carbene derives from the corresponding trihalomethane.

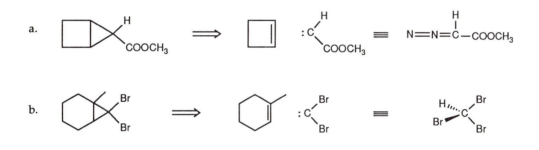

10.38. Given the structure of an epoxide, replace the oxygen atom with a double bond connecting the two carbon atoms that were originally bonded to the oxygen atom.

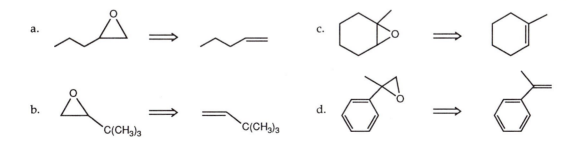

10.39. Electrophilic addition of bromine to a *gem*-disubstituted or trisubstituted double bond in the presence of water yields a vicinal bromohydrin in which the OH group is attached to the tertiary carbon atom. Treating the halohydrin with base generates an alkoxide ion, which displaces the bromide ion to form the epoxide.

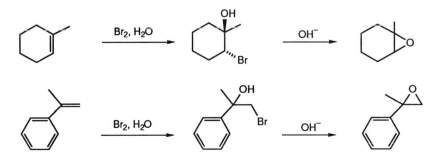

10.40. Epoxidation of a double bond followed by ring opening with water gives a *trans*-diol because the first step proceeds with retention of stereochemistry, and ring opening occurs with inversion of configuration at one of the carbon atoms. Osmium-catalyzed dihydroxylation of an alkene yields the *cis*-diol.

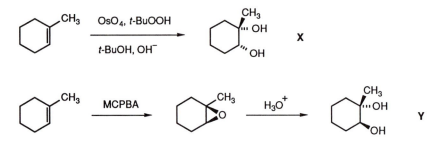

10.41. The cyclic iodine-containing intermediate has a relatively small size ring (five atoms), so formation of a *trans* ring junction in **B'** will be less facile than formation of the *cis* ring junction in **A'**. Compound **A** should therefore be cleaved faster by periodate ion.

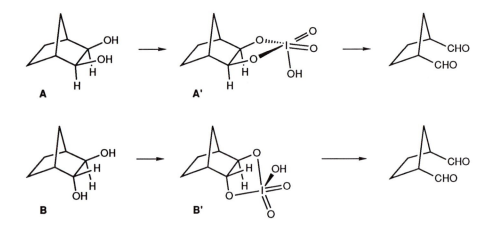

FREE RADICALS: SUBSTITUTION AND ADDITION REACTIONS

11.1. The free radical chlorination of chloromethane occurs in the same way that methane is chlorinated. The mechanism is shown on text pages 466-467. Three separate steps are involved: initiation, propagation, and termination.

initiation

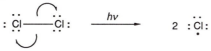

propagation

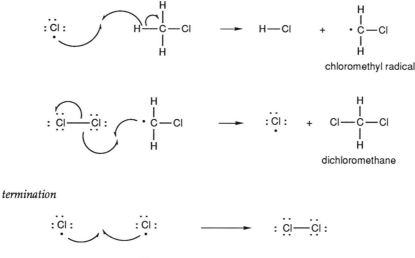

termination

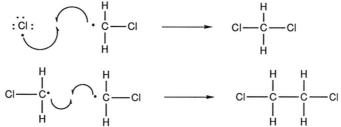

11.2. ΔH values for bromination and iodination of methane are calculated from the formula, $\Delta H°_{rxn} = \Sigma$ (BDE$_{broken}$) − Σ (BDE$_{made}$). The results of these calculations are as follows:

$$Br\cdot + CH_4 \rightarrow \quad HBr + CH_3\cdot \qquad \Delta H°_1 = BDE(CH_3–H) - BDE(H–Br) = (104 - 88) = 16 \text{ kcal/mol}$$
$$CH_3\cdot + Br_2 \rightarrow \quad CH_3Br + Br\cdot \qquad \Delta H°_2 = BDE(Br–Br) - BDE(CH_3–Br) = (46 - 70) = -24 \text{ kcal/mol}$$

$$Br_2 + CH_4 \rightarrow \quad HBr + CH_3Br \qquad \Delta H°_{rxn} = \Delta H°_1 + \Delta H°_2 \qquad\qquad = -8 \text{ kcal/mol}$$

$$I\cdot + CH_4 \rightarrow \quad HI + CH_3\cdot \qquad \Delta H°_1 = BDE(CH_3–H) - BDE(H–I) = (104 - 71) = 33 \text{ kcal/mol}$$
$$CH_3\cdot + I_2 \rightarrow \quad CH_3I + I\cdot \qquad \Delta H°_2 = BDE(I–I) - BDE(CH_3–I) = (36 - 56) = -20 \text{ kcal/mol}$$

$$I_2 + CH_4 \rightarrow \quad HI + CH_3I \qquad \Delta H°_{rxn} = \Delta H°_1 + \Delta H°_2 \qquad\qquad = +13 \text{ kcal/mol}$$

Conclusion: the overall enthalpy change for iodination of methane is greater than zero, so free radical iodination is not spontaneous. Bromination is exothermic, but not as much as chlorination is.

11.3. Follow the procedure outlined in the solution to exercise 2.20; draw the parent carbon skeleton, then attach the chlorine atoms in every possible combination. As you did in exercise 2.20, make certain that you number the carbon chain from each end to avoid drawing two versions of the same compound.

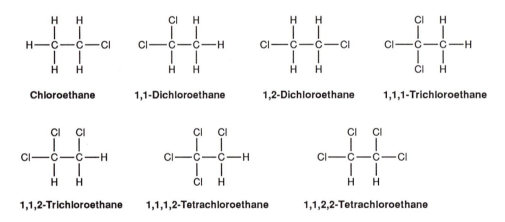

Chloroethane **1,1-Dichloroethane** **1,2-Dichloroethane** **1,1,1-Trichloroethane**

1,1,2-Trichloroethane **1,1,1,2-Tetrachloroethane** **1,1,2,2-Tetrachloroethane**

11.4. Follow the procedure outlined in the solution to exercise 11.2. The BDE for the primary C–H bond in propane is assumed to be the same as that for the primary C-H bond of ethane, which is given in Table 11.2 (text page 469).

a. **propane + Cl • → 1-propyl radical + HCl**
    $\Delta H° = BDE(1°C–H) - BDE(H–Cl) = (98 - 103) \qquad = -5 \text{ kcal/mol}$

b. **propane + Cl • → 2-propyl radical + HCl**
    $\Delta H° = BDE(2°C–H) - BDE(H–Cl) = (95 - 103) \qquad = -8 \text{ kcal/mol}$

11.4.   (continued)

c.   **propane + Cl$_2$ → 1-chloropropane + HCl**
   $\Delta H°_1$ = BDE(C–H) – BDE(H–Cl) = (98 – 103)   = – 5 kcal/mol
   $\Delta H°_2$ = BDE(Cl–Cl) – BDE(C–Cl) = (58 – 81)   = – 23 kcal/mol
   _____
   $\Delta H°_{rxn}$ = $\Delta H°_1$ + $\Delta H°_2$ =   = – 28 kcal/mol

d.   **propane + Cl$_2$ → 2-chloropropane + HCl**
   $\Delta H°_1$ = BDE(C–H) – BDE(H–Cl) = (95 – 103)   = – 8 kcal/mol
   $\Delta H°_2$ = BDE(Cl–Cl) – BDE(C–Cl) = (58 – 80)   = – 22 kcal/mol
   _____
   $\Delta H°_{rxn}$ = $\Delta H°_1$ + $\Delta H°_2$ =   = – 30 kcal/mol

11.5.   2-Methylpropane has nine 1° hydrogen atoms and one 3° hydrogen atom. Statistically, the ratio of monochloro products should be 9:1. Because a 3° center is five times more reactive than a 1° one, the ratio becomes (9 x 1)/(1 x 5) = 9/5

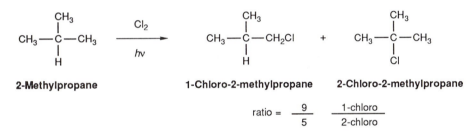

2-Methylpropane   **1-Chloro-2-methylpropane**   **2-Chloro-2-methylpropane**

ratio = $\dfrac{9}{5}$ $\dfrac{\text{1-chloro}}{\text{2-chloro}}$

11.6.   Follow the procedure outlined in the solution to exercise 11.2.

a.   **propane + Br • → 1-propyl radical + H Br**
   $\Delta H°$ = BDE(1°C–H) – BDE(H–Br) = (98 – 88)   =   10 kcal/mol

b.   **propane + Br • → 2-propyl radical + H Br**
   $\Delta H°$ = BDE(2°C–H) – BDE(H–Br) = (95 – 88)   =   7 kcal/mol

c.   **propane + Br$_2$ → 1-bromopropane + HBr**
   $\Delta H°_1$ = BDE(C–H) – BDE(H–Br) = (98 – 88)   =   10 kcal/mol
   $\Delta H°_2$ = BDE(Br–Br) – BDE(C–Br) = (46 – 68)   = – 22 kcal/mol
   _____
   $\Delta H°_{rxn}$ = $\Delta H°_1$ + $\Delta H°_2$ = – 12 kcal/mol

d.   **propane + Br$_2$ → 2-bromopropane + HBr**
   $\Delta H°_1$ = BDE(C–H) – BDE(H–Br) = (95 – 88)   =   7 kcal/mol
   $\Delta H°_2$ = BDE(Br–Br) – BDE(C–Br) = (46 – 68)   = – 22 kcal/mol
   _____
   $\Delta H°_{rxn}$ = $\Delta H°_1$ + $\Delta H°_2$ = – 15 kcal/mol

11.7. Use the numerical data summarized in the solution to exercise 11.6 to label the places on energy curves that correspond to the starting materials, intermediates, and products. The enthalpy change for each reaction is the difference in energy between starting materials and products.

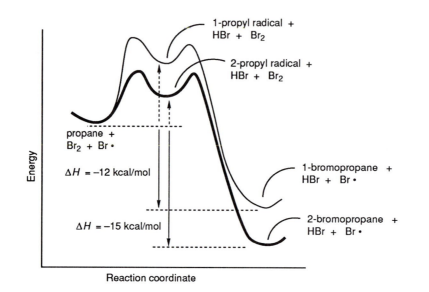

11.8. The order of carbon atom reactivity in radical substitution reactions is benzylic, allylic > 3° > 2° > 1° > methyl. The starting material has 1°, 2°, and 3° carbon atoms. Therefore, reaction is expected to occur at the 3° carbon atom.

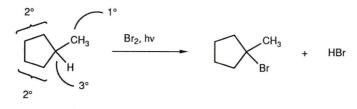

11.9. Allylic radicals formed by hydrogen atom abstraction from 1-hexene, as shown on text page 476, all react with molecular bromine to form the corresponding allylic bromide and a bromine atom. This bromine atom propagates the chain reaction. The major product is the one with the most highly-substituted double bond, *trans*-1-bromo-2-hexene.

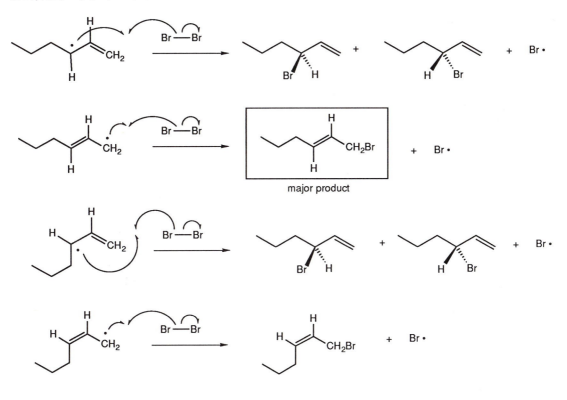

major product

11.10. The hydroperoxy compounds HPETE are specified by the numeric prefix in which the OOH group is attached at the indicated position. The double bond moves away from the position of attachment toward the adjacent bond. The hydrogen atom in arachidonic acid that is allylic to the position at which the hydroperoxy group becomes attached is the one that was removed to form the radical intermediate.

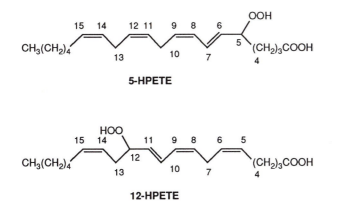

11.10. (continued)

Arachidonic acid

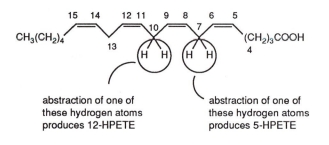

abstraction of one of
these hydrogen atoms
produces 12-HPETE

abstraction of one of
these hydrogen atoms
produces 5-HPETE

11.11. As in the opening of any epoxide ring, a nucleophile reacts at the backside of the carbon atom attached to the epoxy oxygen atom, breaking the carbon-oxygen bond. The oxygen atom reacts with a proton source to form the alcohol group, and the configuration at the carbinol carbon atom is retained.

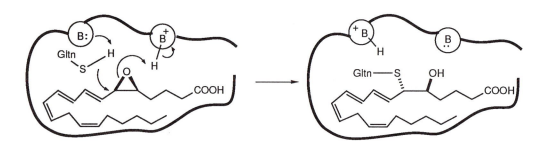

11.12. Raney nickel removes sulfur and selenium atoms, replacing a C–S or C–Se bond with a C–H bond.

a.

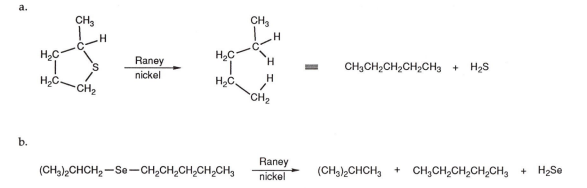

11.13. Hydrogenolysis is used to cleave **benzylic** carbon-oxygen bonds. Bonds between oxygen and other types of sp^3-hybridized carbon atoms are not affected.

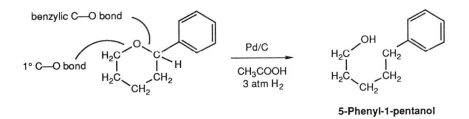

5-Phenyl-1-pentanol

11.14. A phenolate radical has the same type of resonance stabilization that a benzylic radical has because the unpaired electron on oxygen is delocalized over the entire π system.

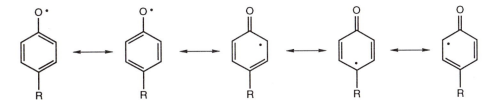

11.15. Even though the triple bond is in a ring, a *trans* double bond is formed if the ring is large enough to accommodate it (> 8 atoms). It may be that transannular effects make the *cis* double bond more stable in this compound, so *cis*-cyclononene may be the product formed. This is a situation where the actual experiment may have to be done to find the result.

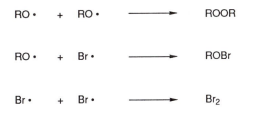

or

11.16. The products of termination steps for radical addition of HBr to an alkene result from combination of the various radical species that are present in the reaction mixture. These species include alkoxy, halogen, and alkyl radicals.

RO• + RO• ⟶ ROOR

RO• + Br• ⟶ ROBr

Br• + Br• ⟶ Br_2

11.16.   (continued)

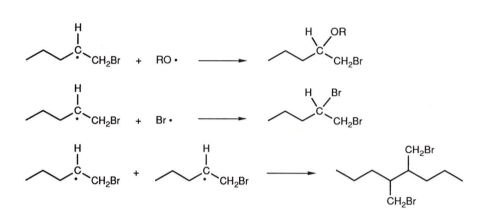

11.17   Follow the procedure outlined in the solutions to exercises 11.6 and 11.7.

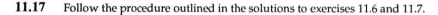

$RCH{=}CH_2$ + $Cl\cdot$ ⇌ $R\overset{\cdot}{C}H{-}CH_2Cl$ ΔH = 65 - 81 = −16 kcal/mol

$RCH{=}CH_2$ + $Br\cdot$ ⇌ $R\overset{\cdot}{C}H{-}CH_2Br$ ΔH = 65 - 68 = −3 kcal/mol

$RCH{=}CH_2$ + $I\cdot$ ⇌ $R\overset{\cdot}{C}H{-}CH_2I$ ΔH = 65 - 53 = + 12 kcal/mol

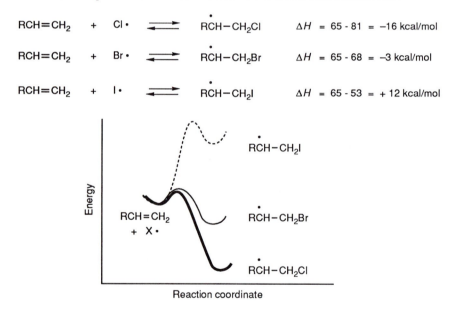

11.18.   The mechanism for polymerization of methyl acrylate is the same as that by which ethylene polymerizes, shown on text pages 489-490.

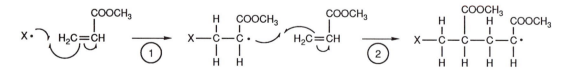

11.18. (continued)

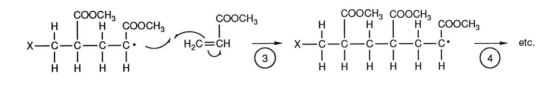

11.19   When an organohalide is treated with tributyltin hydride, reduction or cyclization is likely to occur. If no carbon-carbon double or triple bond is within five atoms of the position at which a halogen atom (Cl, Br, or I) is attached, then the tin hydride reagent reduces the C–X bond, replacing the halogen atom with hydrogen. Fluorine atoms are inert toward reaction with tin hydride reagents.

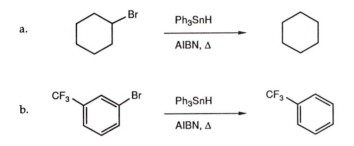

11.20.   The bromoalkene shown in this exercise is of the 5-hexenyl type, so we expect that cyclization will produce a five-membered ring product. The mechanism follows the usual pathway: abstraction of the halogen atom in step (1); cyclization to form the five-membered ring in step (2); and reduction of the carbon-centered radicals in step (3) to form the hydrocarbon product and regenerate the tributyltin radical. Notice that the cyclization step generates *cis* and *trans* isomers.

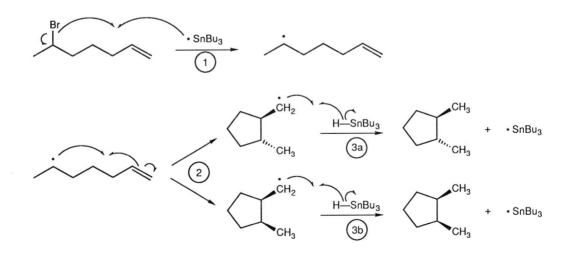

11.21. The mechanism for the given transformation follows the usual course of a radical cyclization process: abstraction of the phenylselenyl group in step (1); cyclization to produce a six-membered ring in step (2) (this substrate is of the 6-heptenyl type); and reduction of the carbon-centered radical to form the C–H bond and regenerate the triphenyltin radical.

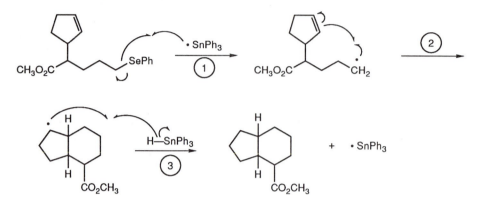

11.22. Step (4) of the scheme shown at the top of text page 503 is a reduction reaction. The vinyl radical reacts with the tin hydride reagent to form the alkene and regenerate the tributyltin radical.

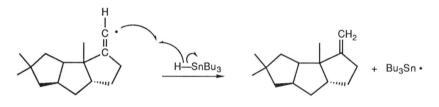

11.23. The mechanism for the given tandem radical cyclization starts with abstraction of the iodine atom. In step (2), cyclization creates the five-membered ring. The alkyl group bearing the radical is attached to the top side of the cyclopentanone ring, so it most likely adds to the cyclopentanone double bond from the top, forming the *cis* ring junction. Step (3) generates another five-membered ring, and the last step is reduction of the carbon-centered radical to form the product and regenerate the tributyltin radical.

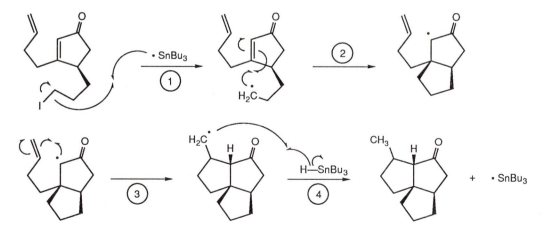

11.24.   Interpret the name according to procedures outlined in Chapter 2. The root word, unsaturation index, and principal functional group are listed below.

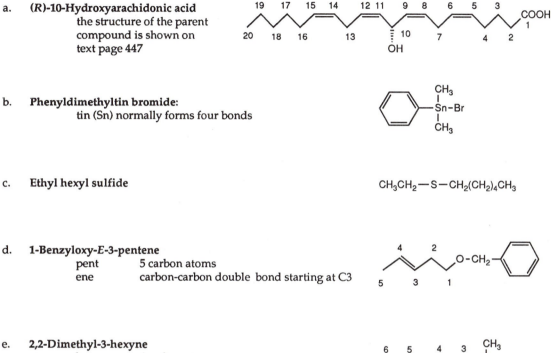

a.   **(R)-10-Hydroxyarachidonic acid**
the structure of the parent
compound is shown on
text page 447

b.   **Phenyldimethyltin bromide:**
tin (Sn) normally forms four bonds

c.   **Ethyl hexyl sulfide**

$$CH_3CH_2-S-CH_2(CH_2)_4CH_3$$

d.   **1-Benzyloxy-E-3-pentene**
       pent   5 carbon atoms
       ene   carbon-carbon double bond starting at C3

e.   **2,2-Dimethyl-3-hexyne**
       hex   6 carbon atoms
       yne   carbon-carbon triple bond starting at C3

$$CH_3-CH_2-C{\equiv}C-\overset{CH_3}{\underset{CH_3}{\underset{|}{\overset{|}{C}}}}-CH_3$$

11.25.   Follow the procedure given in exercise 7.28. For the non-standard types of compounds, compare the names and structures in exercise 11.24.

a.   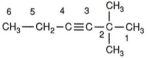   **Tricyclohexyltin hydride**

b.   [cyclobutyl]—S—CH₂CH₂CH₂CH₃

functional group: sulfide
substituents: butyl
                     cyclobutyl

Butyl cyclobutyl sulfide

11.25. (continued)

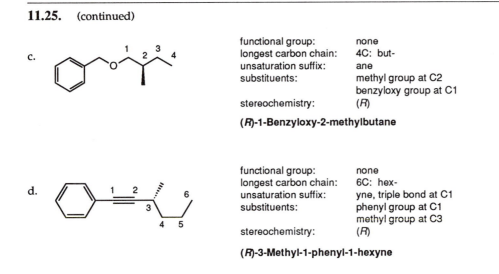

c.

functional group: none
longest carbon chain: 4C: but-
unsaturation suffix: ane
substituents: methyl group at C2
benzyloxy group at C1
stereochemistry: (*R*)

(*R*)-1-Benzyloxy-2-methylbutane

d.

functional group: none
longest carbon chain: 6C: hex-
unsaturation suffix: yne, triple bond at C1
substituents: phenyl group at C1
methyl group at C3
stereochemistry: (*R*)

(*R*)-3-Methyl-1-phenyl-1-hexyne

11.26. Products of a reaction in which reactivity differences among groups is nonexistent are formed in a ratio proportional to the numbers of each type of equivalent site. Isobutane has one 3° carbon atom and nine 1° carbon atoms, so the ratio of monochlorination products is 1:9.

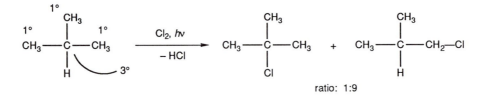

ratio: 1:9

11.27. To predict whether a reaction will yield principally one product, you need to consider relative reactivities of each group in the substrate molecule as well as the reactivity of each reagent. In radical substitution, bromine atoms react at any sp^3-hybridized carbon-hydrogen bond, so we do not have to be concerned about the reagent itself. The substrate reacts such that a benzylic center is more reactive than a 3° one, which in turn is more reactive than a 2° C–H bond. This order of reactivity allows us to predict which carbon atom(s) undergo substitution. If there is a single type of C–H bond, then only one product will form. If there are several of that type of bond, then a mixture of products will be obtained, as in parts b and d, below.

a. A single product is formed because the substrate molecule has completely equivalent 2° carbon atoms.

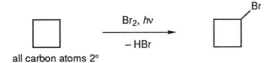

all carbon atoms 2°

11.27.   (continued)

b.   Two products are formed because the substrate molecule has two, nonequivalent 3° carbon atoms.

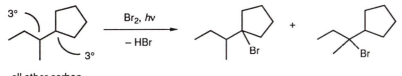

all other carbon
atoms are 1° or 2°

c.   One product predominates because the substrate molecule has a single benzylic carbon atom.

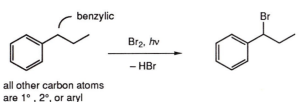

all other carbon atoms
are 1° , 2°, or aryl

d.   Three products are formed because the substrate molecule has three, nonequivalent 2° carbon atoms.

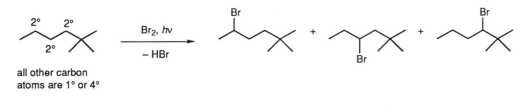

all other carbon
atoms are 1° or 4°

11.28.

a.   NBS and benzoyl peroxide react to form bromine radicals that substitute for hydrogen atoms of hydrocarbon groups.   The order of reactivity is allylic and benzylic > 3° > 2° > 1° > methyl.   Free radical substitution produces both configurations at the carbon atom undergoing substitution.

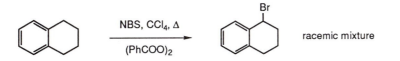

b. Raney nickel removes sulfur and selenium atoms from organic molecules, replacing those heteroatom with two hydrogen atoms.

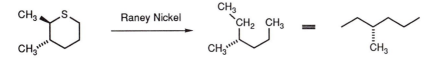

11.28. (continued)

c. Tributyltin hydride reacts with an alkyl halide to generate a carbon radical at the position at which the halogen atom is attached. The resulting radical cyclizes to produce a ring if there is a double bond several atoms away, or it is reduced to form the corresponding hydrocarbon, as it is in this example. The stereochemistry at the halogen-substituted carbon atom is lost during the course of this transformation.

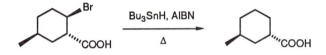

d. HBr, in the presence of peroxides, adds to a double bond in an anti-Markovnikov fashion, so the bromine atom becomes attached to the less highly-substituted end of the double bond. Addition occurs at either face of the double bond, so four stereoisomers are produced.

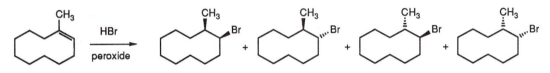

e. Hydrogenolysis cleaves bonds between a benzylic carbon atom and a heteroatom, especially C–O bonds. The oxygen atom is replaced with a hydrogen atom. Other types of carbon-oxygen bonds are not affected.

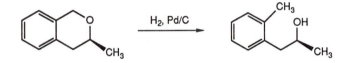

f. See part a of this exercise. The order of reactivity is allylic > 3° > 2° > 1°.

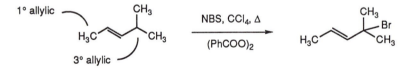

g., h., i. These reactions compare reduction reactions that a triple bond can undergo. Two of these were described in Chapter 10 (text page 418). Catalytic hydrogenation gives either the corresponding alkane or the *cis* double bond, the latter if a poisoned catalyst is employed. A dissolving metal reduction utilizes sodium in liquid ammonia to produce primarily a *trans* alkene.

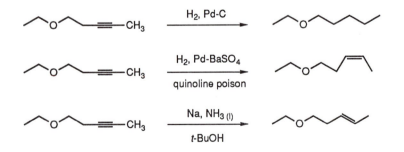

11.29.

a. The energy diagram for the actual mechanism of methane chlorination is shown below, at the left.

b. If chlorination were to occur by the alternative mechanism presented on text page 510, the $\Delta H°$ values calculated from the formula $\Delta H°_{rxn} = \Sigma (BDE_{broken}) - \Sigma (BDE_{made})$ are as follows:

$CH_4 + Cl\cdot$	$\longrightarrow$	$CH_3Cl + H\cdot$	$\Delta H_1 = +20$ kcal/mol
$Cl_2 + H\cdot$	$\longrightarrow$	$HCl + Cl\cdot$	$\Delta H_2 = -45$ kcal/mol
$CH_4 + Cl_2$	$\longrightarrow$	$CH_3Cl + HCl$	$\Delta H_1 = -25$ kcal/mol

The energy diagram for this alternative mechanism is shown below at the right. The one on the left (the actual mechanism) is more likely because the free energy of activation is much less than for the alternative. The overall free energy change is identical, as expected because the reactants and products are the same for each mechanism.

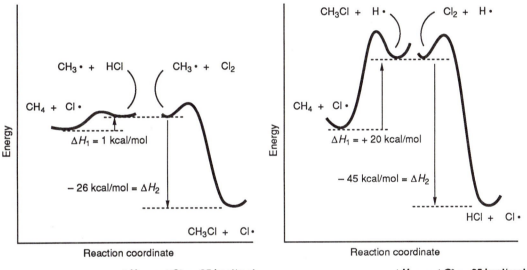

$\Delta H_{total} = \Delta G° = -25$ kcal/mol $\Delta H_{total} = \Delta G° = -25$ kcal/mol

11.30.   In "anti-Markovnikov" addition of HBr by a radical mechanism, the bromine atom becomes attached to the less highly-substituted end of the π bond. With the addition of only one equivalent of HBr, an alkene is produced, which can exist as the (E) or (Z) isomer.

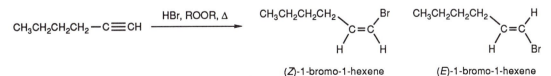

11.31. Follow the procedure outlined in the solution to exercise 11.18.

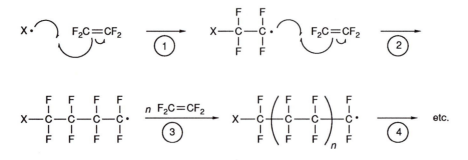

11.32. Assuming that the given monomers combine in a head-to-tail fashion, addition of a radical to the double bond creates a new radical center that adds to the π bond of another monomer. The following show only one of each monomer, but the next unit in a random polymer could be either one of the starting monomers. Butadiene often reacts at each terminus of the conjugated system by 1,4 addition (text page 397).

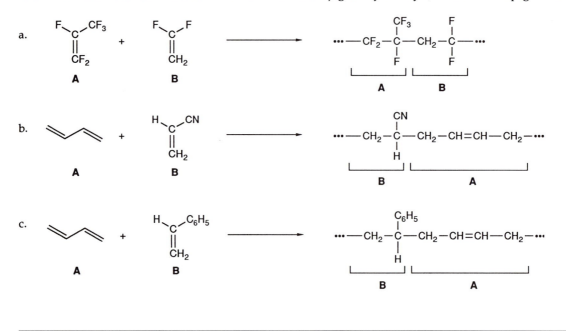

11.33.

a. For radical substitution reactions, specificity is sometimes a problem. The two allylic positions in this substrate are 1° and 3°. Tertiary C–H bonds preferentially undergo free radical bromination.

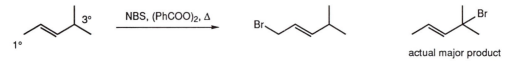

11.33.   (continued)

b.   In radical cyclizations, the heteroatom has to be reactive enough to sustain a chain reaction. Fluorine is unreactive, so no reaction occurs.

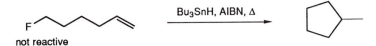

c.   In radical cyclizations, a five-membered ring is formed preferentially, if possible.

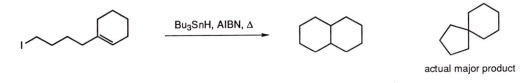

actual major product

11.34.   Calculate $\Delta H°$ values for the given steps by applying the formula, $\Delta H°_{rxn} = \Sigma\ (BDE_{broken}) - \Sigma\ (BDE_{made})$. The calculated values are as follows:

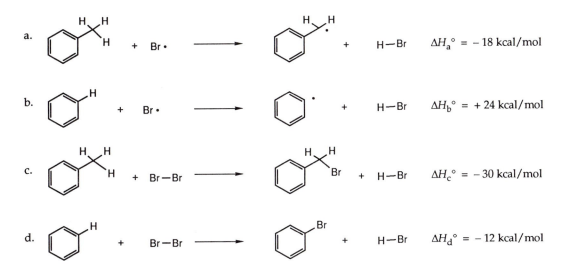

a.   ... + Br· ⟶ ... + H—Br $\Delta H_a° = -18\ \text{kcal/mol}$

b.   ... + Br· ⟶ ... + H—Br $\Delta H_b° = +24\ \text{kcal/mol}$

c.   ... + Br—Br ⟶ ... + H—Br $\Delta H_c° = -30\ \text{kcal/mol}$

d.   ... + Br—Br ⟶ ... + H—Br $\Delta H_d° = -12\ \text{kcal/mol}$

The bromination of toluene proceeds in two exothermic steps. Resonance stabilization of the benzyl radical accounts for the exothermicity of the hydrogen atom abstraction step. A phenyl radical, on the other hand, experiences no resonance stabilization because the unpaired electron is in an orbital that is perpendicular to the π system of the aromatic ring.

Energy diagrams for these two processes are shown on the next page.

11.34.   (continued)

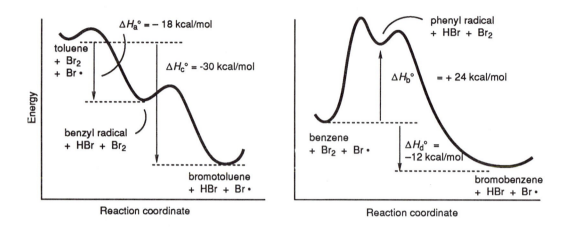

11.35.   Autoxidation occurs when a compound is exposed to molecular oxygen, which itself is a diradical. Abstraction of a hydrogen atom from tetralin produces a benzylic radical in step (1), the formation of which is made possible by resonance stabilization. Recombination of the radicals in step (2) generates the product.

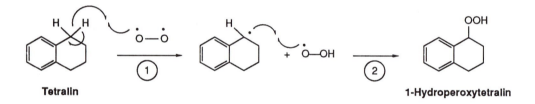

The intermediate radical can also combine with dioxygen, as shown below in step (3).   Hydrogen atom abstraction from another molecule of tetralin in step (4) promotes a chain reaction.

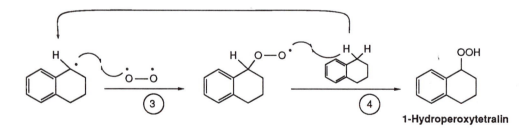

11.36. During pyrolysis of organic molecules, homolysis produces the most stable types of radicals. The order of stability is benzylic, allylic > 3˚ > 2˚ > 1˚.

a. Benzylic C–H and C–C bonds are weakest because the benzylic radical is resonance stabilized.

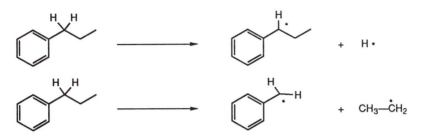

b. 3,4-Dimethylpentane loses a hydrogen atom to form a 3˚ radical and a hydrogen atom. Cleavage of its central carbon-carbon bond generates two 2˚ alkyl radicals.

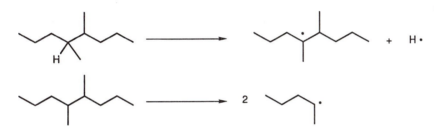

c. Breaking the exocyclic C—C bond of *tert*-butylcyclohexane produces a 2˚ and a 3˚ alkyl radical. Loss of a hydrogen atom from *tert*-butylcyclohexane creates a 3˚ alkyl radical and a hydrogen atom.

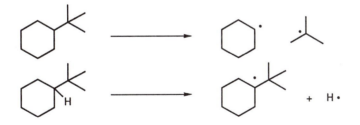

d. Cleaving a C—C bond of 4-methyl-1-pentene creates an allylic and an isopropyl (2˚) radical. Loss of a hydrogen atom generates an allylic or 3˚ alkyl radical and a hydrogen atom.

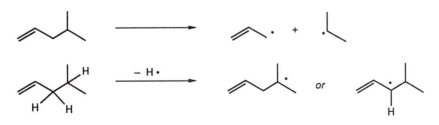

11.37. From the alkene starting material given in this exercise, the PhSe group is abstracted by tirbutyltin radical, which is made by the reaction shown on text page 494. Step (1) produces a carbon-centered radical, and the imine double bond is positioned to form a five-membered ring in step (2). The resulting nitrogen-centered radical leads to formation of a second five-membered ring in step (3). Quenching the carbon radical by tributyltin hydride in step (4) yields the product and regenerates Bu₃Sn• to continue the chain reaction.

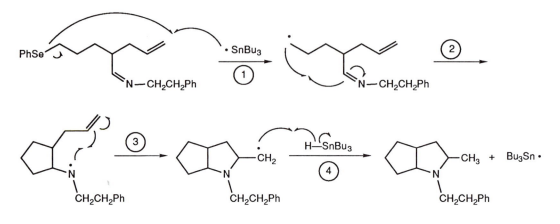

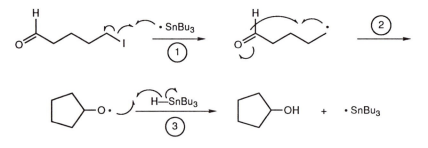

11.38. **a.** The tin radical, generated from Bu₃SnH and AIBN, abstracts the iodine atom from the iodo aldehyde to generate a 1° alkyl radical in step (1). The carbonyl double bond intercepts the radical in step (2), and the resulting oxygen-centered radical is quenched with tin hydride, regenerating Bu₃Sn• to continue the chain reaction.

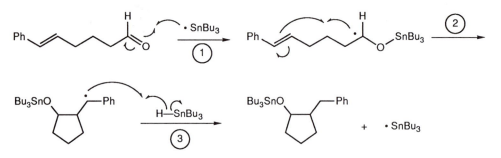

b. Addition of the tin radical to the carbonyl group in step (1) generates a carbon-centered radical (tin is oxyphilic, so it bonds to the oxygen atom) that undergoes addition to the carbon-carbon double bond in step (2). Cyclization is followed by reduction in step (3).

11.38. (continued)

In a separate procedure, hydrolysis yields the alcohol product.

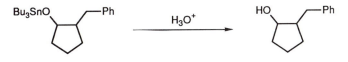

11.39.   The structure of the given substrate is ideal for forming a five-membered ring by radical cyclization via the following pathway.

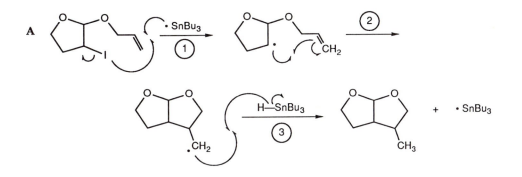

In **B**, the 1° alkyl radical formed in step (2), above, reacts with molecular oxygen to form a carbon-oxygen bond. The hydroperoxy group becomes an alcohol functional group after cleavage of the oxygen-oxygen bond, which is quite weak. Tributyltin hydride is a good reducing agent, so it cleaves O—O bonds.

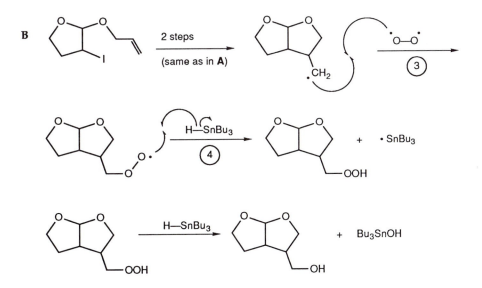

11.40. Galvinoxyl has the structure shown below. The radical is stabilized by extensive delocalization of the unpaired electron over the π system. The *tert*-butyl groups prevent the radical from forming a dimer by keeping two radical centers from getting close enough to form a bond.

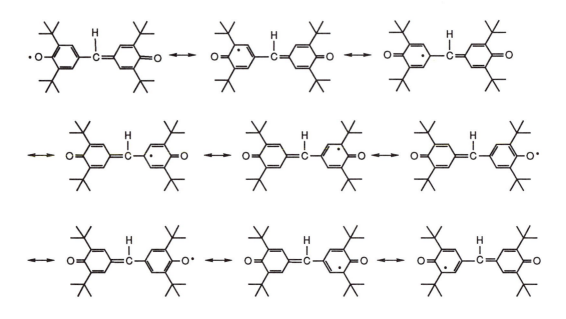

11.41. The Bergman cyclization creates a fused tricyclic system in which the central 6-membered ring is fused to a four-membered ring, which has significant Baeyer strain, and to the rigid benzene ring. The constraints imposed on the carbon skeleton apparently create enough strain in the diradical that ring formation does not occur until the starting material is heated to 100 °C.

11.42. Reaction of an alkyl halide with tributyltin radical generates an alkyl radical that subsequently adds to carbon monoxide. Quenching the acyl radial produces the aldehyde and regenerates the tributyl tin radical to propagate a chain reaction.

11.43. *tert*-Butyl hypochlorite has a weak O—Cl bond that is cleaved upon irradiation, not unlike the photochemical cleavage of Cl_2 or Br_2.

$$(CH_3)_3C—O—Cl \xrightarrow{h\nu} (CH_3)_3C—O \cdot + Cl \cdot$$

A chain reaction subsequently ensues, starting with abstraction of an allylic hydrogen atom from cyclohexene. The allylic radical reacts with *tert*-butyl hypochlorite in step (2) to form a carbon-chlorine bond. This second step also regenerates the *tert*-butoxy radical needed to propagate the chain reaction.

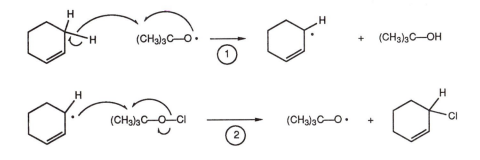

11.44. As with formation of any free radical, pyrolysis of natural rubber generates radicals with a stability order of benzylic, allylic > 3° > 2° > 1°. The results observed for vulcanization of rubber are consistent with those seen for the pyrolysis processes described in the solution to exercise 11.36.

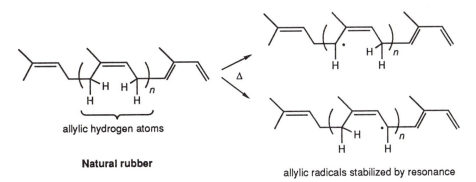

allylic hydrogen atoms

Natural rubber

allylic radicals stabilized by resonance

Reaction of these radicals with sulfur create the crosslinks in vulcanized rubber.

11.45. If we assume that the multistep radical cyclization process described in this exercise will yield a steroid-like material, then we will expect that six-membered rings are formed.

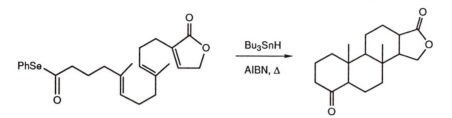

The first step is removal of the PhSe group, which generates an acyl radical.

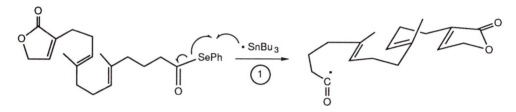

Making the assumption that chair-like conformations exist during each step, we propose the mechanism shown below. The last step regenerates the tributyltin radical to continue propagating the chain reaction.

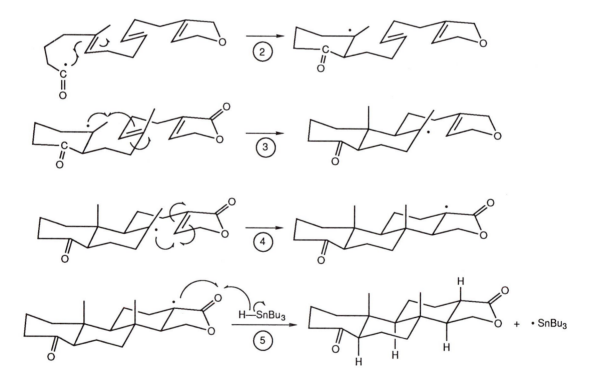

ELECTROPHILIC AROMATIC SUBSTITUTION REACTIONS OF BENZENE AND ITS DERIVATIVES

12.1. Follow the procedure outlined in the solution to exercise 10.34. Each nodal plane in benzene is perpendicular to the plane of the ring, passes through the ring's center, and lies between orbitals with different signs. Both perspective and top views of the ring and orbitals are included in the representations shown below.

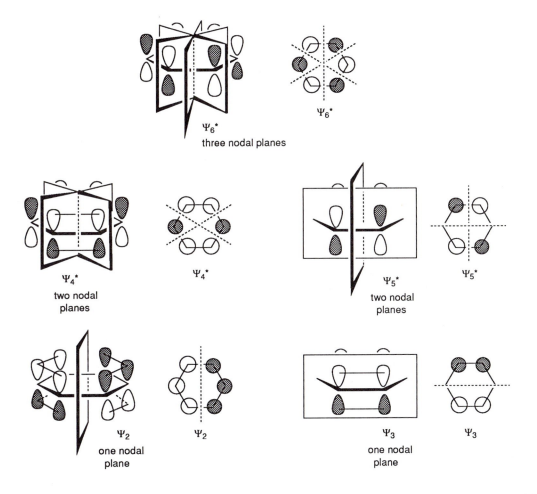

12.2. Relative energy levels of molecular orbitals are generated by extending lines horizontally from each vertex of the ring, as described in the text on page 518. Placing the eight electrons into the orbitals reveals that two of them are unpaired if the molecule is flat. With a model, you can see that the shape of cyclooctatetraene approximates a boat, as shown below at the right.

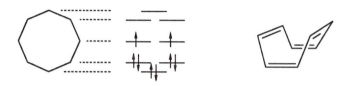

12.3. Relative energy levels of molecular orbitals are generated by extending lines horizontally from each vertex of the ring, as described in the text on page 518. With seven carbon atoms, the neutral cycloheptatrienyl derivative has 7 electrons, the cation has six, and the anion has eight. Only the cation, with 6 electrons, has fully paired electrons and filled orbital levels. It is aromatic; the others are not.

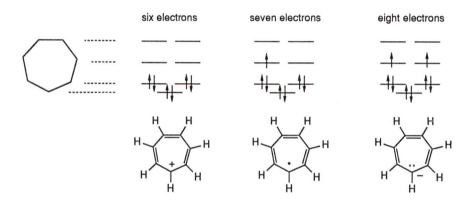

12.4. The root word, given below, specifies the structure of the benzene derivative. Each substitutent is attached according to the numeral (or letter *o, m,* or *p*) associated with it.

a. **3-Chloro-2-methylphenol**

phenol	benzene derivative with a hydroxy group attached to the ring
3-bromo	bromine atom at C3
2-methyl	methyl group at C2

b. *m*-**Hydroxybenzoic acid**

benzoic acid	benzene derivative with a –COOH group attached to the ring
m-hydroxy	OH group at C3

12.4. (continued)

c.   ***o*-Dichlorobenzene**
     two chlorine atoms at adjacent positions on the benzene ring

d.   **2,6-Dimethoxytoluene**
     toluene   benzene derivative with a –CH$_3$
                          group attached to the ring
     2,6-dimethoxy   OCH$_3$ groups attached at C2 and C6

e.   **α,α′-Dibromo-*m*-xylene**
     xylene   benzene derivative with two –CH$_3$
                          groups attached to the ring at C1 and C3
     α,α′-dibromo   a bromine atom is attached to each of the carbon
                          atoms of the methyl substituents

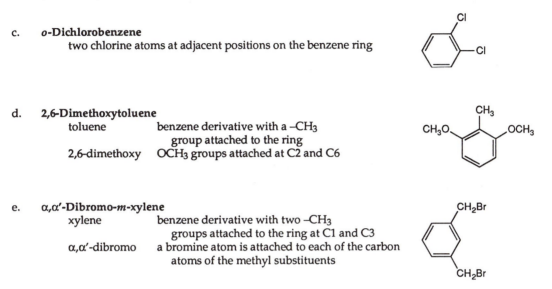

12.5.   Draw the structure of the parent benzene derivative, then attach the substituent(s) at the specified position(s).

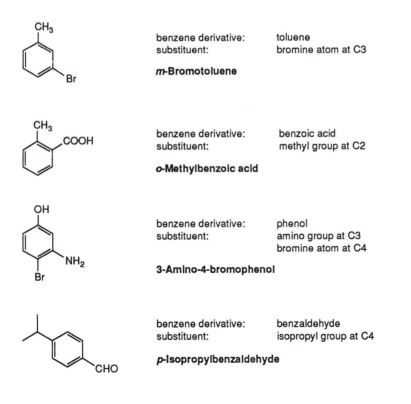

benzene derivative: toluene
substituent: bromine atom at C3

***m*-Bromotoluene**

benzene derivative: benzoic acid
substituent: methyl group at C2

***o*-Methylbenzoic acid**

benzene derivative: phenol
substituent: amino group at C3
                       bromine atom at C4

3-Amino-4-bromophenol

benzene derivative: benzaldehyde
substituent: isopropyl group at C4

***p*-Isopropylbenzaldehyde**

12.6. Chlorination of benzene follows the same mechanism shown for bromination on text pages 524-525. Chlorine and aluminum chloride first react to form the chlorinium and tetrachloroaluminate ions.

AlCl$_3$ +	Cl—Cl	Cl$^+$	AlCl$_4^-$
Lewis acid	Lewis base	electrophile	nucleophile

The remaining steps proceed in the same fashion as described for bromination of benzene.

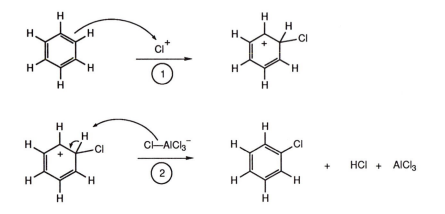

12.7. Nitric acid is a strong mineral acid, so one molecule contributes a proton to another molecule to form a good leaving group (water) that dissociates to form the nitronium ion.

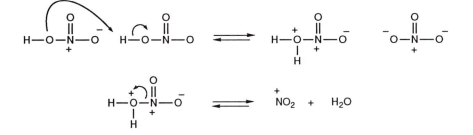

12.8. After reaction of the benzene ring with the nitronium ion, the cationic intermediate is stabilized by delocalization of the positive charge among electrons in the two π bonds.

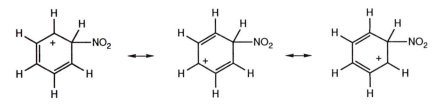

12.9.   Desulfonation of an aryl sulfonic acid starts with ionization of the sulfonic acid group. When the sulfonate salt reacts with sulfuric acid in step (2), the proton is intercepted by a π bond in the ring instead of by the oxygen atom of the sulfonic acid group.

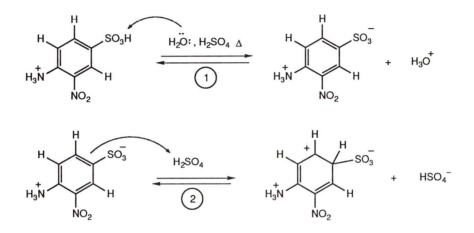

The carbocation formed in the protonation step loses sulfur trioxide, regenerating the aromatic ring. Sulfur trioxide is finally hydrolyzed in the aqueous environment to form dilute sulfuric acid.

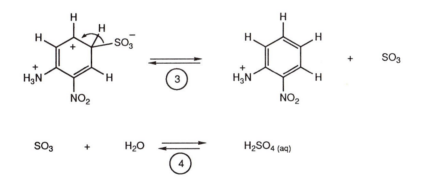

The substrate in this exercise is a derivative of aniline, and the reaction medium is acidic, so the –NH$_2$ group is protonated throughout the preceding steps and exists as the –NH$_3^+$ form. Base (hydroxide ion) is required to remove the proton so that neutral *o*-nitroaniline is isolated.

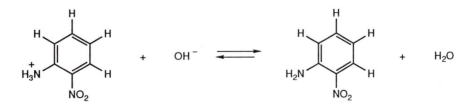

12.10. 1-Chloropropane reacts with aluminum chloride to form a 2° carbocation via rearrangement. The mechanism of this reaction was discussed on text page 281.

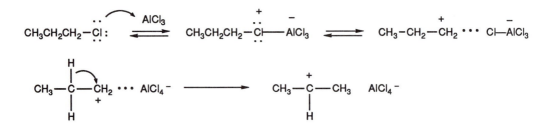

Either 1-propyl(tetrachloroaluminate) or 2-propyl(tetrachloroaluminate) is intercepted by the π bonds of benzene, which leads to formation of isomeric propylbenzene derivatives.

reaction with the 1-propyl intermediate

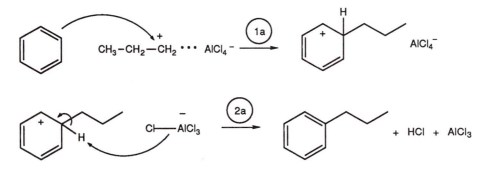

reaction with the 2-propyl intermediate

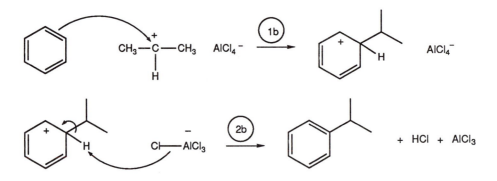

12.11.   When an alkene is used in Friedel-Crafts alkylation, the double bond of the olefin is protonated to form a carbocation, which is intercepted by the π system of benzene. Deprotonation of the cationic intermediate in step (3) regenerates the aromatic system of the product.

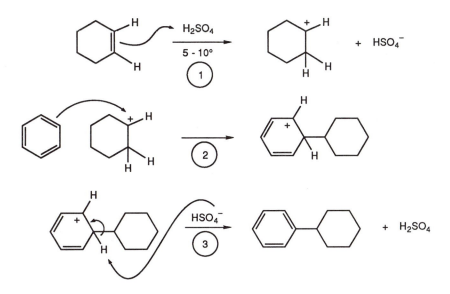

12.12. Friedel-Crafts acylation forms a ketone as product. To prepare an 1° alkylbenzene, an acylated derivative is reduced.

a. The desired product is a ketone, so direct acylation is sufficient.

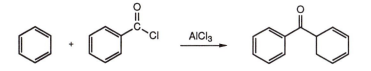

b.   The desired product has an alkyl group attached to the benzene ring through a 1° carbon atom, so acylation followed by reduction is needed.   The substituent is first rewritten as an acyl group. The acid chloride that structurally corresponds to the side chain is used as the starting material, along with benzene.

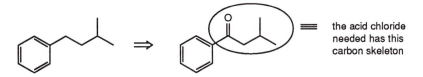

the acid chloride needed has this carbon skeleton

12.12. (continued)

The synthesis comprises Friedel-Crafts acylation followed by reduction. The Clemmensen reduction is particularly good if no other functional groups are present.

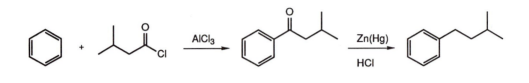

12.13. When asked to predict the major product of electrophilic substitution of a substituted benzene derivative, consider what type of directing group is attached to the ring. A methyl group is an *o,p* director, and steric effects will be minimal, so both isomers of the nitrated product are formed.

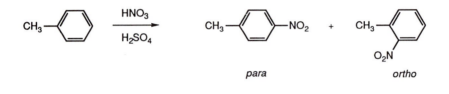

para   ortho

12.14. To generate possible resonance structures for *para* substitution of anisole, attach the electrophile, E⁺, at the position directly across the ring from the methoxy group, then draw the resonance forms by the procedure outlined in example 3.4 (text page 82).

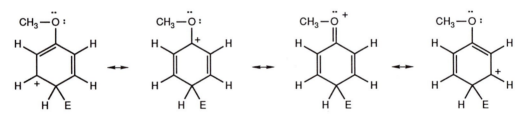

12.15. The phenolic OH group is an *o,p* director, and we are told in this exercise that three nitro groups are introduced into the product. The nitro groups are attached in all *ortho* and *para* positions relative to the OH group.

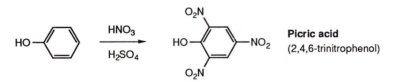

Picric acid
(2,4,6-trinitrophenol)

12.16.   As noted in the text, resonance structures for the intermediates in electrophilic substitution of chlorobenzene are the same as those we draw for anisole (text page 540).

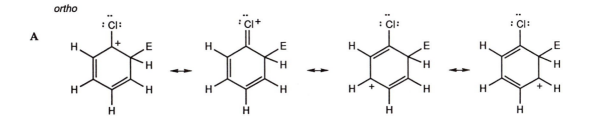

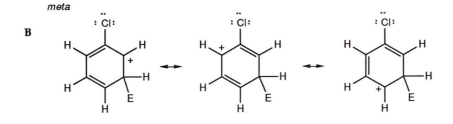

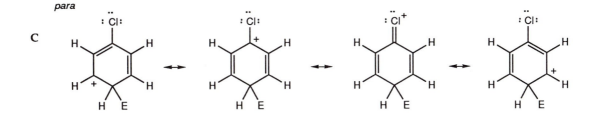

12.17.   Benzoic acid reacts with the brominium ion to form cationic intermediates in which the bromine atom is attached *ortho*, *meta*, or *para* to the COOH group. In the *ortho* and *para*-substituted intermediates, a positive charge appears in one structure (circled, top of next page) at the carbon atom *ipso* to the carboxyl group. The carbon atom of a carbonyl group bears a partial positive charge already, so this form contributes little to resonance stabilization. The *meta* substituted intermediate therefore has the lowest free energy of activation, and *m*-bromobenzoic acid is the predominant product.

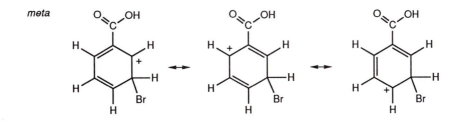

12.17. (continued)

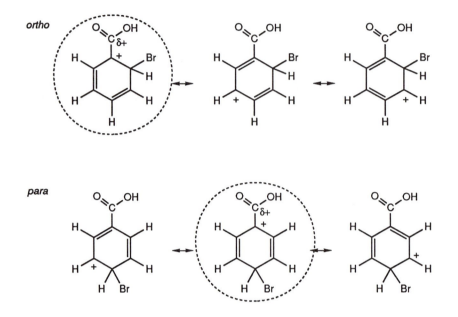

12.18. The *para*-substituted intermediate, formed by reaction of an electrophile with aniline in its protonated from, has a resonance structure that places positive charges on two adjacent atoms (middle, below).

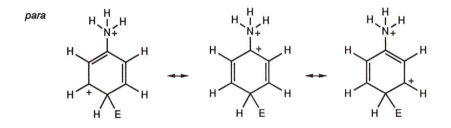

12.19. The *ortho-* and *para*-substituted intermediates formed during electrophilic substitution of trifluoromethylbenzene have a resonance form (circled, below) that places a positive charge on the carbon atom adjacent to the partially positive carbon atom of the trifluoromethyl group. These structures are like those that we drew for benzoic acid (see the solution to exercise 12.17).

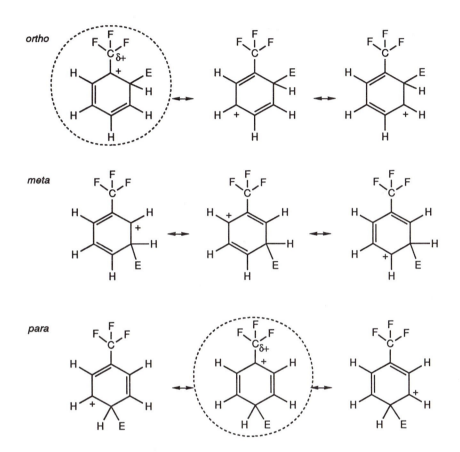

Looking at the influence of inductive effects, we see that a trifluoromethyl group is strongly electron-withdrawing, so it should make the benzene ring electron deficient.

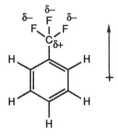

The trifluoromethyl group is a powerful electron-withdrawing group because the highly electronegative fluorine atoms withdraw electron density from the ring by an inductive effect.

12.20.   To predict the product(s) of electrophilic aromatic substitution, consider what effect each substituent has with regard to the position to which it directs the incoming group and how activating or deactivating the substituent is. Do not overlook the fact that certain positions can be equivalent.

a. The carbonyl group is a *meta* director, and the alkyl group of the appended alicyclic ring is an *o, p* director. The influence of the latter predominates, so two products form, *ortho* and *para* to the alkyl substituent.

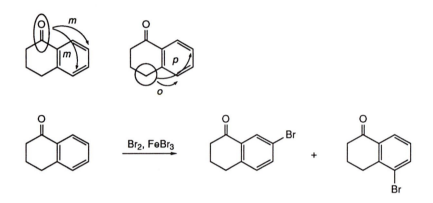

b. Both substituents are *o,p* directors, but the oxygen atom is a more potent director, so the nitro group enters the benzene ring *ortho* and *para* to the oxygen atom.

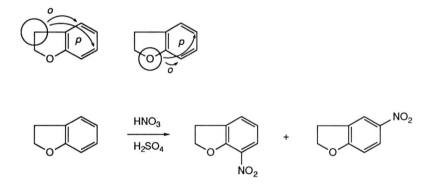

12.21.   A carboxylic acid group is a *meta* director, so direct nitration produces the desired isomer.

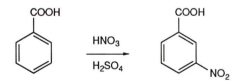

12.22. The substituents in both 3-bromoaniline and 4-bromoaniline (methyl and amino groups) are *o,p* directors. Bromination of aniline (or its acetamide derivative) therefore produces the correct isomer. Direct nitration of bromobenzene also yields the *para*-substituted isomer, and if that pathway is chosen, the nitro group is reduced to the amino group.

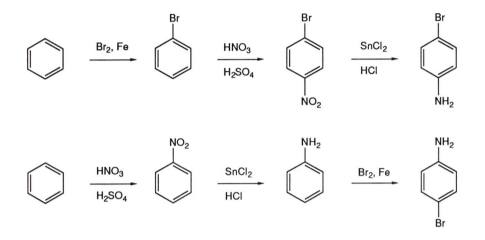

To prepare 3-bromoaniline, one of the substituents needs to be a *meta* director. An amino group is made from a nitro group (a *m* director), so nitrobenzene undergoes bromination to form the correct isomer. Reduction of the nitro group produces the aniline derivative.

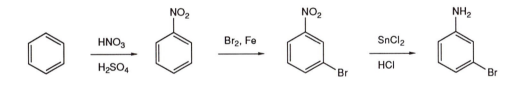

12.23. The Birch reduction of a substituted anisole produces 1,4-cyclohexadiene derivatives in which the carbon atom bearing the methoxy group has a double bond to adjacent carbon atoms in the ring. When a second substituent is attached to the benzene ring, isomers are formed.

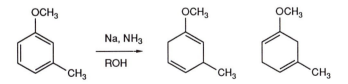

12.24. Oxidation of an alkyl group attached to a benzene ring is most facile if the benzylic carbon atom is tertiary. The reactivity order is $3° > 2° > 1°$. A quaternary center is not oxidized because it lacks a benzylic hydrogen atom, but aldehyde and alkene substituents are also easily oxidized.

a.

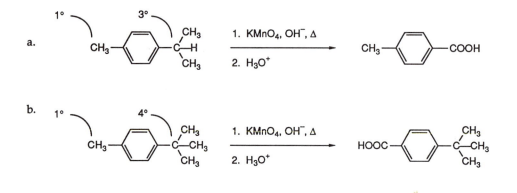

b.

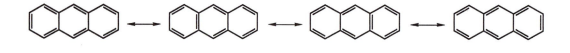

12.25. Resonance forms of anthracene and phenanthrene are generated by moving the π bonds around the ring. In phenanthrene, the C9–C10 bond is a double bond in four of the five resonance forms, so its length will be nearly the same as that in a typical alkene, which means it should be shorter than the other bonds in the molecule.

Anthracene

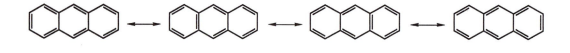

Phenanthrene

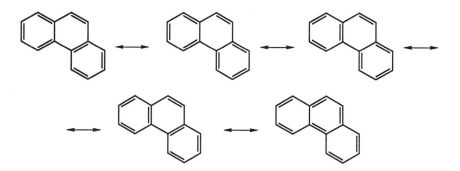

12.26. Six resonance structures can be drawn for the cationic intermediate formed by reaction of an electrophile at C2 of naphthalene. Only two of these have intact benzene rings (boxed, below).

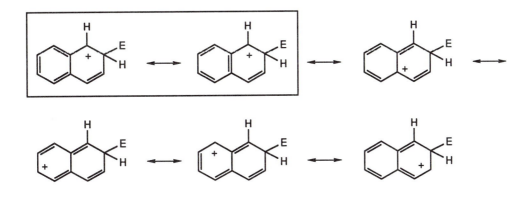

12.27. The methoxy group is a powerful *o,p* -director, but the position *para* to it is blocked in 2-methoxy-naphthalene. Of the two *ortho* positions, one is C1 of the naphthalene ring, which should be more reactive than the other. But C3, the other *ortho* position is less hindered sterically, therefore two products are likely to be formed.

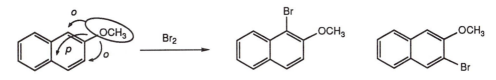

12.28. Like anthracene, phenanthrene is oxidized to a quinone derivative, in this case an *ortho*-quinone.

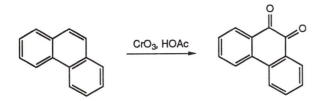

12.29. Because the C9–C10 bond is like that of an alkene, *trans*-addition of bromine will most likely occur.

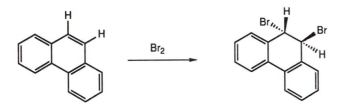

12.30. Interpret the name according to the procedures described in Chapter 2. The root word, given below, indicates the structure of the benzene derivative. Each substitutent is then added according to the numeral (or prefix *o, m,* or *p*) associated with it. The stereochemistry of double bonds is described in Chapter 4 and absolute configuration in Chapter 5.

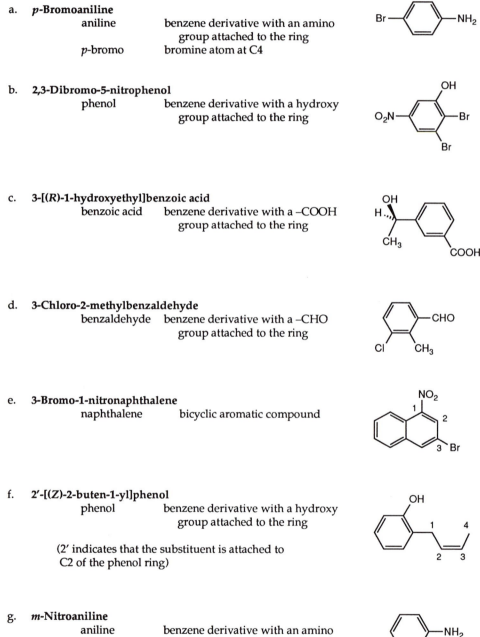

a. ***p*-Bromoaniline**

    aniline   benzene derivative with an amino group attached to the ring

    p-bromo   bromine atom at C4

b. **2,3-Dibromo-5-nitrophenol**

    phenol   benzene derivative with a hydroxy group attached to the ring

c. **3-[(*R*)-1-hydroxyethyl]benzoic acid**

    benzoic acid   benzene derivative with a –COOH group attached to the ring

d. **3-Chloro-2-methylbenzaldehyde**

    benzaldehyde   benzene derivative with a –CHO group attached to the ring

e. **3-Bromo-1-nitronaphthalene**

    naphthalene   bicyclic aromatic compound

f. **2'-[(Z)-2-buten-1-yl]phenol**

    phenol   benzene derivative with a hydroxy group attached to the ring

    (2' indicates that the substituent is attached to C2 of the phenol ring)

g. ***m*-Nitroaniline**

    aniline   benzene derivative with an amino group attached to the ring

    m-nitro   nitro group at C3

12.31.   Write the name of the parent aromatic compound, then include the name of each substituent as a prefix with a numeral (or *o, m,* or *p*) to indicate the position on the ring to which the substituent is attached.

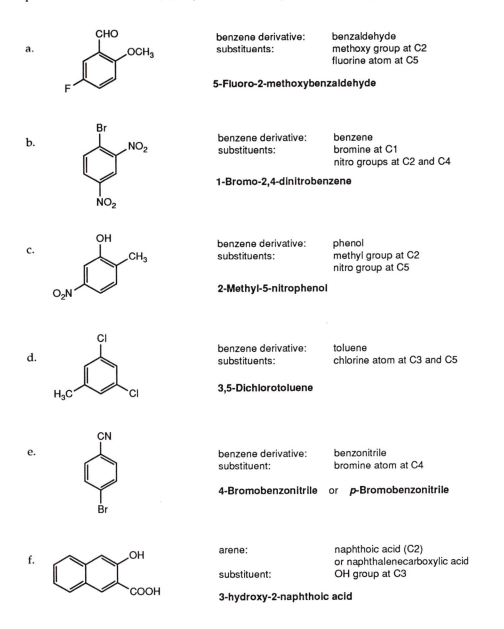

a.

benzene derivative:	benzaldehyde
substituents:	methoxy group at C2
	fluorine atom at C5

5-Fluoro-2-methoxybenzaldehyde

b.

benzene derivative:	benzene
substituents:	bromine at C1
	nitro groups at C2 and C4

1-Bromo-2,4-dinitrobenzene

c.

benzene derivative:	phenol
substituents:	methyl group at C2
	nitro group at C5

2-Methyl-5-nitrophenol

d.

benzene derivative:	toluene
substituents:	chlorine atom at C3 and C5

3,5-Dichlorotoluene

e.

benzene derivative:	benzonitrile
substituent:	bromine atom at C4

4-Bromobenzonitrile   or   ***p*-Bromobenzonitrile**

f.

arene:	naphthoic acid (C2)
	or naphthalenecarboxylic acid
substituent:	OH group at C3

3-hydroxy-2-naphthoic acid

12.32. According to the Hückel formalism, a cyclic compound that is planar and has 4n+2 π electrons is aromatic. Therefore, any system with 2, 6, or 10 electrons is aromatic. Of the compounds shown below, all but the cycloheptatrienyl anion (part c) meet these criteria and are considered aromatic.

a. **Cyclobutadienyl dianion**

2 π bonds = 4 electrons
2– charge = 2 electrons
──────────
6 electrons

b. **Cyclooctatetraenyl dianion**

4 π bonds = 8 electrons
2– charge = 2 electrons
──────────
10 electrons

c. **Cycloheptatrienyl anion**

3 π bonds =   6 electrons
1 unshared pair = 2 electron
──────────
8 electrons

d. **Azulene**

5 π bonds = 10 electrons

e. **Quinoline**

5 π bonds = 10 electrons

12.33. The LCAO representation for azulene is like the one that we draw for naphthalene. Each carbon atom has sp^2 hybridization, so each has a *p* orbital perpendicular to the sigma bonds that define the rings. Overlap occurs between any two adjacent orbitals (both above and below the plane), resulting in complete delocalization of the 10 π electrons.

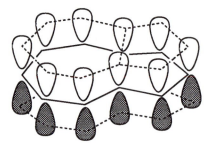

12.34.   To assess how reactive an arene is toward electrophilic substitution, look at the substituents on the ring and decide whether they are *ortho/para* or *meta* directors. For the former, the order of reactivity is alkoxy, hydroxy, amino > alkyl > halogen. All of these make benzene more reactive than a *meta* director does.

a. Bromobenzene versus toluene: both bromine and methyl are *ortho,para* directors, but the methyl group in toluene activates the ring by inductive effects whereas a bromo substituent deactivates the ring by inductive effects. Toluene is the more reactive toward electrophilic aromatic substitution.

b. *p*-Xylene versus *o*-xylene: both compounds have two methyl groups, so the degree of activation should be approximately the same.

c. Nitrobenzene versus chlorobenzene: a chlorine atom is activating by resonance effects whereas a nitro group is deactivating both by resonance and inductive effects. Chlorobenzene is the more reactive toward electrophilic aromatic substitution.

d. Anisole versus toluene: an alkoxy group is more activating than an alkyl group because of resonance effects. Anisole is the more reactive toward electrophilic aromatic substitution.

12.35.   An *ortho/para* directing group falls into one of two categories. It is either an alkyl group or it has an unshared pair of electrons on the atom attached directly to the benzene ring. Substituents a and e (boxed, below) are the only ones that are **not** *o,p* directors.

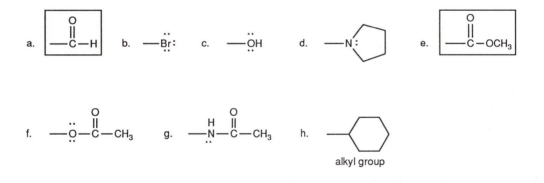

12.36.   To predict the product of electrophilic aromatic substitution, consider to which positions each substituent directs the incoming electrophile and how activating or deactivating the substituent is. Do not overlook the fact that certain positions in the ring are equivalent.

12.36. (continued)

a. The chlorine atoms are equivalent, and they are both *o,p* directors and deactivators. There are two sets of equivalent positions, so the nitro group enters *para* to a chlorine atom.

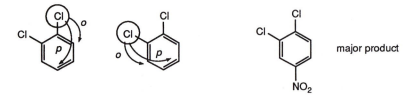

b. The bromine atoms are equivalent, and they are both *o,p* directors and deactivators. Three unique positions exist, so the nitro group enters *para* to a bromine atom.

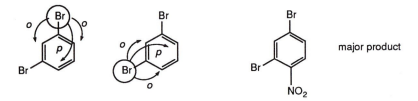

c. The methyl group is an *o,p*-director and activator, and the acetyl group is a *m* director and deactivator, so the methyl group directs substitution. The *para* position is filled, so the nitro group enters *ortho* to the methyl group.

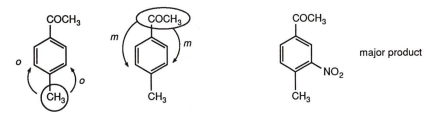

12.37. Follow the procedure outlined in the solution to exercise 12.36.

a. The methyl groups are equivalent, and they are both *o,p* directors and activators. There are two sets of equivalent positions, so the bromine atom enters *para* to a methyl group.

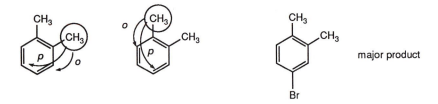

12.37. (continued)

b. The methyl group is an *o,p*-director and activator, and the acetyl group is a *m* director and deactivator, so the methyl group directs substitution. The bromine atom enters *para* to the methyl group.

major product

c. The methyl group is an *o,p*-director and activator, and the chlorine atom is an *o,p* director and deactivator, so the methyl group directs substitution. The bromine atom enters *para* to the methyl group.

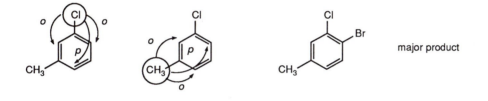

major product

12.38.   Follow the procedure outlined in the solution to exercise 12.36.

a. The methoxy group is an *o,p*-director and activator, and the nitro group is a *m* director and deactivator, so the methoxy group directs substitution. The sulfonic acid group enters *para* to the methoxy group.

major product

b. Both hydroxy and methyl groups are *o,p*-directors and activators, but the hydroxy group is a more potent director by resonance, so it directs substitution. The sulfonic acid group enters *para* to the hydroxy group.

major product

12.38. (continued)

c. Both the methoxy and methyl groups are *o,p*-directors and activators, but the methoxy group is a more potent director by resonance, so it directs substitution. The sulfonic acid group enters *para* to the methoxy group.

major product

12.39. Given that you must start with a disubstituted benzene derivative, remove one of the groups and consider what are the possible starting materials. Then follow the procedure outlined in the solution to exercise 12.36.

a. It is nearly impossible to make a 1,2,3-trisubstituted benzene compound by electrophilic substitution starting with a disubstituted benzene derivative. Steric effects normally favor substitution at the position *para* to an *o,p* directing group (in this case, one of the chlorine atom).

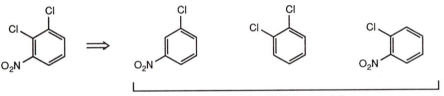

potential starting materials

b. Of the three possible starting materials, we rule out using 1,3-dibromobenzene because there is no straightforward way to introduce an OH group as an electrophile.

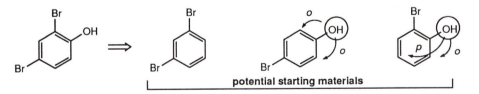

potential starting materials

Starting with either of the phenols shown directly above, we can prepare the desired product by electrophilic bromination. The OH group is the stronger director, so substitution occurs as depicted below.

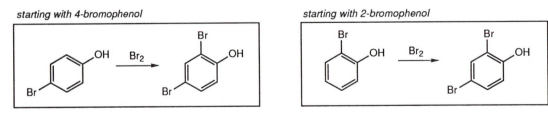

12.40. Follow the procedure outlined in the solution to exercise 12.39.

a.   As noted in the last exercise, it is nearly impossible to make a 1,2,3-trisubstituted benzene compound from a disubstituted benzene compound by electrophilic substitution.

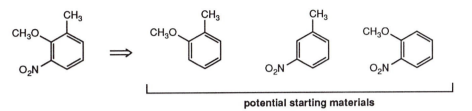

potential starting materials

b.   Of the three possible starting materials, we rule out using *o*-nitrobenzoic acid because there is no good way to introduce a methoxy group as an electrophile. Likewise, we rule out using *p*-nitroanisole because there is no good way to introduce a carboxylic acid group as an electrophile.

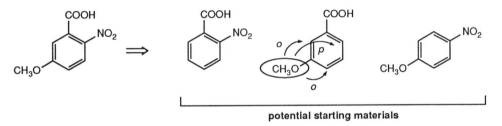

potential starting materials

A nitro group is readily introduced by electrophilic substitution, and the *o,p*-directing nature of the methoxy group in *m*-methoxybenzoic acid directs the nitro group to the desired position.

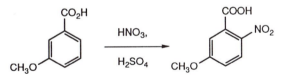

c.   It is nearly impossible to make a 1,3,5-trisubstituted benzene compound by electrophilic substitution when every substituent is an *o,p* director.

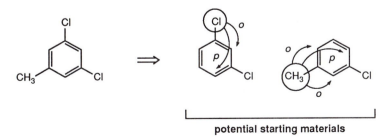

potential starting materials

12.41. Dimerization of styrene begins with protonation of its double bond, which produces the resonance-stabilized benzylic carbocation.

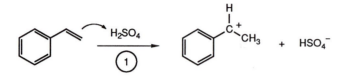

The carbocation is trapped in step (2) by the double bond of a second molecule of styrene, resulting in Markovnikov addition to the alkene double bond.

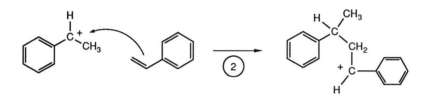

In step 3, one of the benzene rings reacts with the carbocation in the first step of electrophilic aromatic substitution.

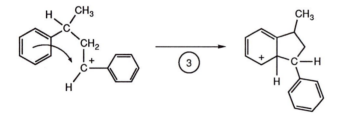

Finally, the hydrogen sulfate ion deprotonates the intermediate, restoring aromaticity to the benzene ring.

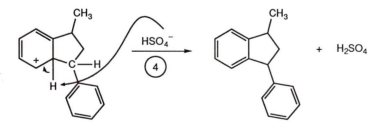

12.42. It may be that strong intramolecular hydrogen bonding stabilizes the benzylic alcohol so that the C—O bond is stronger than normal, and radical formation is energetically unfavorable.

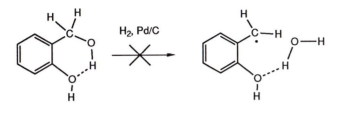

12.43. Nitrosobenzene is unusual because the nitrogen atom of the substitutent group bears an unshared electron pair yet it forms a double bond with the neighboring oxygen atom, too.

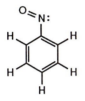

When an incoming electrophile attaches to the ring, the unshared electrons of the nitroso group contributes to resonance stabilization when the electrophile is *ortho* or *para*. Only three good resonance structures can be drawn for the *meta* isomer.

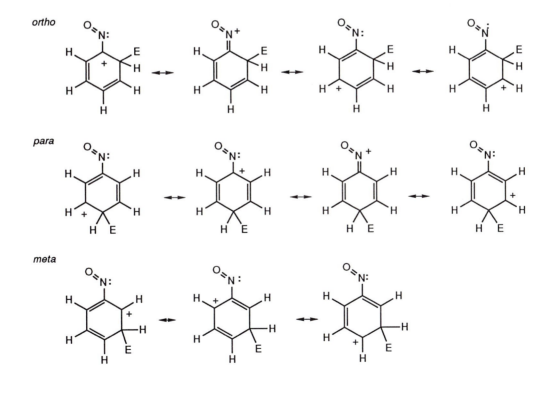

12.44. The mechanism of the Gatterman–Koch reaction follows the same pathway described for Friedel-Crafts acylation (text page 532). The electrophile is formed by protonation of carbon monoxide, which has a nucleophilic carbon atom. The aluminum chloride stabilizes the formyl cation.

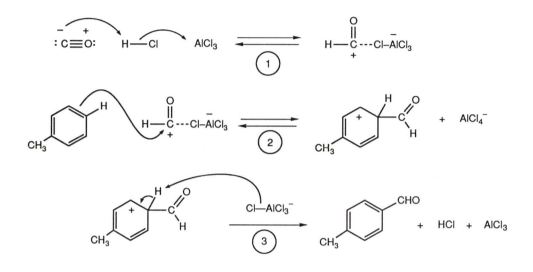

12.45. The synthesis of Bisphenol A consists of two electrophilic alkylation steps. The carbonyl group of acetone is polarized such that protonation occurs at oxygen, generating a carbocation. Phenol is an activated arene, so electrophilic substitution occurs to form a benzylic alcohol.

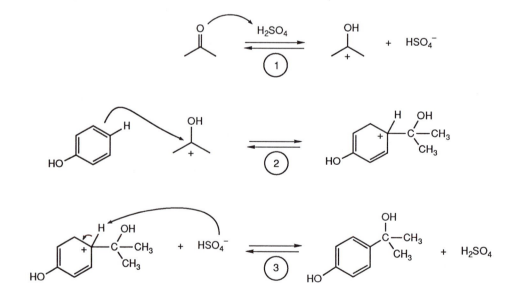

12.45. (continued)

Protonation of the benzylic alcohol OH group forms a good leaving group that leads to dissociation of a water molecule to form a benzylic carbocation.

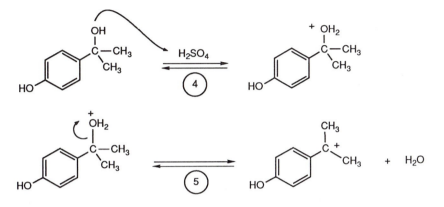

A second electrophilic substitution reaction then takes place to form the product.

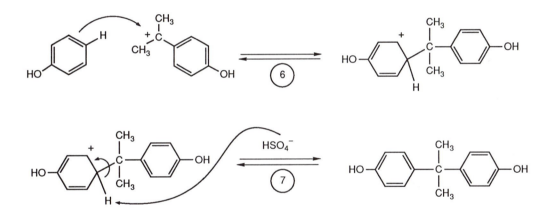

12.46. Follow the procedure outlined in the solution to exercise 12.36 to assess where substitution will take place. In the *para* isomer, all four unsubstituted positions are equivalent, so only one product is formed upon nitration. The *ortho* and *meta* isomers give rise to two and three mononitro products, respectively.

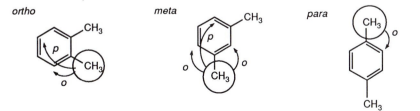

12.46. (continued)

The equation for nitration of *p*-xylene is as follows:

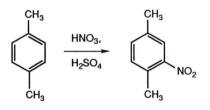

12.47. The competing reactions for the sulfonation of naphthalene exemplify kinetic vs. thermodynamic control. The product formed faster (naphthalene-1-sulfonic acid) has a lower free energy of activation, but it is less stable because of steric interactions between the substituents at C1 and C8. Naphthalene-2-sulfonic acid is the thermodynamic product because it lacks the unfavorable 1,8-steric effects so is more stable. Formation of the 2-substituted isomer requires a higher temperature, however, which means that its formation has a higher free energy of activation.

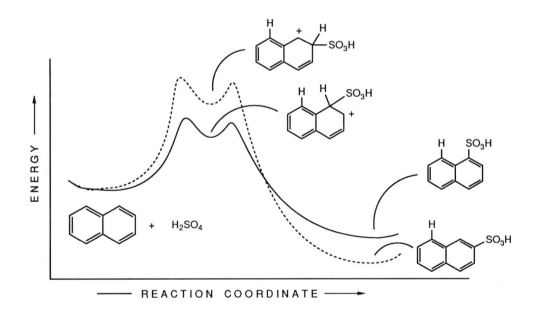

12.48. The general approach for solving problems of this type is described in the examples on text pages 549-553.

a.   The substituents are *para*, yet only the methyl group is a *para* director. The acetyl group is therefore introduced by Friedel Crafts acylation, using toluene as the starting material.

retrosynthesis   *synthesis*

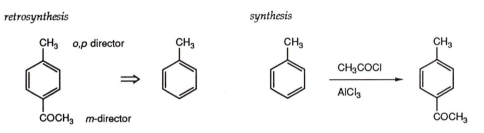

b.   The substituents are *para*, and both are *para* directors. The bromine atom is introduced readily by electrophilic substitution starting with 1-propylbenzene. 1-Propylbenzene, in turn, is made by attaching a propyl group to benzene, but direct Friedel Crafts alkylation is not viable because rearrangement will occur. A propanoyl group is attached without rearrangement, however, and the ketone carbonyl group is subsequently reduced by the Clemmensen procedure.

retrosynthesis

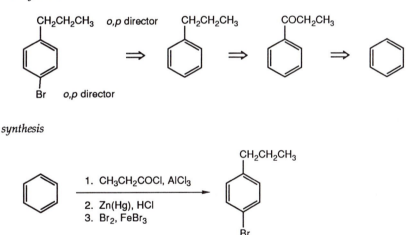

synthesis

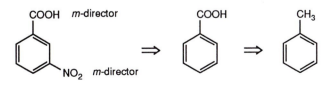

c.   The substituents are *meta* to each other, and both are *meta* directors. A nitro group is introduced readily by electrophilic substitution, and benzoic acid is prepared by oxidizing toluene.

retrosynthesis

12.48. (continued)

synthesis

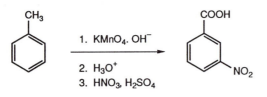

12.49. In electrophilic iodination of activated arenes, NIS is a source of electrophilic iodine. The mechanism follows the same steps as those shown for electrophilic bromination on text pages 524-525.

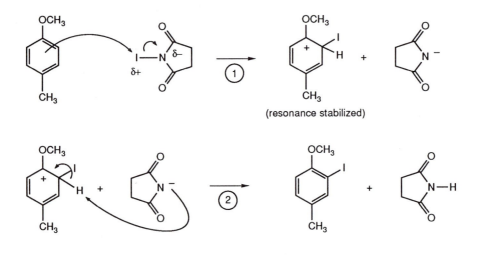

(resonance stabilized)

12.50. The first step is to summarize the reactions that are described in this exercise.

Next, consider what possible structures **X** may have. Based on the molecular formula, compound **X** has to be a bromotoluene derivative, of which there are four.

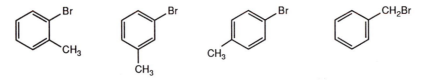

12.50. (continued)

Consider what compounds are made upon nitration of **X** —that is, what are **Y** and **Z**? Every isomer of C_7H_7Br yields two isomeric nitration products.

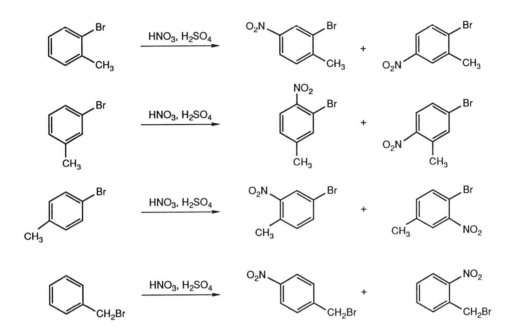

We know that reduction of a nitro to an amino group occurs without altering the substitution pattern of the arene, so **P** and **R** exist in the same isomeric forms as do **Y** and **Z**. Bromination of the aniline derivatives occurs readily. Remember that an amino group activates the ring substantially, so as many as three bromine atoms are substituted if positions *ortho* and *para* to the amine group are open. Only anilines derived from *m*-bromotoluene have a single vacant position *ortho* or *para* to the amine group. Every other aniline undergoes substitution by two bromine atoms, producing $C_7H_6Br_3N$.

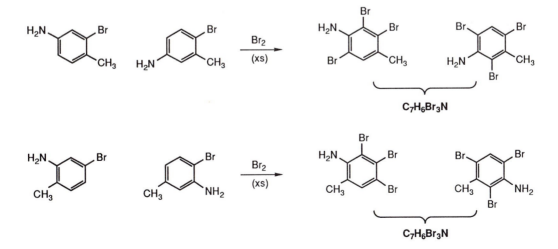

12.50.   (continued)

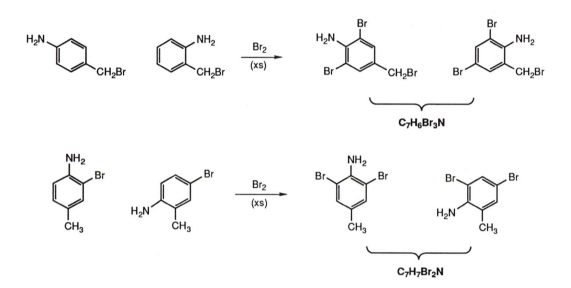

Conclusion: compound **X** is *m*-bromotoluene.

12.51.   When C_{60} undergoes addition, the addend reacts at a double bond between two six-membered rings. Therefore, the expected transformation occurs as follows:

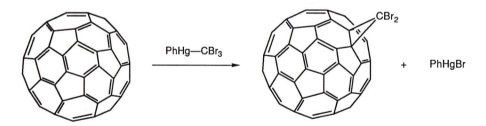

ORGANOMETALLIC CHEMISTRY, THE FORMATION OF C–C BONDS, AND CHEMICAL SYNTHESIS

13.1. The reaction between cyanide ion and a compound with a good leaving group produces the corresponding nitrile as a substitution product. The reaction normally follows an S_N2 pathway, leading to inversion of configuration if the leaving group is attached to a stereogenic carbon atom. An alcohol OH group is first converted to a good leaving group by preparation of its sulfonate ester derivative.

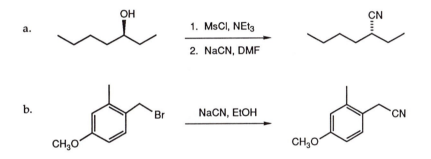

13.2. A Grignard reagent has the same general structure as the corresponding organohalide from which it is made, except that –MgX replaces –X.

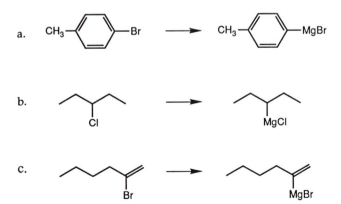

13.3. The Grignard reagent has a nucleophilic carbon atom; it reacts with a ketone as it does with any carbonyl-containing substance, forming a bond with the electrophilic carbon atom. Hydrolysis produces the alcohol.

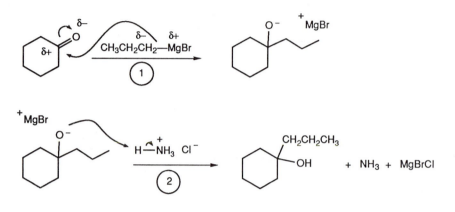

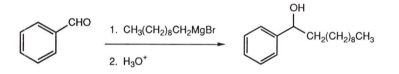

13.4. Grignard reagent reacts with an aldehyde to form the secondary alcohol, after workup.

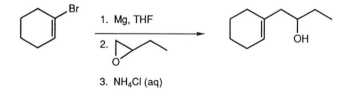

13.5. A Grignard reagent reacts with an epoxide at the less highly-substituted carbon atom of the three-membered ring, forming an alcohol, after workup. In this transformation, a 2° alcohol is the product.

13.6. Groups that interfere with Grignard formation include those with acidic protons and those with a double or triple bond between carbon and a heteroatom (carbonyl and nitrile groups)

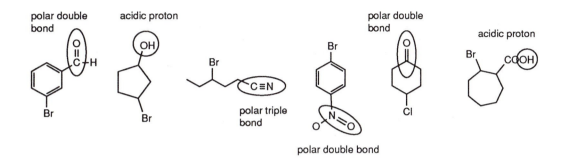

13.7. Groups that interfere with Grignard formation include those with acidic protons and those with a double or triple bond between carbon and a heteroatom (carbonyl and nitrile groups). Carbon-carbon double and triple bonds (except for a terminal alkyne group) do not interfere.

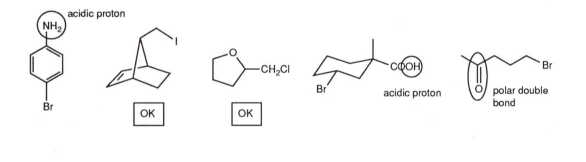

13.8. In the reaction sequence shown in this exercise, *p*-bromophenol first reacts with one equivalent of BuLi. This acid-base reaction forms the phenolate ion. The second equivalent of BuLi reacts via metal-halogen exchange. The resulting organolithium compound adds to a carbon-oxygen double bonds of carbon dioxide, forming a carboxylic acid after workup. This aqueous workup step also reprotonates the phenoxide ion, regenerating the neutral phenol.

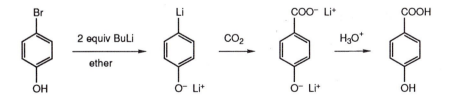

13.9. A terminal alkyne reacts with an alkyl Grignard reagent to generate an alkynyl Grignard reagent. Ethylene oxide undergoes ring-opening upon reaction with the alkynyl Grignard reagent, and that forms a 1° alcohol (after hydrolysis) that has two carbon atoms more than the starting alkyne.

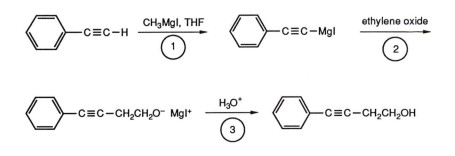

13.10. The organocuprate reagent transfers its alkyl group to replace the halogen atom of the substrate.

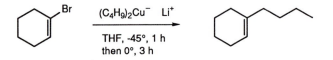

13.11. Higher-order organocuprates react with organohalides to replace a halogen atom with the organic group bonded to copper. A higher-order cuprate reacts with an epoxide by opening the three-membered ring, and the transferred group attaches to the less highly-substituted carbon atom of the original three-membered ring.

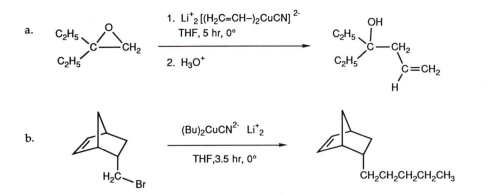

13.12. The oxidation state of the metal ion in a transition metal complex is the difference between the total charges of the ligands and the overall charge on the complex.

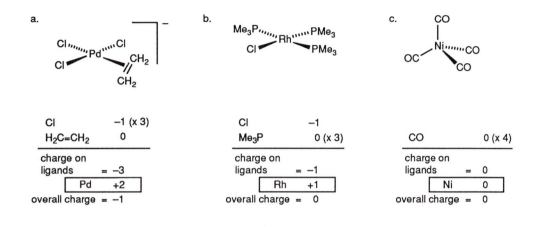

a.			b.			c.		
Cl	−1 (x 3)		Cl	−1		CO	0 (x 4)	
H₂C=CH₂	0		Me₃P	0 (x 3)				

a.

Cl	−1 (x 3)
$H_2C=CH_2$	0

charge on ligands	= −3
Pd	**+2**

overall charge = −1

b.

Cl	−1
Me_3P	0 (x 3)

charge on ligands	= −1
Rh	**+1**

overall charge = 0

c.

CO	0 (x 4)

charge on ligands	= 0
Ni	**0**

overall charge = 0

13.13. Structures **A** through **D** are deduced by applying the descriptions and examples in the accompanying discussion on text pages 599-601. Ethanol replaces one of the phosphine ligands (L) in the first step, producing **A**; then oxidative addition of molecular hydrogen occurs to give **B**. Cyclohexene replaces the ethanol molecule, forming **C**. That step is followed by migratory insertion and re-binding of ethanol to produce **D**. Reductive elimination of cyclohexane regenerates **A**.

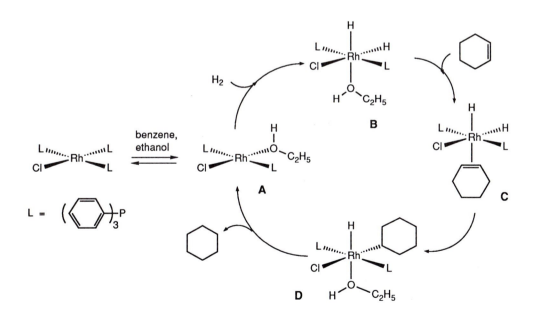

13.14. The configuration of each stereogenic center is specified in the usual way after assigning a priority to each group attached to stereogenic carbon atoms. Configurations in the isotactic polymer occur in two blocks, (R) at one end of the chain, (S) at the other. At the midway point of the chain, priorities of the two groups switch, so the configurations will be opposite one another at equivalent points of the chain.

Configurations in a syndiotactic polymer alternate, except at the very middle of the chain where the priorities switch, which results in two adjacent carbon atoms having the same configuration.

Configurations in an atactic polymer are random.

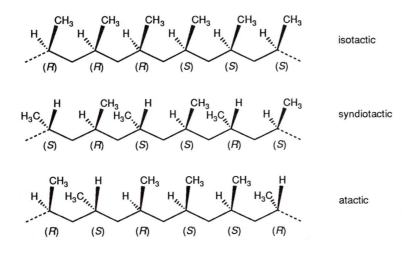

13.15. If an organocuprate reagent reacts by oxidative addition, a four-coordinate copper(III) ion forms that has an iodide ion and three organic groups attached. Reductive elimination creates the carbon-carbon bond, yielding the alkylcopper(I) compound and lithium iodide.

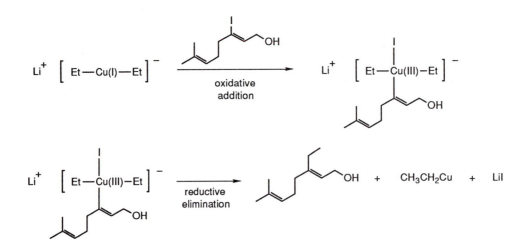

13.16. Structures **A** through **C** are deduced by applying the descriptions and examples in the accompanying discussion on text pages 604-606. After two equivalents of triphenylphosphine dissociate to generate Pd[Ph₃P]₂, oxidative addition of iodobenzene takes place to form **A**. The butyl group is transferred from *B*-butyl-9-BBN, producing **B**. After dissociation of the iodide ion, reductive elimination occurs to form the product, and bis(triphenylphosphine)-palladium(0) is regenerated.

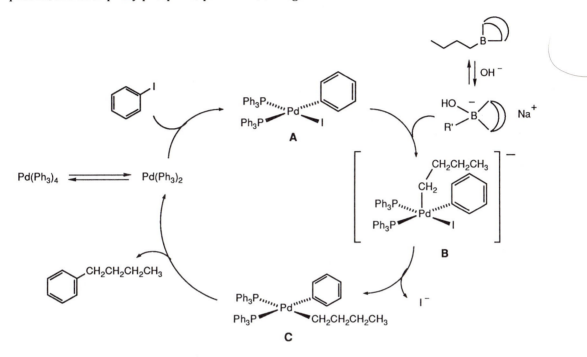

13.17. Catecholborane is prepared by reaction between borane and catechol, which liberates two molecules of dihydrogen.

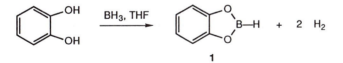

Catecholborane reacts with a terminal alkyne to form the corresponding vinylborane having an (*E*) geometry.

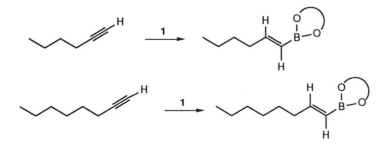

13.18. The equation showing the synthesis of 3-heptanol from ethyl bromide and pentanal is as follows:

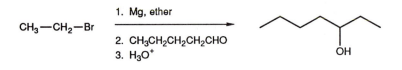

13.19. An alcohol is prepared by addition of a Grignard reagent to a suitable carbonyl compound (see Table 13.1, text page 583). To make a derivative of benzene, consider electrophilic substitution as a way to attach functional groups. There are several ways to synthesize the molecules shown in this exercise; only one method is presented below.

a.   The product is a 2° alcohol, so reaction between a Grignard reagent and an aldehyde is called for. Bromobenzene is made by electrophilic bromination of benzene.

retrosynthesis:

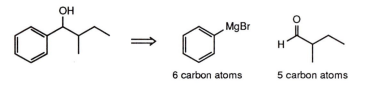

synthesis:

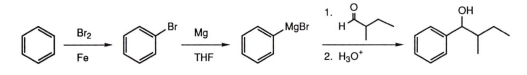

b.   The product is a 1° alcohol, so reaction between a Grignard reagent and either formaldehyde or ethylene oxide yields the desired product.

retrosynthesis:

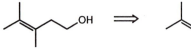

synthesis:

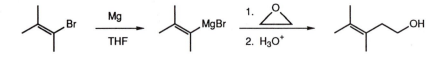

13.20. Retrosynthetic cleavage adjacent to the carbinol carbon atom is reasonable, and that disconnection requires use of an alkyl halide and aldehyde as reactants to make the 2° alcohol product.

retrosynthesis:

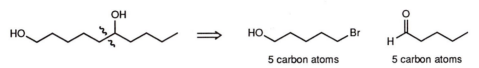

The bromoalkane that is needed has a functional group that is incompatible with formation of a Grignard reagent, so a protecting group is required. The last step in the synthesis is deprotection of the 1° alcohol group.

synthesis:

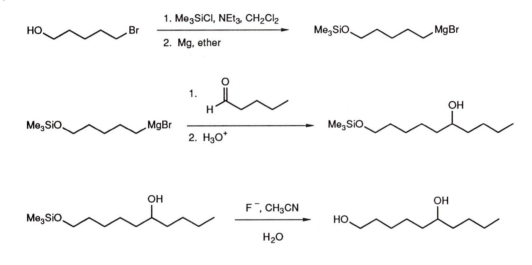

13.21. Interpret the name of the organic portion according to the procedures outlined in Chapter 2. The metal ion associated with the organic portion is included either as a substituent or as the compound root.

a. **(E)-1-Lithio-2-butene**

but	4 carbon atoms
ene	double bond at C2
(E)	*trans* double bond
1-lithio	lithium atom at C1

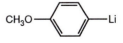

b. **p-Methoxyphenyllithium**

phenyllithium   benzene derivative with a lithium
                 atom attached to the ring

13.21. (continued)

c. **3-Methyl-2-methoxyphenylmagnesium bromide**
phenylmagnesium bromide   benzene derivative with a
bromomagnesium group
attached to the ring

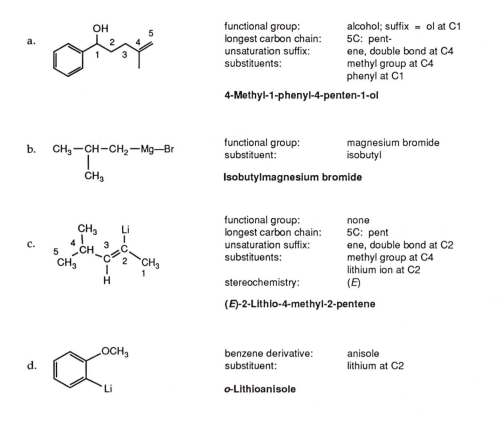

d. **Lithium di-*sec*-butylcuprate**
lithium di(• •• •)cuprate   a compound with the general formula
LiR_2Cu; in this case, R = *sec*-butyl

13.22. Follow the procedure outlined in the solution to exercise 7.28. For the organometallic compounds, compare the names and structures shown in the solution to exercise 13.21.

a.

functional group:	alcohol; suffix = ol at C1
longest carbon chain:	5C: pent-
unsaturation suffix:	ene, double bond at C4
substituents:	methyl group at C4
	phenyl at C1

4-Methyl-1-phenyl-4-penten-1-ol

b. $CH_3—CH—CH_2—Mg—Br$
|
CH_3

functional group:	magnesium bromide
substituent:	isobutyl

Isobutylmagnesium bromide

c.

functional group:	none
longest carbon chain:	5C: pent
unsaturation suffix:	ene, double bond at C2
substituents:	methyl group at C4
	lithium ion at C2
stereochemistry:	(*E*)

(*E*)-2-Lithio-4-methyl-2-pentene

d.

benzene derivative:	anisole
substituent:	lithium at C2

***o*-Lithioanisole**

13.23. To draw the structure of the Grignard reagent derived from a specific organohalide, simply insert "Mg" into the carbon–halogen bond. For parts (d) and (e), below, first draw each structure in its 3-D form to incorporate the stereochemical designation that is given.

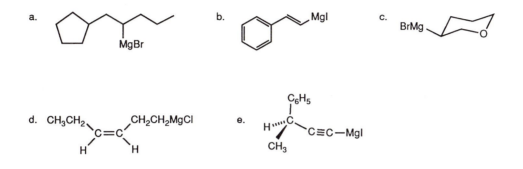

13.24. Groups that interfere with Grignard formation include those with acidic protons and those with carbonyl or nitrile groups. Alkyl fluorides do not normally react with magnesium metal to form Grignard reagents.

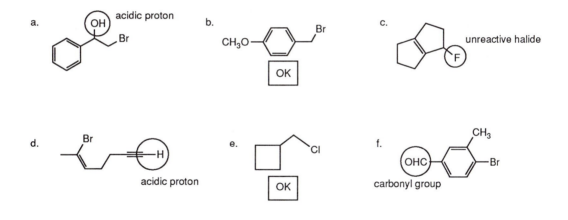

13.25. A Grignard reagent reacts with formaldehyde to form a primary alcohol, after hydrolysis. To decide which alkyl halide is needed, remove the CH$_2$OH group: the resulting organic fragment, bonded to Cl, Br, or I, is an appropriate starting material.

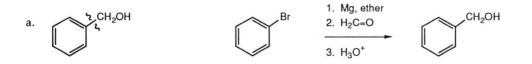

13.25.   (continued)

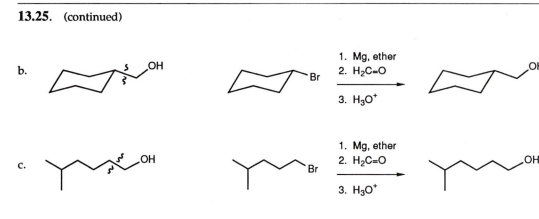

13.26.   A Grignard reagent reacts with any aldehyde besides formaldehyde to form a secondary alcohol, after hydrolysis. To decide which alkyl halide is needed, break the molecule at one of the carbon-carbon bonds adjacent to the OH group. The portion containing the OH group corresponds to the aldehyde needed. The other organic fragment, bonded to Cl, Br, or I, is the starting material from which the Grignard reagent is made. There are often two ways to disconnect the carbon skeleton; only one answer is given here.

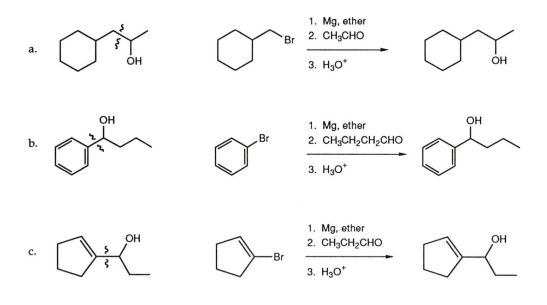

13.27.   A Grignard reagent reacts with a ketone to produce a tertiary alcohol, after hydrolysis. To decide which alkyl halide is needed, break the molecule at one of the carbon-carbon bonds adjacent to the OH group. The portion containing the OH group corresponds to the ketone needed. The other organic fragment, bonded to Cl, Br, or I, is the starting material from which the Grignard reagent is made. There are often three ways to disconnect the carbon skeleton; only one answer is given here.

13.27.   (continued)

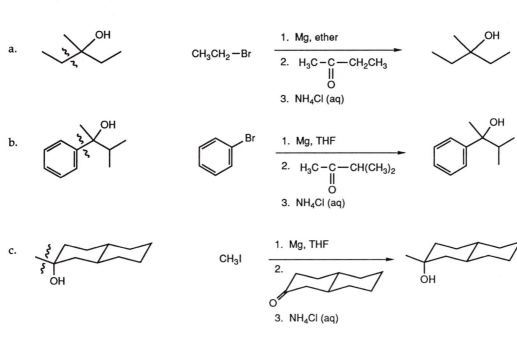

a.

b.

c.

13.28.   Use the identity of the reagents to deduce what the product of each reaction is. Oxidation of 2-hexanol, a secondary alcohol, produces 2-hexanone, **A** (text page 358).   Treating 2-hexanol with PBr₃ leads to replacement of the OH group by a bromine atom, so compound **B** is 2-bromohexane. Reaction of **B** with magnesium metal produces compound **C**, which must be a Grignard reagent. Ketones react with Grignard reagents to form 3° alcohols, after hydrolysis, so compound **D** is the 3° alcohol formed by addition of **C** to the carbonyl group of ketone **A**.

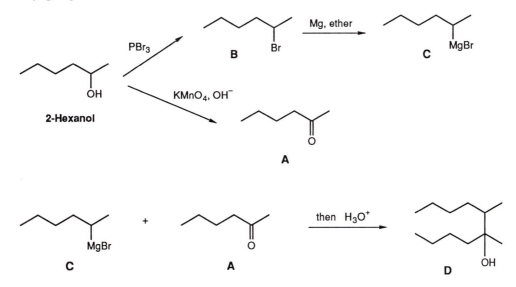

13.29.   Phenylmagnesium bromide, like any Grignard reagent, adds to the carbon-oxygen double bond of aldehydes and ketones. It reacts with substances that have an acidic proton by an acid-base reaction. The products of an acid-base reaction in this case are benzene and the conjugate base of the acidic reactant. If there are no reactive groups, then no reaction occurs.

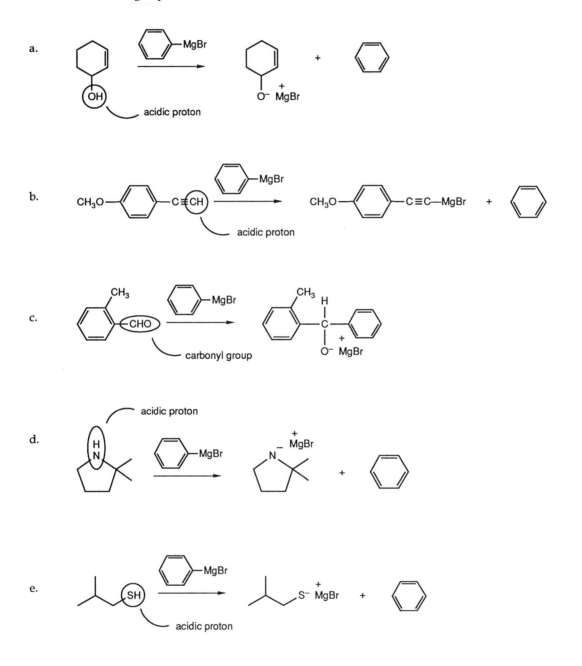

13.30. 1-Bromo-*cis*-2-butene is an allylic halide. Allylic halides form Grignard reagents, but like many allylic species, reaction occurs at two places. The partial negative charge associated with a metal-substituted carbon atom of a Grignard reagent is delocalized over the atoms of the π system, creating two nucleophilic centers.

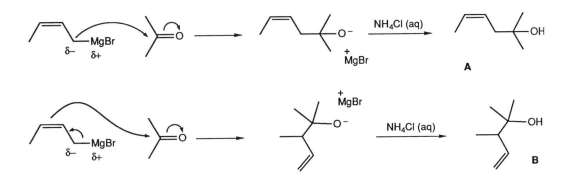

13.31. Protection of an alcohol OH group with the trimethylsilyl group involves stirring the alcohol with chlorotrimethylsilane and a non-nucleophilic base, commonly triethylamine. Deprotecting a trimethylsilyl ether involves stirring the silyl ether with fluoride ion.

Protection of an alcohol OH group as its benzyl ether requires formation of the alkoxide derivative, followed by alkylation with benzyl bromide. Deprotecting the benzyl ether is done by hydrogenolysis (text page 481).

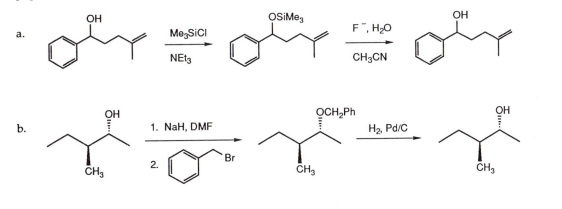

13.32.

a. Bromocyclohexane readily forms a Grignard reagent, which adds to the carbonyl group of acetaldehyde to form a secondary alcohol after acid workup.

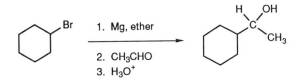

13.32.   (continued)

b. Phosphorus tribromide converts an alcohol to the corresponding bromoalkane, which in turn reacts with magnesium to form a Grignard reagent.   Formaldehyde reacts with a Grignard reagent to produce a 1° alcohol with one carbon more than the starting alkyl halide.

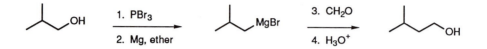

c.   Bromobenzene undergoes halogen-lithium exchange upon treatment with butyllithium.   The phenyllithium so generated reacts with copper(I) iodide to form lithium diphenylcuprate. Iodoethane reacts with this organometallic compound to form ethylbenzene.

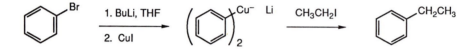

d. *m*-Bromotoluene reacts with magnesium to form the corresponding Grignard reagent, which adds to the carbonyl group of cyclohexanone, yielding the 3° alcohol after workup.

e. 1-Bromohexane reacts with magnesium to form the corresponding Grignard reagent. Grignard reagents do not usually react with alkyl halides, however, so no reaction occurs in step 2. Addition of water protonates the Grignard reagent, producing the hydrocarbon that is the structural analog of the original organohalide. The second alkyl halide is recovered intact.

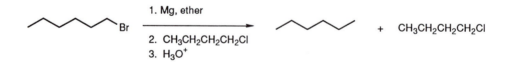

13.33. An alcohol is prepared by reaction between a Grignard reagent and a suitable carbonyl compound (see Table 13.1, text page 583). When a starting material is specified, then the retrosynthetic analysis should be done to generate compounds that are structurally similar to the indicated starting material.

a.   **2-Cyclohexyl-1-ethanol** from **bromocyclohexane.**   The product is a 1° alcohol that has two carbon atoms more than the given starting material. The reaction between ethylene oxide and a Grignard reagent permits such a transformation to be carried out.

retrosynthesis:

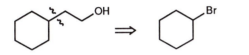

synthesis:

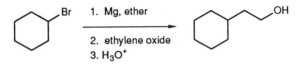

b.   **1-Hexanol** from **1-pentene.**   The product, a 1° alcohol, has one carbon atom more than the starting material, so reaction between a Grignard reaction and formaldehyde is appropriate. The bromide required differs in structure from the specified starting material, 1-pentene, by anti-Markovnikov addition of HBr to the latter.

retrosynthesis:

synthesis:

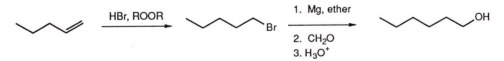

13.33. (continued)

c. **1-Cyclohexylethanol** from **cyclohexanol**. The product is a 2° alcohol with the same ring skeleton that appears in the starting material. An aldehyde corresponding to the acyclic portion of the product is required to react with a Grignard reagent derived from cyclohexanol.

retrosynthesis:

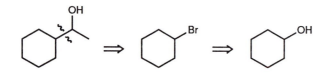

synthesis:

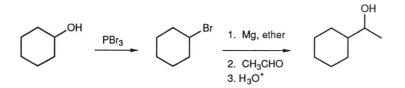

d. **3-Heptanol** from **1-iodobutane**. The product is a 2° alcohol that has three carbon atoms more than the given starting material. Reaction between propanal and the Grignard reagent generated from 1-iodobutane will produce the desired heptanol isomer.

retrosynthesis:

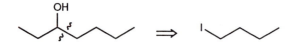

synthesis:

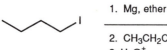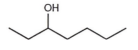

13.34. With a constraint on the number of carbon atoms that each starting material can have, retrosynthesis should break the molecule into pieces that have appropriate sizes. With use of organocuprate reagents, coupling two alkyl groups is often straightforward if one portion is derived from an organolithium compound and the other reactant is a vinyl, aryl, or 1° alkyl iodide. Any ring should be introduced as such.

a.   The retrosynthetic disconnection creates two fragments, one with five carbon atoms and one with six. The righthand portion is denoted as the one from which the organocuprate is derived because the organometallic reactant will then have the metal ion associated with a 2° carbon atom. The lefthand portion is a 1° alkyl iodide.

retrosynthesis:

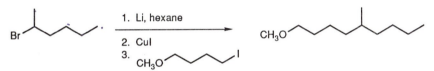

synthesis:

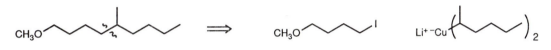

1. Li, hexane

2. CuI

3.

b.   The retrosynthetic disconnection is at the junction between a ring and the attached substituent. Either portion can have the cuprate group, and either can have the iodo substituent.

retrosynthesis:

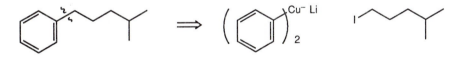

synthesis:

13.34. (continued)

c. The retrosynthetic disconnection is at the junction between a ring and the attached substituent. Either portion can have the cuprate group, and either can have the iodo substituent.

retrosynthesis:

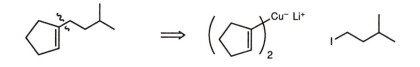

synthesis:

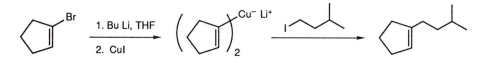

13.35. This transformation, which is like the reaction that takes place between a Grignard reagent and epoxides, provides a way to extend a carbon chain by three carbon atoms. Ring opening occurs because the four-membered ring has significant strain energy. Furthermore, the magnesium ion associates with the oxygen atom of the oxetane ring to form a good leaving group that facilitates ring-opening.

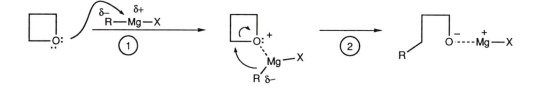

13.36. At the beginning of the reaction, it is likely that a molecule of chlorotrimethylsilane will react with a molecule of 1,4-cyclohexandiol, so the desired monosubstituted product is formed initially. As the concentration of monosubstituted product increases, competition exists between the monosilyl product and the diol for chlorotrimethylsilane. At that point, bis(silylated) product starts forming. In reactions like this, the amount of monosubstituted product is proportional to the number of equivalent functional groups that react with the reagent. Therefore, 50% is the highest yield that is statistically obtainable. Because of the stoichiometry, the other 50% of the mass must consist equally of unreacted starting material and bis(silylated) product.

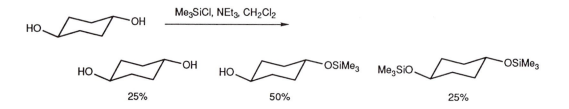

13.37. The stereochemistry of carbon atoms in the three polymer types was described in the solution to exercise 13.14. In a very long chain, the overall effect of chiral centers is negligible because an approximately equal number of each configuration exists.

In isotactic polypropylene, the methyl groups are all on the same side of the chain, but the configuration at any carbon atom is opposite that for the carbon atom at the corresponding position at the other end of the chain. Carbon atoms toward the middle are chiral (four different groups), but their effect on rotation of plane polarized light is likely minimal: for instance in $C_{1000}–CH(CH_3)—C_{1001}$, the two chains are close enough in length as to be essentially the same.

It is possible that atactic polypropylene will have more of one configuration than the other. Statistically, however, the difference approaches zero with increasing chain length.

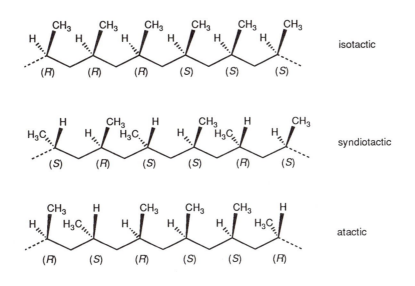

13.38. The oxidation state of the metal ion is the difference between the sum of charges of the ligands and the overall charge on the complex. For these compounds, the overall charge in each case is zero, so the oxidation state is equal in magnitude (but differs in sign) to that on the ligands.

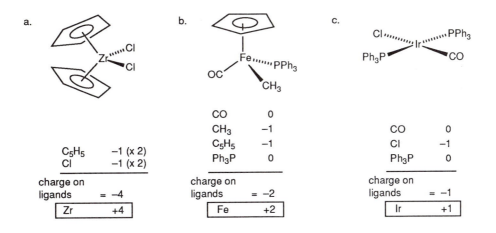

13.39. The oxidation state of the metal ion is calculated as described in the solution to exercise 13.38. If the oxidation state and coordination number stay the same, and two of the ligands react with each other, migratory insertion has occurred. If the oxidation state has become more positive, and the coordination number increases, oxidative addition has taken place. If the oxidation state has become more negative, and the coordination number decreases, reductive elimination has occurred.

a. This is migratory insertion. The oxidation state and coordination number stay the same during the transformation, and the methyl group reacts with one of the CO ligands.

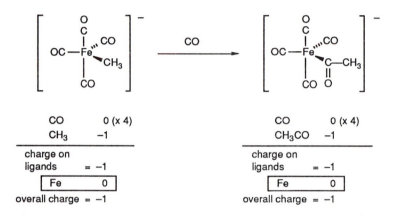

CO	0 (x 4)
CH₃	−1

charge on ligands   = −1

Fe	0

overall charge = −1

CO	0 (x 4)
CH₃CO	−1

charge on ligands   = −1

Fe	0

overall charge = −1

b. This is an example of oxidative addition. The oxidation state becomes more positive, and the coordination number increases.

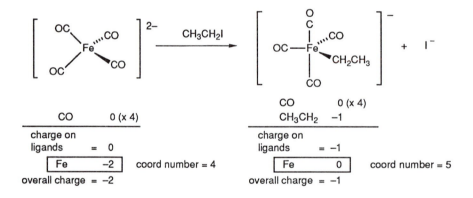

CO	0 (x 4)

charge on ligands   = 0

Fe	−2

coord number = 4

overall charge = −2

CO	0 (x 4)
CH₃CH₂	−1

charge on ligands   = −1

Fe	0

coord number = 5

overall charge = −1

13.39.   (continued)

c. This is an example of oxidative addition. The oxidation state becomes more positive, and the coordination number increases.

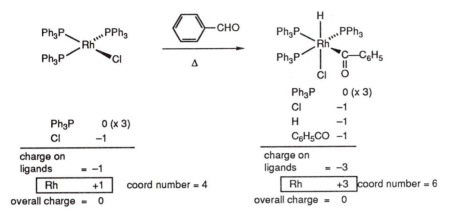

Ph₃P	0 (x 3)
Cl	−1

charge on
ligands   = −1

Rh	+1

coord number = 4

overall charge =   0

Ph₃P	0 (x 3)
Cl	−1
H	−1
C₆H₅CO	−1

charge on
ligands   = −3

Rh	+3

coord number = 6

overall charge =   0

d. This is an example of reductive elimination. The oxidation state becomes more negative, and the coordination number decreases.

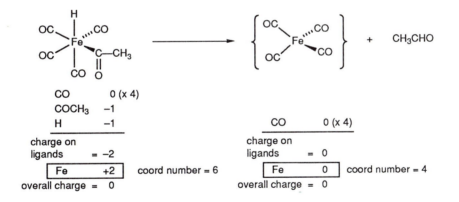

CO	0 (x 4)
COCH₃	−1
H	−1

charge on
ligands   = −2

Fe	+2

coord number = 6

overall charge =   0

CO	0 (x 4)

charge on
ligands   = 0

Fe	0

coord number = 4

overall charge =   0

13.40.   The mechanism of the Stille coupling is the same as that for the Suzuki reaction in many steps. The first step generates bis(triphenylphosphine)palladium(0) by dissociation of ligands.

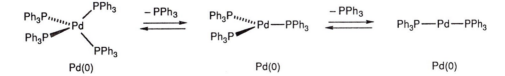

13.40. (continued)

Next, oxidative addition of iodobenzene occurs.

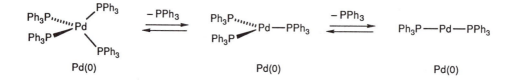

Transmetallation between the organopalladium complex and the organotin compound generates the intermediate that has two organic groups bonded to palladium.

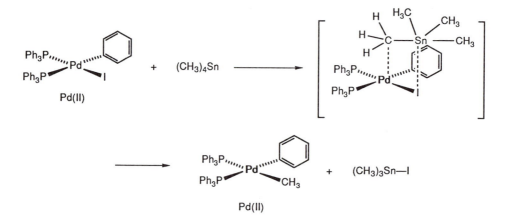

Finally, reductive elimination yields toluene and regenerates bis(triphenylphosphine)palladium(0).

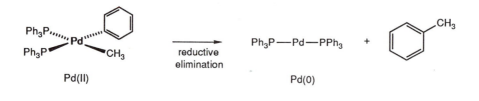

13.41. To assess the atom economy of a pair of reactions, write a complete equation for each (if possible) and consider what byproducts are formed in each. Addition reactions normally have the best atom economy, especially if all of the atoms of a reactant add to the organic substrate. For the transformations in each part of this exercise, the reaction in the box has the better atom economy of each pair being considered.

13.41. (continued)

a. Oxymercuration is an inherently poor procedure because mercury is formed as a product. Disposal of this highly toxic material is costly. Markovnikov addition of water is highly efficient.

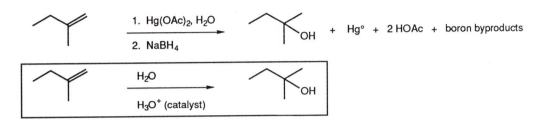

b. Hydroboration is a fairly benign process, but it does produce boric acid as the inorganic product.

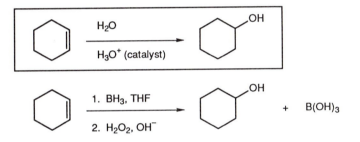

c. The fact that only a third of an equivalent of PBr₃ is needed makes this process more efficient.

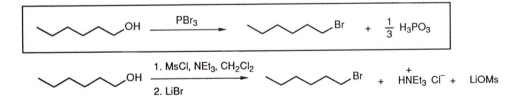

d. Catalytic hydrogenation is an excellent procedure because the catalyst is recovered and recycled.

$$CH_3CH_2CH_2-C\equiv C-H \xrightarrow[\text{2 } t\text{-BuOH}]{\text{2 Na, liquid NH}_3} CH_3CH_2CH_2-CH=CH_2 + 2 \text{ NaO-}t\text{-Bu}$$

$$CH_3CH_2CH_2-C\equiv C-H \xrightarrow[\text{Pd catalyst}]{\text{H}_2\text{, poisoned}} CH_3CH_2CH_2-CH=CH_2$$

13.42. The alcohol functional group defines a place at which to make a logical disconnection in a retro-synthesis. If an OH group is not present, consider how the functional group that does exist can be converted to an alcohol—for instance, carbonyl compounds are reduced to alcohols, and alkenes are hydrated to from alcohols.

If the synthetic goal has a ring as part of the carbon skeleton, consider ways to include the ring as an entity. For instance, benzene derivatives are often made conveniently by electrophilic substitution reactions of suitable precursors. The Diels-Alder reaction is a useful way to make six-membered rings from acyclic starting materials. Cyclohexene and cyclohexadiene derivatives are accessible by Birch reduction of an arene.

There are many ways to synthesize the molecules shown below. Only one route is presented.

a. *Retrosynthesis:*

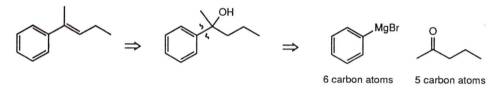

6 carbon atoms   5 carbon atoms

Synthesis: Phenylmagnesium bromide adds to the carbonyl group to 2-pentanone to form the 3° alcohol. Elimination under acidic conditions gives the thermodynamic product, which is the desired (E)-alkene.

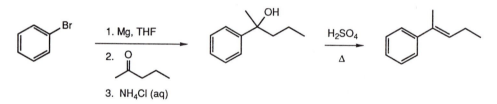

b. *Retrosynthesis:*

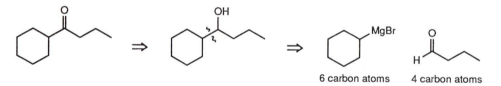

6 carbon atoms   4 carbon atoms

Synthesis: Cyclohexylmagnesium bromide reacts with butanal to form the 2° alcohol. Oxidation to the ketone is carried out with use of a variety of oxidizing agents (see text pages 355-358).

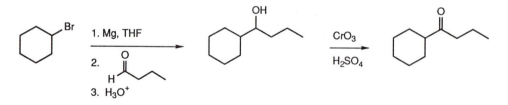

13.42. (continued)

c. *Retrosynthesis:*

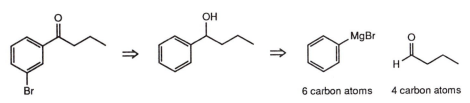

Synthesis: Phenylmagnesium bromide reacts with butanal to form the 2° alcohol. Oxidation to the ketone is carried out with any of a variety of oxidizing agents. The acyl group is a *meta*-director, so electrophilic bromination affords the desired product.

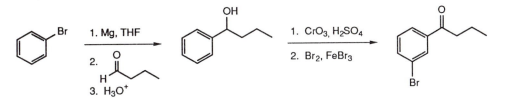

d. *Retrosynthesis:*

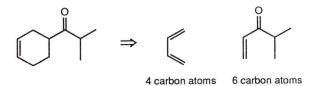

Synthesis: The Diels-Alder reaction is a good way to make a six-membered ring (text page 431). The dienophile must have an electron-withdrawing group attached, which it does here.

13.42. (continued)

e. *Retrosynthesis:*

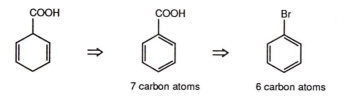

7 carbon atoms   6 carbon atoms

Synthesis: Phenylmagnesium bromide reacts with carbon dioxide to form benzoic acid. Birch reduction converts the benzene ring to the cyclohexadiene derivative (text page 555).

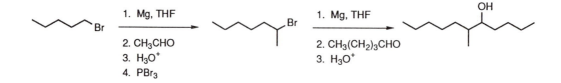

f. *Retrosynthesis:*

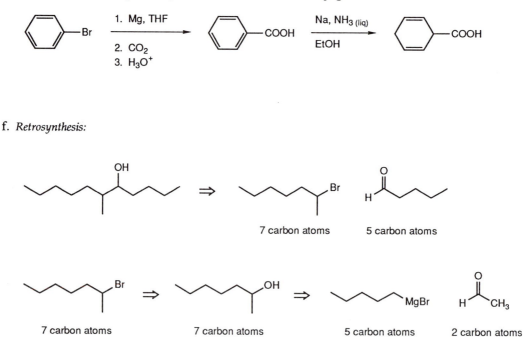

Synthesis: Pentylmagnesium bromide reacts with acetaldehyde to form 2-heptanol, which can be converted to 2-bromoheptane and hence to the corresponding Grignard reagent. Reaction between 2-heptylmagnesium bromide and pentanal yields the desired product.

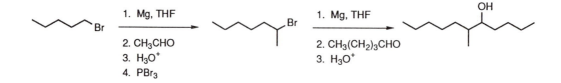

13.42. (continued)

g. *Retrosynthesis:*

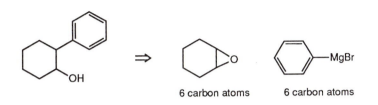

6 carbon atoms   6 carbon atoms

Synthesis: An epoxide reacts readily with a Grignard reagent to form an alcohol.

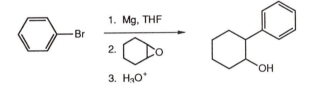

1. Mg, THF

2.

3. H₃O⁺

SPECTROSCOPIC IDENTIFICATION
OF FUNCTIONAL GROUPS

14.1. To assign isotopic distributions of the elements in a compound, first specify its composition. For 2-butanone, the formula is C_4H_8O. Consider next what common isotopes exist for each element: for carbon, the possible isotopes are ^{12}C and ^{13}C; for hydrogen, they are ^1H and ^2H; and for oxygen, they are ^{16}O, ^{17}O, and ^{18}O. Finally, calculate the ratios of the isotopes that combine to give the observed mass. For C_4H_8O, the possible formulas having MW = 74 ($M + 2$) are:

$(^{12}C)_2(^{13}C)_2(^1H)_8{}^{16}O$

$(^{12}C)_3(^{13}C)(^1H)_7(^2H)^{16}O$
$(^{12}C)_3(^{13}C)(^1H)_8{}^{17}O$

$(^{12}C)_4(^1H)_6(^2H)_2{}^{16}O$
$(^{12}C)_4(^1H)_7(^2H)^{17}O$
$(^{12}C)_4(^1H)_8{}^{18}O$

14.2. The molecular ion is identified as the peak at the highest m/z values that fits the pattern(s) illustrated in Figure 14.4 (text page 639). The molecular ion peak for this compound appears at $m/z = 134$. The base peak is the one with a relative intensity of 100%.; for this compound, the base peak has $m/z = 91$. The mass of a fragmentation peak is calculated by subtracting the mass of the fragment lost from the mass of the molecular ion. In this case, the fragment formed by loss of a molecule of water appears at $134 - 18 = m/z$ 116.

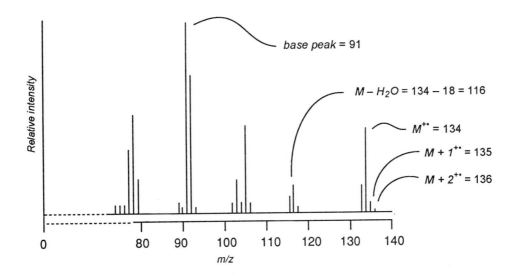

14.3. Apply the data in Figure 14.5 (text page 642) to the observed peak patterns to deduce the presence of the heteroatoms Br, Cl, or S. A noneven value for the molecular weight indicates that an odd number of nitrogen atoms are present.

a. In this spectrum, $M = 76$, so if nitrogen atoms are present, there are an even number. The fact that the peaks at $M + 1$ and $M + 2$ are approximately the same size indicates that a sulfur atom is present.

b. In this spectrum, $M = 170$, so if nitrogen atoms are present, there are an even number. The fact that the peaks at M and $M + 2$ are approximately the same size indicates that a bromine atom is present.

14.4. To assess what type of compound gives rise to the spectra shown in this exercise, start by considering which heteroatoms are present—apply the data in Figure 14.5 (text page 642) to the observed peak patterns. A noneven value for the molecular weight indicates that an odd number of nitrogen atoms are present. Next, calculate m/z values that correspond to the difference between the value of $M^{+\bullet}$ and those of some strong lines in the spectrum.

a. The molecular mass has a noneven value, so this molecule contains an odd number of nitrogen atoms. The difference between m/z 137 and m/z 91 is 46, consistent with loss of either a nitro group or a CH_3S group. No evidence exists that sulfur is present, but nitrogen is. This compound is probably a nitro compound.

b. The molecular mass is even, and there appears to be no evidence for the presence of S or Cl (the MW is too low for Br to be present.) The difference between m/z 72 and m/z 43 is 29, which correlates with loss of a CHO group, so this compound may be an aldehyde. The difference between m/z 72 and m/z 29 is 43, which correlates with loss of a CH_3CO group, so this compound may also be a methyl ketone.

14.5. The peak at 1715 cm^{-1} is clearly the absorption for the C=O stretching vibration of the carbonyl group. Applying the criteria listed in Table 14.5 (text page 656), we rule out the presence of all but the ketone functional group.

Functional group	*base value (cm^{-1})*	*characteristic*
Acid chloride	1800	observed absorption (1715 cm^{-1}) is too low in frequency
Aldehyde	1725	no C–H stretch near 2720 cm^{-1}
Amide	1650-1680	observed absorption (1715 cm^{-1}) is too high in frequency
Carboxylic acid	1710	no broad, strong O—H stretch between 3500 and 2500 cm^{-1}
Ester	1735	no strong, broad band around 1200 cm^{-1}
Ketone	1715	observed absorption is in the correct region

14.6. No strong band appears between 1700 and 1800 cm^{-1}, so the molecule does not contain a carbonyl group. In the region between 2500 and 2000 cm^{-1}, we observe a medium intensity band at about 2250 cm^{-1}. This compound is most likely a nitrile. We rule out the possibility that the compound is a terminal alkyne because no strong band for the C—H stretching vibration is present at 3350 cm^{-1}.

14.7. No strong band appears between 1700 and 1800 cm^{-1}, so the molecule does not contain a carbonyl group. There is also no absorption in the region between 2500 and 2000 cm^{-1}. A strong , broad band is present at 3300 cm^{-1}, so the compound likely has an OH group. Looking at the region around 1100 cm^{-1}, we observe a band at 1150 cm^{-1}, which correlates with the presence of a 1° alcohol functional group. [The bands around 1600 cm^{-1} and between 700 and 900 cm^{-1} indicate that the compound contains a benzene ring—this aspect of the interpretation process is discussed in the text starting on page 667.]

14.8. To predict what peaks will be observed in the infrared spectrum of a given substance, ask yourself what functional groups are present and what corresponding absorption bands each is expected to have.

a.   **3,3-Dimethyl-2-pentanone:** This compound is an aliphatic ketone, so we expect to see a very strong absorption at about 1715 cm^{-1} for the C=O stretching vibration. There are no other diagnostic bands for this functional group.

b.   **4-Methylbenzoic acid:** This compound has a *para*-disubstituted benzene ring, for which we would expect to see bands at about 1600 cm^{-1} and 820 cm^{-1}. The carboxylic acid functional group produces a very strong absorption at about 1690 cm^{-1} for the C=O stretching vibration (aryl-COOH), and a broad, strong band between 3500 and 2500 cm^{-1} for the hydrogen-bonded OH stretching vibration.

c.   **3-(1-Cyano-2-methyl-1-propyl)-1-cyclohexene:** The IR spectrum of this compound will have a medium intensity band at about 2250 cm^{-1} for the CN triple bond stretching vibration. The *cis* double bond in the six-membered ring will produce a medium intensity band at about 1630 cm^{-1} for the C=C stretching vibration and a strong band at 705 cm^{-1} for the out-of-plane bending vibration.

d.   **5-Mercapto-2-methyl-1-pentene:** The IR spectrum of this compound will have a medium intensity band at about 2550 cm^{-1} for the S–H stretching vibration. The terminal alkene will give rise to a medium intensity band at about 1630 cm^{-1} for the C=C stretching vibration as well as two intense bands (one just above 900 cm^{-1} and one just below 1000 cm^{-1}) for out-of-plane bending vibrations.

14.9. No strong band appears between 1700 and 1800 cm^{-1}, so the molecule does not contain a carbonyl group. There is also no absorption in the region between 2500 and 2000 cm^{-1}, nor around 3400 cm^{-1}. A medium intensity band at about 1630 cm^{-1} suggests that the compound contains a double bond or a benzene ring. Looking at the out-of-plane bending region, we see a strong band at 705 cm^{-1}. We conclude that the molecule has a *cis* carbon-carbon double bond.

The absorption at 705 cm^{-1} might also be assigned to a C—Cl stretching vibration. Furthermore, the strong absorption at 960 cm^{-1} may be attributable to the presence of a *trans* carbon-carbon double bond. Remember, absorption bands in the fingerprint region cannot always be assigned unambiguously, so additional data are needed to differentiate these possibilities.

14.10. Differentiating the structures of compounds requires you to recognize which features are dissimilar. If their functional groups differ, then IR spectroscopy is probably best suited. If different heteroatoms are present in each compound, then mass spectrometry constitutes the best method. If the hydrocarbon portion of the molecule varies, then NMR spectroscopy is the best technique.

a.   These compounds have the same overall structures but contain different heteroatoms (Cl and Br). Mass spectra reveal their different molecular weights as well as isotope patterns of ^{35}Cl and ^{37}Cl vs. ^{79}Br and ^{81}Br.

b.   One compound of the given pair is a ketone and the other has alcohol and alkene functional groups. IR spectroscopy is a good way to differentiate these functional groups.

c.   The given compounds are structural isomers, so proton NMR spectra reveal the presence of ethyl groups in one compound versus the methyl and isopropyl groups in the other.

d.   These two compounds have different functional groups, so IR spectra reveal the presence of the nitrile functional group in one and the amine and alkene functional groups in the other.

14.11. Follow the procedure outlined in the solution to exercise 14.1. 3-Pentanone has the composition $C_5H_{10}O$. Possible formulas for $C_5H_{10}O$ that have MW = 87 are:

$(^{12}C)_4(^{13}C)(^1H)_{10}{}^{16}O$

$(^{12}C)_5(^1H)_9(^2H)^{16}O$

$(^{12}C)_5(^1H)_{10}{}^{17}O$

14.12. Follow the procedure outlined in the solution to exercise 14.1. 1-Propanethiol has the composition C_3H_8S. For sulfur, the isotopes are ^{32}S, ^{33}S, ^{34}S, and ^{36}S. Possible formulas for C_3H_8S that have MW = 78 are:

$(^{12}C)_1(^{13}C)_2(^1H)_8{}^{32}S$

$(^{12}C)_2(^{13}C)(^1H)_7(^2H)^{32}S$
$(^{12}C)_2(^{13}C)(^1H)_8{}^{33}S$

$(^{12}C)_3(^1H)_6(^2H)_2{}^{32}S$
$(^{12}C)_3(^1H)_7(^2H)^{33}S$
$(^{12}C)_3(^1H)_8{}^{34}S$

14.13. When more than one heteroatom is present in a molecule, then the molecular ion region has peaks at mass values corresponding to every combination of isotope peaks for each heteroatom. The intensity of each peak is the *product* of the relative intensity of each isotopic contributor. Remember that Cl has two peaks in a ratio of 3:1 with masses of 35 and 37, respectively, and Br has two peaks in a ratio of 1:1 with masses of 79 and 81, respectively.

a. When two chlorine atoms are present, the compositions of formulas (ignoring carbon and hydrogen isotopes) are $C_nH_{2n}(^{35}Cl)_2$, $C_nH_{2n}(^{35}Cl)(^{37}Cl)$, $C_nH_{2n}(^{37}Cl)(^{35}Cl)$, and $C_nH_{2n}(^{37}Cl)_2$, producing peaks at (35 + 35), (35 + 37) & (37 + 35) and (37 + 37) above the mass of the hydrocarbon portion.

These peaks appear in the ratio of (3 x 3): [(3 x 1) + (1 x 3)]: (1 x 1) = 9: 6: 1.

b. When two bromine atoms are present, the compositions of formulas (ignoring carbon and hydrogen isotopes) are $C_nH_{2n}(^{79}Br)_2$, $C_nH_{2n}(^{79}Br)(^{81}Br)$, $C_nH_{2n}(^{81}Br)(^{79}Br)$, and $C_nH_{2n}(^{81}Br)_2$, producing peaks at (79 + 79), (79 + 81) & (81 + 79), and (81 + 81) above the mass of the hydrocarbon portion.

These peaks will appear in the ratio of (1 x 1): [(1 x 1) + (1 x 1)]: (1 x 1) = 1: 2: 1.

c. When a chlorine atom and a bromine atom are present, the compositions of formulas (ignoring carbon and hydrogen isotopes) are $C_nH_{2n}(^{35}Cl)(^{79}Br)$, $C_nH_{2n}(^{35}Cl)(^{81}Br)$, $C_nH_{2n}(^{37}Cl)(^{79}Br)$, and $C_nH_{2n}(^{37}Cl)(^{81}Br)$, producing peaks at (35 + 79), (35 + 81) & (37 + 79), and (37 + 81) above the mass of the hydrocarbon portion.

These peaks will appear in the ratio of (3 x 1): [(3 x 1) + (1 x 1)]: (1 x 1) = 3: 4: 1.

14.14. Mass spectrometry is used to differentiate compounds on the basis of molecular weights, the presence of heteroatoms that have specific mass and isotope patterns (N, S, Cl, and Br, especially), and the masses of fragments that are produced after ionization.

a. These compounds have the same mass, and none of the "special" heteroatoms, so different fragmentation patterns have to be observable. An aldehyde often has a strong peak at $M - 1$, from loss of the aldehyde hydrogen atom, so a peak at $m/z = 71$ would provide good evidence that the sample is an aldehyde.

$$CH_3CH_2-\overset{\overset{\displaystyle O}{\|}}{C}-CH_3 \qquad \text{and} \qquad CH_3CH_2CH_2-\overset{\overset{\displaystyle O}{\|}}{C}-H$$

MW = 72   MW = 72

b. These compounds are *cis, trans* isomers, so they will be very difficult to differentiate using mass spectrometry.

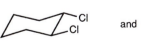 and

MW = 152   MW = 152

14.14. (continued)

c.   These compounds have different masses, and they contain halogen atoms that give rise to different peak patterns at M and $M + 2$.   The alkyl chloride will display a spectrum with peaks at $m/z = 120$ and 122 in a ratio of 3:1, whereas the spectrum of the alkyl bromide will have approximately equal size peaks at $m/z = 122$ and 124.

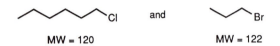

MW = 120   and   MW = 122

d.   These compounds have different masses, and they contain different heteroatoms.   The alcohol will display a spectrum that has a molecular ion peak at $m/z = 122$, whereas the spectrum of the thiol will have a molecular ion peak at $m/z = 124$.   The peak at $m/z = 126$ in the spectrum of the thiol is about the same size as the peak at $m/z = 125$.

MW = 122   and   MW = 124

14.15.   Identify the heteroatoms that are present by looking at the pattern of peaks in the molecular ion region, as illustrated in Figure 14.5 on text page 642.   Determine the MW of the compound (and the presence of nitrogen) by identifying the m/z value of the molecular ion peak using the data in Figure 14.4 (text page 639) and Figure 14.5 (text page 642).   The mass of the species that gives rise to the base peak ion is equal to the m/z value of the base peak.

Compound 15A: no heteroatoms are present (except O, perhaps) because the molecular mass is even ($m/z = 92$) and the sizes of the peaks at $m/z = 92$, 93, 94 decrease in that order (in fact, the peak at $m/z = 94$ is not observed).   You can rule out the possibility that $m/z = 91$ is $M^{+\bullet}$ because the ratio of peaks at $m/z = 91$ and 92 does not correspond to the normal patterns illustrated in Figures 14.4 and 14.5.   The base peak is $m/z = 91$.

Compound 15B: the molecular mass is 76.   The presence of sulfur is indicated by the fact that the peak at $m/z = 78$ ($M + 2$) is about the same size as the peak at $m/z = 77$ ($M + 1$).   The base peak is $m/z = 43$.

Compound 15C: the molecular mass is 170.   The presence of bromine is indicated by the fact that the peak at $m/z = 172$ ($M + 2$) is about the same size as the one at $m/z = 170$ (M).   The base peak is $m/z = 91$.

14.16.   Compound **15B** contains sulfur and its molecular weight is 76, which corresponds to the formula C_3H_8S.   The compound has no sites of unsaturation (text page 370), so the sulfur-containing functional group must be either a thiol or a sulfide.   In the IR spectrum, the thiol group is identified by an absorption band at about 2560 cm^{-1}, which is the S—H stretching vibration.   A sulfide functional group produces no readily identifiable bands in the IR spectrum.

14.17. Follow the procedure outlined in the solution to exercise 14.8 to predict what absorption bands can be expected for each substance, then identify those portions that will be different.

a.   Both ketones and aldehydes have a carbonyl group, but the absorption for the C=O stretching vibration of the aldehyde appears at a slightly higher frequency (1725 cm^{-1} for the aldehyde, 1715 cm^{-1} for the ketone.). The CHO group also has a medium-intensity band at about 2720 cm^{-1}, which is the C—H stretching vibration.

b.   Both ketones and carboxylic acids have a carbonyl group, and the absorption for the C=O stretching vibration appears in about the same place. The COOH group, however, has a broad, strong band that spans the region from about 3500 cm^{-1} to 2500 cm^{-1}, which is the hydrogen-bonded O—H stretching vibration.

c.   Both primary and secondary amines have an amino group, so each compound produces an absorption at about 3300 cm^{-1}, which is the N–H stretching vibration. The pattern of peaks observed for a 1° amine is a weak intensity doublet in the region around 3300 cm^{-1}. A secondary amine has a single weak band in this region (see Figure 14.24 on text page 666).

14.18. Data used to differentiate between types of carbonyl-containing compounds are summarized in Table 14.4 on text page 654.

Compound 18A: The absorption for the C=O stretching frequency at 1750 cm^{-1} has a value higher than that expected for a saturated ketone, so carboxylic acid, ester, and aldehyde functional groups are possibilities. The observation of a very strong band at 1200 cm^{-1} is consistent with the presence of the ester functional group. A carboxylic acid would display a strong, broad band between 3500 and 2500 cm^{-1}, and an aldehyde would have a medium intensity band at about 2720 cm^{-1}.

Compound 18B: The absorption for the C=O stretching frequency is 1710 cm^{-1}, so carboxylic acid, ketone, and aldehyde functional groups are possibilities. The strong, broad band between 3500 and 2500 cm^{-1} is evidence for the presence of a carboxylic acid functional group.

14.19. Compounds that are readily differentiated by infrared spectroscopy normally differ by what functional group is present. If the functional groups in two molecules are the same, significant variation appears only in the fingerprint region of the compounds' spectra.

a. Each compound is a tertiary amide, so differentiating them requires you to distinguish methyl from ethyl groups, which is difficult to do with use of IR spectroscopy.

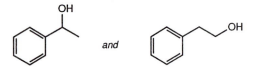

b. One compound is a 2° alcohol, and the other is a primary alcohol. These differ in the position of absorptions produced by the C—O stretching vibration.

c. Each compound is a terminal alkene, so differentiating them requires you to distinguish a methyl from an ethyl group, which is difficult to do with use of IR spectroscopy.

$$CH_3CH_2\overset{\overset{\displaystyle CH_2}{\|}}{C}-CH_2CH_3 \quad and \quad CH_3CH_2\overset{\overset{\displaystyle CH_2}{\|}}{C}-CH_3$$

14.20. Follow the procedure outlined in the solution to exercise 14.17.

a. The C=C stretching vibration for a terminal alkene appears at a lower frequency (1650 cm^{-1}, weak-medium, sharp) than that for the C≡C stretching vibration of a terminal alkyne (2130 cm^{-1}, medium sharp.) In addition, the alkene will have two strong bands, one just above 900 cm^{-1} and the other just below 1000 cm^{-1}, for out-of-plane bending vibrations. The alkyne will have a strong band at about 3300 cm^{-1} for the C—H stretching vibration.

14.20.   (continued)

b.   The nitro group produces two strong bands, one between 1600 and 1500 cm^{-1} and the other between 1390 and 1300 cm^{-1}. The amino group displays two weak–medium bands at about 3400 cm^{-1}.

and

c.   An acid chloride displays a strong band at about 1800 cm^{-1} for the C=O stretching vibration. The alkyl chloride may have a strong band between 600 and 800 cm^{-1}.

and

14.21.   To match structures with spectra, consider what absorptions will be observed for each functional groups listed, then look at each spectrum to see if those features are present.

For the functional groups listed, diagnostic bands are as follows: w = weak, m = medium, s = strong, vs = very strong, br = broad, v br = very broad.

a.   **ROH**   O—H stretching vibration at 3300 cm^{-1} (s, br) ; C—O stretching vibration at 1050–1150 cm^{-1} (s)

b.   **ArOH**   OH stretching vibration at 3400 cm^{-1} (s, br) ; OH stretching vibration at 1220 cm^{-1} (s)

c.   **RCHO**   C=O stretching vibration at 1725 cm^{-1} (vs) ; C—H stretching vibration at 2720 cm^{-1} (m)

d.   **RCOOR′**   C=O stretching vibration at 1740 cm^{-1} (vs) ; C—O stretching vibration at 1220 cm^{-1} (vs)

e.   **RCOOH**   C=O stretching vibration at 3400 cm^{-1} (vs,) ; OH stretching vibration 3500–2500 cm^{-1} (s, v br)

f.   **RNH$_2$**   N—H stretching vibration at 3350 cm^{-1} (w, doublet)

g.   **ArNH$_2$**   N—H stretching vibration at 3400 cm-1 (m, doublet)

h.   **RSH**   S—H stretching vibration at 2530 cm^{-1} (m)

Compound 21A: **1˙ amine:** N—H stretching vibration at 3350 cm^{-1} (w, doublet)

Compound 21B: **ester:**   C=O stretching vibration at 1725 cm^{-1} (vs) ;
                                     C—O stretching vibration at 1150 cm^{-1} (vs)

Compound 21C: **thiol:**   S—H stretching vibration at 2550 cm^{-1} (m)

14.22. A mass spectrum reveals the presence of the heteroatoms N, S, Cl, and Br by the pattern or mass of peaks in the molecular ion region. Certain functional groups are identified by observing peaks that correspond to the fragments produced after ionization (see Appendix B of the text for the appropriate data).

Compound 21A: The presence of N will be apparent from the noneven value for the molecular weight.

Compound 21B: This compound may be identifiable from fragmentation patterns expected for an ester.

Compound 21C: The presence of S is determined by the intensity of the $M + 2$ peak, which should have about the same intensity as the $M + 1$ peak.

14.23. To match structures with spectra, consider what absorptions will be observed for each functional groups listed, then look at each spectrum to see if those features are present.

For the functional groups listed, diagnostic bands are as follows: w = weak, m = medium, s = strong, vs = very strong, br = broad.

a. **ROH**   O—H stretching vibration at 3300 cm^{-1} (s, br) ; C—O stretching vibration at 1050–1150 cm^{-1} (s)

b. **ArOH**   OH stretching vibration at 3400 cm^{-1} (s, br) ; OH stretching vibration at 1220 cm^{-1} (s)

c. **RCOOR′**   C=O stretching vibration at 1740 cm^{-1} (vs) ; C—O stretching vibration at 1220 cm^{-1} (vs)

d. **RCOOH**   C=O stretching vibration at 3400 cm^{-1} (vs,) ; OH stretching vibration 3500–2500 cm^{-1} (s, v br)

e. **RNH$_2$**   N—H stretching vibration at 3350 cm^{-1} (w, doublet)

f. **RNO$_2$**   N—O stretching vibrations, one at 1600–1500 cm^{-1} (vs) and one at 1390–1300 cm^{-1} (vs) .

g. **RCN**   C≡N stretching vibration at 2250 cm^{-1} (m-s)

Compound 23A: **Nitro group:**   N—O stretching vibrations at 1550 cm^{-1} (vs) and 1375 cm^{-1} (vs) .

Compound 23B: **Nitrile group:**   C≡N stretching vibration at 2325 cm^{-1} (m)

Compound 23C: **Alcohol:**   O—H stretching vibration at 3350 cm^{-1} (s, br)
                       2° Alcohol: C—O stretching vibration at 1100 cm^{-1} (s)

14.24. A mass spectrum reveals the presence of the heteroatoms N, S, Cl, and Br by the pattern or mass of peaks in the molecular ion region. Certain functional groups are identified by observing peaks that correspond to the fragments produced after ionization (see Appendix B of the text for the appropriate data).

Compound 23A: The presence of N will be apparent from the noneven value for the molecular weight.

Compound 23B: The presence of N will be apparent from the noneven value for the molecular weight.

Compound 23C: This compound may be identifiable from fragmentation patterns expected for a secondary alcohol. The intensity of the molecular ion peak is likely to be weak.

14.25. Resonance forms that can be drawn for phenol include those that have a double bond between the *ipso* carbon atom and oxygen, which increases the bond strength and thereby raises the carbon–oxygen stretching frequency. These resonance forms are minor contributors because they place a positive charge on the oxygen atom and disrupt the aromaticity of the ring, so the increase in bond order is slight, but apparently significant.

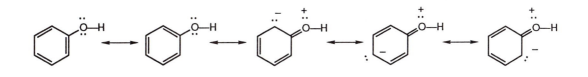

14.26. The mass spectrum described in this exercise clearly indicates that chlorine is present in the structure of compound **X**. The mass of 92 (^{35}Cl) is consistent with the composition C_4H_9Cl, which has no sites of unsaturation. The absence of absorptions in the infrared spectrum in the region around 1600 cm^{-1} confirms that no carbon-carbon double bonds are present. The strong band that appears at 660 cm^{-1} is likely the carbon-chlorine stretching vibration. Possible structures are the following chloroalkanes:

14.27. As described in this exercise, compound **Y** contains only C, H, and O [$C_6H_{14}O$], and we calculate that no sites of unsaturation exist. All functional groups that contain a carbonyl group can be eliminated from consideration, therefore, so the functional group is either an alcohol or ether.

The IR spectrum has a strong absorption at about 3300 cm^{-1}, which indicates that compound **Y** is an alcohol, so we next look for a band between 1200 and 1000 cm^{-1} for the C—O stretching vibration. From observation of a strong band at 1100 cm^{-1}, we conclude that the compound is a 2° alcohol.

Possible structures are therefore as follows:

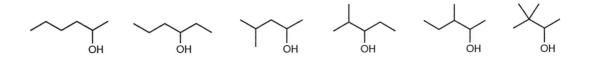

14.28.   To predict what absorptions will be observed in the infrared spectra of reactants and products of a given chemical transformation, identify the functional groups present in starting material and product, and specify what corresponding absorptions are expected for each functional group. The hydrocarbon portions of the molecules will likely be similar.

a.   The starting material is a ketone, which will have a very strong absorption band at 1715 cm^{-1} for the carbonyl C=O stretching vibration. This absorption will disappear as the reaction progresses. Furthermore, the product will show a relatively strong, broad band at 3300 cm^{-1} for the O—H stretching vibration of the alcohol functional group. This 3° alcohol will also have a relatively strong band at about 1050 cm^{-1} for the C—O stretching vibration.

b.   The starting material may have an absorption band in the 600-800 cm^{-1} region for the C—Cl stretching vibration. The benzene ring will produce several absorption bands, but these will be observed in the product as well, so they are not diagnostic. On the other hand, the product will have a very strong band at 1710 cm^{-1} for the carbonyl C=O stretching vibration of the carboxylic acid group, which will also produce a strong and very broad band between 3500 and 2500 cm^{-1} for the hydrogen-bonded O—H stretching vibration.

c.   The starting material is a terminal alkene, which has an absorption in the region around 1600 cm^{-1}. The benzene ring also has absorptions in that region, which may interfere with observation of the alkene C=C stretching vibration. The product, a 2° alcohol, will have a strong, broad absorption band at 3300 cm^{-1} for the alcohol O—H stretching vibration, as well as a relatively strong band at 1100 cm^{-1} for the C—O stretching vibration.

14.29. The reactions shown in exercise 14.28 are ones that you have seen in Chapters 7 through 13. To find out what reagents are needed, consult Appendix D of the text for reference to the appropriate pages. Several methods may be applicable, but only one route is shown below.

a. Conversion of a ketone to a 3° alcohol occurs by the addition of a Grignard reagent to the ketone carbonyl group, followed by hydrolysis.

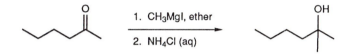

b. An alkyl halide is converted to a carboxylic acid by first forming a Grignard reagent, which is allowed to react with carbon dioxide. Hydrolysis liberates the carboxylic acid product.

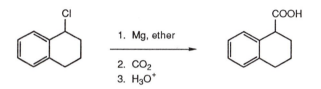

c. An alkene is converted to an alcohol with use of a several reagents. In this reaction, Markovnikov addition of water is required, so the combination of water and sulfuric acid is a good choice. Styrene, the starting material, is prone to cationic polymerization, so a reagent combination that avoids formation of a carbocation is oxymercuration followed by reduction with sodium borohydride.

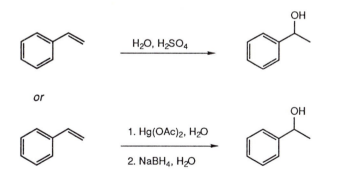

PROTON AND CARBON NUCLEAR MAGNETIC RESONANCE SPECTROSCOPY

15.1. Use the information in Figure 15.6 (text page 696) to make a quick assessment of relative chemical shift values according to the following list: carboxylic acid (δ 12) < aldehyde (δ 10) < aromatic C–H (δ 7-8) < alkene C–H (δ 5-6) < X–C–H (δ 2-5) < C–H (δ 0-2) < TMS (δ 0).   Alcohol and amine protons appear over a substantial range (δ 1–5).   Alkyne C–H chemical shifts tend to be downfield from those of saturated C–H groups.

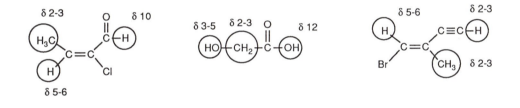

15.2.   To determine which protons are chemically equivalent, decide their relative relationships within the molecule. Protons that are part of a methyl groups are always equivalent, and those in methylene group usually are. If the molecule has a mirror plane, symmetry-related groups of protons are also equivalent.

a.   This molecule has a plane of symmetry that passes through the methyl group, C1 of the ring, and the C3-C4 bond of the ring. The protons attached to C2 are equivalent to each other as well as those attached to C5.

b.   This molecule has a plane of symmetry, so the six methyl protons are all equivalent.

15.2.   (continued)

c.   This molecule has a plane of symmetry, so the four methylene protons are all equivalent.

$$CH_3 - C - (CH_2) - (CH_2) - C - CH_3$$

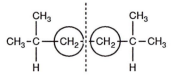

15.3.   The multiplicity of the signal for a given set of protons depends on the number of protons attached to *adjacent* equivalent positions.   If n protons are attached to an adjacent atom, the signal for the observed set of protons appears as a feature with $n + 1$ peaks.

a.   The signal assigned as the resonance of the methine proton appears as a quartet because three protons are attached to the adjacent carbon atom: $3 + 1 = 4 \equiv$ quartet.

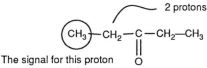

3 protons

The signal for this proton appears as a quartet.

b.   The signal assigned as the methylene proton resonance appears as a triplet because two equivalent protons are attached to the adjacent carbon atom: $2 + 1 = 3 \equiv$ triplet.

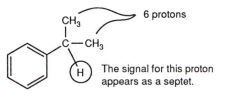

2 protons

The signal for this proton appears as a triplet.

c.   The signal assigned as the resonance of the methine proton appears as a septet because six equivalent protons are attached to adjacent carbon atoms: $6 + 1 = 7 \equiv$ septet.

6 protons

The signal for this proton appears as a septet.

15.4.   The intensity of a signal is proportional to the height of the integration curve over each feature. Measuring the curves for each feature of the spectrum with a ruler, we find their heights to be 20 mm, 8 mm, 8 mm, and 4 mm, from left to right.   Therefore, the ratio between the numbers of protons giving rise to each signal is $5 : 2 : :2 : 1$.

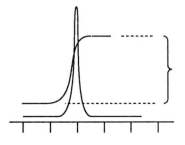

Measure this distance for each feature of the given spectrum. The height in mm is **proportional** to the number of protons that produces this signal.

15.5. Assigning fragment structures to each feature in a proton NMR spectrum is the first step in the interpretation process. First decide what types of protons are present by using data such as that in Table 15.5 (text page 711) or in Figure 15.18 (text page 712). Then, use integrated intensity values to differentiate methyl, methylene, and methine groups, if possible.

a. The chemical shift values for the two features in the given spectrum suggest that the molecule contains aromatic protons as well as protons bonded to an aliphatic carbon atom that has an electron withdrawing group attached. The molecule contains only C and H, so the electron-withdrawing group is most likely the benzene ring. The integrated intensity values of 5 and 3 for the two features correlate with resonances expected for a monosubstituted benzene ring (5 protons) and a methyl group (3 protons). The compound is toluene.

b. The chemical shift values for the three features suggest that the molecule has an aromatic portion as well as two different kinds of protons bonded to aliphatic carbon atoms with one (or two) electron withdrawing groups attached. The molecule is a ketone, which accounts for the presence of one electron-withdrawing group; the benzene ring is another electron withdrawing group. The integrated intensity values of 5, 2, and 3 for the three features correlate with resonances expected for a monosubstituted benzene ring (5 protons), a methylene group (2 protons), and a methyl group (3 protons). The compound is phenylacetone.

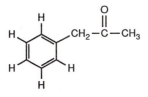

15.6. Follow the procedure outlined in examples 15.1 – 15.4 on text pages 718-727.

a. The medium intensity band in the IR spectrum at 2530 cm^{-1} is most likely the absorption band of an S—H stretching vibration. The appearance of the molecular ion region in the mass spectrum confirms the presence of sulfur: the $M + 1^{+\bullet}$ and $M + 2^{+\bullet}$ peaks are approximately equal in intensity.
The NMR data are summarized as follows:

Chemical shift (δ)	Intensity	Assignment	Multiplicity	J (Hz)	No adjacent protons (n)
3.15	1	C—H	sextet(?)	7	5
1.6	1	S—H	doublet	7	1
1.35	6	—CH$_3$	doublet	7	1

15.6. (continued)

We know that the compound contains an SH group from the IR spectrum. A thiol group must produce an NMR signal with an intensity value of 1, so we assign either the δ 3.15 peak or the δ 1.6 peak as the signal for the thiol proton. An SH group attached to carbon can have at most 3 adjacent protons (CH₃—SH), and the δ 3.15 signal is at least a sextet. Therefore δ 1.6 must be the resonance for the –SH proton. Its appearance as a doublet means that it is bonded to a carbon atom that has a single proton attached. So far, we have H—C—SH.

The δ 3.15 resonance must therefore be attributed to the C—H group adjacent to the thiol functional group. Being attached to a carbon atom that bears a heteroatom, this methine proton produces a signal in the region between δ 2 and δ 5, as expected.

Only 6 protons are left to incorporate into the structure (those that generate the signal at δ 1.35), and they must be equivalent as well as adjacent to a single proton because the δ 1.35 resonance is a doublet. Therefore, we draw the following structure:

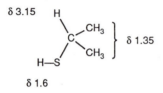

To explain the appearance of the peaks at δ 3.15 (notice that in the table, a question mark follows the assigned multiplicity of "sextet"), we draw a tree structure for splitting of the δ 3.15 signal. The 6 protons of the two methyl groups split the methine proton signal into a septet (6 + 1 = 7). Because the *J* value for coupling of the methine and SH protons is the same as that for coupling of the methine and methyl protons, the second tier of splitting causes the peaks to overlap and appear as an octet of peaks. The very low intensity of the outer peaks explains why only a "sextet" is observed. This example illustrates a valuable lesson—be careful when dealing with splitting patterns that comprise many peaks. Adding two additional peaks to the observed pattern is often necessary (if you were actually recording the spectrum yourself, you could have the NMR spectrometer expand the intensities of the peaks to observe the very weak ones).

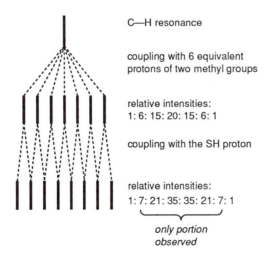

15.6. (continued)

b. The intense band in the IR spectrum at 1725 cm^{-1} is most likely the absorption band for a C=O stretching vibration. The molecule weight of 86 corresponds to the molecular formula $C_5H_{10}O$.

The NMR data are summarized as follows:

Chemical shift (δ)	Intensity	Assignment	Multiplicity	J (Hz)	No adjacent protons (n)
2.45	1	H—C—CO	septet	7	6
2.0	3	CH$_3$—CO	singlet	–	0
0.95	6	—CH$_3$	doublet	7	1

We know that the compound contains a carbonyl group from its IR spectrum. Therefore, the signals at δ 2.45 and at δ 2.0 must be resonances for protons attached to a carbon atom adjacent to the carbonyl group, which is the only electron-withdrawing group in the molecule, as deduced from the molecular formula.

The signal at δ 0.95 must be the resonance of a saturated hydrocarbon group, and the relative intensity is consistent with the presence of two methyl groups. Those methyl groups must be adjacent to the methine group, so this compound is 3-methyl-2-butanone.

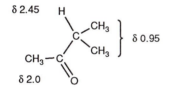

15.7. Use the information in Figure 15.31 (text page 729) to make an assessment of relative chemical shift values according to the following list: carbonyl compounds (δ 150-200) < arenes (δ 100-160) < alkenes (δ 55-160) < nitriles (δ 100-125) < alkynes (δ 20-100) < alkanes (δ 0-60) < TMS (δ 0).

a.   b.   c.

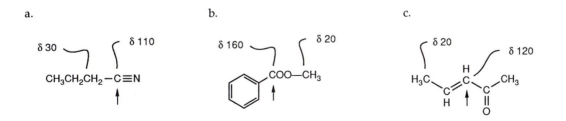

15.8. Assigning fragment structures to each feature in a carbon NMR spectrum is the first step in the interpretation process. First decide what types of carbon atoms are present based on general chemical shift value ranges by using data such as that in Figure 15.37 (text page 735). Then, use the off-resonance data to differentiate 1°, 2°, 3°, and 4° carbon atoms.

Chemical shift (δ)	Assignment	Off-resonance pattern	Type of C atom
164.0	C=O	s	C
146.4	C=C (Ar)	s	C
139.8	C=C (Ar)	s	C
132.6	C=C (Ar)	d	CH
131.6	C=C (Ar)	s	C
127.0	C=C (Ar)	d	CH
126.3	C=C (Ar)	d	CH

Conclusion: this compound is an aromatic carboxylic acid in which the benzene ring is trisubstituted.

15.9. Follow the procedure outlined in the solution to exercise 15.8.

Chemical shift (δ)	Assignment	Off-resonance pattern	Type of C atom
206.2	C=O	s	C
134.1	C=C (Ar)	s	C
129.2	C=C (Ar)	s	C
128.5	C=C (Ar)	d	CH
126.6	C=C (Ar)	d	CH
50.7	C aliphatic	t	CH$_2$
29.0	C aliphatic	q	CH$_3$

Conclusion: this compound is a ketone that has a *para*-disubstituted aromatic ring and a two-carbon aliphatic portion.

15.10. To predict what resonances should be observed in a carbon NMR spectrum as a way to verify structural assignments, use the data compiled in Table 15.11 (text page 741) to decide suitable chemical shift ranges in which the signals will appear.

For the structures shown starting on the next page, values of the chemical shifts ranges are given and followed by a letter that indicates what pattern will be observed in the off-resonance decoupled spectrum (S = singlet, D = doublet, T = triplet, and Q = quartet).

15.10. (continued)

Example 15.1 **Ethyl acetate**

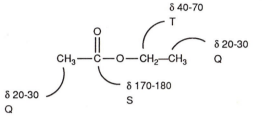

Example 15.2 **Butanoyl chloride**

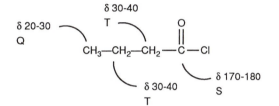

Example 15.3 **Ethyl *trans*-2-butenoate**

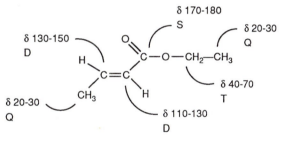

Example 15.4 ***o*-Anisic acid**

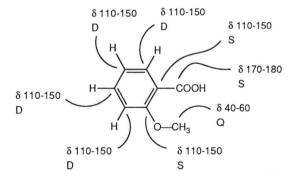

15.10. (continued)

Exercise 15.5a **Toluene**

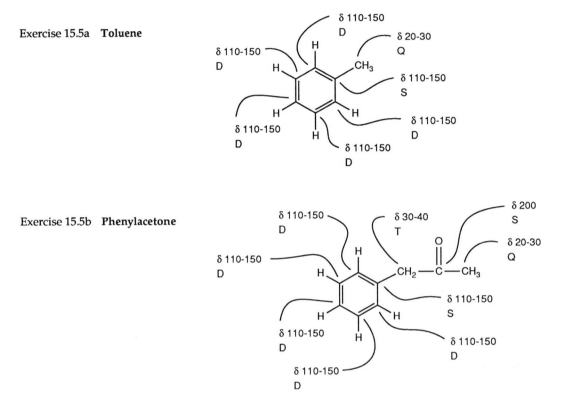

Exercise 15.5b **Phenylacetone**

15.11. Follow the procedure outlined in the solution to exercise 15.2.

a. This molecule has no symmetry, so each set of protons is different. All three protons of the methyl group are equivalent.

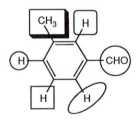

b. This molecule has no symmetry, so each set of protons is different. All three protons of each methyl group are equivalent and both protons of the methylene group are equivalent.

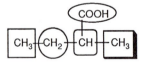

15.11. (continued)

c. This molecule has a plane of symmetry that
    passes through the methyl group and the proton
    para to it. The two protons that are *ortho* to the
    chlorine atoms are equivalent. All three protons
    of the methyl group are equivalent.

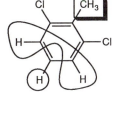

d. This molecule has a symmetry rotation axis that
    is perpendicular to the plane of the double bond.
    The alkene protons are each *cis* to a methyl group, so
    they are equivalent to each other. All six protons
    in the methyl groups are equivalent.

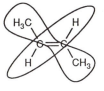

e. This molecule has a plane of symmetry that
    passes through the nitro and hydroxy groups.
    Both protons *ortho* to the nitro group are
    equivalent and both protons *ortho* to the OH group
    are equivalent.

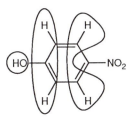

f. This molecule has a plane of symmetry that
    passes through the OH, methylene, and methine
    groups. All six protons of the methyl groups are
    equivalent. Both protons of the methylene group
    are equivalent.

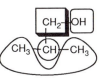

15.12. The chemical shift value expected for a set of protons is affected by the local electronic environment
of the proton. Electronegativity, resonance, and anisotropy effects influence the magnitude of the chemical
shift. Use a compilation such as Table 15.5 (text page 711) or Figure 15.18 (text page 712) to assign chemical
shift values. If the specific environment is not shown in the table or figure, estimate an expected value based
on those given.

a.

$$CH_3CH_2CH_2-\underset{|}{\overset{COOH}{CH}}-CH_3 \qquad CH_3CH_2CH_2-\underset{|}{\overset{OH}{CH}}-CH_3 \qquad CH_3CH_2CH_2-\underset{|}{\overset{Cl}{CH}}-CH_3$$

δ 2.5   δ 3.9   δ 4.0

15.12. (continued)

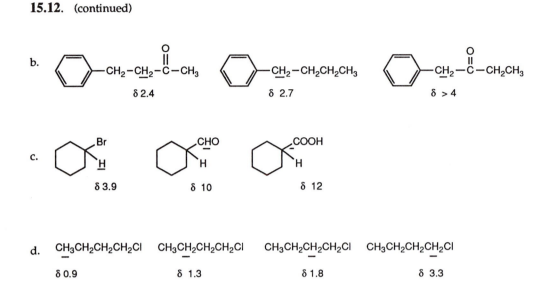

b.

δ 2.4   δ 2.7   δ > 4

c.

δ 3.9   δ 10   δ 12

d.   CH₃CH₂CH₂CH₂Cl   CH₃CH₂CH₂CH₂Cl   CH₃CH₂CH₂CH₂Cl   CH₃CH₂CH₂CH₂Cl

δ 0.9   δ 1.3   δ 1.8   δ 3.3

15.13. To compare chemical shift values for protons in different environments, use a compilation such as Table 15.5 (text page 711) or Figure 15.18 (text page 712). Chemical shift values for protons that have similar environments follow this trend: (upfield) CH₃ < CH₂ < CH (downfield). The circled proton(s) in the following structures have their resonance farthest downfield.

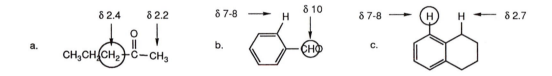

15.14. To compare chemical shift values for carbon atoms in different environments, use a compilation such as Table 15.11 (text page 741).

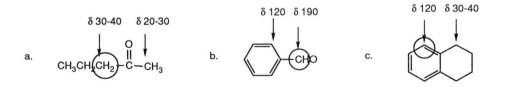

15.15. The multiplicity of the signal for a given set of protons depends on the number of protons attached to *adjacent* equivalent positions. If n protons are attached to an adjacent atom, the signal for the observed set of protons appears as a feature with $n + 1$ peaks.

a.   The signal assigned as the methylene proton resonance appears as a triplet because two equivalent protons are attached to the adjacent carbon atom: $2 + 1 = 3 \equiv$ triplet.

The signal for this proton appears as a triplet.

2 protons

$CH_3CH_2CH_2–Br$

b.   The signal assigned as the methylene proton resonance appears as a sextet because five similar types of protons (i.e., all are bonded to saturated carbon atoms) are attached to the adjacent carbon atoms: $5 + 1 = 6 \equiv$ sextet.

The signal for this proton appears as a sextet.

3 protons   2 protons

$CH_3CH_2CH_2–OH$

c.   The signal assigned as the methine proton resonance appears as a doublet of septets. Six equivalent protons are attached to adjacent methyl groups but a different type of proton is attached to sulfur. Protons on the methyl groups split the methine proton as follows: $6 + 1 = 7 \equiv$ septet. The SH proton splits each of those peaks as follows: $1 + 1 = 2 \equiv$ doublet.

The signal for this proton appears as a doublet of septets.

1 proton

$(CH_3)_2CH–SH$

6 protons

d.   The signal assigned as the alkene proton resonance appears as a doublet because only one proton is attached to the adjacent carbon atom: $1 + 1 = 2 \equiv$ doublet.

1 proton

H   H ← The signal for this proton appears as a doublet.

OCH_3

e.   The signal assigned as the indicated benzene proton resonance appears as a doublet of doublets of doublets. Each of the other three protons on the benzene ring couples with that proton to a different extent because the magnitude of the *J* value varies as a consequence of orientation. One proton is *ortho*, one is *meta*, and one is *para* to the specified proton. For each proton, the splitting follows the $n + 1$ rule: $1 + 1 = 2 \equiv$ doublet.

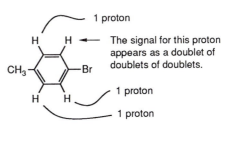

1 proton

H   H ← The signal for this proton appears as a doublet of doublets of doublets.

CH_3 Br

H   H

1 proton

1 proton

f.   This situation is the same as that shown in part e.

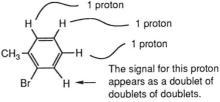

1 proton

H   H   1 proton

1 proton

CH_3 H

Br   H ← The signal for this proton appears as a doublet of doublets of doublets.

15.16. The splitting pattern for a carbon atom resonance in the off-resonance spectrum is created by the protons attached to the carbon atom being examined. The pattern is equal to the number of protons + 1.

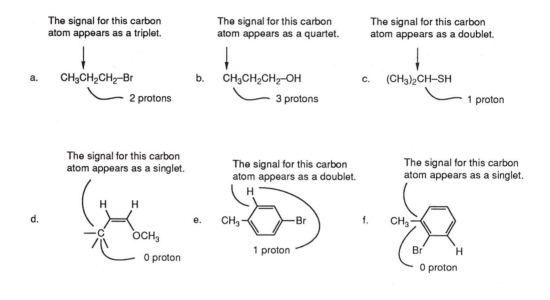

a.

The signal for this carbon atom appears as a triplet.

$CH_3CH_2CH_2–Br$

2 protons

b.

The signal for this carbon atom appears as a quartet.

$CH_3CH_2CH_2–OH$

3 protons

c.

The signal for this carbon atom appears as a doublet.

$(CH_3)_2CH–SH$

1 proton

d.

The signal for this carbon atom appears as a singlet.

0 proton

e.

The signal for this carbon atom appears as a doublet.

1 proton

f.

The signal for this carbon atom appears as a singlet.

0 proton

15.17. The appearance of a proton NMR spectrum is predicted by assigning expected chemical shift values with use of data from Table 15.5 (text page 711) or Figure 15.18 (text page 712). Next, assign the multiplicity of each feature based on the number and type of neighboring protons, using *J* values from Table 15.3 (text page 705). The number of each type of proton allows you to designate the integrated intensity value of each feature.

a.   This molecule has three types of protons. The methine proton couples with six equivalent protons of the geminal methyl groups.

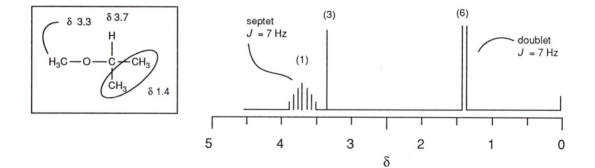

15.17. (continued)

b.   Sizable coupling occurs only between the protons that are *ortho* to each other in the benzene ring. The methyl and methylene groups appear as singlets.

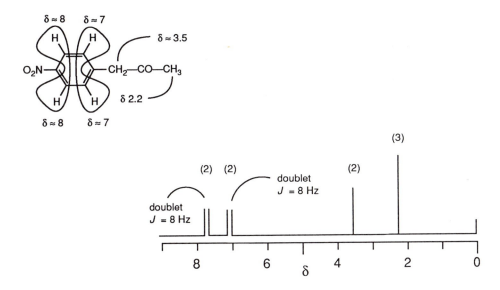

c.   The protons of the methylene group couple both with the aldehyde proton and with those of the methyl group. The protons of the methyl group and the aldehyde proton couple only with those of the methylene group.

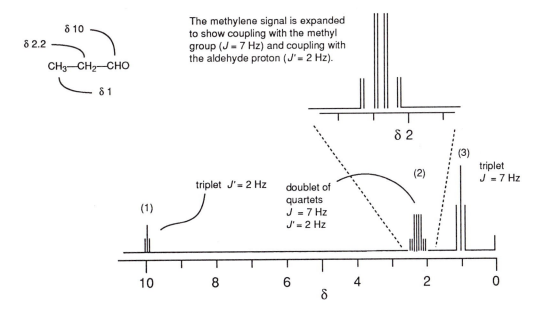

15.18. The broad-band decoupled carbon NMR spectrum appears as a series of single peaks at the expected chemical shift values (Figure 15.37, text page 735). Quaternary carbon atoms invariably produce resonances that have low intensity peaks; the intensities of other resonances vary considerably.

a.

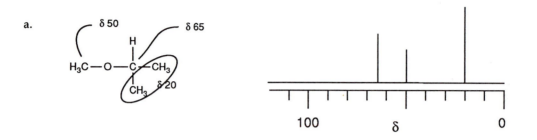

b.

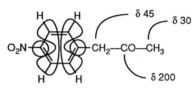

δ 120-140; four peaks, two of which (circled) are quaternary

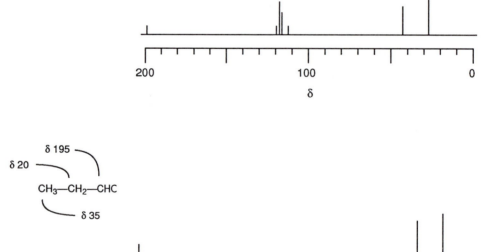

c.

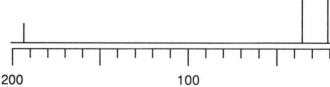

15.19. To differentiate between compounds by proton NMR spectroscopy, you must be able to identify some proton or group of protons that has a chemical shift value apart from those of other resonances observed. If all of the resonances overlap, it may be difficult to see what differences exist. For sets of peaks that are resolved from other resonances, specific spin-spin coupling patterns provide additional data to confirm structures or substructures within the molecule.

a. These molecules will be difficult to differentiate by proton NMR spectra because nearly all of the resonances will appear at values between δ 0.8 and 2.

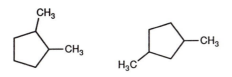

b. The molecules in this exercise are relatively easy to differentiate by proton NMR spectra because the circled protons have chemical shifts that isolate them from the hydrocarbon portions of the molecule. These signals are readily assigned to specific groups, too, so their spin-spin coupling patterns can be observed and compared. Completely different splitting patterns for the appropriate signals in these spectra permit ready identification of the unknown.

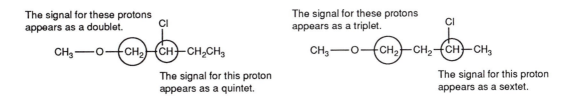

c. The molecules in this exercise are relatively easy to differentiate by proton NMR spectra because the circled proton in the first compound couples only to protons that are *meta* and *para*, so its resonance in the aromatic region of the spectrum will display only small *J* coupling values. In the other isomer, all three aromatic protons have at least one *ortho* coupling, which is normally sizable (approx 8 Hz).

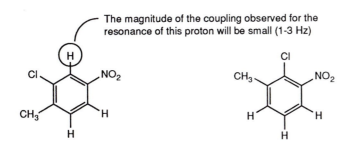

15.20. Follow the procedure outlined in the solution to exercise 15.19.

a. For the alcohol, only one resonance, for the circled proton, will appear in the region farthest downfield (δ 3-4). The ether will have two signals in the region farthest downfield. The splitting pattern of the resonance for the methylene group of the ether will be a readily observable doublet because only one proton is attached to the adjacent carbon atom.

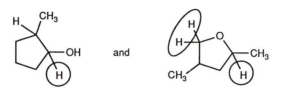

b. For the 2° allylic alcohol, resonances for three protons appear in the alkene region (δ 5-6); only two protons of the 1° allylic alcohol have resonances in the alkene region. Therefore, the integrated intensity in the region δ 5-6 will be different. Moreover, the *cis* alkene has only a single coupling between resonances in the alkene region.

c. Resonances in the aromatic region (δ 7-8) will have an intensity of 3 for the aldehyde, but 4 for the ketone. Also, the aldehyde proton displays a distinctive resonance at about δ 10.

15.21. In carbon NMR spectra, chemical shift values are readily observed in the broad-band decoupled spectrum, and predicted values are assigned using data compiled in Figure 15.37 (text page 735). The number of peaks for a given feature in the off-resonance spectrum is calculated by adding one to the number of protons attached to the carbon atom being observed.

a. For the alcohol, only one carbon atom resonance appears in the region farthest downfield (δ 40-50) whereas the ether will have two such resonances. The signal farthest downfield in the off-resonance spectrum of the alcohol is a doublet.

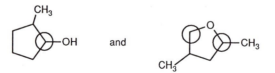

15.21. (continued)

b.   Both allylic alcohols have two resonances in the alkene region (δ 110-150).   In the off-resonance spectrum, however, one of the resonances for the terminal alkene will appear as a triplet.

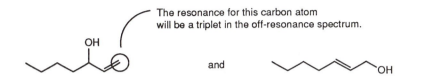

The resonance for this carbon atom will be a triplet in the off-resonance spectrum.

and

c.   For the aldehyde, three resonances in the aromatic region (δ 120-140) appear as doublets in the off-resonance spectrum, whereas for the ketone, four resonances will appear as doublets in the off-resonance spectrum.   Moreover, the aldehyde carbon resonance will appear as a doublet in the off-resonance spectrum.

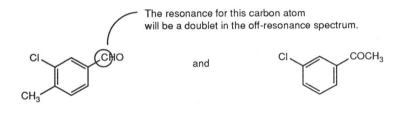

The resonance for this carbon atom will be a doublet in the off-resonance spectrum.

and

15.22.   When performing a chemical reaction, predict what proton NMR resonances you expect to see for the starting material and product using the tables and figures in the chapter.   For the values given below, the first number is the chemical shift relative to the signal for TMS; the letter is the spin-spin splitting pattern (s = singlet, d = doublet, t = triplet, q = quartet, m = multiplet, br s = broad singlet); the last value with the capitalized H is the integrated intensity (2H means two protons, etc.)

a.

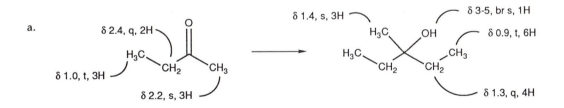

b.   Benzene displays a singlet at δ 7.2.   When an electron-withdrawing group is attached, the resonances are usually farther downfield.   In nitrobenzene, there are three different types of protons, so the signals will appear at > δ 7.2, and they will be split into a complicated pattern with a ratio of 2:2:1.   Upon reduction to aniline, the resonances move upfield, and there will be a new broad absorption for the NH_2 protons.

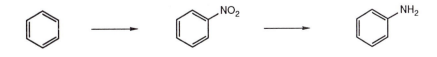

15.22. (continued)

c. Benzene displays a singlet at δ 7.2. In bromobenzene, there are three different types of protons, so the signals will appear slightly downfield from δ 7.2, and they will be split into a complicated pattern with a ratio of 2:2:1. After conversion of bromobenzene to toluene, a new singlet appears at about δ 2.3. The integrated intensity will be 3H compared to 5H for the region δ 7-8.

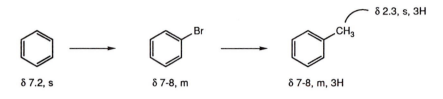

δ 7.2, s   δ 7-8, m   δ 7-8, m, 3H

δ 2.3, s, 3H

15.23. When performing a chemical reaction, predict what carbon NMR resonances you expect to see for the starting material and product using the tables and figures in the chapter. For the values given below, the first number is the chemical shift relative to the signal for TMS, and the letter is the splitting observed in the off-resonance spectrum (s = singlet, d = doublet, t = triplet, q = quartet).

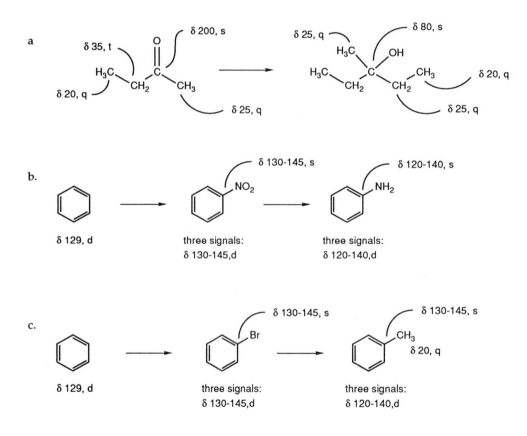

15.24. The reactions shown in exercise 15.24 are ones that you have seen in Chapters 7 through 13 of the text. To find out what reagents are needed, consult Appendix D of the text for reference to the appropriate pages. Several methods may be applicable, but only one route is shown below.

a.   Conversion of a ketone to a 3° alcohol requires reaction with an organometallic compound, followed by aqueous acid workup. A two-carbon fragment is required, so the ethyl Grignard reagent is used.

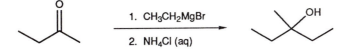

b.   Benzene undergoes nitration upon treatment with a mixture of nitric and sulfuric acids. Reduction of the nitro group forms the amino group.

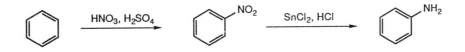

c.   Benzene undergoes bromination with bromine and a Lewis acid catalyst. Bromobenzene reacts with an organocuprate reagent to form a new carbon-carbon bond.

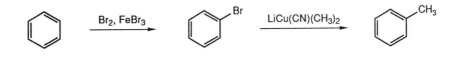

15.25. Compound **X** produces an IR spectrum that has a strong absorption at 1800 cm^{-1}, so it clearly has a carbonyl group. The data used to differentiate between various types of carbonyl-containing compounds are summarized in Table 14.4 on text page 654. The C=O stretching frequency suggests that the compound is an acid chloride.

In the broad band carbon NMR spectrum, we would expect to see a peak at about δ 185-165. In the off-resonance spectrum, this resonance will remain a singlet.

15.26. The proton NMR spectrum has only singlets, so the carbon atoms bearing protons must be isolated from the others by quaternary carbon atoms or by functional groups that do not have hydrogen atoms. The fact that the compound is a ketone means that the remainder of the skeleton has no unsaturation (only one site of unsaturation is present).

The resonance at δ 2.12 (3H) is most likely a methyl group adjacent to the carbonyl group. The resonance at δ 2.32 (2H) is most likely a methylene group adjacent to the carbonyl group. That leaves a singlet at δ 1.02 with an intensity corresponding to 9 protons. Most often a signal at δ 1 with an integrated intensity of 9 is evidence that a *tert*-butyl group is present.

The compound is 4,4-dimethyl-2-pentanone.

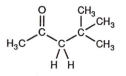

In the broad band carbon NMR spectrum, we expect to see three absorptions between δ 20 and 30 for the methyl and methylene groups, a peak between δ 40 and 60 for the quaternary carbon atom, and a peak between δ 200-220 for the carbonyl group. In the off-resonance spectrum, the peaks at δ 40-60 and δ 200 will appear as singlets. Two of the other peaks will appear as quartets and the remaining one will appear as a triplet.

15.27. The IR spectrum of compound **W** reveals that the compound has no carbonyl or triple bonded functional groups. The strong, broad band at about 3300 cm⁻¹ suggests that an OH group is present. The absorption at 1250 cm⁻¹ is consistent with the notion that the compound is a phenol. Absorptions around 1600 cm⁻¹, in the overtone region between 1700 and 2000 cm⁻¹, and between 700 and 900 cm⁻¹ confirm the presence of a benzene ring.

The proton NMR spectrum also confirms that the compound has a benzene ring. The integrated intensity of 3H in the aromatic region (3 features @ 1H each), along with a singlet at δ 4.9 (1H) and two singlets (each 3H) at δ 2.25 and δ 2.15 suggests that the compound is a trisubstituted benzene, a dimethylphenol.

The strong absorption at 770 cm-1 in the IR region, and the coupling pattern in the aromatic region of the NMR spectrum are consistent with the presence of a 1,2,3-trisubstituted ring. If the compound were 2,6-dimethylphenol, the methyl groups would be equivalent, so only one signal would be observed in the region between δ 2 and 3. Therefore, the compound must be 2,3-dimethylphenol.

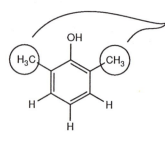

The methyl groups are equivalent by symmetry.

2,3-Dimethylphenol   **2,6-Dimethylphenol**

ASYMMETRIC REACTIONS AND SYNTHESIS

16.1. The stereogenic centers in brucine and quinine, circled in the structures below, are those with sp^3 hybridization and four different groups attached. Nitrogen atoms are normally chiral only when they have sp^3 hybridization and cannot undergo inversion of configuration, for instance when they are at a bridgehead position in a ring. An electron pair counts as one of the four "substituents" of a chiral nitrogen atom.

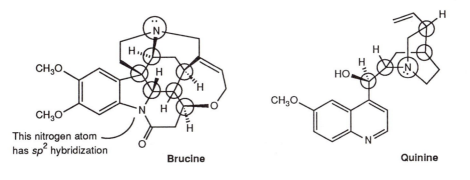

Brucine   **Quinine**

16.2. Resolution of a racemic amine is carried out as follows: the amine is treated in step (1) with the optically active sulfonic acid. Diastereomeric salts are formed and subjected to crystallization in step (2), which leads to the precipitation of one diastereomeric salt, while the other stays in solution. Treating the salt with aqueous NaOH in step (3) regenerates the amine and forms the sodium salt of the sulfonic acid. The sulfonic acid by acid workup.

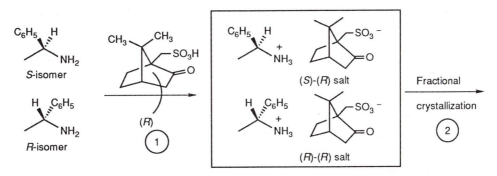

16.2.   (continued)

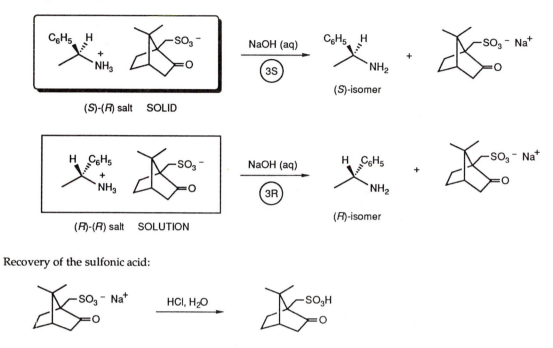

Recovery of the sulfonic acid:

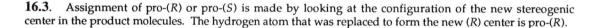

16.3.   Assignment of pro-(*R*) or pro-(*S*) is made by looking at the configuration of the new stereogenic center in the product molecules. The hydrogen atom that was replaced to form the new (*R*) center is pro-(*R*).

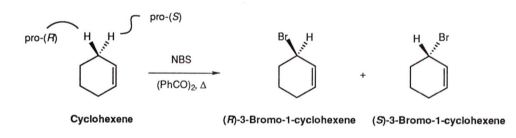

16.4.   Follow the procedure outlined in the solution to exercise 16.3.

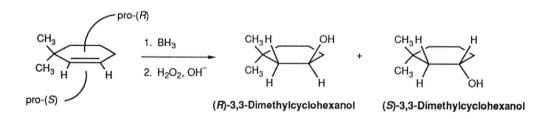

16.5. The %de is the numerical difference between the amounts of each diastereomer produced, expressed as a percentage. The major diastereomer is formed as 80% of the mixture, and the minor one as 20%. The %de is therefore 80 – 20 = 60%.

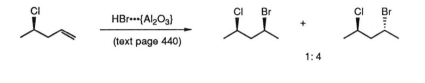

1 : 4

16.6. To calculate the enantiomeric enrichment that is possible from applying the achiral linker strategy, follow procedure described in Example 16.1 on text pages 765-767. The %ee value is initially 88%, so the amounts of each enantiomer are 0.94 (= n) and 0.06 (= $1 - n$). The calculation is as follows:

(R, R)	88.36 %		(R, R)	$\dfrac{88.36}{88.36 + 0.36}$ ≡	99.59 %
(S, S)	0.36 %	fractional crystallization to remove the (R,S)-diastereomer.			
(R, S)	11.28 %		(S, S)	$\dfrac{0.36}{88.36 + 0.36}$ ≡	0.41 %
	100 %				100 %

Conclusion: the purity increases from about 88% of the mixture to over 99%.

16.7. Assign the absolute configuration to each chiral center according to the rules outlined in Chapter 5. Remember that P is chiral when it has four different groups attached. The nitrogen atom in this compound is not chiral because it undergoes pyramidal inversion readily.

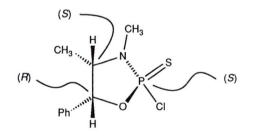

16.8. From the relative intensity values of the peaks, calculated by measuring the heights of the integration curves, determine the percentage of each species. The arithmetic difference between those two values equals the %ee.

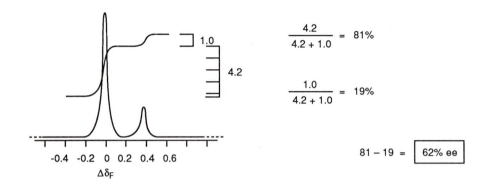

$$\frac{4.2}{4.2 + 1.0} = 81\%$$

$$\frac{1.0}{4.2 + 1.0} = 19\%$$

$$81 - 19 = \boxed{62\% \text{ ee}}$$

16.9. Designing a synthesis of a chiral alcohol via asymmetric hydroboration requires a retrosynthetic analysis that identifies the appropriate alkene precursor. Only one solution is given below for each synthesis; other routes may be possible.

a.   To prepare a cyclic alcohol, start with the corresponding cyclic alkene and introduce the OH group using hydroboration/oxidation. Consult the examples in the text to decide which borane reagent produces the correct stereochemistry.

retrosynthesis:

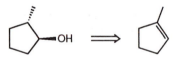

synthesis:

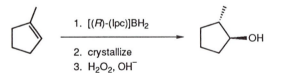

16.9.   (continued)

b.   To prepare an acyclic 2° alcohol, start with a symmetric *cis* alkene and introduce the OH group using hydroboration/oxidation.

retrosynthesis:

synthesis:

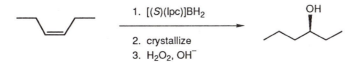

c.   Chiral phenylalkanes are prepared by a Suzuki reaction between a chiral organoborane and bromobenzene. The chiral organoborane is made from the corresponding alkene.

retrosynthesis:

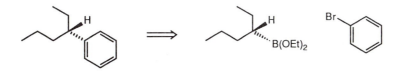

synthesis:

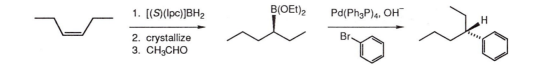

16.10. Allylic alcohols react under Sharpless-Katsuki conditions to form the corresponding chiral epoxide. Details of this procedure are summarized on text pages 777-779. To begin, reorient the epoxy alcohol so that the –CH₂OH group is in the lower right portion of the structure. The allylic alcohol that is needed as the starting material is rendered simply by replacing the O atom in the epoxide structure with a double bond.

a. *retrosynthesis:*

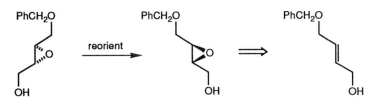

synthesis:

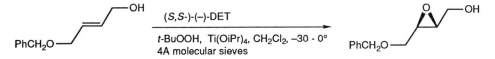

b. *retrosynthesis:*

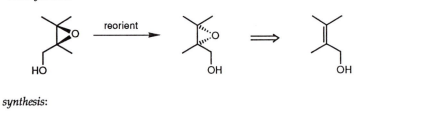

synthesis:

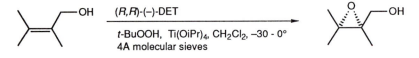

16.11. The epoxy alcohol with the (*R*) configuration is isolated in 30% yield and 95% enantiomeric purity. If you start with 1.00 mol of the allylic alcohol, then (0.95)(0.30) = 0.285 mol of the (*R*) alcohol and (0.05)(0.30) = 0.015 mol of the (*S*) alcohol will react, so 0.215 mol of (*R*) allylic alcohol and 0.485 mol of the (*S*) allylic alcohol remain.

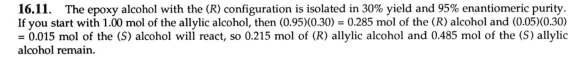

- (*S*) starting material accounts for 0.485/(0.485 + 0.215) = 69.3 % of the mixture.
- (*R*) starting material that remains is 30.7%.
- The enantiomeric excess of (*S*) starting material is 69.3 – 30.7 = 38.6% ee.

16.12.   To predict the stereochemistry of the product from chiral epoxidation, reorient the structure of the substrate and compare with the results shown in Table 16.2 (text page 782).   The (R,R) catalyst produces the (1R,2S) epoxide.

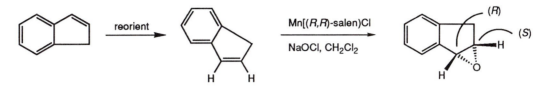

16.13.   Only an allylic alcohol substrate reacts with the reagent combination of Ti(OiPr)₄, DET, and t-butyl hydroperoxide.   The stereochemistry of the product epoxide is deduced from comparisons with the results summarized in the scheme on text page 778.

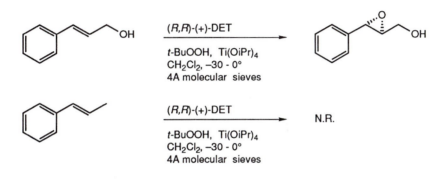

Both compounds shown in this exercise undergo asymmetric dihydroxylation.   Each has a *trans* double bond, so the stereochemistry of the resulting diol is the same and is deduced from comparisons with the figures shown on text page 784.

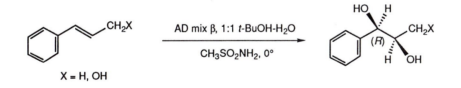

16.14.   Grignard reagents react with epoxides at the less highly-substituted carbon atom.   The carbon atom that remains bonded to oxygen retains its configuration.   The configuration of the other carbon atom is inverted.

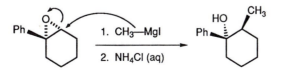

16.15. The steps involved in this synthesis are the same as those described for asymmetric hydroboration on text pages 774-777. The Swern oxidation, step (5), is discussed on text page 350.

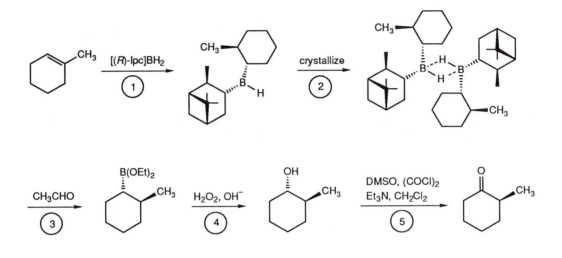

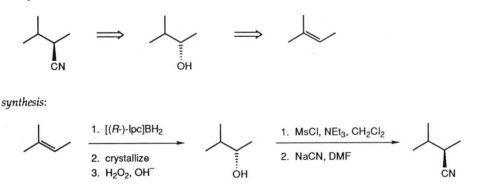

16.16. To plan the synthesis of a chiral molecule, conceive disconnections or exchanges that give compounds with either alcohol or epoxide functional groups because these can be made from an alkene by application of an asymmetric reaction described in the text. Only one solution is given below for each synthesis; other routes may be possible.

a. An aliphatic nitrile is made by nucleophilic substitution, and the S_N2 pathway is stereospecific. The needed substrate is a sulfonate ester, which is made from an enantiomerically pure alcohol. The starting chiral alcohol is made by asymmetric hydroboration of an alkene.

retrosynthesis:

synthesis:

16.16.   (continued)

b.   The dimethoxycyclohexane shown in this exercise is made by alkylating the corresponding diol with methyl iodide. A chiral diol is normally made by direct dihydroxylation or by epoxidation followed by ring-opening. In this case, the two hydroxy groups are *cis*, so dihydroxylation is used.

retrosynthesis:

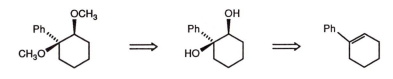

synthesis:

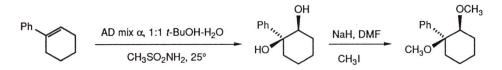

16.17.   Follow the procedure outlined in the solution to exercise 7.28.

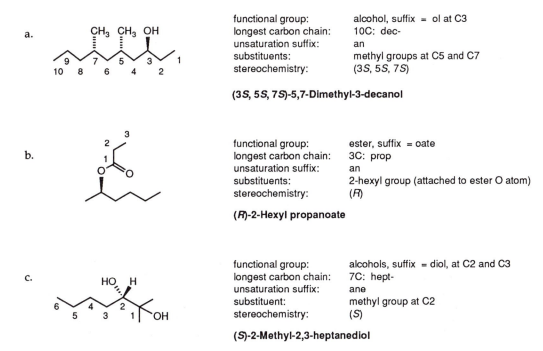

a.

functional group:	alcohol, suffix = ol at C3
longest carbon chain:	10C: dec-
unsaturation suffix:	an
substituents:	methyl groups at C5 and C7
stereochemistry:	(3S, 5S, 7S)

(3S, 5S, 7S)-5,7-Dimethyl-3-decanol

b.

functional group:	ester, suffix = oate
longest carbon chain:	3C: prop
unsaturation suffix:	an
substituents:	2-hexyl group (attached to ester O atom)
stereochemistry:	(R)

(R)-2-Hexyl propanoate

c.

functional group:	alcohols, suffix = diol, at C2 and C3
longest carbon chain:	7C: hept-
unsaturation suffix:	ane
substituent:	methyl group at C2
stereochemistry:	(S)

(S)-2-Methyl-2,3-heptanediol

16.17.   (continued)

d.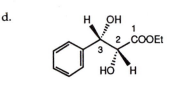

functional group:	ester, suffix = oate
longest carbon chain:	3C: prop-
unsaturation suffix:	an
substituents:	ethyl group (attached to ester O atom)
	hydroxy groups at C2 and C3
	phenyl group at C3
stereochemistry:	(2R, 3S)

Ethyl (2R, 3S)-2,3-dihydroxy-3-phenylpropanoate

16.18.   Interpret the name according to the directions given in Chapter 2.

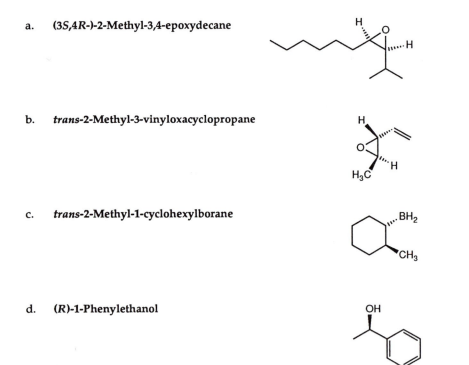

a.   **(3S,4R-)-2-Methyl-3,4-epoxydecane**

b.   ***trans*-2-Methyl-3-vinyloxacyclopropane**

c.   ***trans*-2-Methyl-1-cyclohexylborane**

d.   **(R)-1-Phenylethanol**

16.19.   Resolution of a racemic carboxylic acid is accomplished according to the scheme illustrated on text page 756. The acid is treated with an optically active base (in the scheme at the top of the next page, brucine is represented schematically). The diastereomeric salts are crystallized, which leads to precipitation of one isomer and leaves the other in solution. Treating the salt with aqueous HCl regenerates the carboxylic acid and produces brucine hydrochloride. Brucine is recovered by treating the hydrochloride salt with aqueous NaOH.

16.19.   (continued)

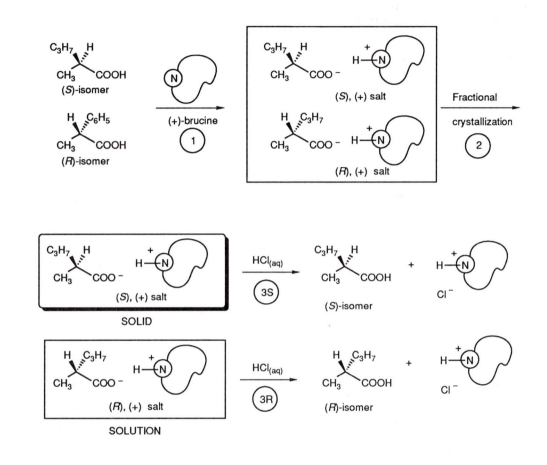

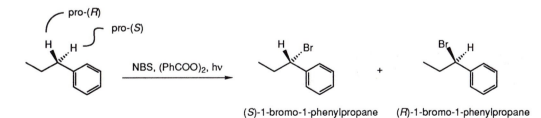

16.20.   Follow the procedures outlined in the solutions to exercises 16.3 and 16.4.

a.   Replacement of a benzylic hydrogen atom yields the bromo derivative. The hydrogen atom that is substituted in forming the (R) enantiomer is pro-(R); the other is pro-(S).

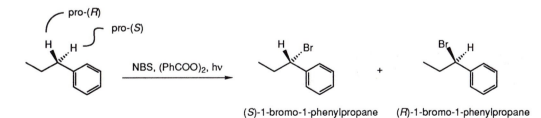

16.20.   (continued)

b.   Reaction of water with a carbocation intermediate forms an alcohol. The top face of the double bond leads to formation of the (*S*)-alcohol, so the top face is pro-(*S*).

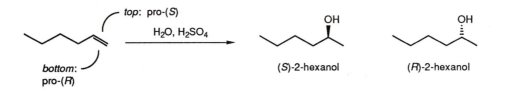

c.   Hydroboration of the alkene, followed by oxidative hydrolysis, produces an alcohol. Addition of the borane to the top face of the double bond leads to formation of the (*S*)-2-methyl compound (the methyl group is "down") so the top face is pro-(*S*).

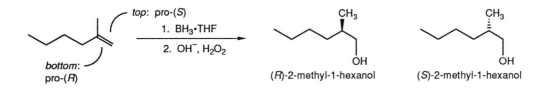

16.21.   Follow the procedure outlined in the solution to exercise 16.20. If a molecule is achiral, replacing either of two like groups normally creates enantiomers, in which case the two like groups are enantiotopic. If the starting material is already chiral, diastereomers are formed upon substitution, and the two like groups are diastereotopic. A methyl group is not prochiral; neither is a carbon atom with two pairs of like groups.

Pro-(*R*) hydrogen atoms are identified as such in the structures below; the other hydrogen atom attached to the same carbon atom is pro-(*S*).

Arachidonic acid is an achiral molecule, so replacing the hydrogen atom of any methylene group creates enantiomers. The protons of each methylene group are therefore enantiotopic.

Oxalosuccinate is chiral, so replacing either hydrogen atom of the methylene group creates a new chiral center, forming diastereomers. The protons are therefore diastereotopic.

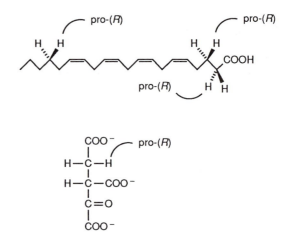

16.21.   (continued)

L-**phenylalanine** is chiral, so replacing either hydrogen atom of the methylene group generates a new chiral center, forming diastereomers. The protons are therefore diastereotopic.

Citric acid is achiral, but replacing a hydrogen atom in one of the methylene groups makes the central carbon atom chiral, in addition to creating a new chiral center. This molecule provides an example where the protons are diastereotopic, even though the original molecule is achiral. This situation results because the quaternary carbon atom is also prochiral.

Lauric acid is achiral, so replacing the hydrogen atom of any methylene group creates enantiomers. Each circled methylene group has enantiotopic protons.

The protons of the methyl group are not prochiral because replacing any one of them leaves two protons attached to that carbon atom, which is still achiral.

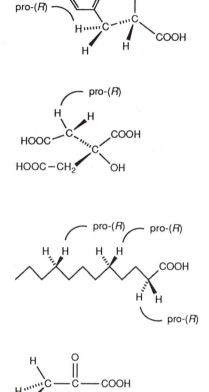

16.22.   Given both the names and structures, we interpret the stereochemistry according to the associated prefix, namely (Z) and (2R, 3S). The top face of the double bond in aconitate is the one to which the incoming H and OH groups add. We have to modify our designation of pro-(R) or pro-(S) to include a numeral that designates which group correlates with the stereochemical descriptor. In this example, the top face of the double bond can be called the pro-(2R) face with respect to hydration.

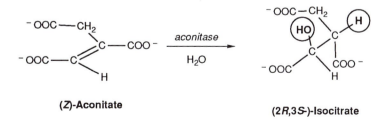

(Z)-**Aconitate**   (2R,3S)-**Isocitrate**

16.23. Using the procedures outlined in Chapter 15, we know that signals in the given spectrum that lie between δ 7 and 8 are aromatic proton resonances; the ones at δ 5.2 comprise the resonance for the methine proton, with its spin-spin coupling. The peaks at δ 4.05 represent the methylene group resonances with spin-spin coupling.

The methine carbon atom is a stereogenic center, which makes the methylene protons diastereotopic. Diastereotopic protons engage in geminal spin-spin coupling with a J value of approximately 10 Hz. Each methylene proton is therefore coupled to the methine proton and to its diastereotopic methylene proton partner, which gives rise to the observed unsymmetric pattern. Furthermore, the chemical shifts of H_a and H_b are close enough that second order effects are present here, which complicates that region greatly.

The methine proton, on the other hand, is coupled separately to each of the diastereotopic methylene protons, which accounts for its appearance as two doublets. The magnitude of the coupling to each diastereotopic proton is different because the conformational relationship is different. The vicinal coupling constant depends on the dihedral angle between neighboring protons. Even though the dihedral angle between vicinal protons varies because of conformational mobility in an acyclic compound, the relationship between H_a/H_c and between H_b/H_c is always different in any given conformation.

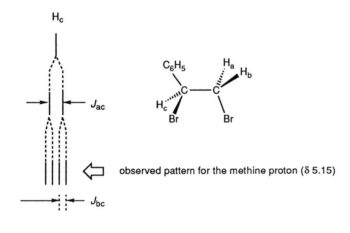

observed pattern for the methine proton (δ 5.15)

16.24. The structures of (+)-**2** and (–)-**2** are diastereomeric, differing only in the configuration of the carbon atom to which the amino group is attached (assignments of (+) and (–) are arbitrary):

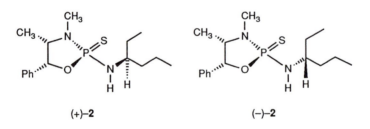

(+)–**2**   (–)–**2**

16.24.   (continued)

The %ee value is calculated from the ratio of the integrated area of the two curves (again, assigned below arbitrarily to the two isomers). The assumption is made that no change in the ratio occurs upon making derivatives of the enantiomeric amine products.

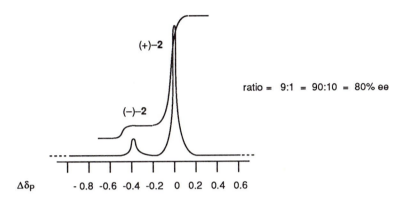

ratio = 9:1 = 90:10 = 80% ee

16.25.   To draw the structure of an enantiomeric product, switch the positions of any two groups attached to every chiral center. The amount of the minor isomer subtracted from the amount of the major isomer is the %ee value.

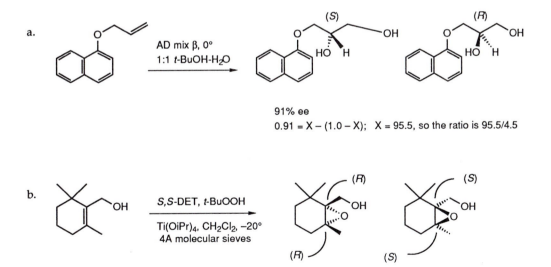

a.   AD mix β, 0°
1:1 *t*-BuOH-H₂O

91% ee
0.91 = X − (1.0 − X);   X = 95.5, so the ratio is 95.5/4.5

b.   *S,S*-DET, *t*-BuOOH

Ti(OiPr)₄, CH₂Cl₂, −20°
4A molecular sieves

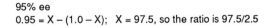

95% ee
0.95 = X − (1.0 − X);   X = 97.5, so the ratio is 97.5/2.5

16.25.   (continued)

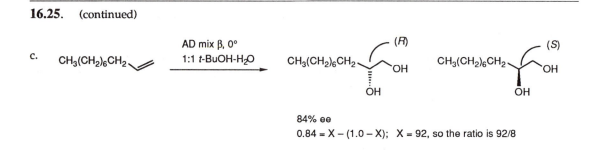

c.

84% ee
0.84 = X – (1.0 – X);   X = 92, so the ratio is 92/8

16.26.   The %de or %ee of a reaction is equal to the amount of the major diastereomer produced minus the amount of the minor one.

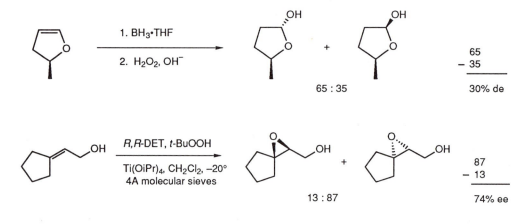

16.27.   To plan the synthesis of a chiral molecule, conceive disconnections or exchanges that identify compounds with the alcohol functional group. A chiral alcohol is often readily prepared by asymmetric hydroboration followed by oxidative hydrolysis. Only one solution is given below for each synthesis; other routes may be possible.

a.   To prepare a cyclic compound that contains a chiral center, choose a starting material with an intact ring, then append any substituents. The substituent here is a vinyl group, and formation of a carbon-carbon bond employs the Suzuki coupling reaction between an organoborane and vinyl iodide. The organoborane is made by hydroboration/oxidation of the cyclic alkene.

retrosynthesis:

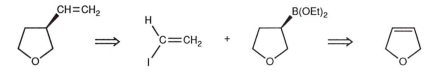

16.27.   (continued)

synthesis:

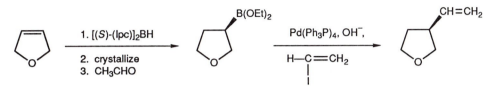

b.   An organoazide is made by nucleophilic substitution between azide ion and an aliphatic compound with a good leaving group. The mesylate ion is a good leaving group that can be made from an alcohol. The alcohol is made by asymmetric hydroboration followed by oxidation/hydrolysis.

retrosynthesis:

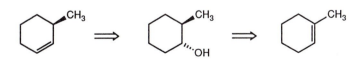

synthesis:

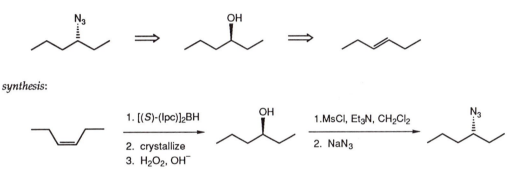

c.   The given alkene is made by an elimination reaction starting with *trans*-2-methyl-1-cyclohexanol. An E2 reaction requires that the proton and leaving group must have an *anti* relationship, and this requirement prevents the double bond from forming at the chiral carbon atom that bears the methyl group. The hydroboration step sets the stereochemistry of the carbon atom with the methyl group.

retrosynthesis:

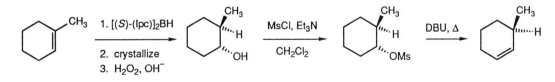

synthesis:

16.28. The Sharpless-Katsuki epoxidation method is specific for allylic alcohols. The stereochemistry of the diethyl tartrate isomer used in the procedure dictates the stereochemistry of the product epoxide according to the scheme illustrated on text page 778.

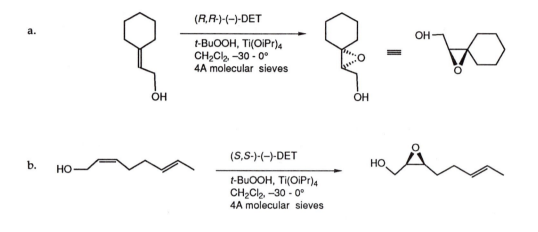

16.29. Only an allylic alcohol substrate reacts with the reagent combination of Ti(OiPr)$_4$, DET, and *t*-butyl hydroperoxide. The stereochemistry of the product epoxide is deduced from the results illustrated in the scheme on text page 778.

All of the substrates undergo asymmetric dihydroxylation. The stereochemistry is deduced from a comparison with the structures shown on text page 784.

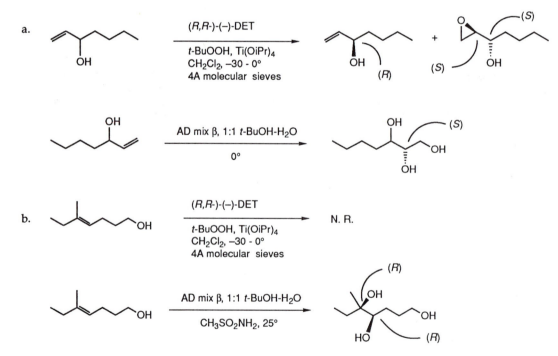

16.29.   (continued)

c.

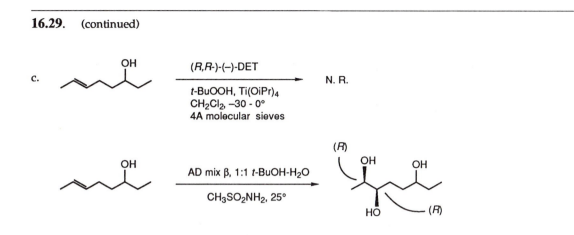

16.30.   Planning the synthesis of a chiral molecule requires that you conceive disconnections or exchanges that give compounds with either alcohol or epoxide functional groups. The rationale for this approach derives from the fact that these functional groups can be made by reactions between alkenes and reagents described in this chapter. Only one solution is given below for each synthesis; other routes may be possible.

a.   The ketone is made by oxidizing the corresponding alcohol, and the chiral alcohol is made by hydroboration/oxidation to set the stereochemistry of the ethyl group.

retrosynthesis:

synthesis:

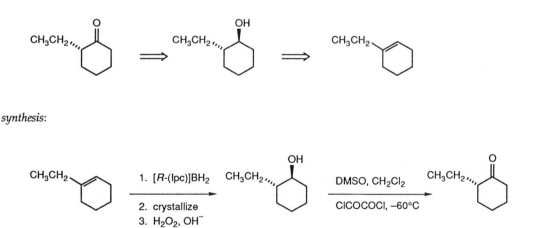

16.30. (continued)

b.   The given azidocyclohexane is made from the analogous alcohol by nucleophilic substitution via the mesylate derivative. The chiral alcohol that is needed is made by hydroboration/oxidation, which also sets the stereo-chemistry of the phenyl group. The cyclohexene derivative is prepared by adding the phenyl Grignard reagent to cyclohexanone, followed by dehydration. Cyclohexanone is made by hydration of cyclohexene, followed by oxidation.

retrosynthesis:

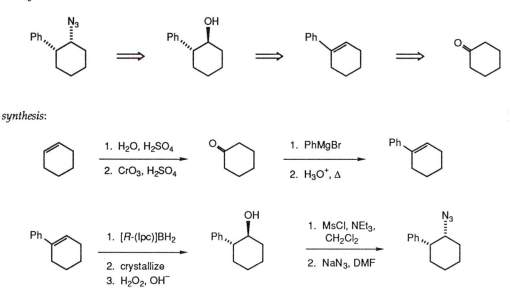

synthesis:

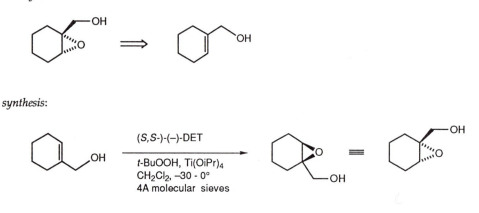

c.   The product is a seven-carbon epoxide, so a single step suffices. We choose the allylic alcohol as the starting material so that we can make use of Sharpless-Katsuki epoxidation.

retrosynthesis:

synthesis:

16.30.   (continued)

d.   A β-phenyl alcohol is equivalent to an epoxide that has been cleaved by hydrogenolysis, the benzylic C–O bond being more susceptible to cleavage than the other (text page 481). The chiral epoxide is made from the corresponding alkene, which in turn is made by coupling an organoborane and bromobenzene (Suzuki coupling).

retrosynthesis:

synthesis:

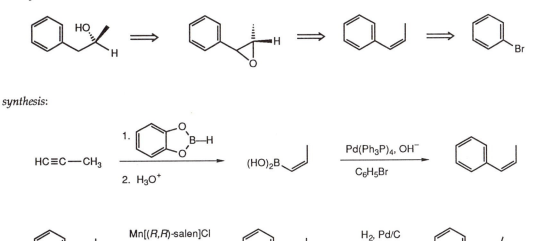

16.31.   First, summarize the reactions described in the accompanying paragraph.

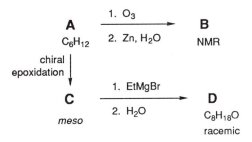

Compound **A** is an alkene because it has one site of unsaturation and reacts with ozone and undergoes epoxidization.

From the proton NMR spectrum, we conclude that compound **B** is an aldehyde because it has a resonance at δ 9.9. The appearance of this downfield signal as a triplet means that a methylene group is adjacent to the CHO group. The splitting pattern observed for the methylene group indicates that it is bonded both to a methyl and aldehyde groups. Compound **B** is therefore propanal, CH_3—CH_2—CHO.

16.31.   (continued)

The fact that only a single compound (**B**) is obtained from ozonolysis of **A** means that compound **A** is symmetric, and *cis*- or *trans*-3-hexene are the only options for its identity.

Epoxidation of the double bond in **A** forms a *meso* product, **C**.   Compound **A** must therefore be *cis*-3-hexene.   Compound **D** is racemic because addition of an ethyl group occurs at either end of the epoxide ring. This step destroys one of the stereogenic centers, and enantiomers are formed.

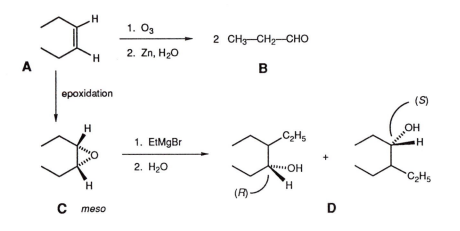

16.32.   As described in the solution to exercise 16.31, we first outline the reactions that occur.

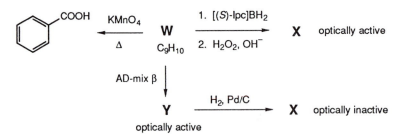

Permanganate ion converts an alkyl substituent attached to a benzene ring to a carboxylic acid group. Only one COOH group is present after oxidation of compound **W**, so the ring be monosubstituted to begin. Possible structures for **W** are therefore as follows.

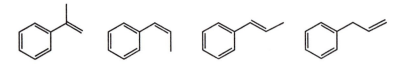

Even though 2-phenyl-1-propene does not have a benzylic hydrogen atom, $KMnO_4$ oxidizes the 2-propenyl to a methyl ketone group, then to the –COOH group.

16.32.   (continued)

Next, we consider the consequences of carrying out the given reactions with each of these possible starting materials. Hydroboration/oxidation of 3-phenyl-1-propene yields a primary alcohol in which there is no stereogenic carbon atom. Because compound **X** is optically active, we rule out the possibility that **W** is that compound.

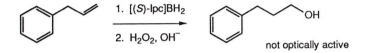

The other three compounds are converted to products that have a stereogenic center, although the disubstituted alkenes may generate mixtures of regioisomers (1-phenyl-2-propanol and 1-phenyl-1-propanol).

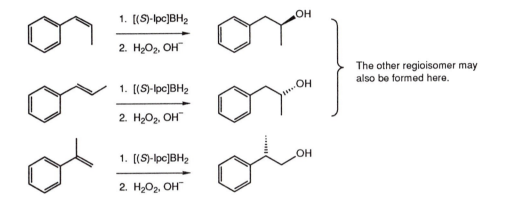

Next, we consider the results expected for asymmetric dihydroxylation of these three substances. All three will produce optically active diols that correspond to compound **Y** in the original scheme. Only hydrogenolysis of the dihydroxylated product from 2-phenyl-1-propene produces a racemic mixture of alcohols that correspond to **X**, however.

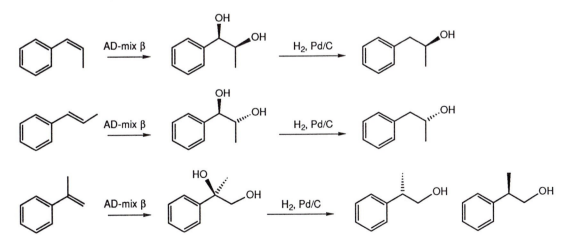

16.32.   (continued)

Therefore, compound **W** is 2-phenyl-1-propene. Its expected proton and ^{13}C NMR spectra are shown below. For the proton spectrum, numbers in parentheses represent the integrated intensities of the peaks. For the carbon NMR spectrum, letters in parentheses represent the expected pattern of peaks in the off-resonance spectrum (s = singlet, d= doublet, t = triplet, and q = quartet).

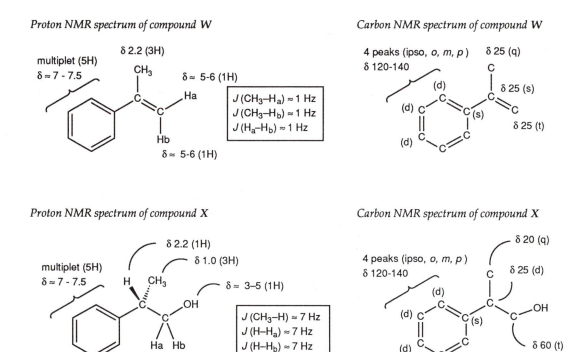

Proton NMR spectrum of compound W

Carbon NMR spectrum of compound W

Proton NMR spectrum of compound X

Carbon NMR spectrum of compound X

16.33.   The first step in the transformation presented in this exercise converts the alcohol to its mesylate derivative. If methylmercaptide ion were to react at the carbon atom bearing the leaving group, then the stereochemistry of the epoxide ring would not change.

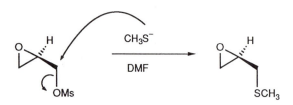

16.33. (continued)

The fact that the configuration of C2 is inverted means that the epoxide ring must be involved. We know that epoxides are susceptible to ring opening by nucleophiles (text page 308). Therefore, if the methylmercaptide ion reacts at the epoxide ring terminus, the resulting alkoxide ion formed in step (1), below, reacts in step (2) at the carbon atom bearing the mesylate group, regenerating an epoxide ring with inversion of configuration at the central carbon atom.

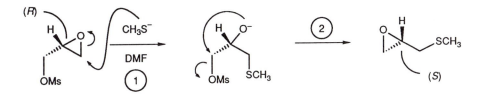

16.34. An asymmetric synthesis takes place when the possible transition states leading to enantiomeric products differ in their free energies of activation, as illustrated in Figure 16.1 (text page 759). An enzyme active site is a chiral catalyst because it is constructed from chiral building blocks, the α-amino acids,. The prochiral starting material therefore interacts with the chiral active site to produce diastereomeric transition states. In many enzyme-catalyzed reactions, this energy difference is great enough to generate enantiomerically pure products.

16.35. The active site of the enzyme that converts citrate to aconitate and isocitrate is chiral, but citrate is prochiral, having two ⁻OOC–CH₂– groups attached to the central carbon atom. When citrate is bound within the active site, the prochiral groups are differentiated by their proximity to the base that promotes dehydration:

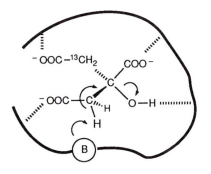

Many types of interactions—hydrogen bonds, electrostatic attractions, and van der Waals forces (schematically shown as dashed lines in the figure)—are used to orient the substrate molecule within the active site. But after citrate is bound, only one of the two methylene groups is positioned near the base, B, that promotes dehydration. Thus, even after the reaction reaches equilibrium, the ^{13}C-labeled carbon atom cannot have reacted in the same way that the unlabeled methylene group does.

NUCLEOPHILIC ADDITION REACTIONS
OF ALDEHYDES AND KETONES

17.1. Cyanide ion is a good nucleophile, and it adds to the ketone carbonyl group by reaction with the electrophilic carbon atom. The resulting alkoxide ion in this exercise reacts with HCN to form the cyanohydrin of cyclohexanone and to regenerate cyanide ion.

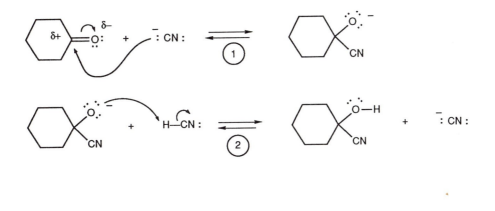

17.2. Recall from Chapter 4 (text page 132) that a cyclopropane ring is highly strained. Even with bent bonds, the ∠C–C–C values are approximately 102°, which is compressed from the normal value of 109.5°. If a carbonyl group is present in a three-membered ring, then the disparity between theoretical (120°) and actual (102°) bond angles is even greater. By reacting with water to produce an sp^3-hybridized carbon atom, cyclopropanone is converted to a compound having only the "normal" strain of cyclopropane.

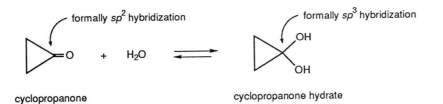

cyclopropanone   cyclopropanone hydrate

17.3.   Markovnikov addition of water to the double bond of 1-bromocyclohexene produces a *gem*-bromohydrin [steps (1) - (3)]. As described in the text on page 810, *gem*-halohydrins are unstable and lose the elements of HX to form the carbonyl product.

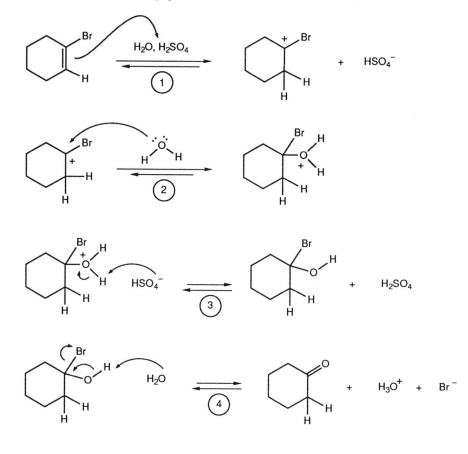

17.4.   In the first step of the reaction between benzaldehyde and ethylene glycol, the carbonyl group of benzaldehyde is protonated by its reaction with sulfuric acid. This makes the carbon atom more electrophilic. In step (2), an oxygen atom of ethylene glycol acts as a nucleophile and intercepts the carbocation. Hydrogen sulfate ion removes a proton from the protonated oxygen atom in step (3). In the fourth step, the hydroxyl group that was generated from the original carbonyl group is protonated by its reaction with sulfuric acid, generating a good leaving group. Loss of water in step (5) generates another carbocation, which is intercepted by the other oxygen atom of ethylene glycol in step (6). Finally, hydrogen sulfate ion removes the proton from the protonated acetal to yield the product, the ethylene acetal of benzaldehyde. Another name for this product is 2-phenyl-1,3-dioxolane (see the solution to exercise 17.29f).

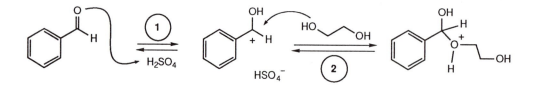

17.4.   (continued)

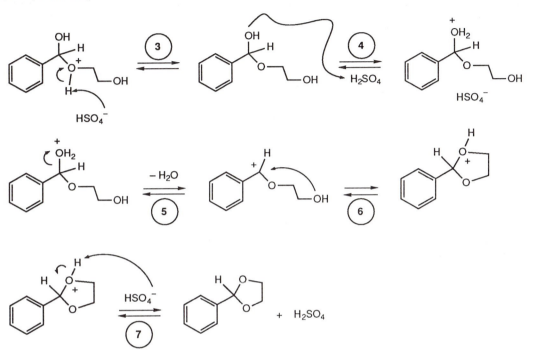

17.5.   In the first step of acetal formation from a vinyl ether, the alkene double bond is protonated by its reaction with sulfuric acid. The proton adds in a Markovnikov fashion, generating a resonance-stabilized carbocation. In step (2), the nucleophilic oxygen atom of the alcohol intercepts the carbocation. Hydrogen sulfate ion removes a proton from the protonated oxygen atom in step (3) to form the acetal product.

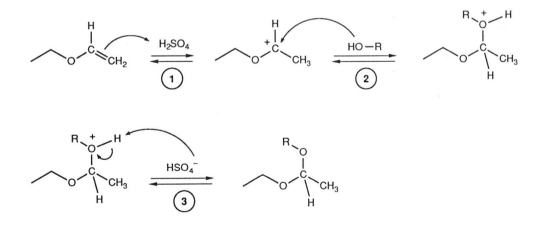

17.6. The mechanism for thioacetal formation follows the same steps described in the solution to exercise 17.4. The carbonyl group of cyclohexanone reacts with the sulfuric acid, which enhances the electrophilicity of the carbon atom. Then, the nucleophilic sulfur atom of ethanedithiol intercepts this carbocation in step (2). Another carbocation is subsequently generated by protonation and loss of water in steps (4) and (5), and that intermediate is intercepted by the other sulfur atom.

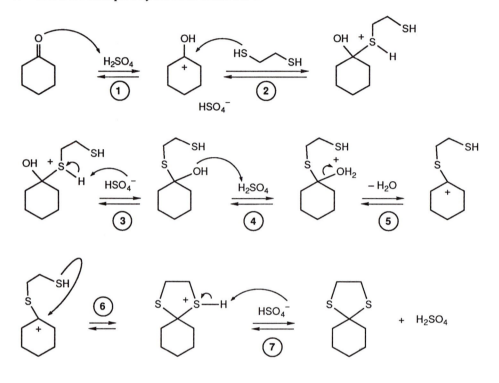

17.7. The mechanism for thioacetal formation proceeds via the same steps that are shown in the solution to exercise 17.6, except that the catalyst is a Lewis acid rather than a protic acid. The OH group of the hemiacetal binds zinc chloride and forms a good leaving group. Dissociation of the leaving group produces an electrophilic carbon atom that is intercepted by the nucleophilic sulfur atom of ethanedithiol. The hydroxide ion associated with zinc chloride removes the proton in the last step to liberate the thioacetal.

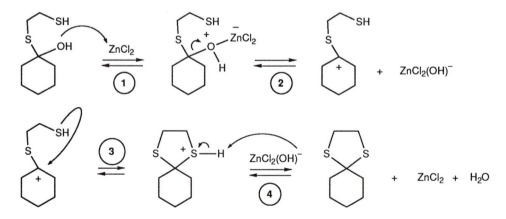

17.8.   The first two steps constitute Friedel-Crafts acylation (text page 532). Step 3 is thioacetalation, which proceeds by the mechanism illustrated in the solution to exercise 17.7. Raney nickel leads to replacement of the sulfur atoms with hydrogen atoms, converting the thioacetal carbon atom to a methylene group.

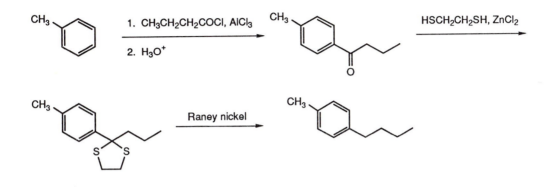

17.9.   The mechanism of acetal hydrolysis is the reverse of acetal formation. From the hemiacetal stage, protonation in step (1) converts the HOCH$_2$CH$_2$O— group into a good leaving group. An unshared pair of electrons of the hemiacetal OH group promotes dissociation of ethylene glycol in step (2), and a molecule of water deprotonates the cationic intermediate, forming the carbonyl group in step (3).

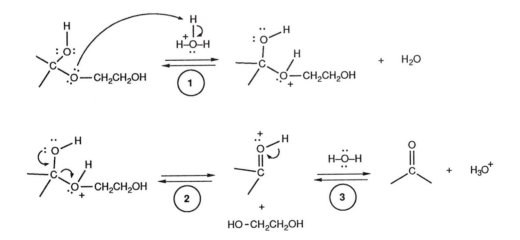

17.10. The mechanism for hemithioacetal hydrolysis is the reverse of hemithioacetal formation. The sulfur atom forms a complex with the chloromercury(II) ion, creating a good leaving group. Carbonate ion removes a proton from the OH group in step (2) to form the carbon-oxygen double bond by displacing the mercapto(mercury) group.

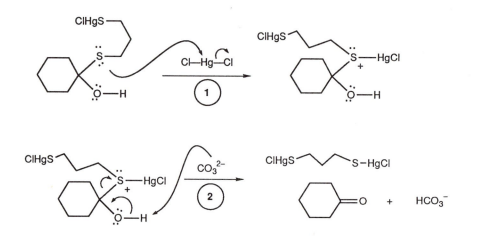

17.11. The mechanism for the hydrolysis of a THP ether is the same as the first steps in acetal hydrolysis. In the first step, the exocyclic oxygen atom is protonated and forms a good leaving group. An electron pair on the endocyclic oxygen atom displaces the molecule of alcohol in step (2). The carbon atom of the oxycation is trapped by water in step (3). Finally, the protonated 2-hydroxytetrahydropyran intermediate undergoes an acid-base reaction to form the neutral hemiacetal.

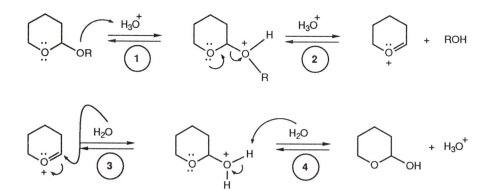

17.12. To draw the structure of a given carbohydrate derivative, copy the Fischer projection of the appropriate D sugar (the identity of which is underlined in the names given below) from Table 17.1 (text page 823). If the prefix has the word "deoxy", change the OH group to a hydrogen atom at the indicated position (the aldehyde carbon atom is C1). If the sugar has the L configuration, change the configuration of every stereogenic center by switching the positions of H and OH on each stereogenic carbon atom.

a. L-<u>arabinose</u>

b. D-2-deoxy<u>galactose</u>

c. L-3,4-dideoxy<u>talose</u>

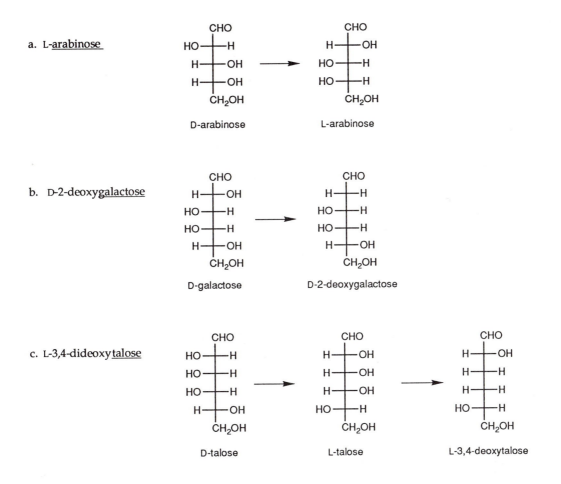

17.13. To draw the structure of a given carbohydrate derivative, copy the Fischer projection of the appropriate D sugar from Table 17.1 (text page 823). Draw a vertical line from the aldehyde carbon atom (which becomes the anomeric carbon atom), and connect the line to the oxygen atom at C4 if the suffix is furanose and to the oxygen atom at C5 if the suffix is pyranose. The oxygen atom of the original aldehyde group is changed to an OH group: the OH group is drawn to the left if the compound is the β anomer and to the right if the compound is the α anomer.

a. β-D-<u>arabino</u>furanose

b. α-D-<u>galacto</u>pyranose

c. α-D-<u>ribo</u>furanose

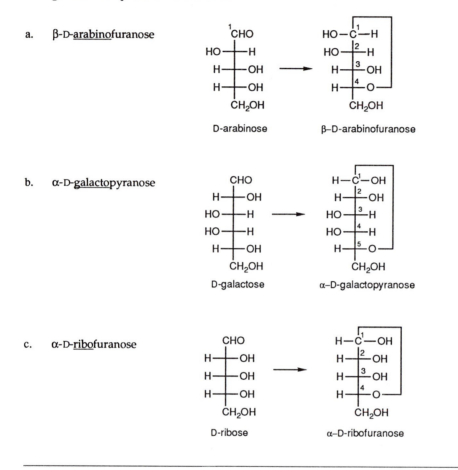

17.14. To draw the Fischer projection of β-D-lyxofuranose, follow the procedure outlined in the solution to exercise 17.13. Remember that the β anomer has the OH group pointing left; the α anomer has the OH group on the right. To convert from the Fischer to the Haworth projection, draw the ring in its perspective form, and place groups on the right side of the Fischer projection below the plane of the ring in the Haworth form.

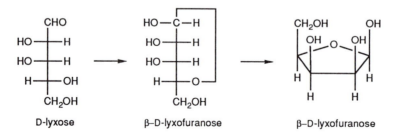

17.15.   The mechanism for hydrolysis of a glycoside is the same as that for the conversion of an acetal to the corresponding hemiacetal derivative, for instance as outlined in the solution to exercise 17.9. For α-D-gluco-pyranose, protonation by H_3O^+ in step (1) converts the alcohol portion of the hemiacetal to a good leaving group. An unshared pair of electrons on the anomeric OH group promotes dissociation of the C5 OH group in step (2). Water is the base that deprotonate the cationic intermediate, generating the carbonyl form of the sugar in step (3).

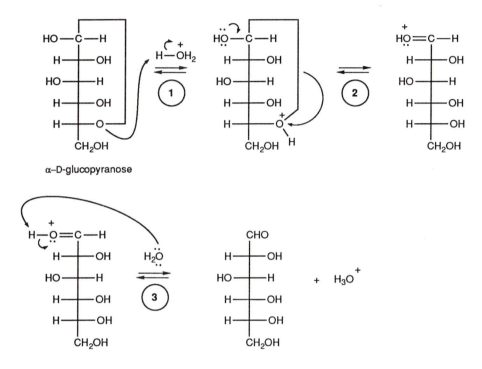

α–D-glucopyranose

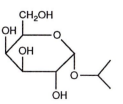

$+$ H_3O^+

17 16.   To interpret the name of a carbohydrate with the suffix "–oside", first recognize that the sugar is an acetal derivative. The unnumbered prefix is the alkyl group bonded to the oxygen atom attached at the anomeric carbon atom. The Greek letter indicates whether this alkoxy group is "up" (β) or "down" (α). The portion of the name "furan" (5 atoms) or "pyran" (6 atoms) tells what size ring is formed.

a.   **Ethyl β-D-fructofuranoside**

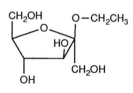

b.   **Isopropyl α-D-galactopyranoside**

17.17. D-Xylose exists in equilibrium with its five-membered ring hemiacetal (glycoside) form (compare its structure with that of D-ribose). When a glycoside is treated with methanol and a catalytic amount of HCl, the methoxy acetal derivative is formed. Treating that compound with sodium hydride and methyl iodide in DMF leads to methylation of every other –OH group.

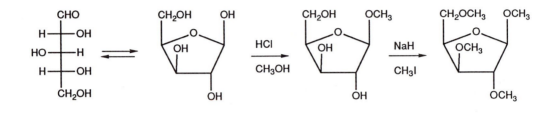

17.18. Hydrolysis of ethyl 2,3,4,6-tetra-O-methyl-α-D-glucopyranoside begins with protonation of the oxygen atom of the anomeric alkoxy group (–OEt). The other oxygen atom bonded to the anomeric carbon atom participates with its pair of electrons to expel the leaving group in step (2). The resulting cation is intercepted by a molecule of water in step (3), and deprotonation of the next intermediate yields the hemiacetal form of the carbohydrate. No other alkoxy group reacts this way because each is ether oxygen atom, not an acetal oxygen atom. The hemiacetal derivative formed in step (4) subsequently undergoes ring opening and epimerization at the anomeric carbon atom, like the hemiacetal form of any sugar.

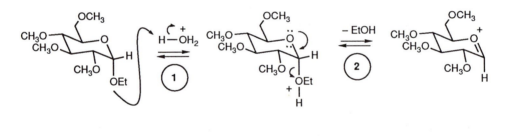

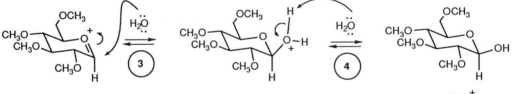

17.19. The amomer of a carbohydrate is an isomer in which only the configuration of the hemiacetal carbon atom is inverted. Two anomeric carbon atoms are present in lactose; the one that exists in the hemiacetal form is at the lower right end of the structure. β-Lactose has the anomeric OH group in the equatorial position, as shown below at the right.

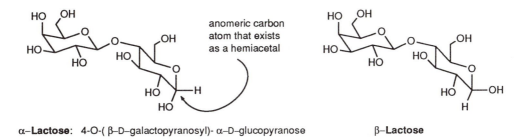

α–**Lactose**: 4-O-(β–D–galactopyranosyl)- α–D–glucopyranose   β–**Lactose**

17.20. For cell wall cleavage to occur in the reaction catalyzed by lysozyme, an acidic group (Glu-35) is required to protonate the oxygen atom of the acetal bond. After dissociation of an alcohol fragment, a carbocation forms. This intermediate is stabilized partially by the negative charge of carboxylate ion Asp-52. If both carboxylic acid side chains (Asp and Glu) are protonated, then no carboxylate ion is present to stabilize the carbocation intermediate. If both side chains are deprotonated, then an acidic group is not available to initiate the first step of the reaction. At the optimum pH values, both necessary reactive centers, a carboxylic acid group and a carboxylate ion, are present in the active site of the enzyme.

17.21. Organometallic reagents react with ketones and aldehydes by addition of the organic group ("R" of RLi or RMgX) to the carbon-oxygen double bond, which results in formation of a new bond between the R group and the carbonyl carbon atom. An alcohol is the product after aqueous acid workup.

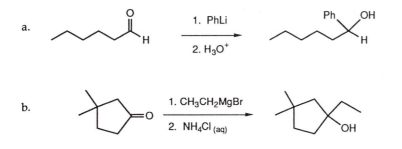

17.22. An alcohol is the product of metal hydride reduction of an aldehyde or ketone carbonyl groups. An aldehyde forms a primary alcohol, and a ketone forms a secondary alcohol.

a.

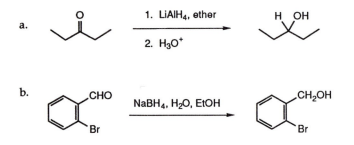

b.

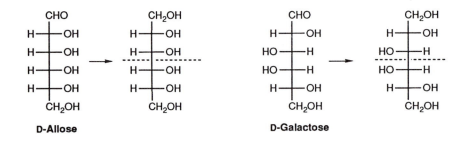

17.23. If a D-aldohexose is known to yield a *meso* alditol upon reduction, the OH groups attached to C2 & C5 and to C3 & C4 must have a mirror-image relationship to each other (the mirror plane is shown as a dashed line in the structures shown below). Only two aldohexoses (six-carbon aldehyde sugars) form *meso* alditols upon reduction: D-allose and D-galactose.

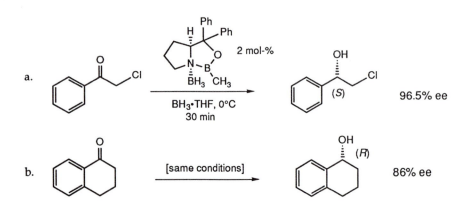

17.24. The structure of the product from reduction of a ketone by the (*S*)-oxazaborolidine reagent shown on text page 845 in combination with borane•THF (catalytic conditions) is deduced by looking at the examples given in the text. The absolute configuration of the alcohol is normally (*R*) when the (*S*)-oxazaborolidine is used, but the chloro ketone forms a product with the (*S*) configuration because the priorities of the substituents change when chlorine is present.

a.

b.

17.24.   (continued)

c.

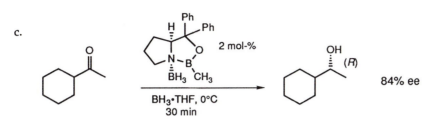

17.25.   Friedel-Crafts acylation of benzene with benzoyl chloride yields benzophenone. Clemmensen reduction of the ketone produces the hydrocarbon diphenylmethane.

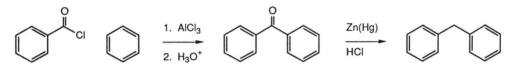

A retrosynthetic precursor of diphenylmethane is the corresponding benzylic alcohol, which is made by a Grignard reaction between phenylmagnesium bromide and benzaldehyde followed by hydrolysis.

retrosynthesis:

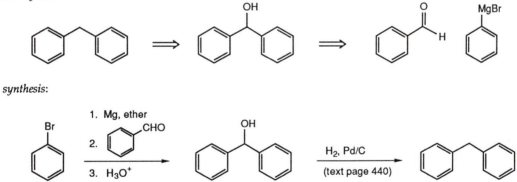

synthesis:

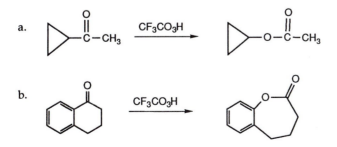

17.26.   The Baeyer-Villiger reaction occurs when a ketone or aldehyde is treated with peracid. The product is the ester formed by inserting the "new" oxygen atom into the bond between the carbonyl group and the substituent that is more highly substituted (aryl migrates more readily than alkyl).

a.

b.

17.27. The symmetry of D-glucose is such that two five-membered ring lactones form when the chain termini are converted to carboxylic acid functional groups.

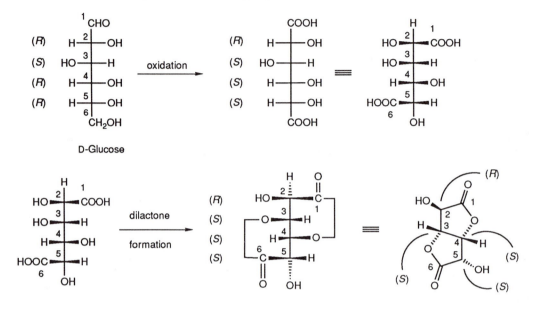

D-Glucose

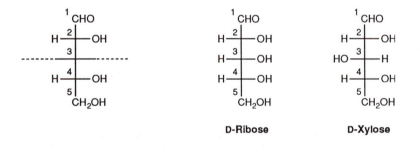

17.28. If a D-aldopentose yields a *meso* diacid upon oxidation, the OH groups attached at C2 and C4 must have a mirror relationship to each other with respect to a mirror plane at C3. Only two aldopentoses yield *meso* aldaric acids: D-ribose and D-xylose.

D-Ribose   D-Xylose

17.29. Interpret the name according to procedures outlined in Chapter 2.

a. **(Z)-3-chloro-2-pentenal**

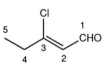

17.29.   (continued)

b.   **2-chloro-3-nitrobenzaldehyde**

c.   **α,α′-difluoroacetone**

d.   **3′,4′-dimethoxyacetophenone**

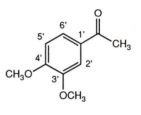

e.   **dichloroacetaldehyde diethyl acetal**
     An acetal is a derivative of the aldehyde group; in this
     compound, an ethyl group is attached to each oxygen
     atom of the acetal functional group.

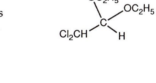

f.   **2-(1,1-dimethylpropyl)-1,3-dioxolane**
     1,3-Dioxolane is the five-membered ring heterocycle with
     oxygen atoms at positions 1 and 3.

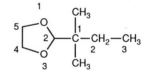

17.30.   Interpret the spectra according to directions given in Chapter 15 (follow the procedure outlined in the solution to exercise 15.17).

a.

δ 10	aldehyde proton (H$_a$)	doublet	J_{ab} = 2 Hz
δ 5-6	alkene proton (H$_b$)	doublet	J_{ab} = 2 Hz
δ 2-2.5	methylene group	quartet	J = 7 Hz
δ 1	methyl group	triplet	J = 7 Hz

17.30. (continued)

b.

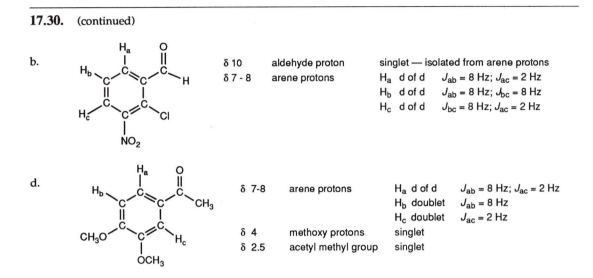

δ 10   aldehyde proton   singlet — isolated from arene protons

δ 7 - 8   arene protons

H_a d of d J_{ab} = 8 Hz; J_{ac} = 2 Hz

H_b d of d J_{ab} = 8 Hz; J_{bc} = 8 Hz

H_c d of d J_{bc} = 8 Hz; J_{ac} = 2 Hz

d.

δ 7-8   arene protons

H_a d of d J_{ab} = 8 Hz; J_{ac} = 2 Hz

H_b doublet J_{ab} = 8 Hz

H_c doublet J_{ac} = 2 Hz

δ 4   methoxy protons   singlet

δ 2.5   acetyl methyl group   singlet

17.31. Follow the procedure given in the solution to exercise 7.28.

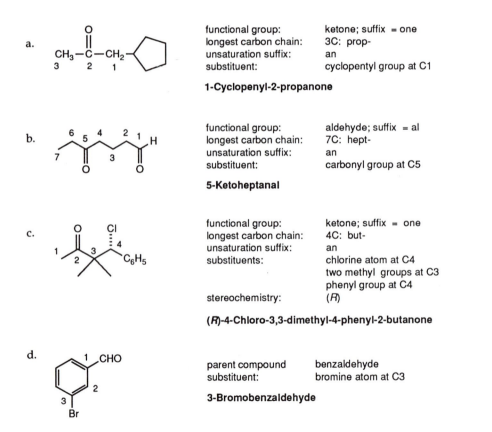

a.

functional group: ketone; suffix = one

longest carbon chain: 3C: prop-

unsaturation suffix: an

substituent: cyclopentyl group at C1

1-Cyclopenyl-2-propanone

b.

functional group: aldehyde; suffix = al

longest carbon chain: 7C: hept-

unsaturation suffix: an

substituent: carbonyl group at C5

5-Ketoheptanal

c.

functional group: ketone; suffix = one

longest carbon chain: 4C: but-

unsaturation suffix: an

substituents: chlorine atom at C4

two methyl groups at C3

phenyl group at C4

stereochemistry: (*R*)

(*R*)-4-Chloro-3,3-dimethyl-4-phenyl-2-butanone

d.

parent compound   benzaldehyde

substituent: bromine atom at C3

3-Bromobenzaldehyde

17.31.   (continued)

e.

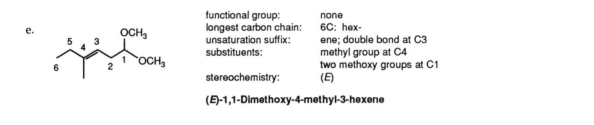

functional group:	none
longest carbon chain:	6C: hex-
unsaturation suffix:	ene; double bond at C3
substituents:	methyl group at C4
	two methoxy groups at C1
stereochemistry:	(*E*)

(*E*)-1,1-Dimethoxy-4-methyl-3-hexene

17.32.   In carbohydrate names, the D or L designation specifies the stereochemistry of the stereogenic carbon atom farthest from the carbonyl group. An aldehydic sugar is an aldose, and a keto sugar is a ketose. The total number of carbon atoms identifies the carbohydrate as a tetrose, pentose, or hexose.

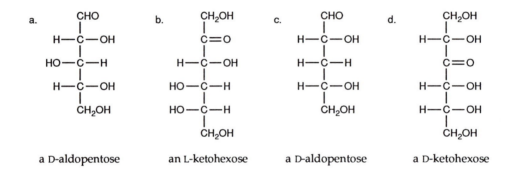

a D-aldopentose   an L-ketohexose   a D-aldopentose   a D-ketohexose

---

17.33.   To name a carbohydrate, first identify the root word by comparing the structure of the compound in its Fischer projection with structures shown in Table 17.1 on text page 823.

- If the compound is cyclic, a suffix is appended to the root word to denote whether the compound exists in a five- (**furanose**) or six-membered (**pyranose**) ring.
- If the compound is an acetal, which is identified by the presence of an –OR group attached to the anomeric carbon atom, then the suffix is changed from **-ose** to **-oside** and the name of the substituent group of the acetal appears as a separate, unnumbered prefix.
- For cyclic forms, β is used to denote the configuration of the anomeric carbon atom if the oxygen-containing group (OH or OR) is above the plane, or α if it is below the plane of the ring.

17.33. (continued)

a.

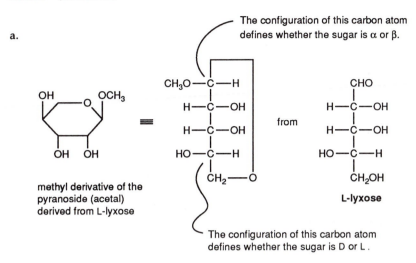

OH ⟋O⟍ OCH₃

OH OH

methyl derivative of the
pyranoside (acetal)
derived from L-lyxose

The name of this carbohydrate is **methyl β-L-lyxopyranoside**.

b.

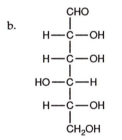

The name of this carbohydrate is **D-gulose**.

c.

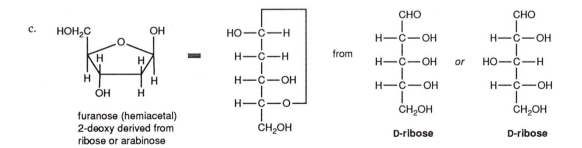

HOH₂C ⟋O⟍ OH

H   H
H   H
OH   H

furanose (hemiacetal)
2-deoxy derived from
ribose or arabinose

The name of this carbohydrate is **β-D-2-deoxyribofuranose**. This compound can also be named as **β-D-2-deoxyarabinose**, because the OH group that has been removed could have either configuration initially.

17.33.   (continued)

d.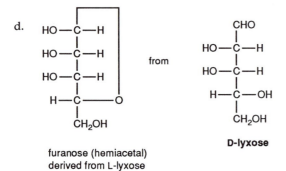

furanose (hemiacetal)
derived from L-lyxose

D-lyxose

The name of this carbohydrate is β-D-**Lyxofuranose**

e.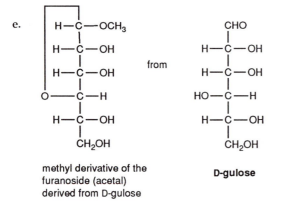

methyl derivative of the
furanoside (acetal)
derived from D-gulose

D-gulose

The name of this carbohydrate is **methyl α-D-gulofuranoside**

17.34.   A reducing sugar is one that has either an aldehyde or hemiacetal functional group (keto sugars are also reducing in many instances because they isomerize to aldoses). The latter exists in equilibrium with the aldehyde form. Compounds with an aldehyde group react with a mild oxidizing agent. For compounds shown in exercise 17.33, the reducing sugars are **b, c,** and **d.** Compounds **a** and **e** are acetals, so they are not reducing sugars.

17.35. Structures of the common carbohydrates are shown in Table 17.1 on text page 823. Switching the horizonal substituents (H and OH) at every stereogenic center in a sugar structure generates the L isomer. Any substitutions in the sugar structure are indicated by use of appropriate prefixes.

a. **L-glucose.** This compound is the mirror image of D-glucose, the Fischer projection of which is shown in Table 17.1.

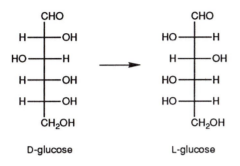

b. **D-2-deoxythreose.** Numbering starts at the aldehyde group of an aldose. Deoxy means that an oxygen atom has been removed from the position indicated by the numeral, in this case C2. The configuration of C3 remains the same as that in the parent sugar.

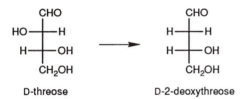

c. **L-3-deoxymannose.** This sugar lacks the OH group at C3, and the stereochemistry of each stereogenic center has been inverted relative to those in the parent carbohydrate.

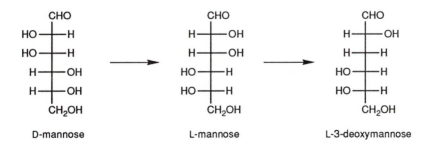

17.36. Follow the instructions outlined in Example 17.1 on text page 826. The root word of the sugar, below, is indicated by underlining.

a. β-D-<u>gluco</u>furanose

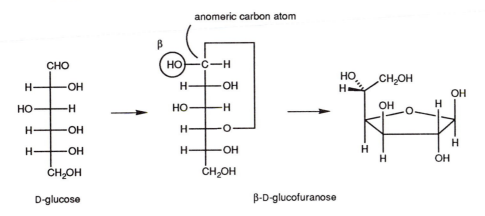

b. α-D-2-deoxy<u>manno</u>pyranose

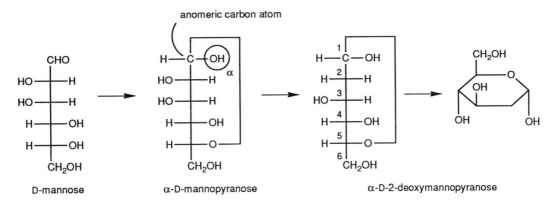

c. α-D-<u>talo</u>pyranose-6′-phosphate

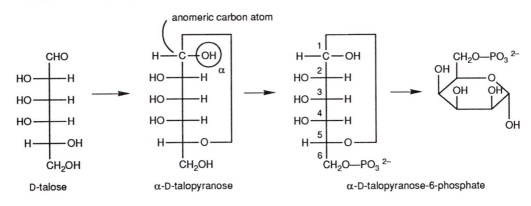

17.36. (continued)

d. methyl ß-D-<u>galacto</u>furanoside

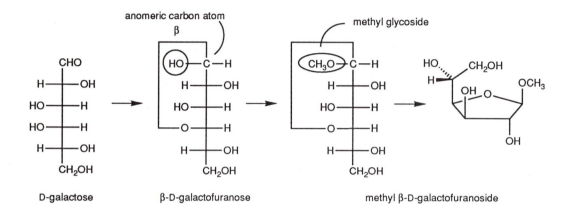

17.37. An anomer of a sugar differs only in the configuration of its anomeric carbon atom (the one with two oxygen atoms attached). The configuration of every stereogenic center is inverted in the enantiomer of a sugar.

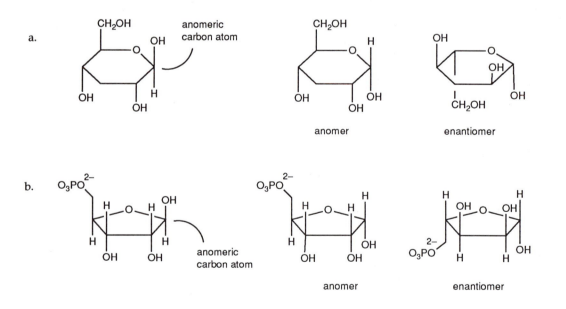

17.38. To name the carbohydrate derivatives, follow the procedure outlined in the solution to exercise 17.33. Draw the structures of the product(s) of each transformation on the basis of the reactions discussed in the chapter.

a.   D-Mannose is treated with isopropanol and a catalytic amount of HCl, which produces the acetal derivatives. Both anomers (α and β) are formed. The compounds are **isopropyl α-D-mannopyranoside** and **isopropyl β-D-mannopyranoside**

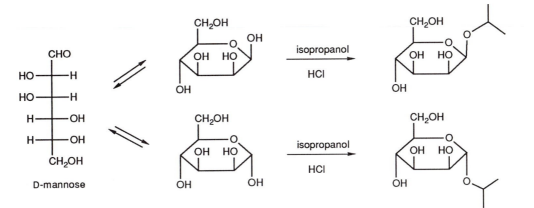

b.   Treating the acetal derivative of the sugar with sodium hydride and ethyl iodide in DMF leads to alkylation of all the OH groups (only one anomer is shown for this step, but the other anomer undergoes the same reaction). The products are **isopropyl 2,3,4,6-tetra-O-ethyl-α-D-mannopyranoside** and its β anomer.

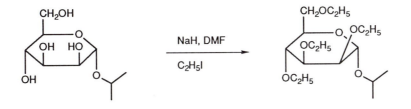

c.   Dilute sulfuric acid catalyzes the hydrolysis of the acetal to form the hemiacetal, which for a sugar is in equilibrium with the open (linear) form. The products are **2,3,4,6-tetra-O-ethyl-α-D-mannopyranose** and its β anomer.   The open chain forms is **2,3,4,6-tetra-O-ethyl-D-mannose.**

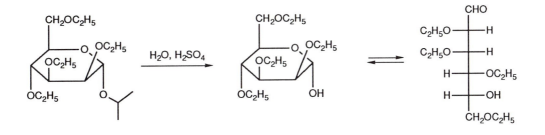

17.38. (continued)

d. Sodium borohydride in aqueous ethanol reduces the aldehyde group to the corresponding primary alcohol. The ether and alcohol functional groups are not affected by this reagent. The product is **2,3,4,6-tetra-*O*-ethyl-D-mannitol.**

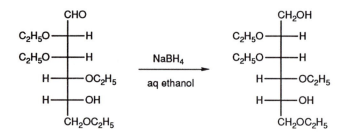

17.39. A ketone reacts with a variety of reagents by addition of a nucleophile to the carbon oxygen double bond. The benzene ring of propiophenone is susceptible to reaction by electrophilic substitution pathways as described in Chapter 12.

a. The combination of Zn(Hg) and HCl constitutes the Clemmensen reduction, which converts a ketone carbonyl group to a methylene group.

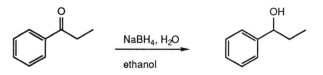

b. The combination of NaBH₄ in aqueous ethanol reduces the ketone carbonyl group to form the corresponding 2° alcohol.

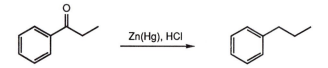

c. A ketone reacts with HOCH₂CH₂OH and *p*-TsOH to form the ethylene acetal derivative.

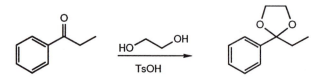

17.39. (continued)

d.   An aromatic ketone reacts with HNO_3 and H_2SO_4 to form the nitrated derivative. A ketone carbonyl group is a *meta* director (text page 544).

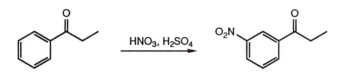

e. A ketone is reduced by $LiAlH_4$ in ether to form the 2° alcohol after workup with aqueous acid.

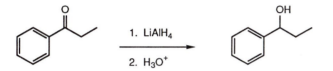

17.40.

a. Grignard and organolithium reagents add to a ketone to form a 3° alcohol after workup with aqueous acid.

b.   Acetals are unstable in aqueous acid, undergoing hydrolysis to produce two equivalents of an alcohol and a carbonyl compound. In this case, the two "alcohols" are part of a single molecule, so the products are a diol and a ketone.

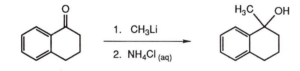

c. Grignard reagents react with aldehydes to form 2° alcohols after hydrolytic workup. This step creates a new stereogenic center, but it does not affect the chirality of other carbon atoms in the molecule. In this reaction, diastereomers are produced.

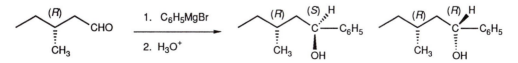

17.40.   (continued)

d.   A Grignard reagent reacts with the acidic proton of a carboxylic acid functional group, generating the carboxylate salt. Acid workup regenerates the carboxylic acid group, so no apparent reaction occurs.

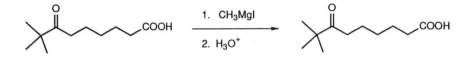

e.   An aryl iodide undergoes substitution upon reaction with an organocuprate reagent, and the iodine atom is replaced by the alkyl or aryl group attached to copper (the propyl group in this example). A ketone carbonyl group is normally inert toward reaction with an organocuprate reagent.

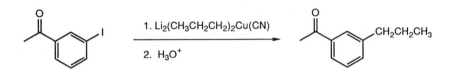

17.41.   When a limit of six carbon atoms is set on the size of permissible starting materials, retrosynthetic analysis of aromatic compounds requires breaking the aryl-alkyl bond:

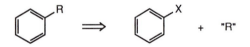

When a benzylic alcohol is the desired product, the reactants that are needed include the phenyl (or aryl) Grignard reagent and an aliphatic aldehyde or ketone.

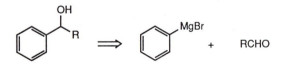

17.41. (continued)

a. A 2° benzylic alcohol is made by reaction between phenylmagnesium bromide and an aldehyde.

retrosynthesis:

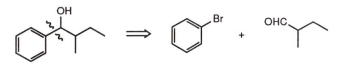

synthesis:

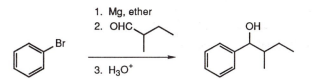

b. A 3° benzylic alcohol is made by the reaction between an aryl Grignard reagent and a ketone.

retrosynthesis:

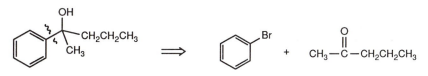

synthesis:

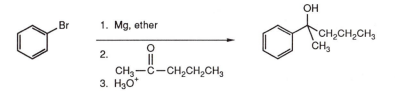

c. A 3° alcohol is prepared by reaction between a Grignard reagent and a ketone. The aldehyde group at the end of the chain can be incorporated into the starting material, but it has to be protected before a Grignard reagent is made from that portion of the molecule.

retrosynthesis:

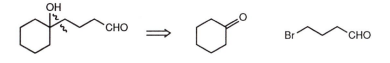

17.41. (continued)

synthesis:

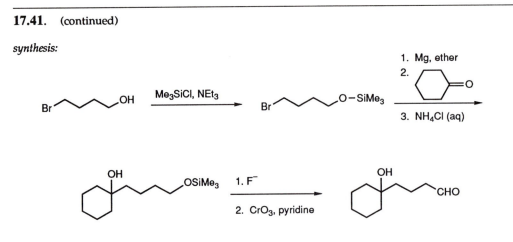

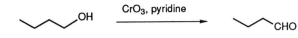

17.42. The most general method to make an aldehyde is oxidation of the corresponding primary alcohol. Ozonolysis is another good strategy if the starting material has more carbon atoms than the desired product.

a. From 1-butanol: oxidation

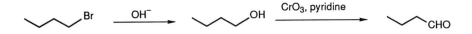

b. From 1-bromobutane: substitution followed by oxidation

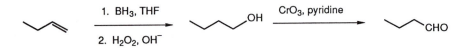

c. From 1-butene: hydroboration followed by oxidation

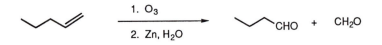

d. From 1-pentene: ozonolysis to remove the "extra" carbon atom

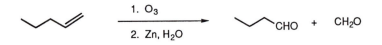

17.43. If a D-aldopentose yields a chiral alditol, the OH groups attached to C2 and C4 must be on opposite sides of the vertical in a Fischer projection. Thus, two aldopentoses, arabinose and lyxose form chiral products upon reduction.

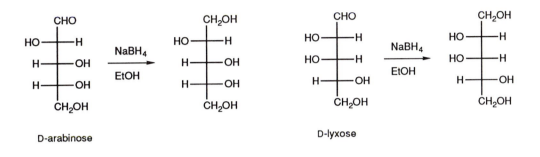

17.44. If an L-aldohexose forms a *meso* aldaric acid, the OH groups on C2 & C5 and on C3 & C4 must be on the same side of the vertical in a Fischer projection. Only two aldohexoses yield *meso* alditols: L-allose and L-galactose.

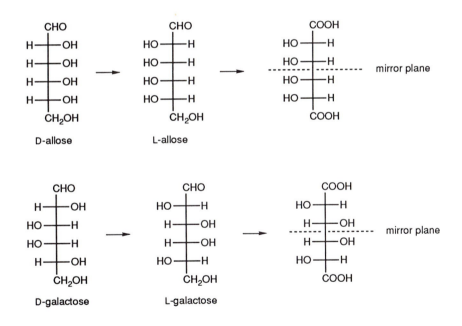

17.45. In the **pinacol rearrangement**, 2,3-dimethyl-2,3-butanediol reacts with acid and is protonated of one of the OH groups in step (1). Dissociation of a molecule of water in step (2) forms a carbocation, and a methyl group migrates in step (3) to form the resonance stabilized hydroxy-substituted cation. In the fourth and last step, deprotonation yields 3,3-dimethyl-2-butanone.

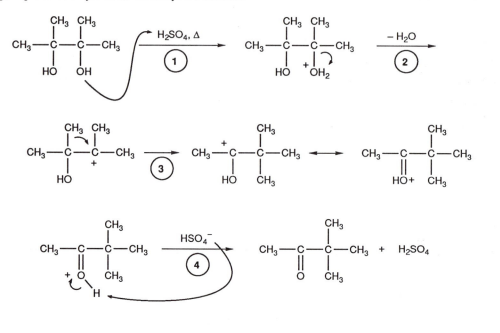

<hr>

17.46.

a. An organolithium compound adds to a ketone group to form a 3° alcohol after mild acid hydrolysis in step 2. NaH converts the alcohol OH group to its salt, the alkoxide ion. That species is a nucleophile and reacts with an alkyl halide by an S_N2 pathway to form the ether.

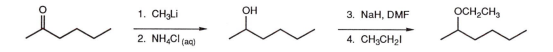

b. Benzene undergoes Friedel-Crafts acylation with propanoyl chloride to form the ketone after hydrolytic workup in step 2. The carbonyl group is subsequently reduced with use of lithium aluminum hydride to yield the 2° alcohol after hydrolysis.

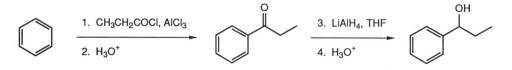

17.46.   (continued)

c. An aldehyde reacts with an alcohol in the presence of acid and a dehydrating agent to form the acetal derivative. Stereogenic centers within the molecule are normally not affected.

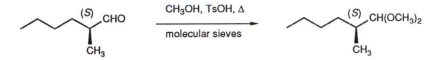

d. A γ-hydroxy aldehyde cyclizes to its hemiacetal form. In the presence of an alcohol and an acid catalyst, the corresponding acetal is produced. The chirality of the carbinol carbon atom is not affected by hemiacetal formation.

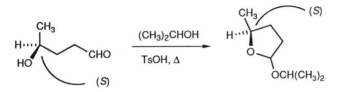

e. Hydroboration of a terminal alkene yields a 1° alcohol after oxidative hydrolysis. Hydrolysis of the acetal functional group in step 3 produces a δ-hydroxy ketone, which cyclizes under the acidic conditions to form the hemiacetal.

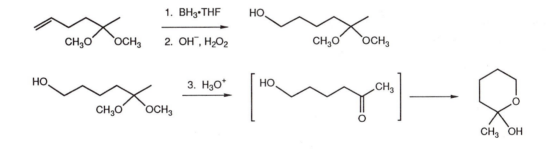

17.47.   a. PCC (pyridinium chlorochromate) oxidizes a 2° alcohol to the corresponding ketone. Fluoride ion unmasks the alcohol functional group, leading to spontaneous cyclic hemiacetal formation.

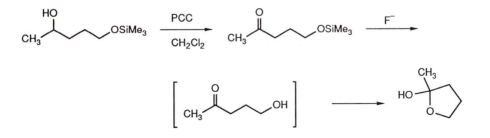

17.47.   (continued)

b. The oxazaborolidine reagent produces a chiral alcohol from an aryl alkyl ketone. Reaction of the alcohol with trimethylchlorosilane yields the trimethylsilyl ether.

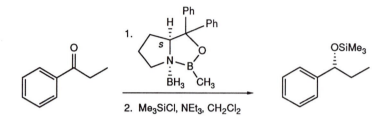

c. The specified enzyme along with NADH reduces a dialkyl ketone to the (*S*) alcohol. LDA is a strong base that deprotonates the alcohol OH group to form the alkoxide ion. When a good leaving group is present in the same molecule, intramolecular nucleophilic substitution takes place, forming the cyclic ether.

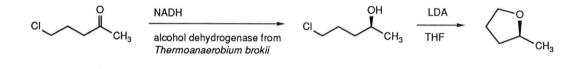

17.48.   When cyanide ion is present with a ketone or aldehyde, the nucleophilic carbon atom of CN^- reacts with the electrophilic carbon atom of the carbonyl group. A silicon–cyanide bond is weaker than a silicon–oxygen bond, so the alkoxy group reacts with cyanotrimethylsilane to displace cyanide ion and form the trimethylsilyl derivative of the cyanohydrin. Cyanide ion is thereby regenerated and so fulfills its role as a catalyst in this transformation.

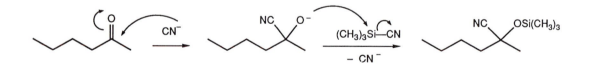

a.   Silicon has an extremely high affinity for fluorine. Therefore, reaction of the trimethylsilyl cyanohydrin with HF leads to removal of the trimethylsilyl group. A cyanohydrin, however, is stable toward acid, so the reaction stops there.

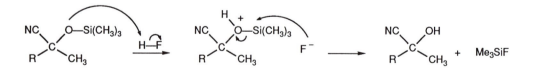

17.48.   (continued)

b.   A trimethylsilyl ether is stable toward base, so no reaction occurs upon treatment with OH⁻.

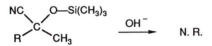

c.   Silicon has an extremely high affinity for fluorine. Therefore, fluoride ion reacts with the trimethylsilyl group, generating the conjugate base of the cyanohydrin. Displacement of cyanide ion leads to regeneration of the ketone carbonyl group.

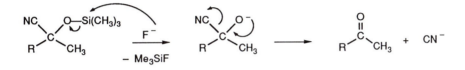

17.49.   The two reactions described in this exercise constitute the first steps of the Haworth synthesis of naphthalenes. Succinic anhydride reacts as an acylating agent with benzene to form the phenyl keto acid. Clemmensen reduction converts the carbonyl to a methylene group without affecting the carboxylic acid functional group.

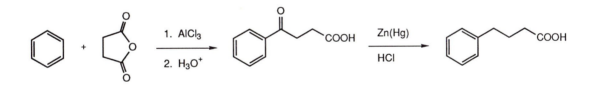

17.50.   The reaction between a ketone and a peracid constitutes the Baeyer-Villiger reaction, which yields an ester as the product. The migration aptitude of the carbonyl substituent follows the order: aryl > 4° > 3° > 2° > 1° carbon atom.

a.

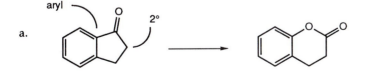

17.50. (continued)

b.

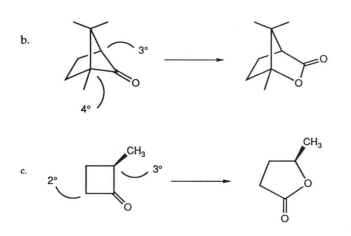

c.

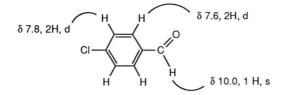

17.51. a. The IR spectrum of compound **51A** displays a strong absorption band at 1700 cm^{-1} that is assigned as a carbonyl stretching vibration. Looking at the appropriate portions of the spectrum summarized in Table 14.4 (text page 654), we conclude that the compound is an aldehyde because a band for the C–H stretching vibration is observed at 2710 cm^{-1}.

The proton NMR spectrum confirms that an aldehyde group is present, and it further shows that the compound has a *para*-disubstituted benzene ring. From the molecular formula of C_7H_5ClO, we conclude that the compound is 4-chlorobenzaldehdye.

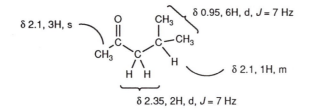

b. The IR spectrum of compound **51B** displays a strong absorption band near 1720 cm^{-1} that is assigned as a carbonyl stretching vibration. Looking at the appropriate portions of the spectrum summarized in Table 14.4 (text page 654), we conclude that the compound is a ketone because diagnostic bands for other types of carbonyl groups are absent.

The proton NMR spectrum has resonances attributable to the presence of a methyl group adjacent to a carbonyl group, a methylene group adjacent to a carbonyl group, and an aliphatic isopropyl group. Based on the observed spin-spin coupling patterns and the molecular formula of $C_6H_{12}O$, we conclude that the compound is 4-methyl-2-pentanone.

δ 2.1, 3H, s

δ 0.95, 6H, d, *J* = 7 Hz

δ 2.1, 1H, m

δ 2.35, 2H, d, *J* = 7 Hz

17.52.   a. A *trans*-2-alkylcycloalkane is prepared by opening the ring of an epoxycycloalkane

Retrosynthesis:

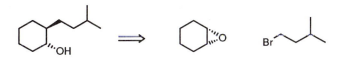

Synthesis:

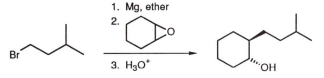

b. A lactone is made by Baeyer-Villiger oxidation of a cyclic ketone.

Retrosynthesis: *Synthesis*

17.53.   Each product illustrated in this exercise is a 2° alcohol with greatly different steric bulk on each side of the carbinol center. Enantioselective reduction of a ketone therefore comes to mind as a synthetic procedure.

a.   The needed ketone, 1-phenyl-1-butanone is made by Friedel-Crafts acylation of benzene. Reduction with use of a chiral oxazaborolidine yields the alcohol product.

retrosynthesis:

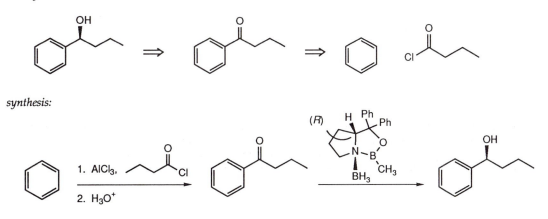

synthesis:

17.53. (continued)

b.   The needed ketone is made by a Grignard reaction followed by oxidation of the resulting 2° alcohol. Reduction of the ketone carbonyl group makes use of a chiral oxazaborolidine.

retrosynthesis:

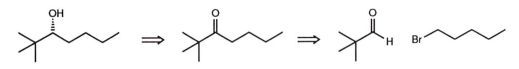

synthesis:

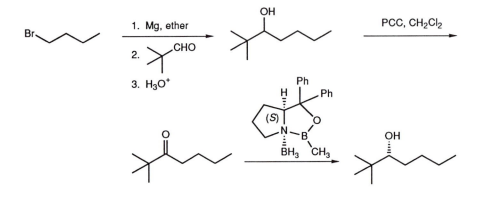

17.54.   The first step is to summarize the reactions described in the accompanying paragraph of this exercise.

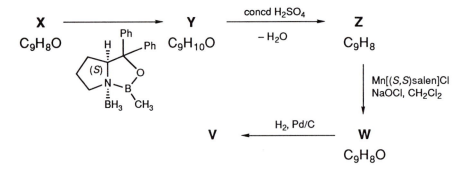

Next, use the spectroscopic data to identify one or more of the unknowns. The IR spectrum of **X** displays a strong absorption band at 1705 cm⁻¹ that is assigned as a carbonyl stretching vibration. Looking at the appropriate portions of the spectrum summarized in Table 14.4 (text page 654), we conclude that the compound is a ketone because diagnostic bands for the other types of carbonyl groups are absent.

17.54. (continued)

The proton NMR spectrum shows that the compound is a disubstituted aromatic compound (δ 7-8), and there are two methylene groups (δ 3.1, d of d, 2H and δ 2.65, d of d, 2H). From the pattern in the aromatic region, we know that the molecule does not have a *para*-disubstituted benzene ring. Calculating the sites of unsaturation, we find that there are six—four are part of the benzene ring, and one is the ketone carbonyl group, so there must be one additional double bond or ring. There are no signals in the NMR spectrum attributable to a carbon-carbon double bond, so there must be another ring. The compound is 1-indanone.

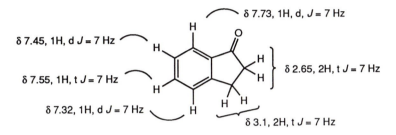

Looking at the reactions carried out starting with compound **X**, we summarize the results as follows:

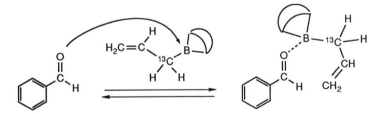

X   **Y**   **Z**   **W**   **V**   *meso*

17.55. When *B*-allyl-9-BBN is treated with benzaldehyde, an acid-base reaction takes place between the boron atom, a Lewis acid, and the oxygen atom of benzaldehyde, a Lewis base.

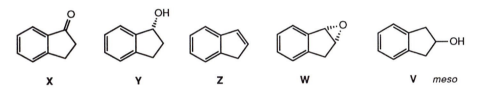

In the next step, electrons move through a concerted six-membered ring transition state to form the new carbon-carbon bond between the terminus of the allyl group and the carbonyl carbon atom.

ADDITION–ELIMINATION REACTIONS OF ALDEHYDES AND KETONES

18.1. The reaction between cyclohexanone and *tert*-butylamine begins with activation of the carbonyl group by its reaction with sulfuric acid. The cationic intermediate is intercepted by the nucleophilic nitrogen atom of the amine, which is subsequently deprotonated to yield the hemiaminal in step (3).

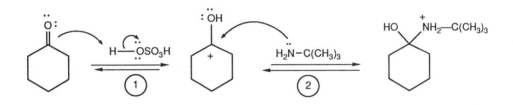

Protonation of the OH group in step (4) generates a good leaving group, and the unshared pair of electrons on nitrogen displaces a molecule of water in step (5).

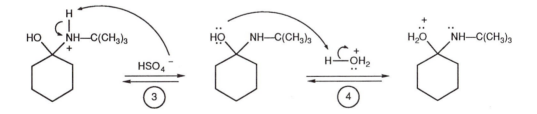

Finally, the iminium ion is deprotonated to form the neutral imine.

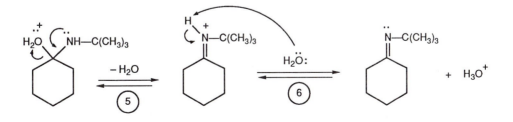

18.2. The mechanism for the reaction between 3-pentanone and hydroxylamine follows the same steps shown in the solution to exercise 18.1.

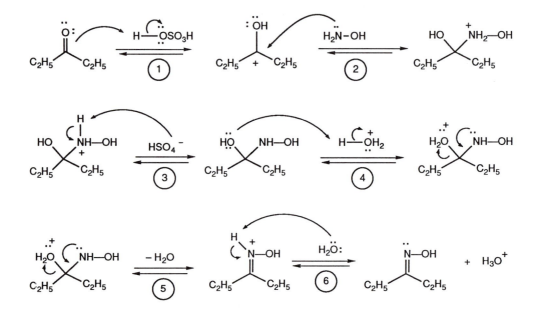

18.3. Hydrolysis of an imine follows the reverse of the steps shown in the solution to exercise 18.1. The amine produced from the reaction shown in this exercise is 2-amino-2-methylpropane (*tert*-butyl amine).

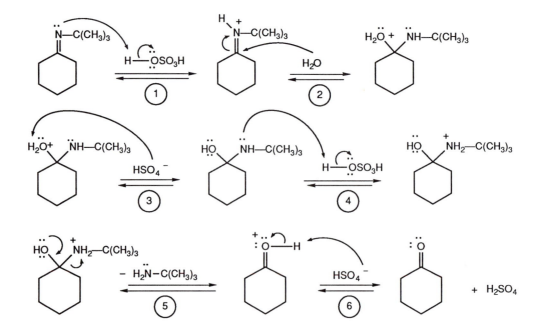

18.4.   The first step of the reaction scheme shown in this exercise is a Wolff-Kishner reaction, which converts the ketone group to a methylene group. The second step is hydrolysis of the acetal, which unmasks a ketone group that had been protected. Remember that acetal protecting groups are stable under basic and nucleophilic conditions (text page 817).

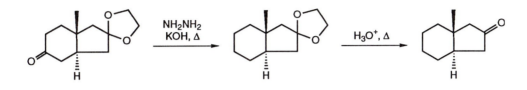

18.5.   The first step in each reaction scheme of this exercise is the formation of an imine. The combination of hydrogen and a metal catalyst reduces the carbon-nitrogen double bond to produce the corresponding amine; sodium borohydride in aqueous ethanol does the same transformation.

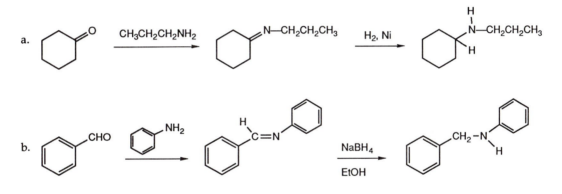

18.6.   In forming an imine crosslink, the amino group of lysine undergoes addition to the carbonyl group of an allysine residue. Proton transfer yields the hemiaminal in step (2). When the electron pair on nitrogen moves to create the carbon-nitrogen double bond in step (3), the OH group removes the proton from nitrogen.

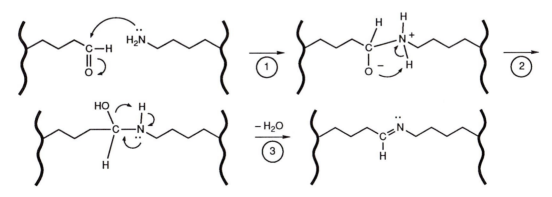

18.7. Hydrolysis of the imine link between the opsin lysine residue and retinal begins with the protonated iminium form of this adduct. A molecule of water adds to the carbon-nitrogen double bond in the first step to form the hemiaminal after proton transfer in step (2). An electron pair on the oxygen atom moves in step (3) to create the carbon-oxygen double bond, which displaces the lysine residue as a leaving group. Proton transfer in step (4) completes the transformation.

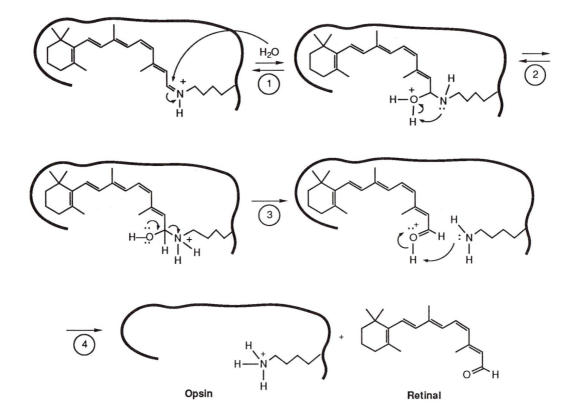

Opsin   Retinal

18.8. If two water molecules enter the active site of *glycogen phosphorylase*, PLP might activate the water molecules sufficiently that the acetal oxygen atom could be protonated. Cleavage of glycogen would then occur as before (steps 2 and 3), but under these conditions, a water molecule intercepts the glucose cation in step (4) to form glucose instead of its phosphorylated derivative.

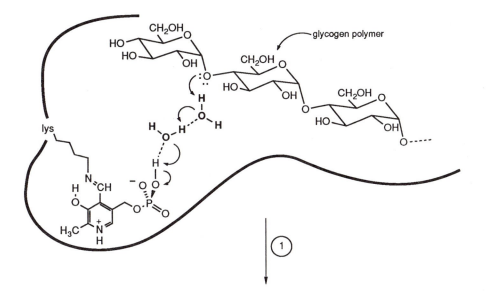

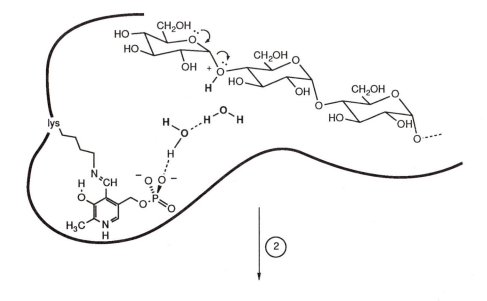

18.8. (continued)

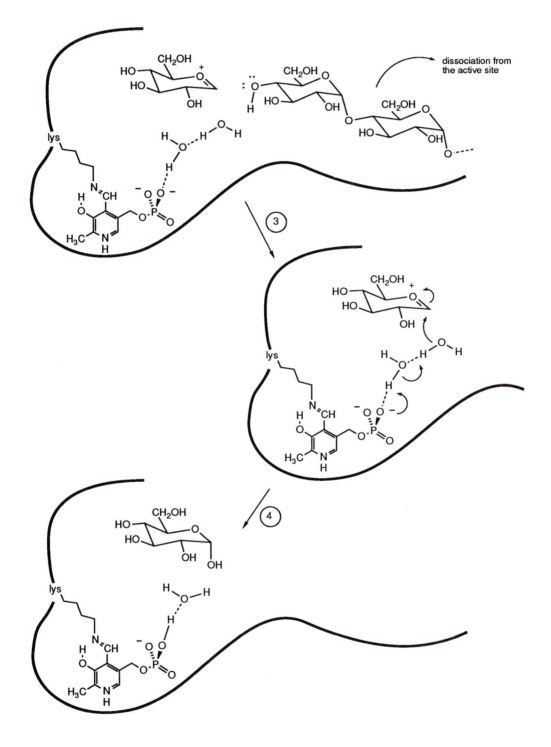

18.9. In transimination, an amine nitrogen atom undergoes nucleophilic addition to the carbon-nitrogen double bond of the imine. In the second step, an acid-base reaction leads to transfer of a proton from one nitrogen atom to the other.

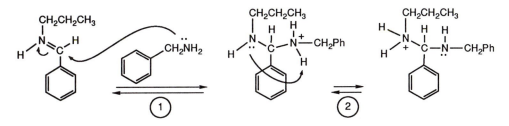

In step (3), the electron pair displaces the amine to form a new carbon-nitrogen double bond.

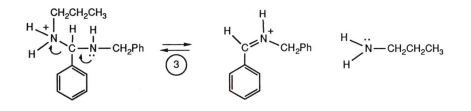

18.10. The mechanism for imine hydrolysis, which is the reaction that takes place to liberate the ketoacid intermediate from pyridoximine phosphate during amino acid metabolism, has already been shown in the solution to exercise 18.3. The steps comprise the following:
  (1) addition of water to the imine double bond
  (2) proton transfer
  (3) displacement of the amine as a leaving group
  (4) proton transfer
In the scheme shown below, (PxP) is the pyridine portion of pyridoxyl phosphate.

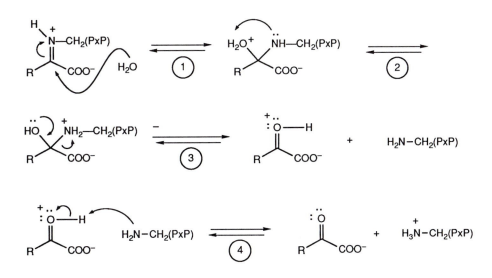

18.11. The conversion of the PMP derivative of α–ketoglutarate to regenerate PLP and glutamate takes place by a pathway that starts with reduction of the carbon-nitrogen double bond via a series of electron and proton transfer processes in steps (1) and (2).

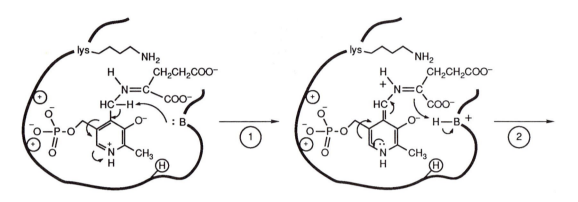

After the bond has been reduced, transamination occurs between the amino acid-PLP complex and the side chain of the lysine residue (cf. exercise 18.9). This phase leads to displacement of glutamate and regeneration of the enzyme–PLP complex.

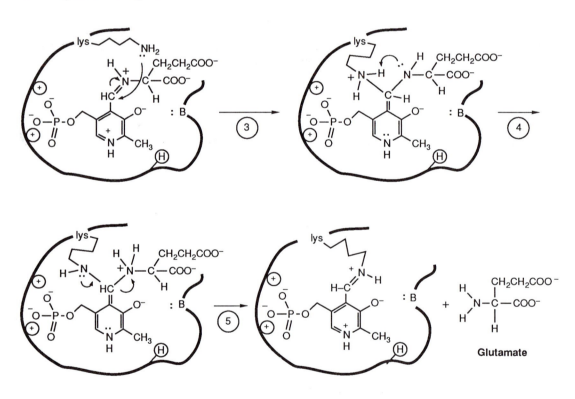

18.12. The steps leading to formation of GABA after decarboxylation of glutamate are the same as those illustrated in the solution to exercise 18.10. In the scheme shown below, (PxP) is the pyridine portion of pyridoxyl phosphate.

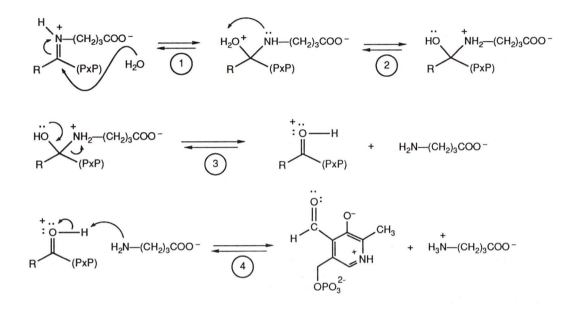

18.13. The product of the reaction between a ketone and a 2° amine is an enamine. An enamine undergoes hydrolysis in aqueous acid to yield a ketone and a 2° amine.

a.

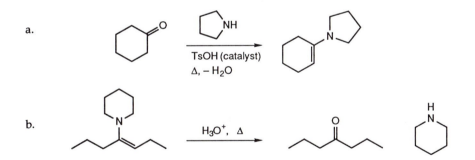

b.

The mechanism for enamine hydrolysis is the reverse of steps needed to make an enamine.

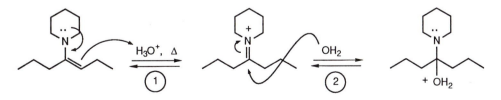

18.13. (continued)

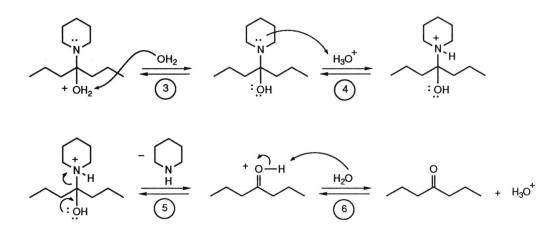

18.14. A diaminoalkane reacts with a ketone in the same way that a diol reacts with a carbonyl group to form a cyclic acetal (compare this reaction with the equation at the top of text page 819). An amine nitrogen atom adds to an activated carbonyl group that has been formed by protonation in step (1). Elimination of a molecule of water followed by cyclization yields a heterocycle that has two nitrogen atoms attached to the carbon atom of the original carbonyl group.

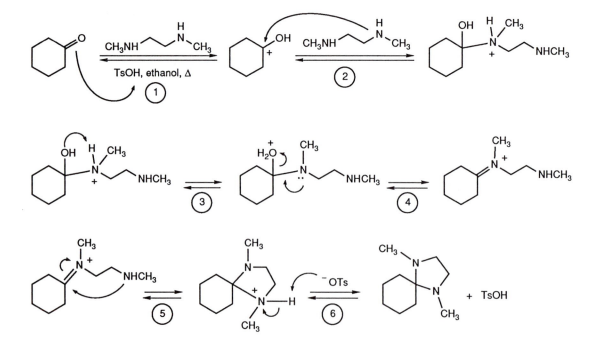

18.15. An alkyl halide reacts with triphenylphosphine to form an alkyltriphenylphosphonium salt. Treating that salt with a strong base generates the corresponding ylide.

a.

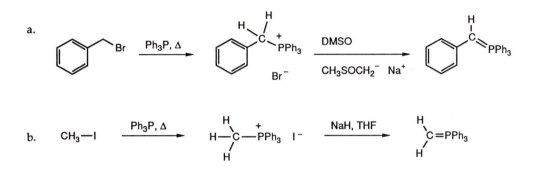

b.

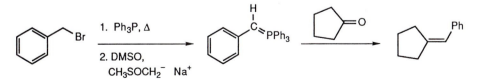

18.16. The logical disconnection that makes use of a Wittig reaction to prepare an alkene is the bond between the carbon atoms connected by the double bond. If the double bond is exocyclic to a ring, the ring component normally is the carbonyl partner. In most cases, the ylide reactant comprises the end of the alkene double bond with the less highly-substituted carbon atom.

a. *retrosynthesis*:

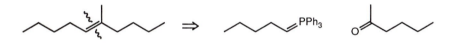

synthesis:

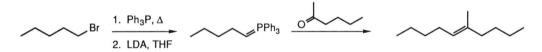

b. *retrosynthesis*:

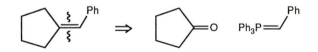

synthesis:

18.17. To accomplish two-carbon homologation by a Wittig procedure, the ylide must be derived from a compound that has one carbon atom more than the reagent prepared from chloromethyl methyl ether. The methoxy group must therefore be attached to a chloro<u>ethyl</u> rather than a chloro<u>methyl</u> group.

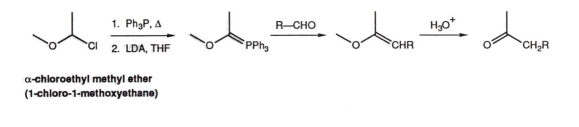

**α-chloroethyl methyl ether
(1-chloro-1-methoxyethane)**

18.18. If a compound has a double bond, then either a Wittig or Horner-Emmons reaction can be used to construct the carbon skeleton. If the double bond is conjugated with a carbonyl or nitrile group (that is, between the carbon atoms α and β to the carbonyl or nitrile group), then the phosphonate ylide is normally chosen.

a. *retrosynthesis*:

synthesis:

b. *retrosynthesis*:

synthesis:

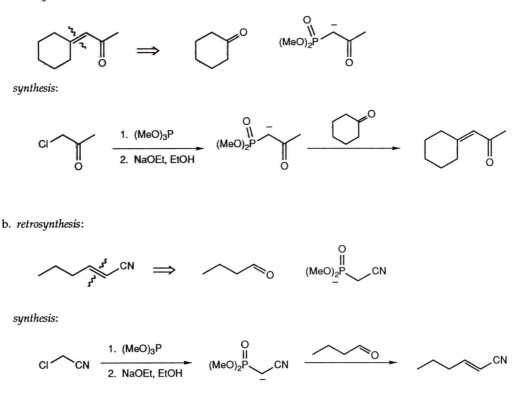

18.19. Cyclopropyl diphenylsulfonium ylide reacts with the carbonyl group of decanone to form epoxide **A**.

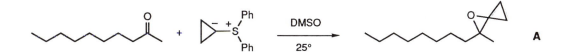

This epoxide, when treated with acid, is first protonated at its oxygen atom. The strained three-membered ring opens to form a 3° carbocation. Recall that carbocations readily rearrange (text page 280). If one of the alkyl groups migrates, then a more stable carbocation is formed. Deprotonation of this species yields the cyclobutanone.

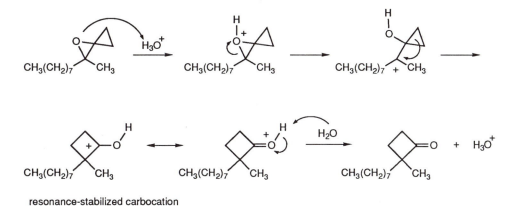

resonance-stabilized carbocation

18.20. Follow the procedure given in the solution to exercise 7.28.

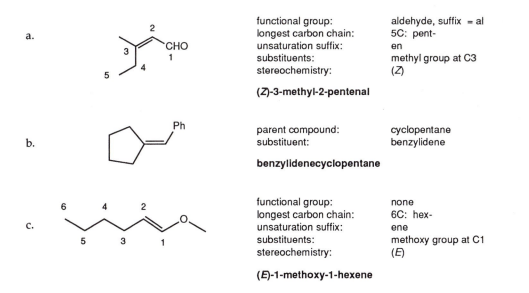

a.

functional group:	aldehyde, suffix = al
longest carbon chain:	5C: pent-
unsaturation suffix:	en
substituents:	methyl group at C3
stereochemistry:	(Z)

(Z)-3-methyl-2-pentenal

b.

parent compound:	cyclopentane
substituent:	benzylidene

benzylidenecyclopentane

c.

functional group:	none
longest carbon chain:	6C: hex-
unsaturation suffix:	ene
substituents:	methoxy group at C1
stereochemistry:	(E)

(E)-1-methoxy-1-hexene

18.20.   (continued)

d.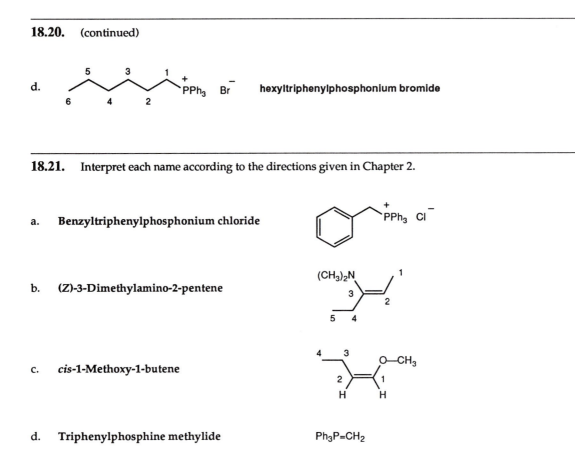

5 3 1
6 4 2
$^+$PPh$_3$ Br$^-$ **hexyltriphenylphosphonium bromide**

18.21.   Interpret each name according to the directions given in Chapter 2.

a.   **Benzyltriphenylphosphonium chloride**

$^+$PPh$_3$ Cl$^-$

b.   **(Z)-3-Dimethylamino-2-pentene**

(CH$_3$)$_2$N   1
3   2
5   4

c.   ***cis*-1-Methoxy-1-butene**

4   3   O—CH$_3$
2   1
H   H

d.   **Triphenylphosphine methylide**

Ph$_3$P=CH$_2$

18.22.   Semicarbazide is a hydrazine derivative with an aminocarboxy substituent. The NH$_2$ group at the hydrazine end of the molecule undergoes addition–elimination with the aldehyde carbonyl group (the other NH$_2$ group is attached directly to a carbonyl group, so it is more like an amide, which does not react with ketones and aldehydes to form C=N bonds).

The first step of the mechanism is protonation of the aldehyde oxygen atom, which activates the carbonyl group by creating an electrophilic center. Semicarbazide intercepts this carbocation in step (2), and deprotonation of that intermediate occurs in step (3). Protonation of the OH group in step (4) generates a good leaving group. Deprotonation of the next intermediate with displacement of a molecule of water generates the semicarbazone product.

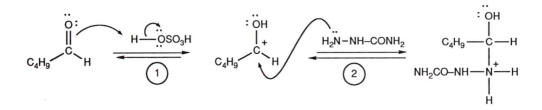

18.22. (continued)

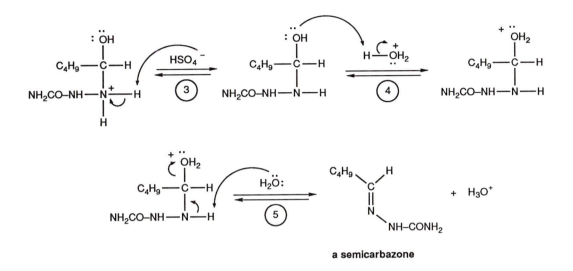

a semicarbazone

18.23. An imine is formed by reaction between a primary amine and an aldehyde or ketone. The nitrogen atom replaces the oxygen atom in the carbonyl component, with loss of a molecule of water.

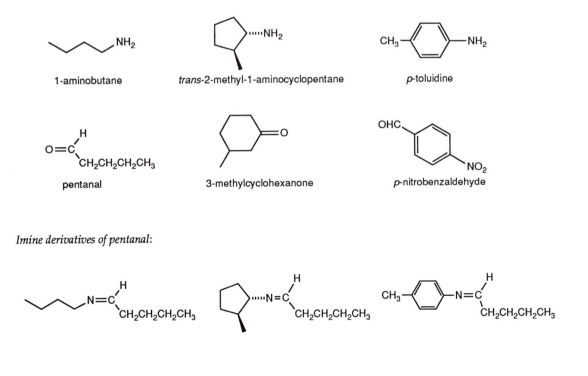

1-aminobutane

trans-2-methyl-1-aminocyclopentane

p-toluidine

pentanal

3-methylcyclohexanone

p-nitrobenzaldehyde

Imine derivatives of pentanal:

18.23.   (continued)

Imine derivatives of 3-methylcyclohexanone:

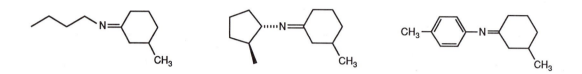

Imine derivatives of p-nitrobeznaldehyde:

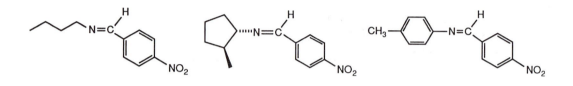

18.24.   An enamine is formed by reaction between a secondary amine and an aldehyde or ketone. The nitrogen atom replaces the oxygen atom in the carbonyl component, and a carbon-carbon double bond is formed between the original carbonyl carbon atom and one of the adjacent carbon atoms.

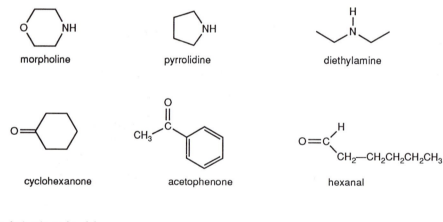

morpholine   pyrrolidine   diethylamine

cyclohexanone   acetophenone   hexanal

Enamine derivatives of cyclohexanone:

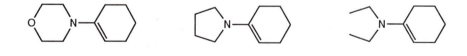

18.24.   (continued)

Enamine derivatives of acetophenone:

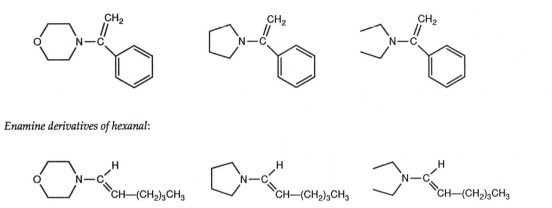

Enamine derivatives of hexanal:

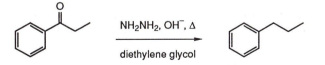

18.25.

a.   Addition of an organolithium reagent to a ketone produces a 3° alcohol.

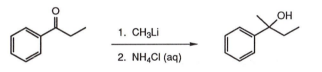

b.   Sodium borohydride reduces a ketone to a 2° alcohol.

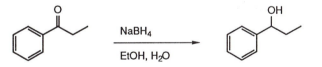

c.   The Wolff-Kishner reaction transforms the carbonyl group of a ketone to a methylene group.

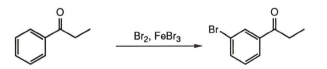

d.   Bromine in the presence of a Lewis acid leads to bromination of the aromatic ring.

18.25. (continued)

e.   A Wittig reaction forms a double bond where the ketone C=O group was, replacing the oxygen atom with the alkylidene group attached to phosphorus.

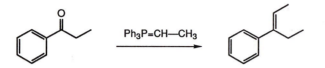

f.   A sulfur ylide converts a carbonyl group into an epoxide. The Grignard reagent opens the epoxide ring at the less highly-substituted end to form the 3° alcohol.

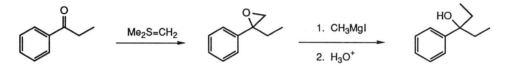

g.   The secondary amine morpholine reacts with a ketone and an acid catalyst to form an enamine.

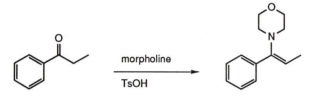

h.   Hydroxylamine reacts with a ketone carbonyl group to form an oxime.

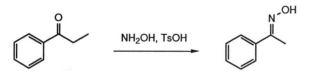

18.26.

a.   A secondary amine reacts with a ketone to form an enamine if the carbon atom adjacent to the carbonyl group has at least one proton. An acid is normally added to catalyze the condensation reaction. The stereochemistry of the double bond is (*E*).

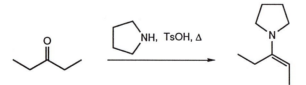

18.26. (continued)

b. Dimethylsulfonium methylide reacts with a ketone to form an epoxide. A racemic mixture is obtained.

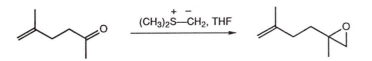

c. A phosphonium ylide reacts with an aldehyde to create a new carbon-carbon double bond. Both *cis* and *trans* isomers of the new double bond are formed.

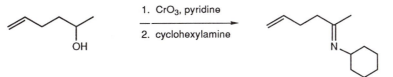

d. Hydrolysis of an acetal under acidic conditions unmasks the aldehyde carbonyl group (text page 817). Subsequently, the 1° amino group condenses with the carbon-oxygen double bond to form the imine, in this example a cyclic one. The stereogenic center that bears the amino group is not affected, so its configuration stays the same.

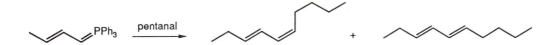

e. Oxidation of a 2° alcohol generates a ketone, which reacts with a 1° amine to form an imine.

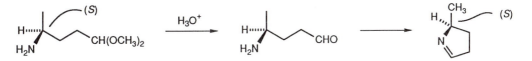

f. Friedel-Crafts acetylation of naphthalene with acetyl chloride yields the methyl ketone, which reacts with piperidine, forming the enamine.

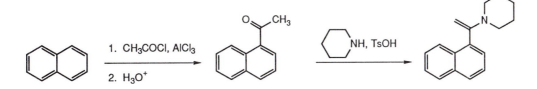

18.26. (continued)

g. Hydroboration of an alkyne followed by oxidative hydrolysis produces the corresponding aldehyde. Reaction with the methylene Wittig reagent forms the terminal alkene having one additional carbon atom.

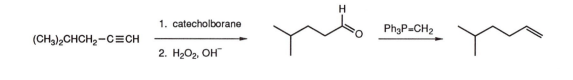

18.27. In the reaction that is commonly used to make pyrazole and its derivatives, hydrazine reacts with one of the carbonyl groups of a diketone in step (1), which yields the protonated hydrazone after two proton transfer steps and loss of a molecule of water. In step (5) deprotonation leads to formation of the hydrazone, and this is the end of the first stage of the reaction.

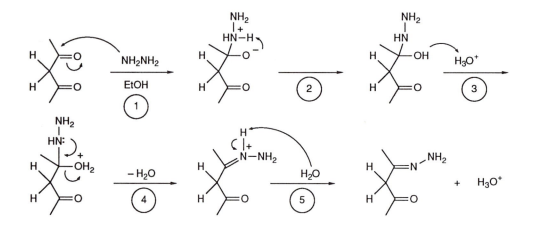

The next four steps repeat the same process at the other carbonyl group (the second stage).

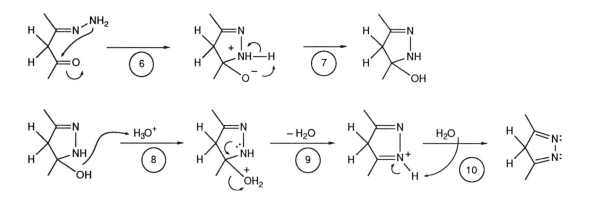

18.27. (continued)

Finally, two proton transfer steps (11) lead to tautomerism and formation of the product. Tautomerism creates a fully conjugated aromatic system in which the two π bonds and an electron pair on a nitrogen atom constitute the 6 π electron system.

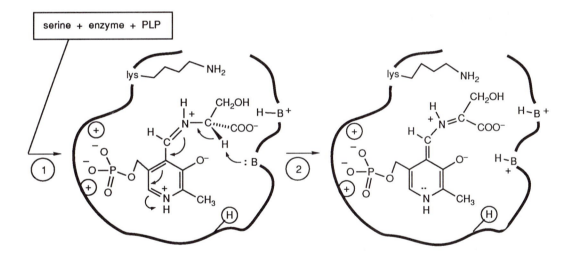

18.28. The enzyme-catalyzed dehydration of serine occurs in three stages. In the first two steps, the amino acid becomes attached via an imine linkage to the PLP coenzyme within the enzyme active site, and a base removes the α-proton, as seen before (text page 891).

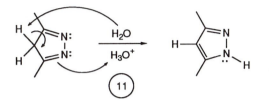

In the subsequent step, electrons move from the nitrogen atom of the dihydropyridine, expelling the OH group from serine as it becomes protonated by an acidic group in the enzyme active site. Protonation generates a molecule of water, which acts as the leaving group in this step.

18.28. (continued)

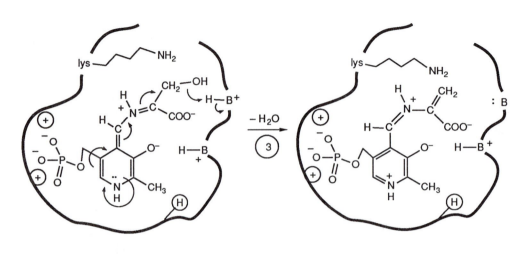

18.29. If the final step in the synthesis of Valium is formation of the imine bond, then the immediate precursor must be the compound in which the carbon-nitrogen double bond has been disconnected. Those two atoms are attached to the elements of water. The carbon atom is the one that is attached to O, and the nitrogen atom is attached to two atoms of H:

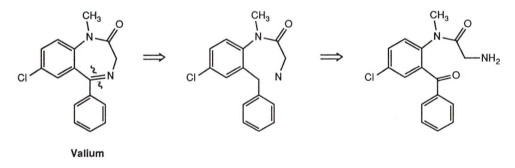

Valium

The mechanism for ring-closure starts with addition of the amino group to the double bond of the carbonyl group, followed in step (2) by proton transfer from the nitrogen atom to the oxygen atom.

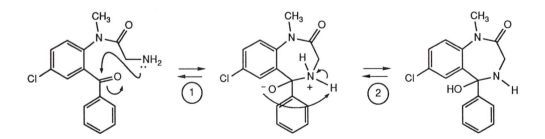

18.29.   (continued)

In step (3), the OH group is protonated to form a good leaving group, and the electron pair on the nitrogen atom displaces the water molecule, forming the C=N bond.

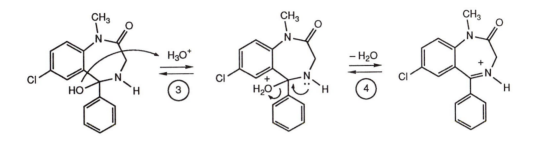

In the final step, the nitrogen atom is deprotonated to yield the neutral product.

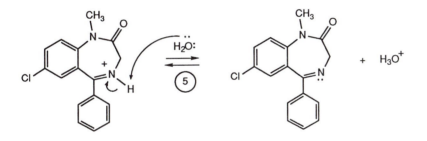

18.30.   The two steps needed to convert glutamate-5-semialdehyde to proline involve cyclization, which creates the imine linkage, and reduction of the imine to form the amine.

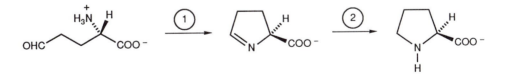

18.30.   (continued)

The mechanism is straightforward. After deprotonation of the amino group, nucleophilic addition of the nitrogen atom to the aldehyde carbonyl group takes place, followed by dehydration to form the imine bond. The imine group, like a carbonyl group (see text page 846), is reduced by NADH to form the 2° amine.

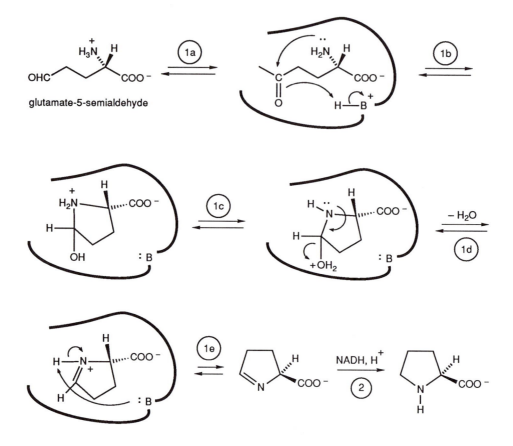

glutamate-5-semialdehyde

18.31.   IR spectra reveal what functional groups are present in a molecule. To deduce what differences will be observed in the IR spectrum during the course of a specific reaction, identify the functional group in the starting material and product and decide what absorption bands will be observed for each.

a.

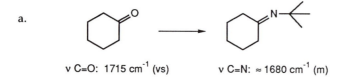

ν C=O: 1715 cm^{-1} (vs)   ν C=N: ≈ 1680 cm^{-1} (m)

18.31. (continued)

b.

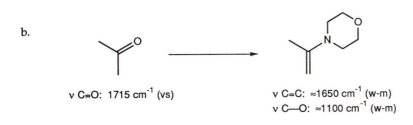

v C=O: 1715 cm⁻¹ (vs)   v C=C: ≈1650 cm⁻¹ (w-m)
                              v C—O: ≈1100 cm⁻¹ (w-m)

18.32. When performing a reaction, you can predict what resonances will be observed in the proton NMR spectra of the starting material and product by using data from the tables and figures in Chapter 15. For the resonances given below, the first value is the chemical shift relative to the signal for TMS; the number with the capitalized H is the integrated intensity (2H means two protons, etc.); and the lower case letter is the spin-spin coupling pattern (s = singlet, d = doublet, t = triplet, q = quartet, m = multiplet, br s = broad singlet). The *J* value for the coupling is the final parameter given.

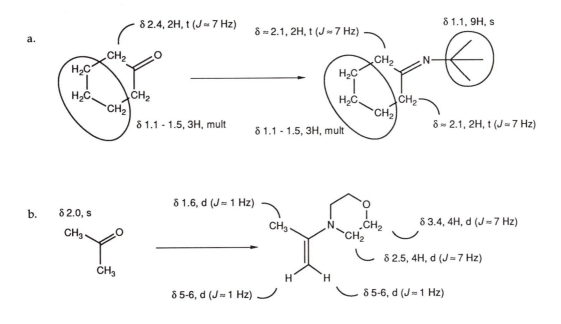

18.33. When performing a reaction, you can predict what resonances will be observed in the carbon NMR spectra of the starting material and product by using data from the tables and figures in Chapter 15. For the resonances given in the equations shown below, the first value is the chemical shift relative to the signal for TMS; the letter is the splitting observed in the off-resonance spectrum (s = singlet, d = doublet, t = triplet, q = quartet).

18.33.   (continued)

a.

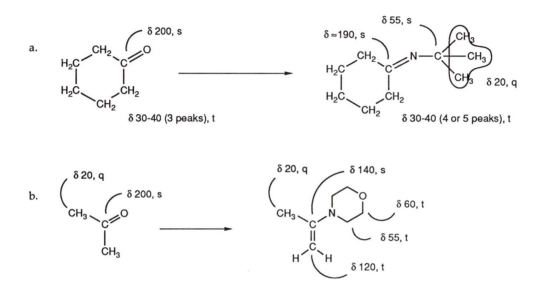

b.

18.34.

a. The conversion of a ketone to an imine requires reaction between the ketone and a 1° amine in the presence of an acid catalyst.

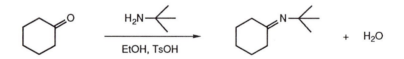

b. The conversion of a ketone to an enamine requires reaction between the ketone and a 2° amine in the presence of an acid catalyst.

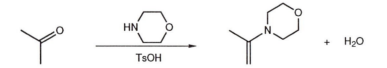

18.35. If a Wittig reaction is to be used to prepare an alkene, the retrosynthetic disconnection is at the double bond. One end of the alkene comes from the ylide reagent and the other from the carbonyl component (aldehyde or ketone).

a. This Wittig reaction uses the combination of an ylide derived from toluene with acetone. It is also possible to treat the ylide derived from 2-bromopropane with benzaldehyde.

retrosynthesis:

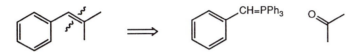

synthesis:

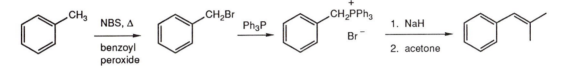

b. This Wittig reaction requires the combination of two aliphatic components. The methylene ylide is ideal for adding one carbon atom to form a terminal alkene.

retrosynthesis:

synthesis:

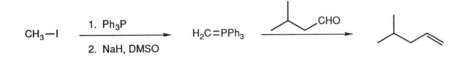

c. This Wittig reaction requires the combination of a cyclic ketone and an ylide with two carbon atoms.

retrosynthesis:

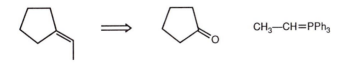

18.35. (continued)

synthesis:

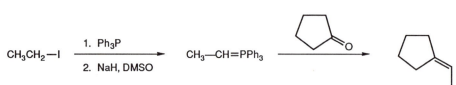

d. Recognize that an aldehyde is the product of a Wittig reaction that employs the methoxymethylene ylide. The initial alkene product is a vinyl ether that is hydrolyzed to unmask the carbonyl group. Cyclohexanone is made from cyclohexene by hydration of the double bond, followed by oxidation of the 2° alcohol.

retrosynthesis:

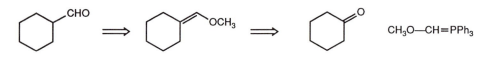

synthesis:

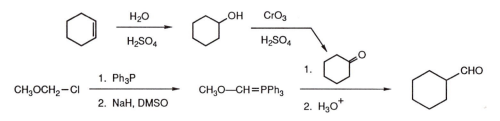

18.36.

a. The imine bond is a logical disconnection site. The aldehyde is made from benzene by attaching a two-carbon fragment, which is done by reaction between a Grignard reaction and ethylene oxide. The alcohol product is oxidized to the aldehyde. Aniline is made from benzene by nitration followed by reduction.

retrosynthesis:

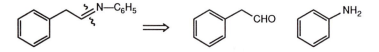

18.36. (continued)

synthesis:

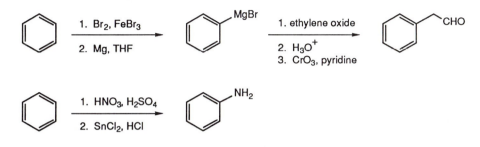

b.   When a desired product has no functional group, you have to insert one (or more) in places where they are readily converted to hydrocarbon units. A Wittig reaction followed by catalytic hydrogenation of the alkene double bond is an easy way to make an alkane.

The starting materials in this exercise are limited to those with four carbon atoms, so because the product has 12 carbon atoms, disconnection to produce three, four-carbon fragments is reasonable. The symmetry of the molecule also makes this approach logical because the two ends are identical.

retrosynthesis:

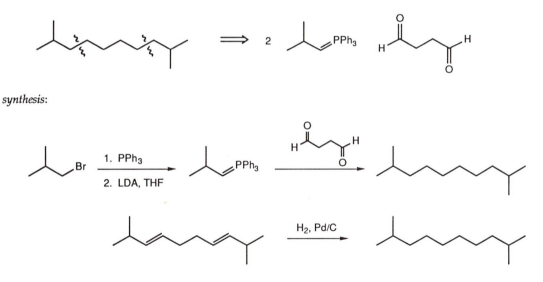

synthesis:

c.   A carbon-carbon double bond provides a logical disconnection site that can be reconstructed by application of a Wittig reaction. The aldehyde is made from 2-bromobutane by addition of one carbon atom, a disconnection that makes use of a Grignard reaction with formaldehyde, followed by oxidation.

retrosynthesis:

18.36. (continued)

synthesis:

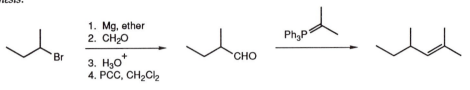

18.37. A chiral diol is made from the corresponding alkene by asymmetric dihydroxylation.

a. The alkene starting material is made from cyclopentanone via Wittig olefination.

retrosynthesis:

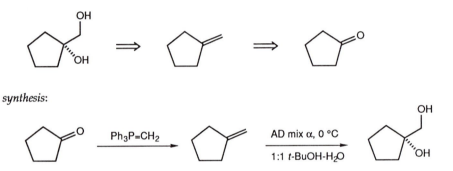

synthesis:

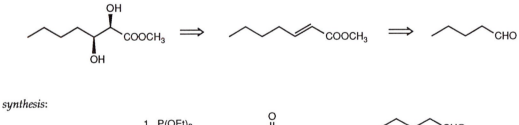

b. The α,β-unsaturated ester is made from an aldehyde via Horner-Emmons olefination.

retrosynthesis:

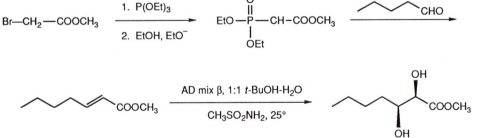

synthesis:

Br—CH₂—COOCH₃ → [1. P(OEt)₃ 2. EtOH, EtO⁻] → EtO—P(=O)(OEt)—CH—COOCH₃ → ...

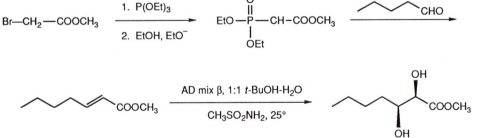

18.38.   Reductive amination, by its very name, implies that two reactions are taking place: formation of a carbon-nitrogen bond and reduction. The transformation takes place when a carbonyl compound is treated with a 1° amine, so imine formation most likely constitutes the first step. Even at pH 7, formation of a imine is quite facile, even though water is not a strong acid. (If the solution becomes too basic, however, the proton concentration is low enough that reaction does not occur).

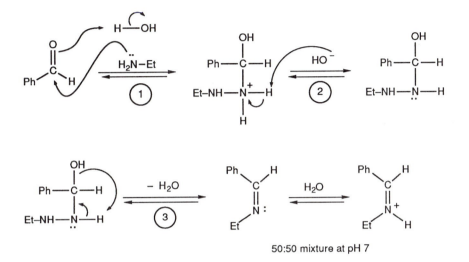

50:50 mixture at pH 7

The reason that cyanoborohydride ion is able to reduce the C=N bond of an imine relates to the property that an imine exists partially protonated in the given pH range.   An iminium ion is more susceptible to reaction with a nucleophile than is either the neutral imine or the related carbonyl compound. The protonated form of an imine has a resonance contributor in which the carbon atom bears a positive charge.

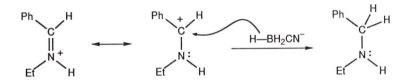

18.39.   Deprotonation of an alkyltriphenylphosphonium salt generates an ylide, which has a nucleophilic carbon atom. Alkylation occurs in step (2) by reaction between the nucleophilic carbon atom and the electrophilic carbon atom of the alkyl halide. Deprotonation with NaH in step (3) produces an ylide, too. Its reaction with benzaldehyde forms the carbon-carbon double bond of the product.

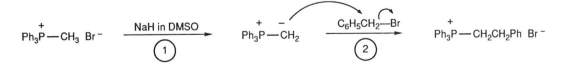

18.39.   (continued)

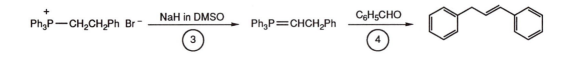

18.40.   A six-membered ring is formed by a Diels-Alder reaction between a conjugated diene and a dienophile. The dienophile in this scheme is vinyltriphenylphosphonium bromide. The resulting cyclohexenyl phosphonium salt is subsequently deprotonated by reaction with LDA in step (2), which forms an ylide. Reaction of the ylide with propanal forms the exocyclic carbon-carbon double bond.

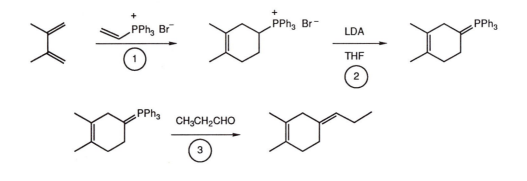

18.41.   First, summarize the reactions described in the accompanying paragraph.

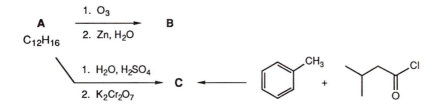

The identity of **C** is easy to reckon because it is formed by Friedel-Crafts acylation between toluene and 3-methylbutanoyl chloride. Compound **C** is 1-(4-methylphenyl)-3-methyl-1-butanone.

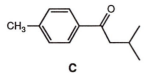

C

18.41. (continued)

Compound **B** is *p*-tolualdehyde, the structure of which is determined from the illustrated IR and proton NMR spectra.

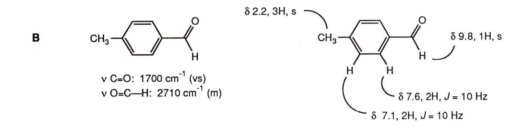

Working backwards from the structure of compound **C**, we conclude that dichromate ion was used to oxidize a 2° alcohol to make the ketone group in **C**, and the alcohol must have been made from an alkene by hydration. To check this, we ask whether **A** can be converted to **B** by ozonolysis. The answer is "yes".

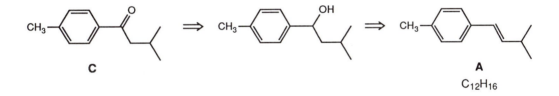

Finally, to convert compound **B** to compound **A**, a Wittig reaction is employed.

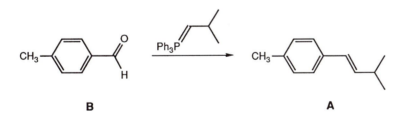

18.42. First, summarize the reactions described in the accompanying paragraph. The IR spectrum of compound **Z** indicates that it is an aldehyde. Its synthesis by a Wittig reaction followed by hydrolysis suggests that the reactions that convert **X** to **Y** to **Z** constitute the homologation process described on text page 902. Compound **X** is therefore an aldehyde or ketone.

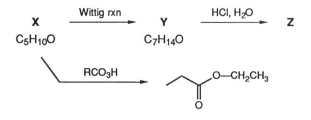

If compound **X** is a ketone, then its reaction with a peracid constitutes a Baeyer-Villiger reaction. The product of that transformation, ethyl propanoate, reveals that compound **X** is a ketone. In fact, it must be 3-pentanone.

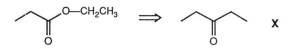

The reactions that convert **X** to **Y** to **Z** are as follows:

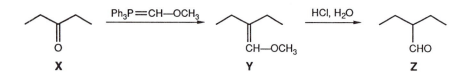

18.43. Dimethylsulfonium methylide is a nucleophile, so it reacts with an alkyl halide to form the homologous sulfonium salt. The dimethylsulfonium group is a good leaving group (text page 351), so reaction of the alkylated sulfonium salt with strong base produces the alkene by an E2 pathway.

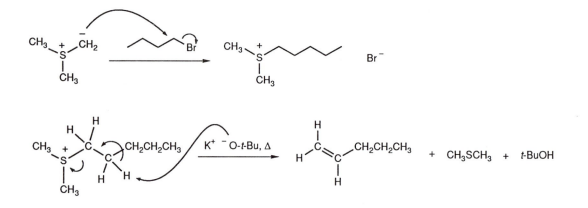

ADDITION–ELIMINATION REACTIONS OF CARBOXYLIC ACID AND DERIVATIVES

19.1. An organolithium reagent reacts with a carboxylic acid to form the corresponding ketone after acid workup. The alkyl group of the organometallic compound replaces the OH portion of the COOH group. Cerium chloride facilitates this transformation.

19.2. A carboxylic acid reacts with an alcohol in the presence of a mineral acid to form an ester. A dehydrating agent such as molecular sieves is sometimes added to shift the equilibrium toward formation of the ester product.

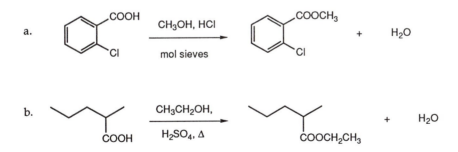

463

19.3. If one of the alcohol groups of a diol is oxidized before the other one is, then cyclization occurs, either at the aldehyde or carboxylic acid stage. At the aldehyde stage, a hemiacetal forms and is oxidized to the lactone.

via the aldehyde

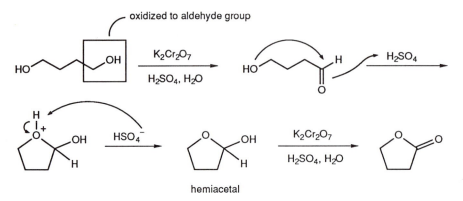

via the carboxylic acid

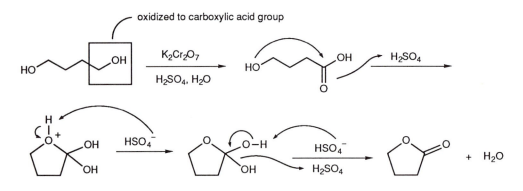

19.4. A hydroxy carboxylic acid undergoes intra- or intermolecular esterification. If there are enough atoms separating the alcohol and carboxylic acid groups, a lactone is formed. For an α-hydroxy carboxylic acid, however, a dimer is produced. This transformation proceeds via several steps that involve addition of the alcohol OH group to the carboxylic acid carbonyl group. Elimination of water from the tetrahedral intermediate produces the ester group.

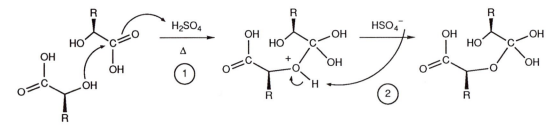

19.4.   (continued)

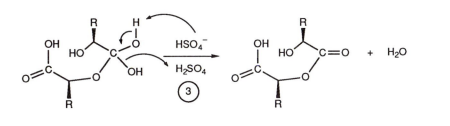

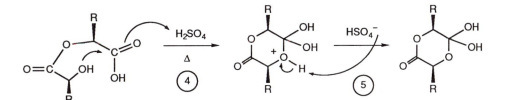

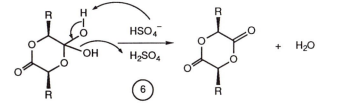

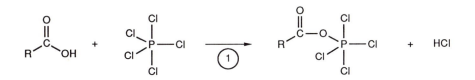

19.5.   A carboxylic acid reacts with PCl_5 in the first step by displacing a chloride ion. The proton of the carboxylic acid group combines with the chloride ion to form HCl.

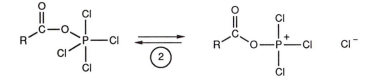

A four-coordinate acyloxy(trichloro)phosphonium ion forms in step (2) by dissociation of chloride ion.

19.5. (continued)

In step (3), Cl⁻ adds to the carbonyl group of the phosphonium salt to form a tetrahedral intermediate. This intermediate collapses in step (4) to produce the acid chloride and phosphorus oxychloride.

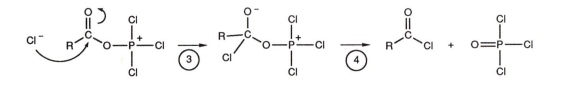

19.6. A 3° alcohol cannot be used for direct (acid-catalyzed) esterification because of a competing elimination pathway (E1) that takes place in acid (text page 326). A route that makes use of an acid chloride is therefore required.

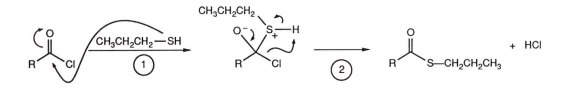

19.7. A thiol is a good nucleophile, so it reacts with an acid chloride by addition to the carbonyl group. A tetrahedral intermediate that is formed subsequently collapses to yield the thioester and HCl.

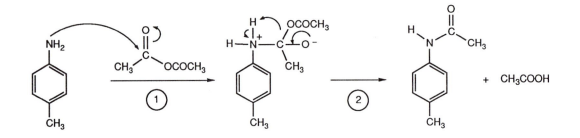

19.8. Anilines have a nucleophilic nitrogen atom that adds to the carbonyl group of acetic anhydride to form a tetrahedral intermediate. Regeneration of the carbon-oxygen double bond displaces acetate ion, which removes a proton from nitrogen to form the products, acetic acid and an acetamide derivative.

19.9. An amide is made by a reaction between an amine and either an ester or an acid chloride. If a hydroxy group is present in the desired product, as in part b, then the corresponding lactone is a good choice as a starting material. If an acid chloride is used to make an ester, a base like pyridine or triethylamine is normally included in the reaction mixture to react with the HCl that is formed. An ester or lactone needs no added base because the leaving group is an alkoxide ion, which reacts with a proton attached to nitrogen.

a.

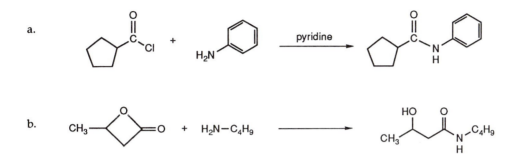

b.

19.10. Hydrazine reacts with an acid chloride in the same way that an amine does. The second nitrogen atom acts as a base, so aqueous OH⁻ has to be added at the end of the reaction to obtain the neutral hydrazide derivative.

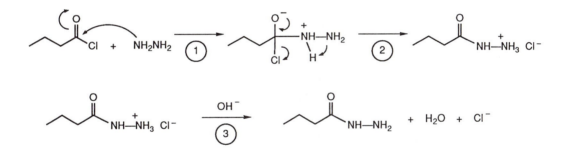

19.11. Hydrolysis of a lactone follows the same mechanism involved in the hydrolysis of any ester. Hydroxide ion adds to the carbonyl group to form a tetrahedral intermediate. That intermediate collapses to regenerate the carbon–oxygen double bond, which leads to opening the lactone ring. Acid workup yields the hydroxy acid.

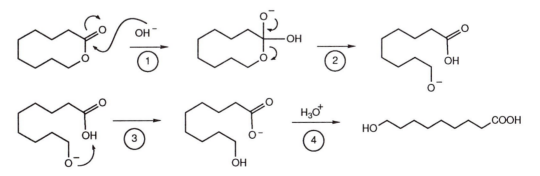

19.12. In any acid-catalyzed hydrolysis reaction of a carboxylic acid derivative, the carbonyl group is protonated, forming an activated intermediate that is trapped by reaction with a molecule of water. Protonation of the X group in step (4) subsequently generates a good leaving group that is displaced when the carbon-oxygen double bond is regenerated.

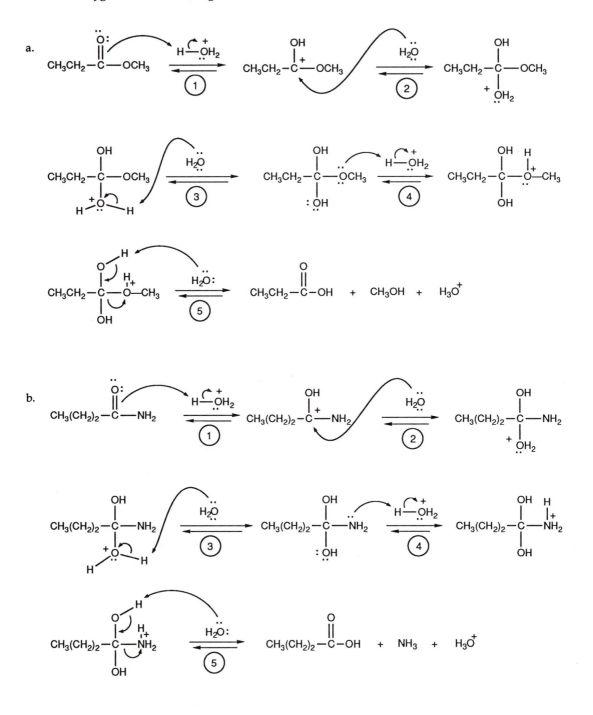

19.13.   The chymotrypsin-catalyzed hydrolysis of *p*-nitrophenyl acetate starts with binding of the aromatic molecule in the active site of the enzyme. In step (1), the serine hydroxyl group adds to the carbonyl group of the substrate, generating a tetrahedral intermediate. The tetrahedral intermediate subsequently collapses, displacing *p*-nitrophenol.

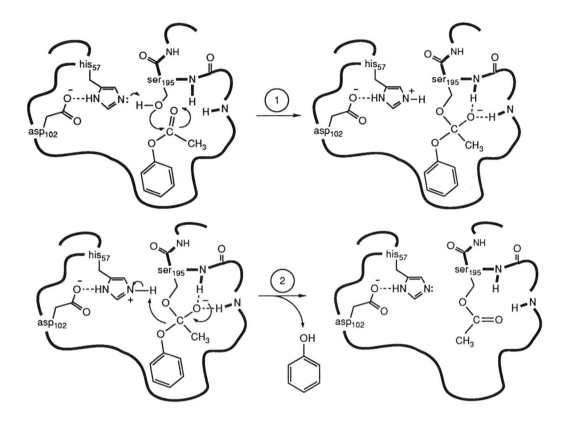

A molecule of water enters the active site and adds to the carbonyl group of the acetylated serine residue in step (3). A tetrahedral intermediate is formed once again.

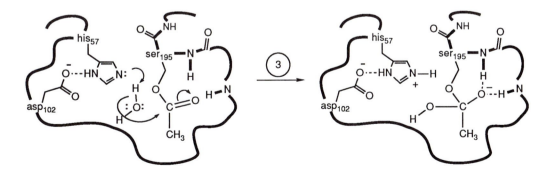

19.13. (continued)

The anionic intermediate collapses in step (4), forming acetic acid and regenerating the unoccupied chymotrypsin active site.

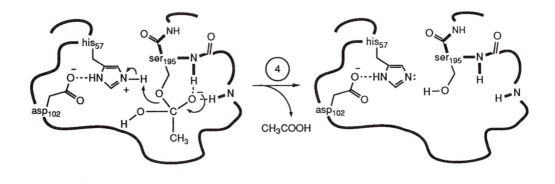

19.14. Transesterification occurs under basic conditions by nucleophilic addition of alkoxide ion to the carbonyl group, which forms a tetrahedral intermediate. When the carbon-oxygen double bond is regenerated in step (2), the original alkoxy group is displaced. The released alkoxy group reacts with the alcohol solvent to generate more alkoxide ion (boxed reaction, below).

If a low MW alcohol such as methanol is formed, its removal from the reaction mixture by distillation drives the equilibrium to generate the desired product.

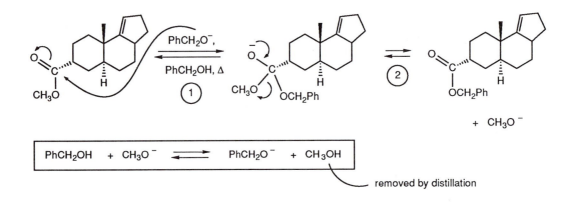

19.15. The mechanism of this transformation follows the usual one for transesterification: the oxygen atom of the alcohol OH group adds to the carbonyl group of the ester, which is followed by regeneration of the carbonyl group and expulsion of methanol.

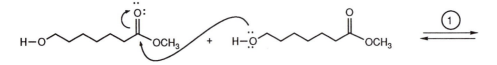

19.15.   (continued)

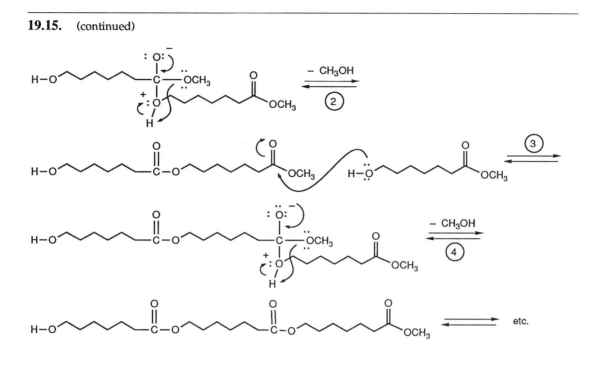

19.16.   An acid-catalyzed reaction of a nitrile begins with protonation of the nitrogen atom. The carbocation formed in step (1) is intercepted by a nucleophile, in this case methanol. Proton transfer from oxygen to nitrogen forms the product salt.

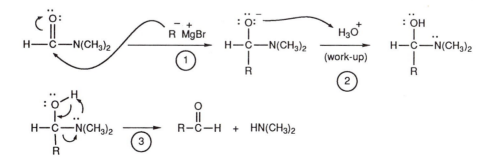

19.17.   In the reaction between a Grignard reagent and an amide, the nucleophilic R group of RMgX adds in the first step to the carbonyl group, forming a tetrahedral intermediate. Protonation during workup yields the geminal amino alcohol (a hemiaminal) which is unstable and eliminates a molecule of the amine.   In step (3), dimethylamine is displaced during formation of the carbon-oxygen double bond of the aldehyde.

19.18. In the reaction between a Grignard reagent and nitrile, the nucleophilic R group of RMgX adds in the first step to the carbon-nitrogen triple bond, forming an iminate ion. Protonation at nitrogen in step (2) forms the imine, which is unstable and undergoes hydrolysis in steps (3)–(6) to form the ketone.

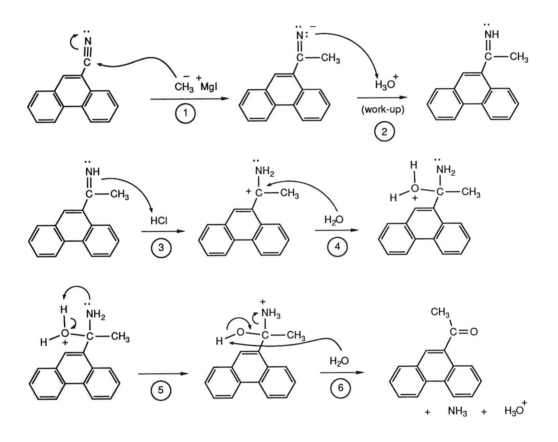

19.19. A ketone is generally prepared from a carboxylic acid derivative by reaction with an appropriate organometallic reagent. The reaction between an acid chloride and an organocuprate reagent is an especially good reagent combination for preparing a ketone. A lithium diorganocuprate is made from copper(I) iodide and an organolithium compound.

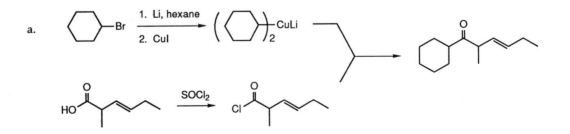

19.19.   (continued)

b.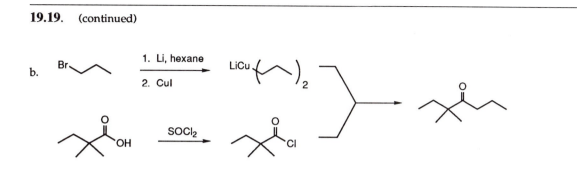

19.20.   A lactone is an ester, so according to the discussion in the text, two equivalents of a Grignard reagent add to a lactone, forming a 3° alcohol at the original carbonyl carbon atom. The alcohol portion of the lactone is unmasked by the reaction.

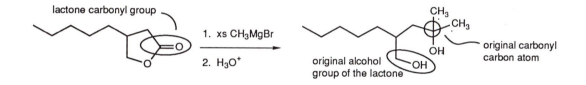

19.21.   Aldehydes undergo nucleophilic addition of H⁻ to the carbonyl group. The resulting alkoxide ion is protonated during workup with aqueous acid to form the alcohol.

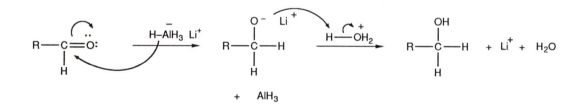

19.22.   Aldehydes are prepared from carboxylic acids or their derivatives by two common routes: 1) complete reduction to the alcohol followed by partial oxidation, or 2) partial reduction of an ester or acid chloride (this procedure works best with aromatic acid chlorides).

a.

b.

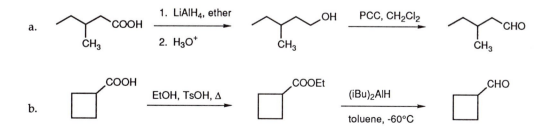

19.23. In the Killiani-Fischer synthesis, the aldehyde group of the sugar is converted to its cyanohydrin derivative. Hydrolysis of the nitrile group leads to formation of the corresponding lactone, which occurs through trapping the iminium intermediate by an alcohol OH groups.

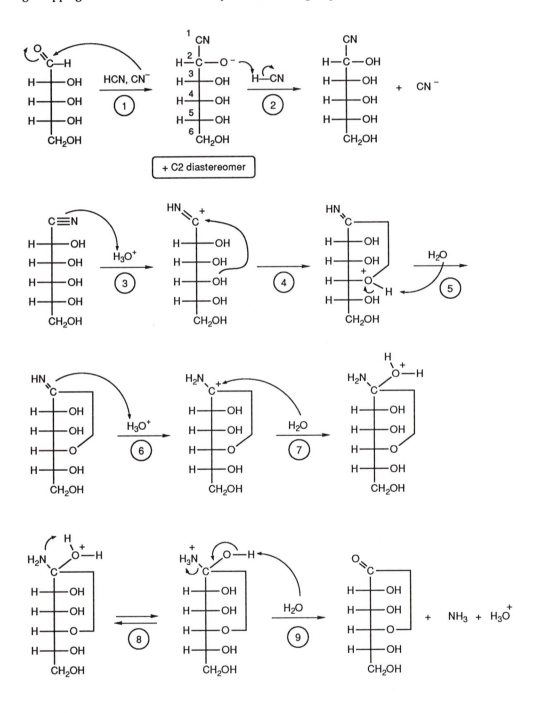

19.23. (continued)

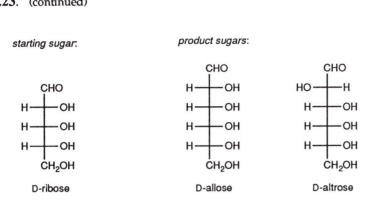

19.24. Interpret the name according to the directions given in Chapter 2.

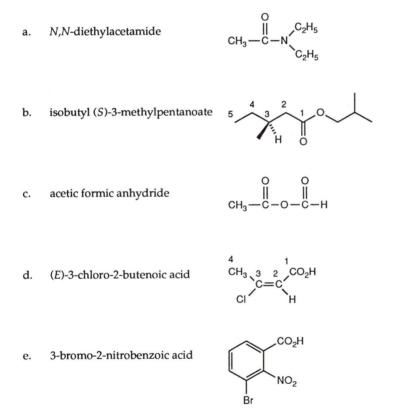

a.   *N,N*-diethylacetamide

b.   isobutyl (*S*)-3-methylpentanoate

c.   acetic formic anhydride

d.   (*E*)-3-chloro-2-butenoic acid

e.   3-bromo-2-nitrobenzoic acid

19.25.   Follow the procedure outlined in the solution to exercise 7.28.

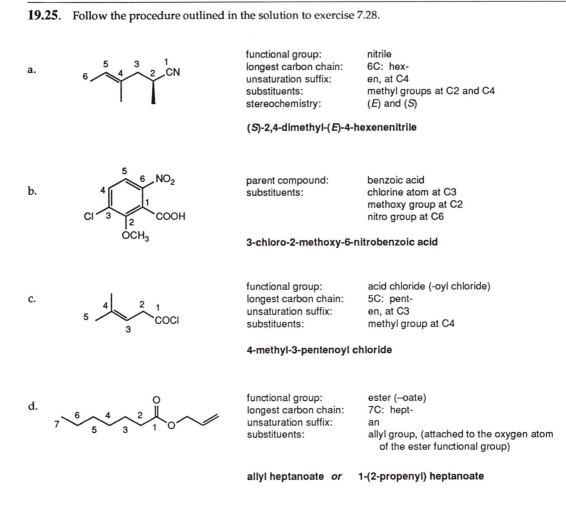

a.

functional group:	nitrile
longest carbon chain:	6C: hex-
unsaturation suffix:	en, at C4
substituents:	methyl groups at C2 and C4
stereochemistry:	(*E*) and (*S*)

(*S*)-2,4-dimethyl-(*E*)-4-hexenenitrile

b.

parent compound:	benzoic acid
substituents:	chlorine atom at C3
	methoxy group at C2
	nitro group at C6

3-chloro-2-methoxy-6-nitrobenzoic acid

c.

functional group:	acid chloride (-oyl chloride)
longest carbon chain:	5C: pent-
unsaturation suffix:	en, at C3
substituents:	methyl group at C4

4-methyl-3-pentenoyl chloride

d.

functional group:	ester (–oate)
longest carbon chain:	7C: hept-
unsaturation suffix:	an
substituents:	allyl group, (attached to the oxygen atom of the ester functional group)

allyl heptanoate   *or*   1-(2-propenyl) heptanoate

19.26.   a. A carboxylic acid reacts with a 1° or 2° alcohol and an acid catalyst to form an ester.

$$CH_3CH_2CH_2CH_2-\overset{O}{\overset{\|}{C}}-OH \quad \xrightarrow[\text{TsOH, }\Delta]{(CH_3)_2CHCH_2OH} \quad CH_3CH_2CH_2CH_2-\overset{O}{\overset{\|}{C}}-O-CH_2CH(CH_3)_2 \; + \; H_2O$$

b.   Lithium aluminum hydride reduces a carboxylic acid to form a 1° alcohol after acid workup.

$$CH_3CH_2CH_2CH_2-\overset{O}{\overset{\|}{C}}-OH \quad \xrightarrow[\text{2. }H_3O^+]{\text{1. LiAlH}_4} \quad CH_3CH_2CH_2CH_2-CH_2-OH$$

19.26.   (continued)

c.   A carboxylic acid reacts with thionyl chloride to form an acid chloride. An acid chloride reacts with an amine to form an amide.

$$CH_3CH_2CH_2CH_2-\overset{\overset{\displaystyle O}{\|}}{C}-OH \xrightarrow[\text{2. } CH_3CH_2CH_2CH_2NH_2]{\text{1. } SOCl_2, \Delta} CH_3CH_2CH_2CH_2-\overset{\overset{\displaystyle O}{\|}}{C}-NHCH_2CH_2CH_2CH_3$$

d.   Lithium aluminum hydride reduces a carboxylic acid to form a 1° alcohol after acid workup. PCC oxidizes a 1° alcohol to the corresponding aldehyde.

$$CH_3CH_2CH_2CH_2-\overset{\overset{\displaystyle O}{\|}}{C}-OH \xrightarrow[\substack{\text{2. } H_3O^+ \\ \text{3. PCC, } CH_2Cl_2}]{\text{1. } LiAlH_4} CH_3CH_2CH_2CH_2-CHO$$

e.   A carboxylic acid reacts with thionyl chloride to form an acid chloride. An acid chloride is reduced at low temperature to an aldehyde with use of a lithium tri(*t*-butoxy)aluminum hydride.

$$CH_3CH_2CH_2CH_2-\overset{\overset{\displaystyle O}{\|}}{C}-OH \xrightarrow[\text{2. } LiAlH(O\text{-}t\text{-}Bu)_3, -78\ °C]{\text{1. } SOCl_2, \Delta} CH_3CH_2CH_2CH_2-CHO$$

f.   A carboxylic acid reacts with thionyl chloride to form an acid chloride, which in turn reacts with an alcohol to form an ester. An ester reacts with a Grignard reagent to yield a 3° alcohol that has two like groups attached to the carbinol carbon atom.

$$CH_3CH_2CH_2CH_2-\overset{\overset{\displaystyle O}{\|}}{C}-OH \xrightarrow[\substack{\text{2. } CH_3OH \\ \text{3. } CH_3MgI, \text{ ether} \\ \text{4. } NH_4Cl\ (aq)}]{\text{1. } SOCl_2, \Delta} CH_3CH_2CH_2CH_2-\overset{\overset{\displaystyle OH}{|}}{\underset{\underset{\displaystyle CH_3}{|}}{C}}-CH_3$$

19.27.   IR spectra reveal which functional groups are present in a molecule. To deduce what differences are be observed in the IR spectra of reactants and products of a specific reaction, identify the functional group in the starting material and in the product and decide what absorption bands will be observed for each.

a.   A carboxylic acid has a carbonyl group, and the alcohol product does not, so there should be a noticeable difference in the appearance of the region around 1700 cm^{-1}. A carboxylic acid also has a strong, very broad band for the OH stretching vibration. The alcohol has an absorption from the O–H stretching vibration, too, but it is narrower in width and at higher frequency.

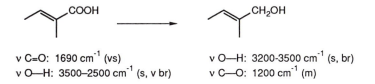

ν C=O: 1690 cm^{-1} (vs)
ν O—H: 3500–2500 cm^{-1} (s, v br)

ν O—H: 3200-3500 cm^{-1} (s, br)
ν C—O: 1200 cm^{-1} (m)

19.27. (continued)

b. A carboxylic acid and its acid chloride derivative each have a carbonyl group, but the stretching frequency observed for an acid chloride is at 1800 cm^{-1}, the highest frequency of any common C=O stretching vibration.

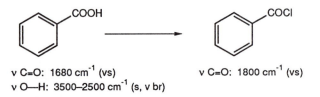

v C=O: 1680 cm^{-1} (vs)
v O—H: 3500–2500 cm^{-1} (s, v br)

v C=O: 1800 cm^{-1} (vs)

c. A ketone and an ester each have a carbonyl group, but the ester also has a very strong absorption band at about 1200 cm^{-1} caused by the C—O stretching vibration. The product also has a carbon-carbon double bond, which will have an absorption band at about 1650 cm^{-1}. The band for the C=O stretching vibration may obscure the C=C stretching vibration, however.

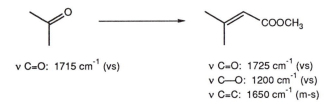

v C=O: 1715 cm^{-1} (vs)

v C=O: 1725 cm^{-1} (vs)
v C—O: 1200 cm^{-1} (vs)
v C=C: 1650 cm^{-1} (m-s)

19.28. When performing a reaction, you can predict which resonances will be observed in the proton NMR spectra of the starting material and product by using data from the tables and figures in Chapter 15. For the resonances given below, the first value is the chemical shift relative to the signal for TMS, the number with the capitalized H is the integrated intensity (2H means two protons, etc.), and the lower case letter is the spin-spin coupling pattern (s = singlet, d = doublet, t = triplet, q = quartet, m = multiplet, br s = broad singlet). The *J* value for the spin-spin coupling is the final parameter given.

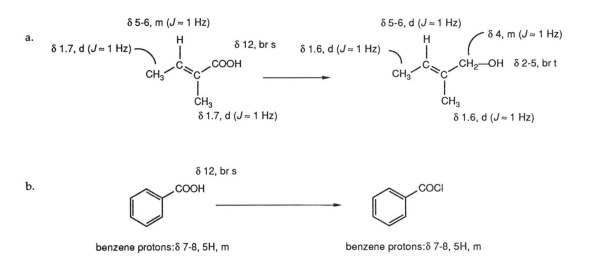

19.28.   (continued)

c.

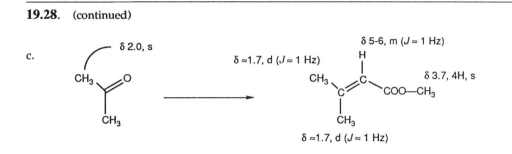

δ 2.0, s

$\delta \approx 1.7$, d ($J \approx 1$ Hz)

δ 5-6, m ($J \approx 1$ Hz)

δ 3.7, 4H, s

$\delta \approx 1.7$, d ($J \approx 1$ Hz)

19.29.   a. A carboxylic acid is converted to an alcohol by reduction with lithium aluminum hydride followed by workup with aqueous acid.

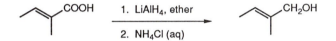

1. LiAlH$_4$, ether

2. NH$_4$Cl (aq)

b.   A carboxylic acid is converted to its acid chloride derivative by heating it with thionyl chloride.

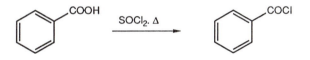

SOCl$_2$. Δ

c.   A ketone is converted to an ester that has two additional carbon atoms with use of a phosphonate ylide made from methyl bromoacetate and trimethylphosphite (Horner-Emmons reaction, text page 904).

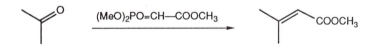

(MeO)$_2$PO=CH—COOCH$_3$

19.30.   In the Killiani-Fischer synthesis, a carbohydrate is converted to two isomeric carbohydrates that each have one additional carbon atom. Chiral centers present in the starting compound retain their configurations in the products; epimers are formed at the new stereogenic carbon atom that is formed.

a.

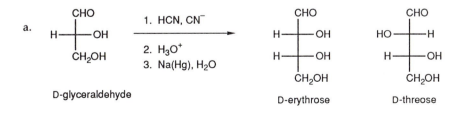

1. HCN, CN$^-$

2. H$_3$O$^+$

3. Na(Hg), H$_2$O

D-glyceraldehyde

D-erythrose

D-threose

19.30.   (continued)

b.

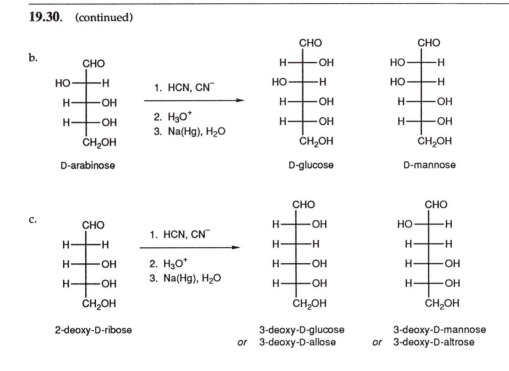

19.31.   The transformation of the chloroalkene to the cyclized product starts with nucleophilic addition of the Grignard reagent to the carbonyl group. Treating the alcohol with formic acid leads to formation of a carbocation that is trapped in step (4) by the double bond. The 2° carbocation formed after cyclization occurs is subsequently trapped by formate ion in step (5), yielding the ester. The final steps constitute the acid-catalyzed hydrolysis of an ester [steps (6) - (9)]. The presence of a chlorine atom on the carbinol carbon atom leads to formation of the ketone carbonyl group by elimination of HCl in the last step.

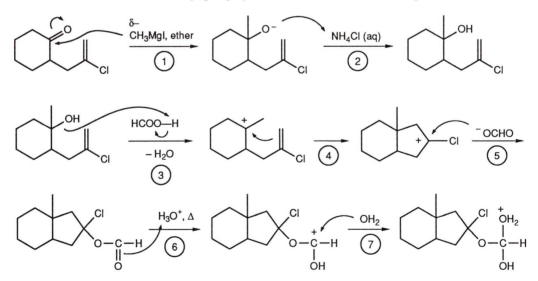

19.31.   (continued)

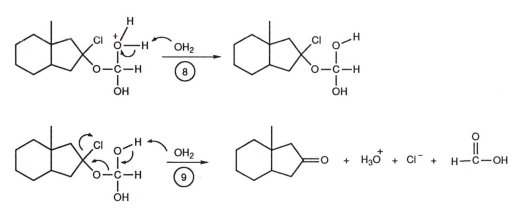

19.32.   The rates of addition-elimination reactions of carbonyl compounds are affected by the electrophilicity of the carbonyl carbon atom and by steric effects of any substituents.

a.   The carbonyl carbon atom of a carboxylic acid is less electrophilic than its proton, the removal of which effectively prevents most addition reactions to the carbonyl group. An amide has a poorly electrophilic carbonyl group because of resonance stabilization by the nitrogen atom. An aldehyde is not very sterically hindered, but it is less reactive than an anhydride toward addition-elimination because the anhydride has a better leaving group (acetate ion) attached to the carbonyl carbon atom.

b.   An acid chloride is the most reactive of the acid derivatives. Esters are reactive but steric effects of the alkoxy group affect the rate of addition-elimination, so a methyl ester reacts faster than a *tert*-butyl ester.

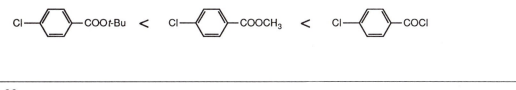

19.33.

a.   An amine reacts with an ester, displacing a molecule of alcohol and forming the amide.

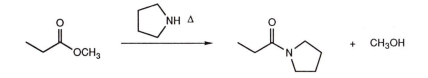

19.33.   (continued)

b.   A phenol reacts with an acid chloride, with the oxygen atom displaces chloride ion to form the aryl ester. Triethylamine is a base that reacts with the HCl that is generated.

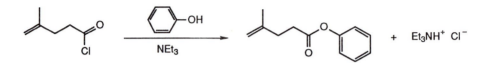

c.   An alkene double bond is cleaved with hot, concentrated permanganate ion (text page 443). An alkene carbon atom that bears a hydrogen atom becomes a carboxylic acid group.

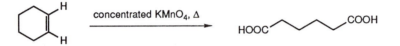

d.   Nitriles are hydrolyzed under basic conditions, but an α-hydroxy nitrile (a cyanohydrin) is converted to a ketone or aldehyde by loss of HCN (text page 807).

e.   An aldehyde reacts with hydroxylamine to form an oxime (text page 874). Dehydration transforms an oxime to a nitrile.

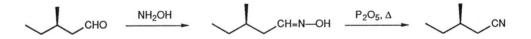

f.   Lithium aluminum hydride reduces an amide to an amine. Acetic anhydride reacts with an amine to produce an acetamide derivative. The acetal protecting group is inert toward reaction with hydride reagents, and the aqueous acid workup (step 2) is not vigorous enough to hydrolyze the acetal group.

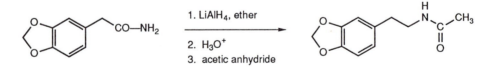

g.   Thionyl chloride converts a carboxylic acid to its acid chloride derivative. A lithium dialkylcuprate replaces the chloride ion with an alkyl group, in this case methyl. The resulting ketone undergoes a Wittig reaction to form the corresponding alkene.

19.33. (continued)

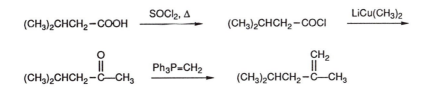

h.   Dichromate ion converts a 1° alcohol to a carboxylic acid, which reacts with methanol and acid to form the methyl ester.

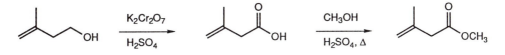

19.34.   The first step in acid chloride formation with use of oxalyl chloride is an acid-base reaction that produces a carboxylate ion. That anion undergoes addition-elimination (steps 2 and 3) to form a mixed anhydride. In step (4), chloride ion adds to the carbonyl group that was part of the original carboxylic acid. Finally, the tetrahedral intermediate collapses, displacing carbon dioxide, carbon monoxide and chloride ion and forming the acid chloride.

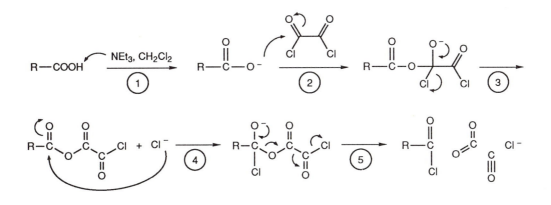

19.35.   There are many ways to make carboxylic acids; the reactions shown below summarize some synthetic routes.

a.   **Benzyl alcohol:** a 1° alcohol is oxidized to a carboxylic acid by dichromate ion (text page 853).

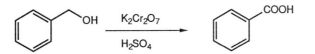

19.35. (continued)

b.   **Bromobenzene:** after conversion to its Grignard derivative, reaction with carbon dioxide (text page 582) followed by hydrolysis yields the acid.

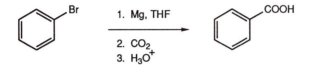

c.   **Ethylbenzene:** an alkyl group attached to a benzene ring is oxidized by permanganate ion (text page 556) to a carboxylic acid group, after acid workup.

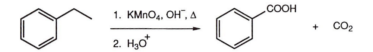

d.   **N-Methylbenzamide:** an amide is hydrolyzed to the analogous carboxylic acid (text page 942).

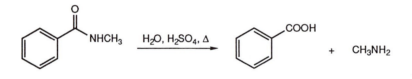

e.   **Benzonitrile:** a nitrile is hydrolyzed to the analogous carboxylic acid (text page 953).

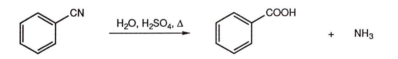

f.   **Styrene:** a carbon-carbon double bond is cleaved upon ozonolysis (text page 441) or by permanganate ion (text page 443). A carboxylic acid is formed when the alkene carbon atom bears a hydrogen atom.

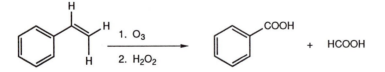

19.36. a. The reaction of a Grignard reagent with carbon dioxide is one of the simplest ways to prepare a carboxylic acid.

retrosynthesis:

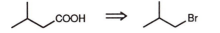

synthesis:

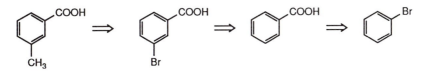

b. The *meta*-substitution pattern of the benzene ring in the product suggests that the acid group should be used to introduce a substituent that can be replaced by the methyl group (direct introduction of a methyl group by Friedel-Crafts alkylation is unlikely because a *meta* director normally deactivates the ring too much). Bromination attaches a bromine atom, which can be replaced by reaction with an organocuprate reagent. The carboxylic acid group is first protected as its ester before treatment with the Gilman reagent to avoid an acid-base reaction. Deprotection in the last step regenerates the carboxylic acid group.

retrosynthesis:

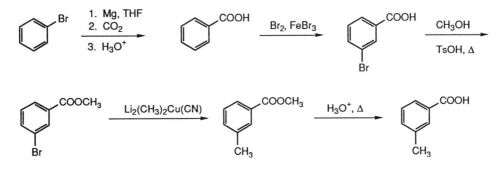

synthesis:

19.36. (continued)

c. The carboxylic acid is made by reaction of a Grignard reagent with carbon dioxide. The aldehyde group must be protected and deprotected during the synthesis.

retrosynthesis:

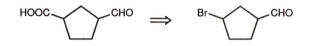

synthesis:

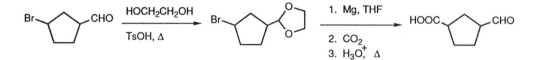

19.37. Use the data in Table 14.4 (text page 654) to differentiate carboxylic acid derivatives by the position and shapes of their IR absorption bands.

37A	RCOOR'	vC=O:	1720 cm⁻¹ (vs)	vC—O:	1230 cm⁻¹ (vs)
37B	RCOCl	vC=O:	1800 cm⁻¹ (vs)		
37C	RCOOH	vC=O:	1690 cm⁻¹ (vs)	vO—H:	3400-2500 cm⁻¹ (m, br)

19.38. Use the data in Figure 15.37 (text page 735) to differentiate carboxylic acid derivatives by their carbon NMR chemical shift values.

38A	RCOOR'	C=O:	δ 168
38B	RCONH₂	C=O:	δ 175
38C	RCN	C≡N:	δ 120

19.39. A cyclic anhydride reacts the same as an acyclic one except that the leaving group remains part of the product molecule. Thus, the alcohol OH group of propanol adds to one of the carbonyl groups of the anhydride in step (1) to form a zwitterionic tetrahedral intermediate. When this intermediate collapses in step (2) and regenerates the carbonyl group, the carboxylate ion that is formed deprotonates the alkoxy substituent in step (3) to form the acid-ester.

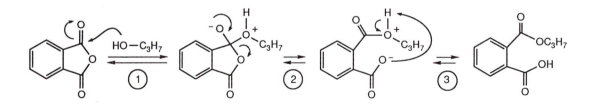

19.40. Hydride and organometallic reagents differ in their reactivity toward the various types of carbonyl compounds.

a. Sodium borohydride is capable of reducing a ketone functional group, but not an ester. Therefore, the 2° alcohol is produced. Cyclization may occur to form the lactone.

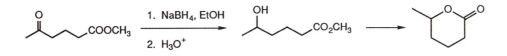

b. Lithium aluminum hydride reduces every type of carbonyl group. Therefore, both the ketone and carboxylic acid groups are reduced, and the diol is obtained.

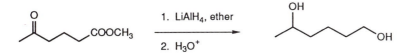

c. Lithium aluminum hydride reduces an ester to form two molecules of alcohol. A lactone is therefore converted to a diol. Because an ester is reduced by breaking the single bond between the carbonyl carbon atom and the oxygen atom, the stereochemistry of the chiral center is not affected during this particular transformation.

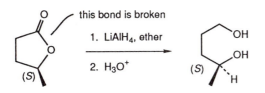

d. A Grignard reagent adds to an ester to form a 3° alcohol. With a lactone, the group that was the alkoxy portion of the ester is liberated by the reaction. The stereochemistry of the chiral center is not affected during this particular transformation.

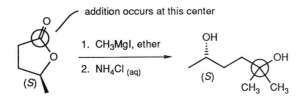

19.41. Friedel-Crafts acylation is a straightforward way to make an aryl alkyl ketone such as 1-phenyl-1-butanone Using organometallic reagents, low temperature addition of a Grignard reagent to an acid chloride is often successful. Reaction of an acid chloride with an organocuprate reagent is usually more reliable.

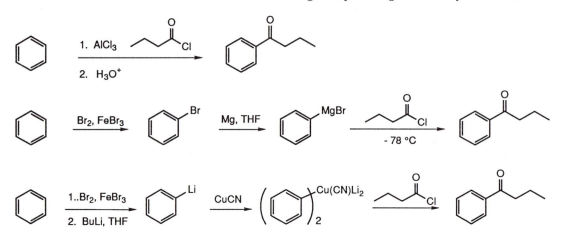

19.42. a. A chiral diol is made by asymmetric dihydroxylation of the corresponding alkene. The α,β-unsaturated ester is made from the acid.

retrosynthesis:

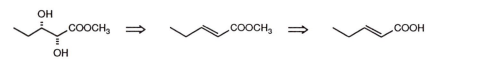

synthesis:

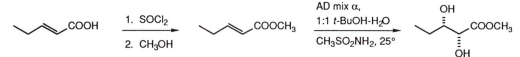

b. A chiral epoxide is made by asymmetric epoxidation of the corresponding alkene. In this case, the product derives from an allylic alcohol, which is prepared by reduction of the corresponding acid (there are better reagents than LiAlH₄ for this transformation; they are discussed in Chapter 22).

retrosynthesis:

synthesis:

19.43. Copolymerization of diphenyl carbonate and Bisphenol A is an example of transesterification, the mechanism of which is shown on text page 952. In the first step, one of the oxygen atoms of Bisphenol A adds to the carbonyl group of diphenyl carbonate. The tetrahedral intermediate that is formed collapses, expelling an equivalent of phenol.

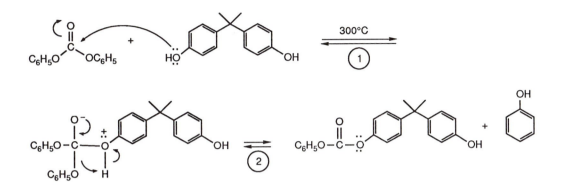

In the next stage, the process is repeated by another equivalent of Bisphenol A.

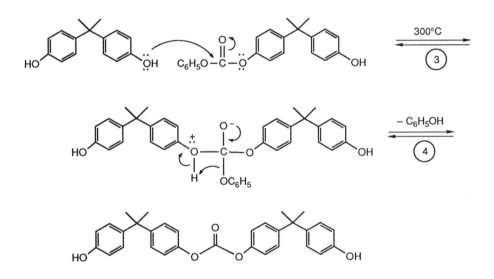

19.44. Propiolactone has a strained four-membered ring. An impurity such as hydroxide ion reacts with the carbonyl group of the lactone, leading to ring opening. The first stage of the overall transformation to polymer is hydrolysis of the lactone, producing the conjugate base of the hydroxy acid.

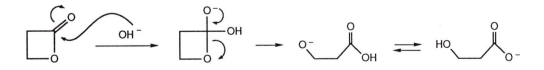

19.44.   (continued)

The conjugate base of the hydroxy acid is in equilibrium with the alkoxy acid, and this alkoxide ion is a nucleophile that reacts with another equivalent of the lactone. Nucleophilic addition to the lactone carbonyl group is repetitive process that leads to polymer formation.

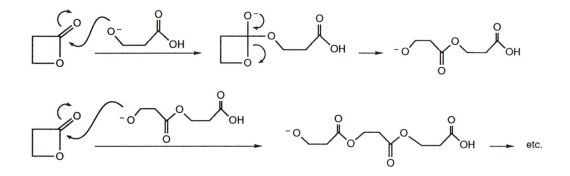

The IR spectrum has absorptions for the ester group (with a small amount of the carboxylic acid end group). The proton NMR spectrum consists of two triplets for the nonequivalent methylene groups (and minor peaks for the protons at the chain ends).

IR spectrum: proton NMR spectrum:

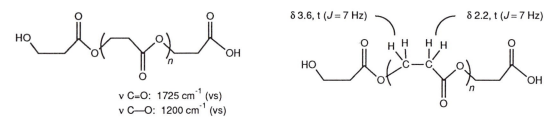

ν C=O: 1725 cm^{-1} (vs)
ν C—O: 1200 cm^{-1} (vs)

δ 3.6, t (J = 7 Hz) δ 2.2, t (J = 7 Hz)

19.45.   Two transformations take place in the formation of valium from an amide precursor:

(1) hydrolysis of the amide group to liberate an amino group, and
(2) intramolecular imine formation that generates the benzodiazepam ring.

In the first stage of the transformation, the catalytic triad of the amidase (indicated in the cartoon by the bold amino acid abbreviations) reacts with the amide linkage of the valium prodrug. A tetrahedral intermediate is formed in the first step

19.45.   (continued)

In step (2), the tetrahedral intermediate collapses with expulsion of the valium precursor.   The acylated enzyme is subsequently returned to its active form by hydrolysis of the acyl group attached to the serine side chain.

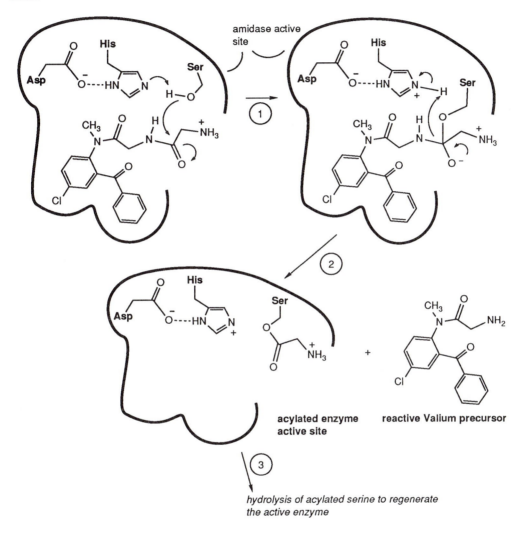

The mechanism for ring formation was illustrated in the solution to exercise 18.29.

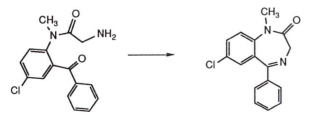

This transformation of the active precursor to Valium proceeds by the mechanism illustrated in the solution to exercise 18.29.

19.46. The Wohl degradation converts one carbohydrate to another having one less carbon atom. Cleavage occurs from the aldehyde end.

The first step is addition-elimination between the aldehyde group and hydroxylamine, producing an oxime (text page 874).

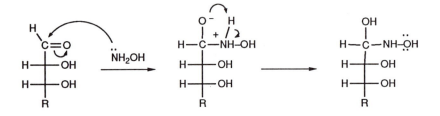

The second step is an addition-elimination reaction between the oxygen atom of the oxime and acetic anhydride, forming an ester derivative of the oxime.

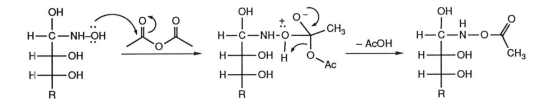

The last process comprises two steps: elimination that generates a new π bond of the nitrile functional group, and elimination that forms the aldehyde functional group. The cyanohydrin, which is produced by the first elimination step, is unstable under the basic conditions of the reaction. So as soon as the nitrile group is formed, deprotonation occurs to generate the aldehyde. The carbon chain of the carbohydrate is thereby decreased in length by one carbon atom.

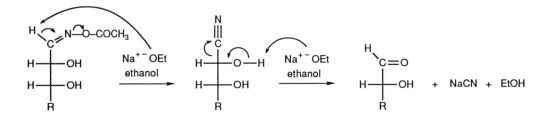

19.47.   The Wohl degradation, described in exercise 19.46, converts a carbohydrate to the next lower homolog. Configurations of any chiral centers that remain are not affected.

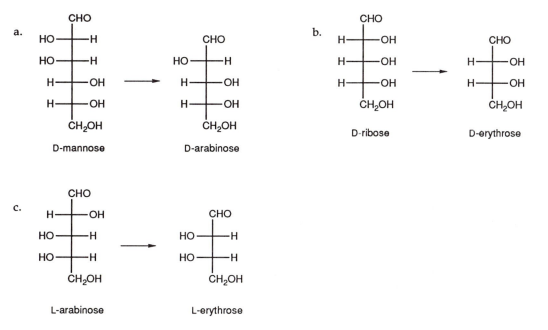

19.48.   The Wohl degradation, described in exercise 19.46, converts a carbohydrate to its next lower homolog.   Configurations of any chiral centers that remain are not affected.   If we consider which carbohydrate is used to prepare a particular sugar, the starting carbohydrate can have either configuration at the carbon atom that becomes the new aldehyde group (that is, C2), because chirality of that center is destroyed during the transformation.

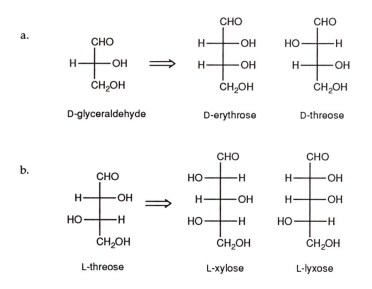

19.48. (continued)

c.
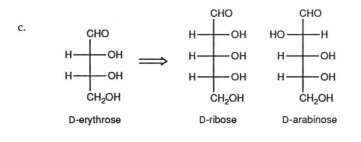

D-erythrose   D-ribose   D-arabinose

19.49. a. Compound **49A** has a strong absorption at 1700 cm^{-1}, so it clearly has a carbonyl group. The data used to differentiate various carbonyl-containing compounds are summarized in Table 14.4 on text page 654. This spectrum shows a very strong broad band between 3500 cm^{-1} and 2500 cm^{-1}, which is consistent with the notion that this compound is a carboxylic acid.

From the proton NMR spectrum, we confirm that the compound is a carboxylic acid by observation of a broad singlet at about δ 12. Absorptions between δ 5 and δ 6 indicate that the compound has a double bond, and the singlet near TMS shows that there is a methyl group.

Because the coupling between the alkene protons is small, it is difficult to tell whether this compound has a *cis* or *trans* geometry. The strong absorption band in the IR spectrum at 960 cm^{-1} (compare Figure 14.30, text page 671) suggests that this compound is the *trans* isomer.

δ	integr	splitting	J (Hz)	assignment
12	1H	s	–	COOH
6.3	1H	s	–	C=C–H
5.6	1H	s		C=C–H
2.0	3H	s	–	CH$_3$

b. Compound **49B** has a strong absorption at 1740 cm^{-1}, so it clearly has a carbonyl group. The data can be used to differentiate various carbonyl-containing compounds are summarized in Table 14.4 on text page 654. The strong band at 1200 cm^{-1} is consistent with the presence of an ester functional group. A carboxylic acid would have a very strong broad band between 3500 cm^{-1} to 2500 cm^{-1}.

For the proton NMR spectrum, we summarize the data as shown at the top of the next page.

19.49. (continued)

δ	integr	splitting	J (Hz)	assignment
4.2	2H	q	7	O–CH$_2$; coupling with CH$_3$
3.7	2H	t	7	Cl–CH$_2$; coupling with CH$_2$
2.5	2H	q	7	CH$_2$ adjacent to a carbonyl group; coupling with CH$_2$
2.1	2H	t	7	CH$_2$; coupling with two CH$_2$ groups
1.2	3H	t	7	CH$_3$; coupling with CH$_2$

Compound **49B** is ethyl 4-chlorobutanoate.

$$Cl-CH_2-CH_2-CH_2-\overset{\overset{\displaystyle O}{\|}}{C}-O-CH_2-CH_3$$

19.50. As shown following the paragraph that introduces this exercise, initiation generates the tributyltin radical from hexabutylditin, which in turn abstracts a halogen atom from the substrate. Cyclization occurs by addition-elimination with the carbonyl group of the thio- or selenoester. The PhSe• or PhS• radical produced by elimination from the tetrahedral radical intermediate propagates the reaction by abstraction of a halogen atom from another equivalent of starting material.

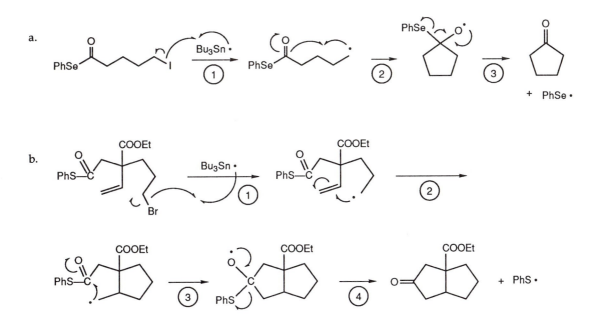

19.51. In the tetracarbonylcobalt(-1) catalyzed conversion of an alkyl halide to a carboxylic acid, the first step is oxidative addition: the cobalt goes from an oxidation state of –1 to +1, and the coordination number increases. From the standpoint of RX, the reaction follows an S_N2 pathway.

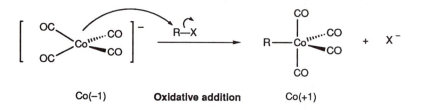

In the second step, CO binds to the cobalt ion, and the alkyl group migrates onto a coordinated CO molecule.

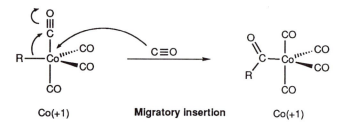

In the third step, hydroxide ion undergoes nucleophilic addition to the carbonyl group of the acyl ligand, forming a tetrahedral intermediate.

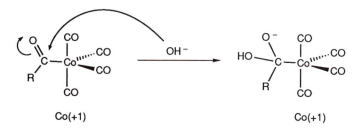

In the fourth step, the tetrahedral intermediate collapses to form the carboxylic acid functional group. The tetracarbonylcobaltate(-1) ion functions as the leaving group. For cobalt, this is reductive elimination because the oxidation state of the cobalt ion goes from +1 to –1, and the coordination number decreases.

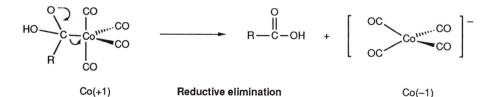

THE ACID–BASE CHEMISTRY OF CARBONYL COMPOUNDS

20.1. The dianion of a phosphoric acid derivative is stabilized by delocalization of unshared and π electron pairs among the atoms that have sp^2 hybridization.

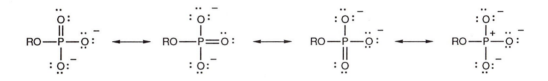

20.2. A plot of pK_a values versus the number of halogen atoms is used to estimate the approximate pK_a value of difluoroacetic acid.

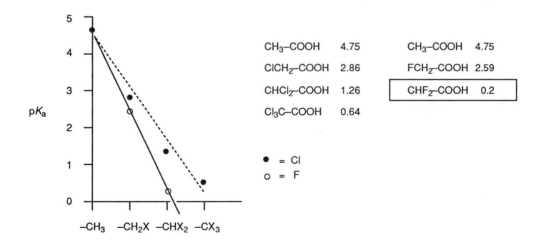

CH_3-COOH	4.75	CH_3-COOH	4.75
$ClCH_2-COOH$	2.86	FCH_2-COOH	2.59
$CHCl_2-COOH$	1.26	CHF_2-COOH	0.2
$Cl_3C-COOH$	0.64		

● = Cl
○ = F

20.3.   Dissociation of a proton from oxalic and malonic acids produces a monoanion that forms a five- or six-membered ring by hydrogen bonding. Acids that have additional carbon atoms between carboxylic acid groups would have to form larger rings, which is entropically less favorable.

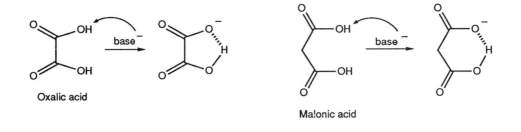

20.4.   The most electrophilic atom in butanoamide is a proton bonded to nitrogen. In *N,N*-dimethylbutano-amide, the most electrophilic center is the carbonyl carbon atom. Butanoamide undergoes an acid-base reaction and forms a resonance-stabilized anion, whereas the dimethyl compound undergoes nucleophilic addition to the carbonyl group, leading to hydrolysis.

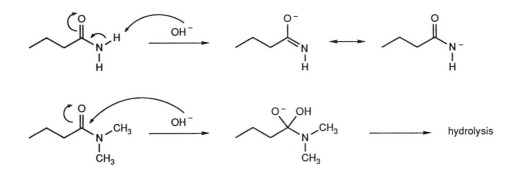

20.5.   The enol form of phenol is aromatic (6 π electron). Recall from Chapter 12 (text page 516) that the empirical resonance energy of a benzene derivative is substantial (> 30 kcal/mol).

2,4-cyclohexadienone (keto form)   phenol (enol form)

20.6.   In the conversion of glyceraldehyde-3-phosphate to dihydroxyacetone phosphate, protonation of the carbonyl group leads to isomerization and formation of the enediol.   Deprotonation of the OH group at C2 yields the ketone.   Deprotonation of the OH group at C1 regenerates the aldehyde form.

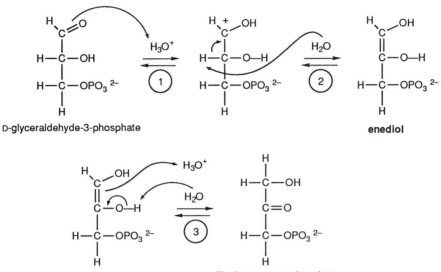

20.7.   The proton attached to the α carbon atom of phenylalanine ethyl ester is slightly acidic ($pKa \approx 25$). This proton is removed by base to form the enolate derivative.   Reprotonation occurs at either face of the enolate double bond, generating enantiomers.

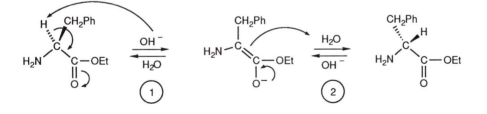

20.8.   Each round of iodination of a methyl ketone consists of two steps: a) formation of the enolate ion, and b) reaction of the anion with iodine, displacing iodide ion.

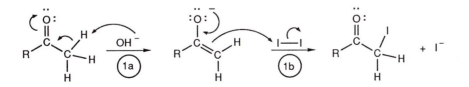

20.8. (continued)

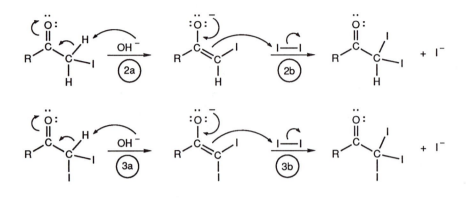

20.9. The two sets of reaction conditions in this exercise illustrate the difference between kinetic and thermodynamic deprotonation. At the lower temperature, the kinetic product forms. At the higher temperature, the thermodynamic product is generated.

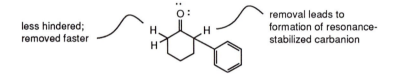

The thermodynamic product is a result of extensive delocalization of the negative charge.

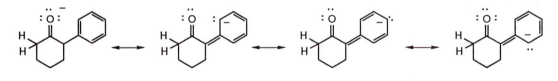

20.10. Both compounds in this exercise are prepared by alkylation of a ketone enolate ion with a reactive 1° alkyl halide.

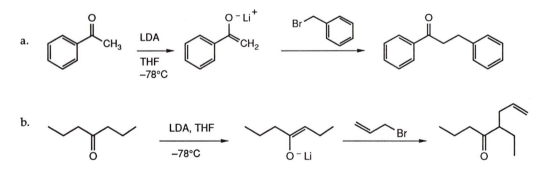

20.11. An enolate ion is formed by deprotonation of a ketone at its alpha carbon atom. This nucleophilic species reacts with diphenyl disulfide at the sulfur-sulfur bond, displacing phenylthiolate ion as a leaving group.

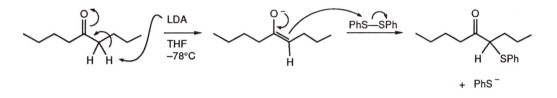

20.12. Hydrolysis of an iminium ion follows the same steps as hydrolysis of an imine, except that the group is already activated for reaction with the nucleophilic water molecule because the nitrogen atom bears a positive charge.

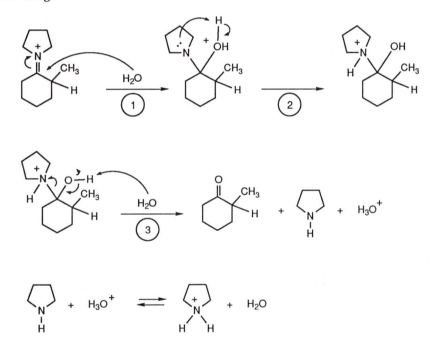

20.13. A ketone reacts with pyrrolidine to form an enamine. When treated with a reactive alkyl halide, an enamine forms the alkylated product as its iminium salt. Hydrolysis generates the alkylated ketone.

a.

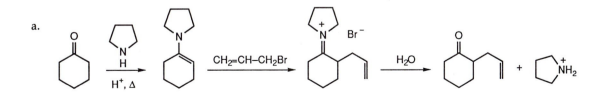

20.13. (continued)

b.

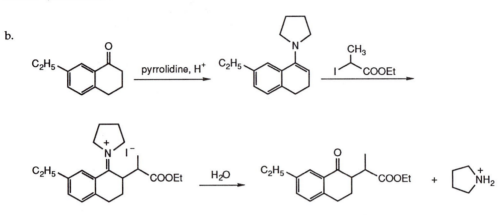

20.14. The dimethylhydrazone derivative of a ketone is deprotonated by LDA at the less highly substituted carbon atom. Alkylation followed by hydrolysis produces the alkylated ketone.

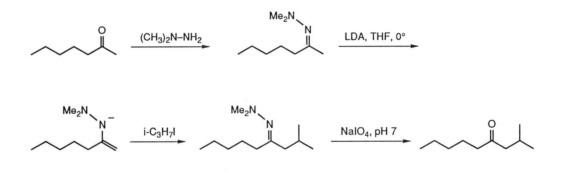

20.15. A lactone forms an enolate derivative in the usual way, by deprotonation at the position alpha to the carbonyl group. Alkylation is an S_N2 process with the enolate ion serving as the nucleophile.

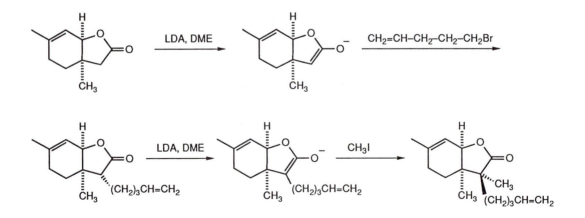

20.15. (continued)

This molecule is saddle-shaped, so the bottom face of the lactone ring is less hindered, and alkylation occurs from below the ring in each step.

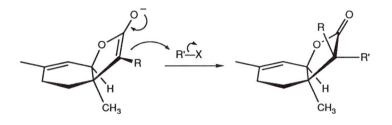

20.16. The dianion of a carboxylic acid is alkylated at the carbon atom alpha to the carboxylate group. Acid workup regenerates the carboxylic acid. An ester reacts with LDA at low temperature to form a monoanion; alkylation is followed by hydrolysis with hot, aqueous acid to liberate the carboxylic acid product.

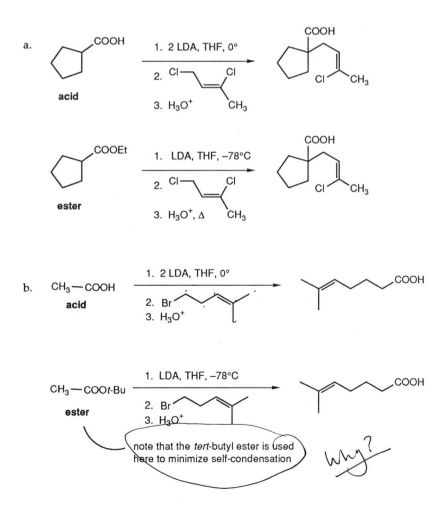

20.17. Resonance forms for the anions of diethyl malonate and ethyl cyanoacetate are like those drawn for ethyl acetoacetate: the negative charge and the electron pair are delocalized onto the cyano or carbonyl group.

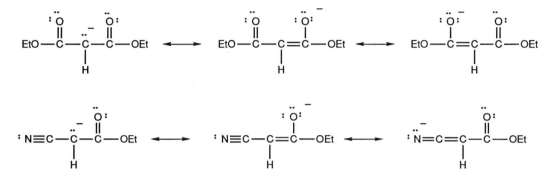

20.18. An active methylene compound has the form Z—CH₂—Z', where Z and Z' are electron withdrawing substituents, often carbonyl groups. Any compound that has two or more "Z-groups" may be called an "active methylene compound", even if it is actually a methine derivative.

a.   *retrosynthesis*:

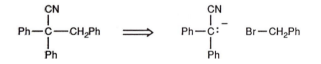

synthesis:

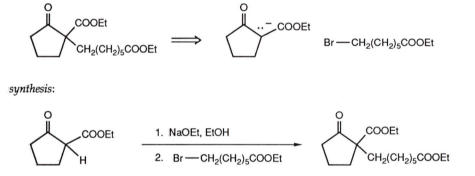

b.   *retrosynthesis*:

$$Ph-\underset{\underset{Ph}{|}}{\overset{\overset{CN}{|}}{C}}-CH_2Ph \implies Ph-\underset{\underset{Ph}{|}}{\overset{\overset{CN}{|}}{C}}:^- \quad Br-CH_2Ph$$

synthesis:

$$Ph-\underset{\underset{Ph}{|}}{\overset{\overset{CN}{|}}{C}}-H \quad \xrightarrow[\text{2. PhCH}_2\text{Br}]{\text{1. NaOEt, EtOH}} \quad Ph-\underset{\underset{Ph}{|}}{\overset{\overset{CN}{|}}{C}}-CH_2Ph$$

20.19. Reaction of an ester enolate ion with dimethyl disulfide produces the α-sulfenylated species. That compound readily undergoes deprotonation to form a second enolate ion. Alkylation proceeds smoothly by an S_N2 pathway.

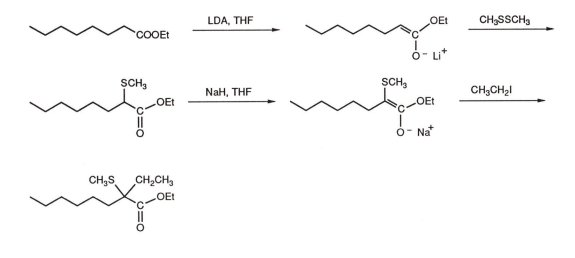

20.20. The hydrolysis of esters is discussed in the text starting on page 941. The carbonyl group is first protonated to generate a carbocation, and a molecule of water intercepts the positively-charged center. Transfer of a proton in step (3) is followed by regeneration of the carbonyl group in step (4), which forms the acid and expels a molecule of ethanol.

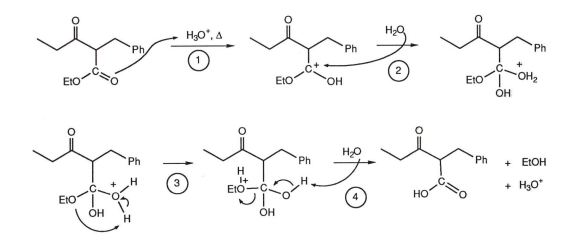

20.21. To contemplate the retrosynthesis of a carbonyl compound in which alkylation is a key step, attach an ester group at the position alpha to the carbonyl group of the given product. If the product is a carboxylic acid, think of the acid group as its ester derivative first. Then consider which alkylating agent is needed to prepare the carbon atom skeleton.

20.21.   (continued)

a.   This carboxylic acid is considered as a diester precursor; the cyclobutane ring is made by double alkylation of diethyl malonate with a dihaloalkane.

retrosynthesis:

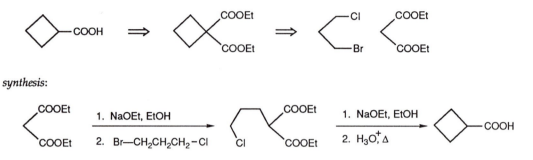

synthesis:

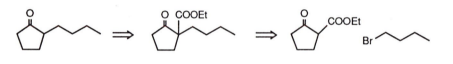

b.   The given ketone is considered as a keto ester precursor.

retrosynthesis:

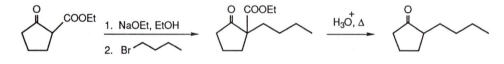

synthesis:

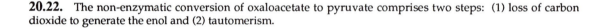

20.22.   The non-enzymatic conversion of oxaloacetate to pyruvate comprises two steps: (1) loss of carbon dioxide to generate the enol and (2) tautomerism.

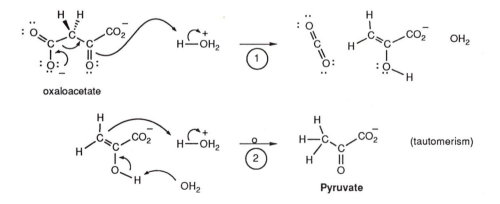

20.23. To draw the structure of the conjugate base of a carbonyl compound, first, interpret the name of the starting compound (the acid). Then remove the most acidic proton, which is normally the one alpha to a carbonyl (or other unsaturated) group. If there are two electron withdrawing groups, the most acidic proton is the one attached to the atom between them.

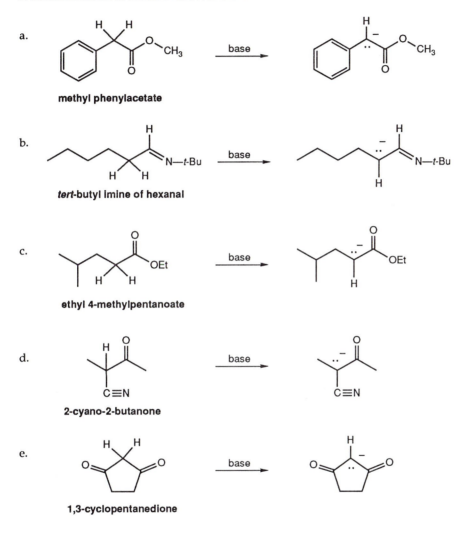

a.

methyl phenylacetate

b.

tert-butyl imine of hexanal

c.

ethyl 4-methylpentanoate

d.

2-cyano-2-butanone

e.

1,3-cyclopentanedione

20.24. To draw the structure of an enol form, first interpret the name of the starting carbonyl compound, then change the C=O group to a C—OH group. Remove a proton from the carbon atom adjacent to the carbonyl group, and insert a double bond between the original carbonyl carbon atom and the adjacent one.

a.

20.24.   (continued)

b.

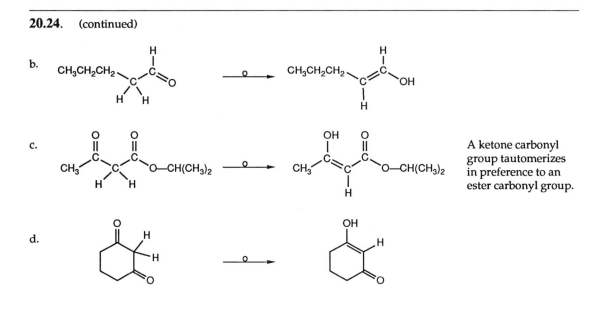

c.

A ketone carbonyl group tautomerizes in preference to an ester carbonyl group.

d.

20.25.   Follow the procedure described in the solution to exercise 3.4.

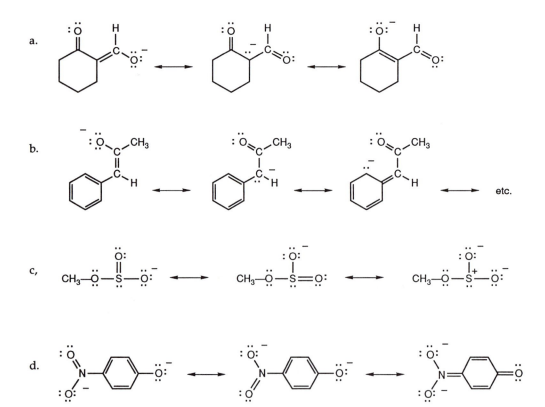

a.

b.

etc.

c,

d.

20.26. Use the pK_a values given in Table 6.3 (text page 224) to evaluate the acidity of each type of proton. Generally, a proton attached to an oxygen atom is the most acidic. If unsaturation is present, the acidity increases (i.e., a phenol OH group is more acidic than an alcohol OH group) . If no proton is attached to a heteroatom, the most acidic proton is the one bonded to a carbon atom adjacent to one or more carbonyl groups.

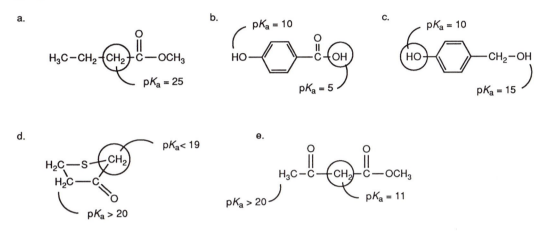

20.27. Ethyl acetoacetate has two carbonyl groups, so two carbanions can be stabilized by delocalization with the π bond of the adjacent carbonyl group. Besides protons attached to the methylene group situated between the two carbonyl groups, protons attached to the terminal methyl group are also relatively acidic.

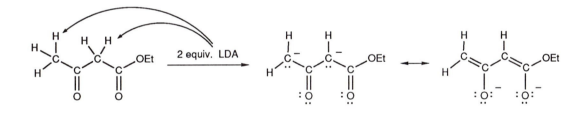

20.28. Assess the acidity of each compound by looking for structural differences based on relative electro-negativities of substituents, the identity of the atom to which the acidic proton is attached, and possibilities for delocalization.

a.   Based on electronegativity alone, the methoxy group is the most electron-withdrawing, so the carboxylic acid with that substituent should be the most acidic. The unsubstituted compound is the least acidic.

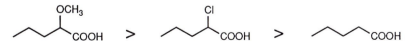

20.28.   (continued)

b.   A proton attached to sulfur is generally more acidic than one attached to oxygen. The nitro group on the benzene ring, being electron withdrawing, makes the proton of the carboxylic acid group even more acidic. Even though a methyl group is electron donating, substitution of a benzoic acid in the *ortho* position makes that compound more acidic than benzoic acid itself.

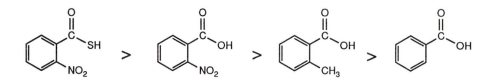

c.   Protons adjacent to a carbonyl group are relatively acidic, but for esters and amides, delocalization of electrons among the atoms of the carbonyl group and oxygen or nitrogen offsets resonance stabilization that the carbonyl group provides to a neighboring anionic center.

$$CH_3CH_2CH \overset{O}{\underset{}{C}} CH_3 \quad > \quad CH_3CH_2CH \overset{O}{\underset{}{C}} OCH_3 \quad > \quad CH_3CH_2CH \overset{O}{\underset{}{C}} N(CH_3)_2$$

20.29.   A base reacts with an acid by removing a proton from the latter, generating the conjugate base and conjugate acid of the reactants. The equilibrium lies to the side that has the weaker acid (text page 225).

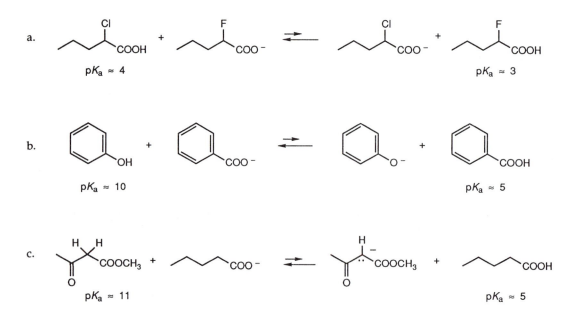

20.29.   (continued)

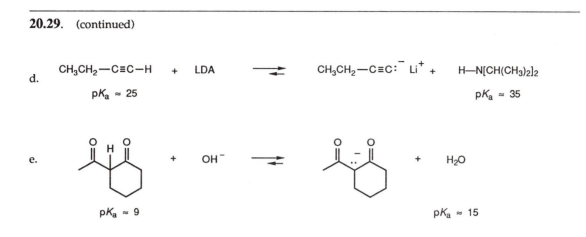

d.

$CH_3CH_2-C\equiv C-H$ + LDA $\rightleftharpoons$ $CH_3CH_2-C\equiv C:^- \; Li^+$ + $H-N[CH(CH_3)_2]_2$

$pK_a \approx 25$ $pK_a \approx 35$

e.

$pK_a \approx 9$ + OH^- $\rightleftharpoons$ + H_2O

$pK_a \approx 15$

20.30.

Type of carbanion	general structure of starting material required	approximate pK_a of starting material	base and solvent required	general structure of product after reaction with an alkyl halide, RX
ketone enolate	$R-CH_2-\overset{\overset{O}{\|\|}}{C}-R'$	20	LDA, THF	$R-\overset{\overset{R}{\|}}{C}H-\overset{\overset{O}{\|\|}}{C}-R'$
ester enolate	$R-CH_2-\overset{\overset{O}{\|\|}}{C}-OR'$	25	LDA, THF	$R-\overset{\overset{R}{\|}}{C}H-\overset{\overset{O}{\|\|}}{C}-OR'$
carboxylic acid dianion	$R-CH_2-\overset{\overset{O}{\|\|}}{C}-OH$	25	LDA, THF	$R-\overset{\overset{R}{\|}}{C}H-\overset{\overset{O}{\|\|}}{C}-OH$
nitrile carbanion	$R-CH_2-C\equiv N$	25	LDA, THF	$R-\overset{\overset{R}{\|}}{C}H-C\equiv N$
1,3-diketone enolate		9	EtO$^-$, EtOH	
β-ketoester enolate		11	EtO$^-$, EtOH	
ketimine carbanion *	$R-CH_2-\overset{\overset{NR''}{\|\|}}{C}-R'$	> 20	LDA, THF or RMgX, THF	$R-\overset{\overset{R}{\|}}{C}H-\overset{\overset{NR''}{\|\|}}{C}-R'$
aldehyde imine carbanion *	$R-CH_2-\overset{\overset{NR''}{\|\|}}{C}-H$	> 20	LDA, THF or RMgX, THF	$R-\overset{\overset{R}{\|}}{C}H-\overset{\overset{NR''}{\|\|}}{C}-H$

* A C=N bond is less able to stabilize an α-carbanion than a C=O bond is; the pK_a value of an imine should therefore be slightly greater than that for an aldehyde or ketone.

20.31.   a.   A 2° amine reacts with a ketone to form an enamine.   An enamine reacts with a reactive alkyl halide at the carbon atom adjacent to the original carbonyl group.   Hydrolysis of the alkylated imine regenerates the ketone.

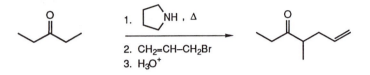

b.   LDA deprotonates the position adjacent to a carbonyl group.   However, an aryl halide does not undergo an S_N2 reaction with the enolate ion, so the second step does not take place.

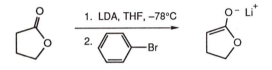

c.   LDA deprotonates the position adjacent to a nitrile group.   The benzylic halide undergoes an S_N2 reaction with the carbanion serving as a nucleophile, and the product is the alkylated nitrile.

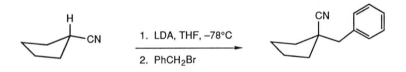

d.   A 1° amine reacts with an aldehyde to form an imine, which is deprotonated by reaction with a Grignard reagent to form the carbanion.   Alkylation at the carbon atom adjacent to the original carbonyl group takes place, and hydrolysis regenerates the aldehyde functional group.

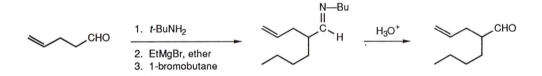

e.   A diketone reacts with alkoxide ion to form a carbanion.   Alkylation occurs at the carbon atom between the carbonyl groups.

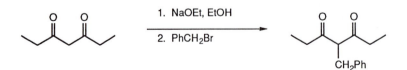

20.31.   (continued)

f.   A keto ester reacts with alkoxide ion to form a carbanion.   Alkylation occurs at the carbon atom situated between the carbonyl groups. Hydrolysis under the influence of heat leads to elimination of the ester group by its conversion to the carboxylic acid followed by decarboxylation.

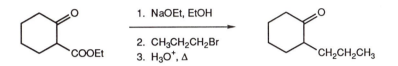

g.   LDA deprotonates the position adjacent to an ester (lactone) carbonyl group. The second step proceeds by attachment of a phenylselenyl group at the alpha position. Oxidation of the Se atom leads to elimination, forming the unsaturated lactone.

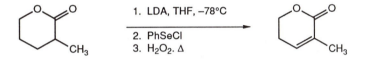

h.   The combination of halogen and hydroxide ion constitutes the haloform reaction, which converts a methyl ketone to the salt of the corresponding carboxylic acid with one less carbon atom.

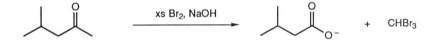

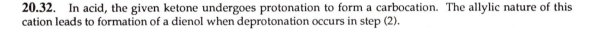

20.32.   In acid, the given ketone undergoes protonation to form a carbocation. The allylic nature of this cation leads to formation of a dienol when deprotonation occurs in step (2).

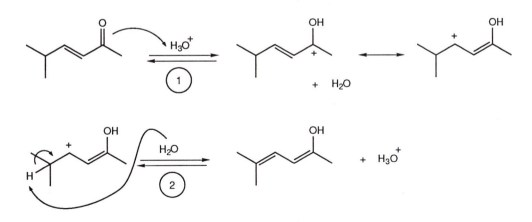

20.32. (continued)

Tautomerism regenerates the carbonyl compound in which the carbon-carbon double bond is no longer conjugated with the carbonyl group.

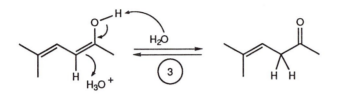

20.33. IR spectra reveal which functional groups are present in a molecule. To deduce what differences are be observed in the IR spectra of reactants and products of a specific reaction, identify the functional group in the starting material and in the product and decide what absorption bands will be observed for each.

a.   A ketone and a carboxylic acid each have a carbonyl group, but the acid has a very strong, broad absorption band between 3500 and 2500 cm^{-1} that is attributed to the O—H stretching vibration.

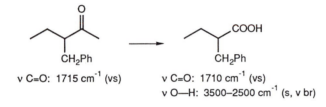

v C=O: 1715 cm^{-1} (vs)   v C=O: 1710 cm^{-1} (vs)
                                    v O—H: 3500–2500 cm^{-1} (s, v br)

b.   An ester and a carboxylic acid each have a carbonyl group, but the acid has a very strong, broad band between 3500 and 2500 cm^{-1} that is attributed to the O—H stretching vibration. The ester has a very strong band at about 1200 cm^{-1} caused by the C—O stretching vibration. Two ester groups are present in the starting compound, so there may be two C=O stretching vibrations around 1700 cm^{-1}.

CH_3—$\overset{\textstyle COOCH_2CH_3}{\underset{\textstyle COOCH_2CH_3}{C}}$ $\longrightarrow$ CH_3—CH_2—COOH

v C=O: 1735 cm^{-1} (vs)   v C=O: 1710 cm^{-1} (vs)
v C—O: 1200 cm^{-1} (vs)   v O—H: 3500–2500 cm^{-1} (s, v br)

c.   The first two compounds in this scheme each have a saturated ketone functional group that displays a very strong absorption around 1715 cm^{-1}. Upon substitution by a heteroatom at the adjacent carbon atom, the carbonyl stretch will be shifted to between 1730-1740 cm^{-1} (this effect was not discussed in Chapter 14, so do not be alarmed that you did not know how a heteroatom affects a C=O stretching frequency.) More noticeable will be new absorptions from stretching and bending vibrations associated with a monosubstituted benzene ring.

20.33.   (continued)

Upon reaction to form the α,β-unsaturated ketone, the C=O stretching vibration will be shifted to about 1690 cm^{-1}, and the absorption bands associated with the benzene ring will be gone.   The C=C bond has stretching vibrations that lead to absorption bands at about 1650 cm^{-1} (probably obscured by the very strong C=O stretching vibration) and in the out-of-plane bending region at about 900 and 1000 cm^{-1}.

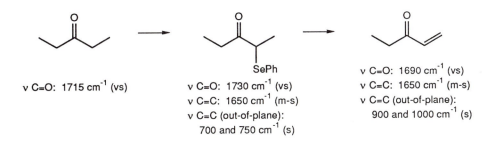

v C=O: 1715 cm^{-1} (vs)

v C=O: 1730 cm^{-1} (vs)
v C=C: 1650 cm^{-1} (m-s)
v C=C (out-of-plane):
   700 and 750 cm^{-1} (s)

v C=O: 1690 cm^{-1} (vs)
v C=C: 1650 cm^{-1} (m-s)
v C=C (out-of-plane):
   900 and 1000 cm^{-1} (s)

20.34.   When performing a reaction, you can predict which resonances will be observed in the proton NMR spectra of the starting material and product by using data from the tables and figures in Chapter 15.   For the resonances given below, the first value is the chemical shift relative to the signal for TMS, the number with the capitalized H is the integrated intensity (2H means two protons, etc.), and the lower case letter is the spin-spin coupling pattern (s = singlet, d = doublet, t = triplet, q = quartet, m = multiplet, br s = broad singlet).   The J value for the spin-spin coupling is the final parameter given.

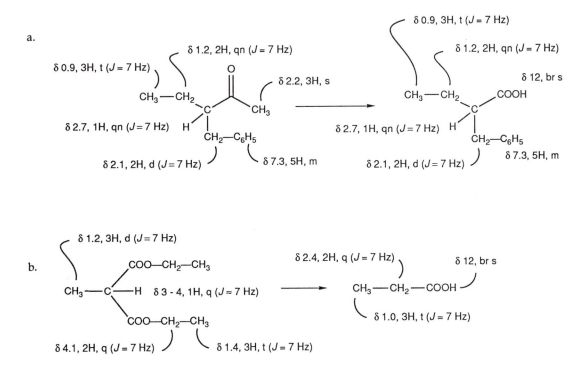

20.34. (continued)

c.

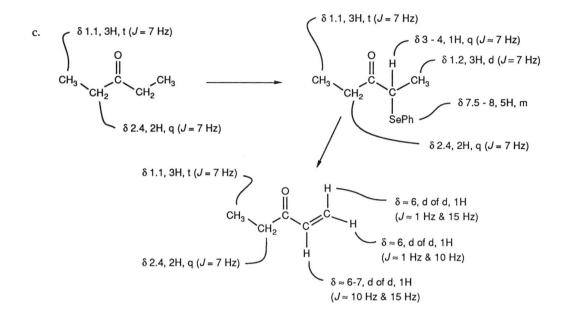

20.35. a. A methyl ketone is converted to a carboxylic acid via the haloform reaction followed by acid workup.

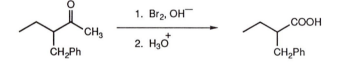

b. A malonate ester undergoes hydrolysis and decarboxylation upon heating with aqueous acid.

$$CH_3-\overset{COOCH_2CH_3}{\underset{COOCH_2CH_3}{|}} \quad \xrightarrow{\overset{+}{H_3O},\ \Delta} \quad CH_3-CH_2-COOH \ + \ 2\ EtOH \ + \ CO_2$$

c. An enolate ion, formed from reaction between a ketone and LDA in an inert solvent, reacts with phenylselenyl bromide to yield an α-phenylselenyl ketone. Oxidation to the selenoxide leads to elimination, producing the α.β-unsaturated ketone.

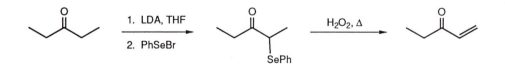

20.36. MCPBA converts a cyclic ketone to a lactone, which is an example of the Baeyer-Villiger reaction (text page 851). The more highly-substituted carbon atom is the one that migrates. If the oxidation is done first, then enolate formation occurs only on one side of the lactone carbonyl group. If alkylation is carried out first, the positions adjacent to the carbonyl group are equivalent. After alkylation, the ketone undergoes the Baeyer-Villiger reaction by insertion of an oxygen atom into the side that was alkylated.

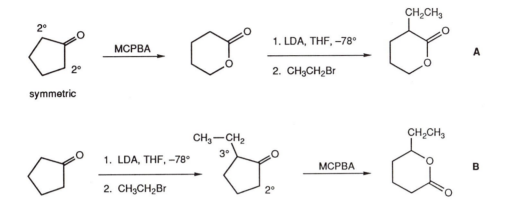

20.37.

a. A primary alcohol is oxidized to the analogous carboxylic acid with dichromate ion. Treating a carboxylic acid with at least two equivalents of methyllithium (especially in the presence of cerium(III) chloride) produces the methyl ketone.

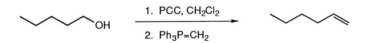

b. The best way to make a terminal alkene is by Wittig olefination. Partial oxidation of the 1° alcohol to the aldehyde provides a way to obtain the starting material needed for the Wittig reaction.

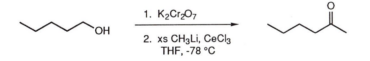

c. An α,β-unsaturated ester that has more carbon atoms than the starting material is best made by the Horner-Emmonds reaction. Partial oxidation of the 1° alcohol to the aldehyde provides a way to obtain the starting material needed for the alkene-forming reaction.

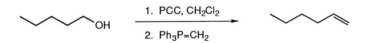

20.37. (continued)

d. An α,β-unsaturated ester that has the same number of carbon atoms as the starting material is best made by an elimination reaction from the corresponding ester. Oxidation of the 1° alcohol to the acid is followed by esterification (step 2). Enolate ion formation, reaction with PhSeBr, and elimination affords the product.

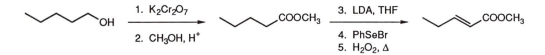

e. An α-benzyl aldehyde is made by reaction between the enamine derivative of the aldehyde and benzyl bromide, followed by hydrolysis. Partial oxidation of the alcohol to the aldehyde provides a way to obtain the starting material needed for alkylation.

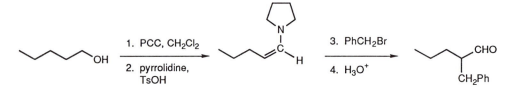

f. An epoxide is made from the corresponding alkene. As shown in part b, the required alkene is made by a Wittig reaction.

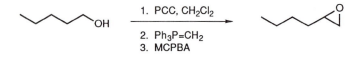

20.38. Follow the procedure outlined in the solution to exercise 20.21.

a. This carboxylic acid is considered as a diester precursor. Two alkylation steps are needed prior to hydrolysis and decarboxylation.

retrosynthesis:

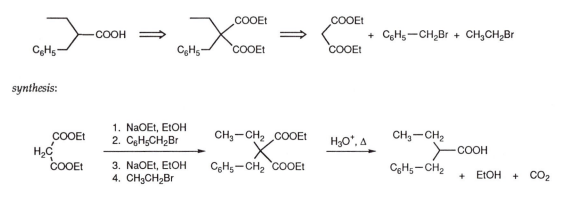

synthesis:

20.38. (continued)

b. This ketone is considered as a keto ester precursor.

retrosynthesis:

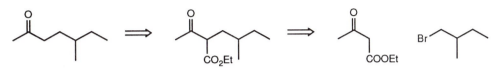

synthesis:

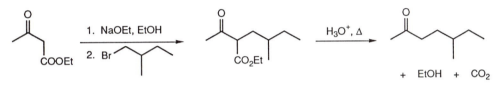

20.39. The instructions given in this exercise indicate that each compound is a keto ester, so we know that one substituent will be attached to the oxygen atom of the ester functional group.

a. For compound **39A** the proton NMR spectrum assignments are tabulated below. The structure is formed by putting together the fragments.

δ	integr	splitting	J (Hz)	assignment
7.3	5H	m	various	monosubstituted phenyl ring
4.1	2H	q	7	O–CH$_2$; coupled with a CH$_3$ group
3.8	1H	t	7	CH adjacent to two carbonyl groups; coupled with a CH$_2$ group.
3.2	2H	d	7	CH$_2$ adjacent to one carbonyl group; coupled with a CH group
2.2	3H	s	–	CH$_3$ adjacent to a ketone group
1.2	3H	t	7	CH$_3$; coupled with a CH$_2$ group

20.39. (continued)

b. For compound **39B** the proton NMR spectrum assignments are tabulated below. The structure is formed by putting together the fragments.

δ	integr	splitting	J (Hz)	assignment
3.8	3H	s	-	O–CH₃
3.5	2H	s	–	CH₂ between two carbonyl groups
2.6	2H	q	7	CH₂ adjacent to a carbonyl group; coupled with a CH₃ group
1.1	3H	t	7	CH₃; coupled with a CH₂ group

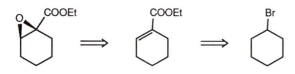

20.40.

a. The chiral epoxide shown in this exercise is made from the unsaturated ester. The exocyclic carbon atom is introduced via a Grignard reaction followed by esterification.

retrosynthesis:

synthesis:

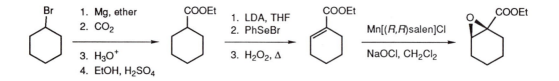

20.40. (continued)

b. The chiral alcohol shown in this exercise is made by asymmetric reduction of the chloroketone. Friedel-Crafts acylation provides a route to the needed starting material.

retrosynthesis:

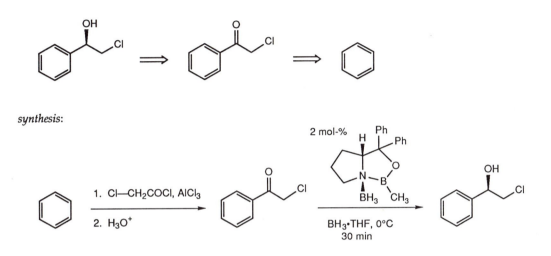

synthesis:

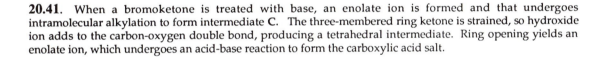

20.41. When a bromoketone is treated with base, an enolate ion is formed and that undergoes intramolecular alkylation to form intermediate **C**. The three-membered ring ketone is strained, so hydroxide ion adds to the carbon-oxygen double bond, producing a tetrahedral intermediate. Ring opening yields an enolate ion, which undergoes an acid-base reaction to form the carboxylic acid salt.

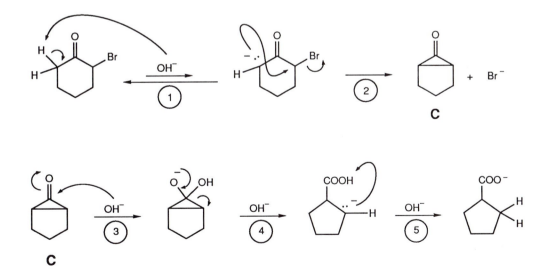

20.42. During enamine synthesis, double bond formation at the carbon atom bearing the methyl group (**B**) creates steric hindrance between the methyl group and hydrogen atoms of the pyrrolidine ring.

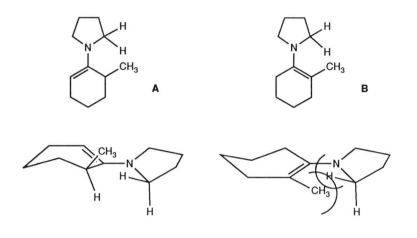

20.43. 1,3-Cyclohexanedione is an active methylene compound, so the first equivalent of Grignard reagent removes a proton from the CH₂ group, effectively "protecting" one carbonyl group from addition because it exists in its enolate form. After addition of the second equivalent of the Grignard reagent to the "normal" carbon-oxygen double bond, no reactive electrophilic center remains. Hydrolysis yields the enol, which eliminates a molecule of water under the acidic conditions of workup, forming the unsaturated ketone.

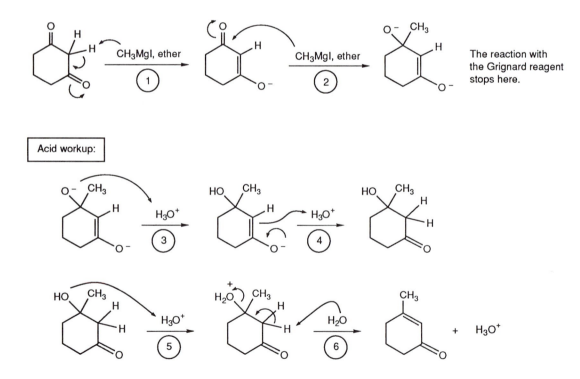

NUCLEOPHILIC ADDITION REACTIONS OF ENOLATE IONS WITH CARBONYL GROUPS

21.1. Two products of the aldol reaction of propanal are the β-hydroxyaldehdye and the α,β-unsaturated aldehyde. The mechanism follows the same steps as those illustrated for acetaldehyde on text pages 1034-1035.

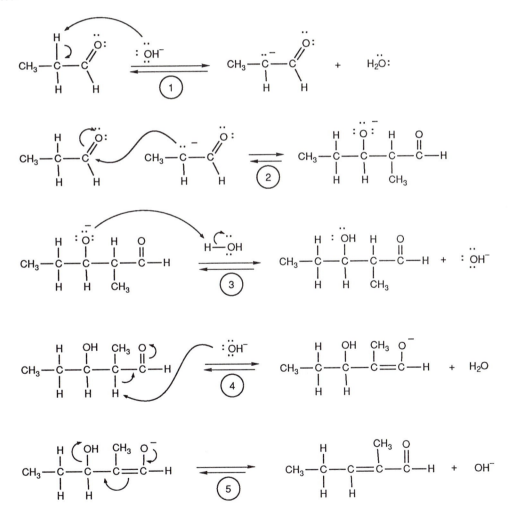

21.2. An aldol reaction performed with acid catalysis requires two initial steps: (1) formation of the enol derivative of the aldehyde, and (2) carbonyl activation of a second equivalent of aldehyde.

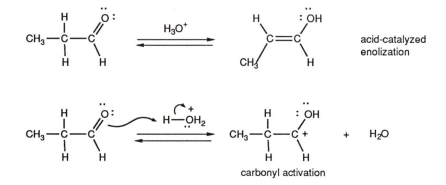

acid-catalyzed
enolization

carbonyl activation

The remaining steps occur as described for acetaldehyde on text pages 1036-1037.

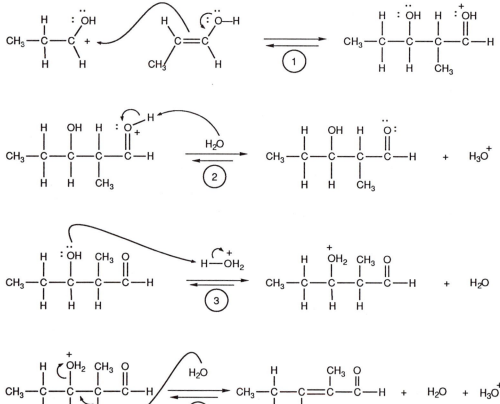

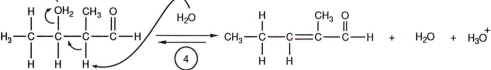

21.3. A retroaldol reaction occurs by the reverse of the steps shown in the solution to exercise 21.2.

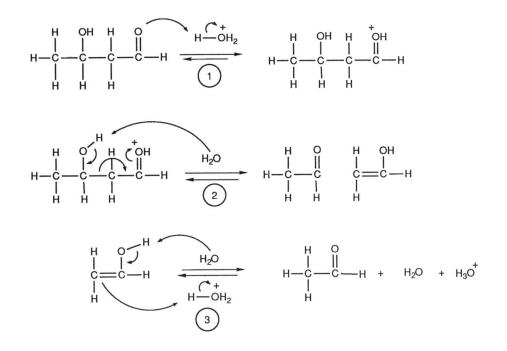

21.4. The mechanism for the aldol reaction of acetone is the same as that for the acid-catalyzed aldol reaction of acetaldehyde. In this instance, the acid is H—(P), where (P) is the conjugate base of the acidic polymer. The enol form of acetone (**B**) is generated first; it reacts with the cation formed by protonation of acetone by the polymer (**A**).

enolization:

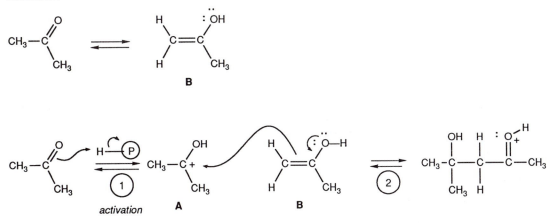

21.4.   (continued)

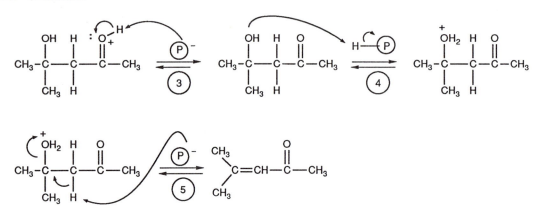

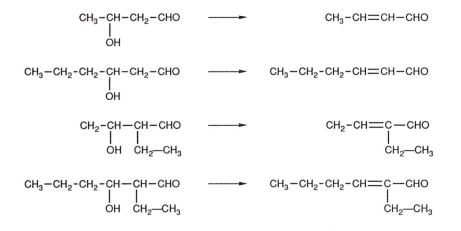

21.5.   The unsaturated aldehydes formed by a crossed aldol reaction between acetaldehyde and butanal have a double bond between C2 and C3 generated by removing the elements of water from the hydroxy aldehydes shown in Table 21.1 (text page 1039).

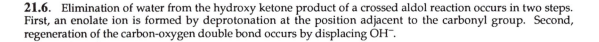

21.6.   Elimination of water from the hydroxy ketone product of a crossed aldol reaction occurs in two steps. First, an enolate ion is formed by deprotonation at the position adjacent to the carbonyl group. Second, regeneration of the carbon-oxygen double bond occurs by displacing OH⁻.

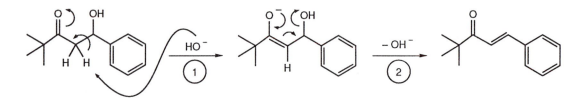

21.7. When one reactant in a crossed aldol reaction is an unsymmetric ketone, the less-highly substituted alpha carbon atom is the one that adds to the aldehyde carbonyl group. Dehydration forms the unsaturated ketone.

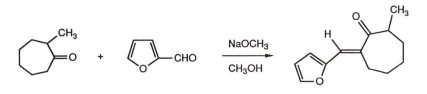

21.8. A directed aldol reaction that employs a boron enolate as one reactant leads to formation of a β-hydroxy ketone. When the stereochemistry of the boron enolate is known, you can predict the stereochemistry of the product by looking at the chair transition state of the reaction according to the schemes shown on text page 1044.

a. The boron enolate is produced from the ketone. Reaction with the aldehyde carbonyl group generates the product.

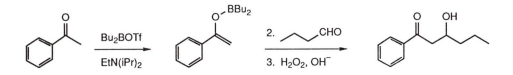

b. The stereochemistry of the boron enolate is given, so looking at the structures of the chair transition states, we predict that the *threo* product is the major one.

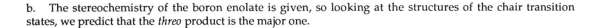

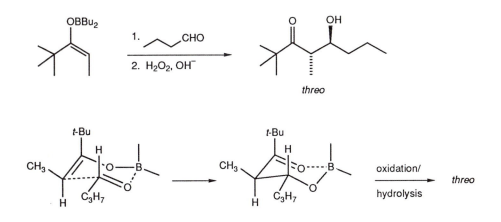

21.9. A carbanion derived from a 3° amide or from an imine have a negative charge at the carbon atom adjacent to the unsaturation. This nucleophilic center adds to the ketone carbonyl group, forming a new carbon-carbon bond. The initial product after workup with aqueous acid is a 3° alcohol in each case, so elimination occurs to afford the unsaturated carbonyl compound. In part b, workup unveils the aldehyde group by hydrolysis of the C=N bond, too.

a.

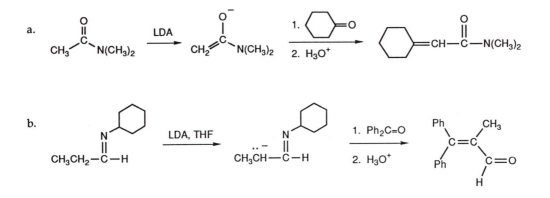

b.

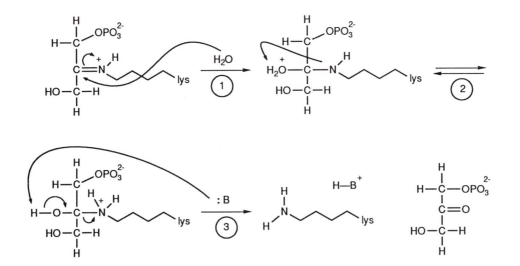

21.10. An iminium salt has an electrophilic carbon atom, so it is susceptible to reaction with a nucleophile such as water. Proton transfer occurs in step (2), forming the protonated amine, a good leaving group. A base in the enzyme active site subsequently deprotonates the OH group at C2, leading to formation of the carbonyl group in the product.

21.11.   As in many transformations of carbohydrates in which carbon-carbon bonds are made or broken, an aldol (or retroaldol) reaction is involved. Biological aldol reactions often make use of Schiff-base catalysis, in this case by reaction between the amino group of a lysine residue and the carbonyl group of 2-keto-3-deoxy-6-phosphogluconate.

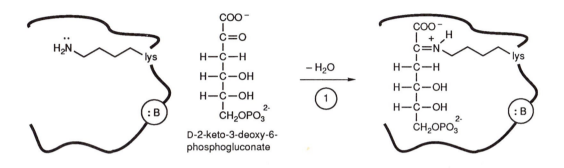

In step (2), a base removes the proton from the OH group attached to C4, cleaving the C3-C4 bond of the substrate molecule. Reprotonation of the enamine in step (3) is followed by hydrolysis of the Schiff-base in step (4) (see the solution to exercise 21.10 for the detailed mechanism) yields pyruvate.

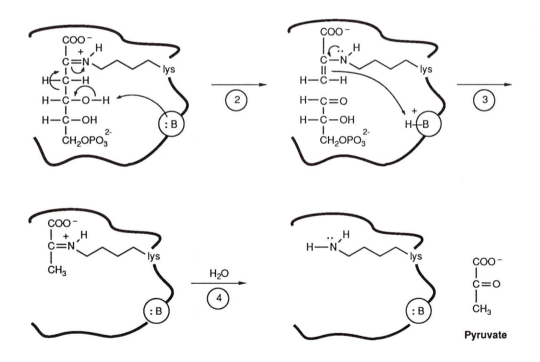

21.12. If carbon-carbon bond formation is non-stereospecific, then two new chiral centers (C3 and C4) are generated. Four diastereomers are therefore produced, the structures of which are shown below.

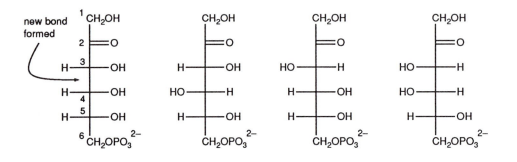

21.13. FDP aldolase catalyzes the reaction between DHAP and an aldehyde to form a new carbon-carbon bond. The aldehyde carbon atom (denoted by *) becomes C4 of the resultant carbohydrate, which has the D–configuration.

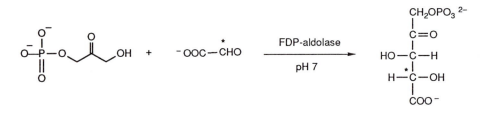

21.14. As described in Chapter 20 on text page 1013, ethyl acetoacetate is alkylated twice at its methylene group under the conditions given in the first step of this scheme. Dialkylation creates a quaternary center, so when that compound is treated with NaOEt in ethanol, the product undergoes a retro-Claisen reaction, generating two esters.

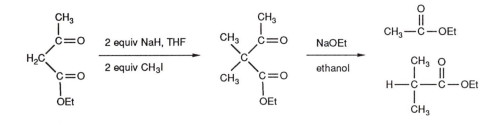

21.15. The transformation shown in this exercise comprises three stages: a Dieckmann (Claisen) condensation, hydrolysis of an ester, and decarboxylation of a β-keto acid. Steps (1) - (5) constitute a Claisen condensation: formation of an enolate ion, addition to the other ester carbonyl group, displacement of ethoxide ion, deprotonation [step (4)] and acid workup, which forms the β-keto ester.

21.15. (continued)

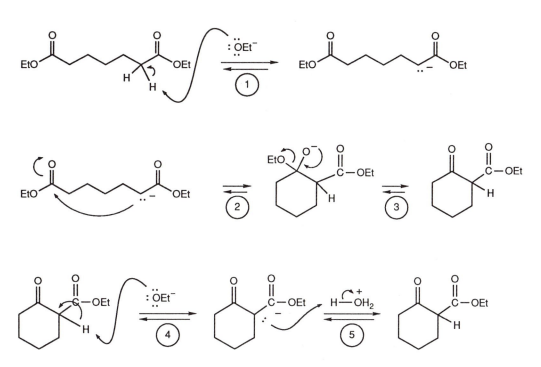

In the next stage, the β-keto ester undergoes hydrolysis by protonation of the ester carbonyl group, reaction with water, and displacement of ethanol. Decarboxylation occurs via a cyclic transition state in step (10), followed by tautomerism to yield the final product.

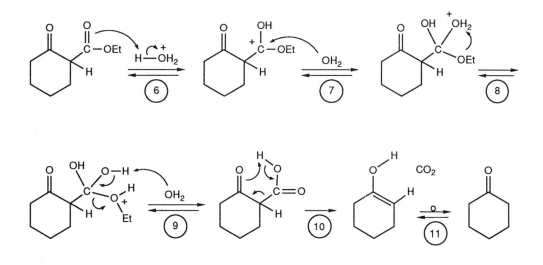

21.16. The transformation shown in the equation is the Claisen condensation of a diester in which only one ester functional group can be deprotonated at its alpha position. The first step creates an enolate ion, which adds to the other ester carbonyl group. The addition step is followed by displacement of ethoxide ion and deprotonation [step (4)]. Acid workup, step (5), yields the β-keto ester product.

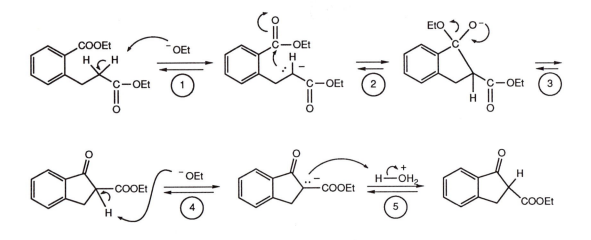

21.17. In a crossed Claisen condensation between an ester and an acid chloride, the first step generates the enolate ion from the ester component. Addition of that enolate ion to the carbonyl group of the acid chloride produces a tetrahedral intermediate, which subsequently collapses to regenerate the carbonyl group with expulsion of chloride ion.

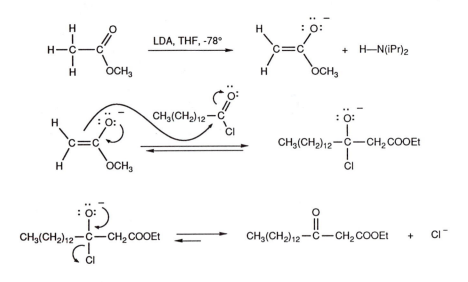

21.18.   Both kinetic and thermodynamic enolate ions can be generated by deprotonation of a ketone with ethoxide ion as the base. Addition of the enolate ion to the carbonyl group of ethyl formate followed by displacement of ethoxide ion completes the first stage of this crossed Claisen condensation. Only the product formed via the kinetic enolate ion can be deprotonated, however; the other product, from reaction of the thermodynamic enolate ion, undergoes the reverse steps (a retro-Claisen reaction) to regenerate starting materials. Eventually, all of the starting ketone reacts at the less highly-substituted alpha position.

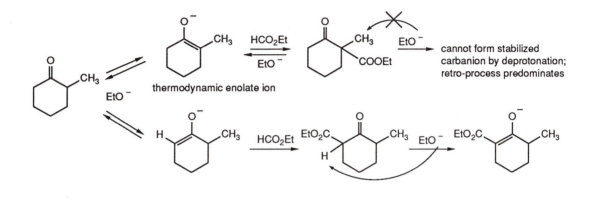

21.19.   The use of diethyl carbonate and base leads to formation of a β-keto ester, which is readily alkylated upon treatment with base and an alkyl halide. Hydrolysis and decarboxylation forms the α-alkylated ketone. This three-step procedure provides a good way to alkylate the position adjacent to a ketone carbonyl group without the problem of self-condensation.

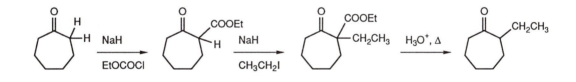

21.20.   Transesterification occurs by the addition of the thiol group of ACP to the carbonyl group of acetyl-CoA. A base is required to regenerate the carbonyl group in step (3), and the base itself is regenerated by reaction in step (4) with the thiolate ion formed after collapse of the tetrahedral intermediate.

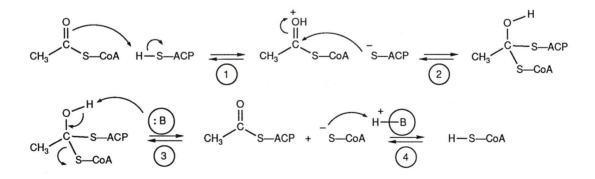

21.21. The biosynthesis of a fatty acid proceeds as follows: an acyl thioester condenses with an enolate ion generated during decarboxylation of malonyl—ACP, forming a β-keto thioester. The ketone group of the β-keto thioester is reduced to an alcohol, which subsequently dehydrates to form the α,β-unsaturated thioester. The double bond of the unsaturated thioester is reduced to form the saturated thioester, which undergoes the next round comprising the same series of reactions. This builds the fatty acid two carbon atoms at a time.

$$CH_3-CO-S-ACP$$

1. **Claisen condensation** with ⁻OOC—CH₂-CO-S—ACP
2. Reduction of β-keto group
3. Elimination of water to generate the double bond
4. Reduction of the double bond

$$CH_3CH_2CH_2-CO-S-ACP$$

1. **Claisen condensation** with ⁻OOC—CH₂-CO-S—ACP
2. Reduction of β-keto group
3. Elimination of water to generate the double bond
4. Reduction of the double bond

$$CH_3CH_2CH_2CH_2CH_2-CO-S-ACP$$

1. **Claisen condensation** with ⁻OOC—CH₂-CO-S—ACP
2. Reduction of β-keto group
3. Elimination of water to generate the double bond
4. Reduction of the double bond

$$CH_3CH_2CH_2CH_2CH_2CH_2CH_2-CO-S-ACP$$

etc.

21.22. Acetoacetate loses carbon dioxide spontaneously to form the enolate derivative of acetone. If a proton source is available, the enolate ion is protonated to yield the neutral ketone.

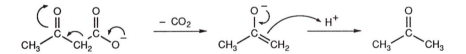

An alternate pathway involves protonation of the carboxylate ion to form the acid, which undergoes decarboxylation via a cyclic transition state to form the corresponding enol. Tautomerism generates acetone.

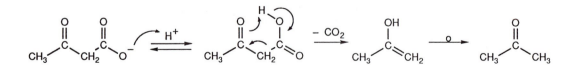

21.23. Follow the procedure given in the solution to exercise 7.28.

a.

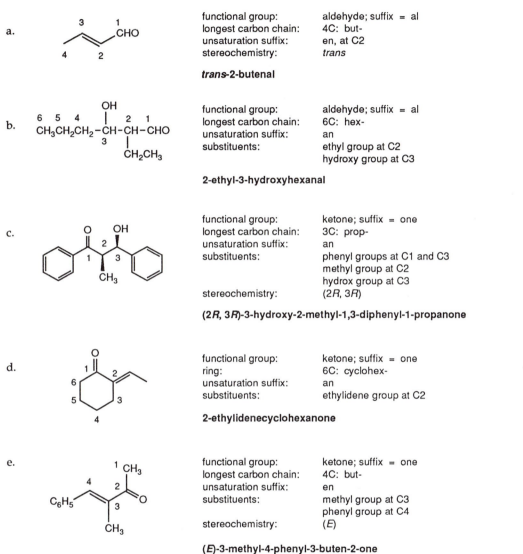

functional group: aldehyde; suffix = al
longest carbon chain: 4C: but-
unsaturation suffix: en, at C2
stereochemistry: *trans*

trans-2-butenal

b.

functional group: aldehyde; suffix = al
longest carbon chain: 6C: hex-
unsaturation suffix: an
substituents: ethyl group at C2
hydroxy group at C3

2-ethyl-3-hydroxyhexanal

c.

functional group: ketone; suffix = one
longest carbon chain: 3C: prop-
unsaturation suffix: an
substituents: phenyl groups at C1 and C3
methyl group at C2
hydrox group at C3
stereochemistry: (2R, 3R)

(2R, 3R)-3-hydroxy-2-methyl-1,3-diphenyl-1-propanone

d.

functional group: ketone; suffix = one
ring: 6C: cyclohex-
unsaturation suffix: an
substituents: ethylidene group at C2

2-ethylidenecyclohexanone

e.

functional group: ketone; suffix = one
longest carbon chain: 4C: but-
unsaturation suffix: en
substituents: methyl group at C3
phenyl group at C4
stereochemistry: (E)

(E)-3-methyl-4-phenyl-3-buten-2-one

21.24. Follow the procedure outlined in Chapter 2 and in the solution to exercise 20.23.

a.   the enolate ion of *t*-butyl acetate

21.24. (continued)

b. methyl 2-methyl-3-ketobutanoate

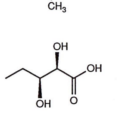

c. (2R,3S)-dihydroxypentanoic acid

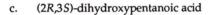

d. 3-^{14}C-dihydroxyacetone phosphate

$$CH_2O-PO_3^{2-}$$
$$|$$
$$C=O$$
$$|$$
$$^{14}CH_2OH$$

21.25. Aldehydes undergo self-condensation to form an aldol product. If two reactants are present, consider which one(s) can form an enol or enolate, and consider what product(s) will form if the enolate derivative reacts with the other carbonyl containing compound.

a. In this crossed aldol reaction, only the acetophenone derivative forms an enolate ion. Addition of the enolate ion to formaldehyde yields the primary alcohol. Elimination of water produces the α,β-unsaturated ketone.

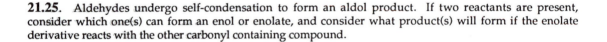

b. In this crossed aldol reaction, only the acetophenone derivative forms an enolate ion. Addition of the enolate ion to the benzaldehyde derivative yields the α,β-unsaturated ketone (chalcone) after elimination of water.

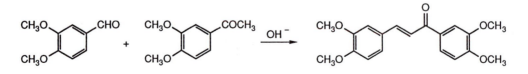

21.25. (continued)

c.   In this directed aldol reaction, the boron enol adds to the aldehyde carbonyl group, forming the β-hydroxy ketone. Assuming that a chair-shaped transition state is involved, the (E)-enol forms the *threo* product.

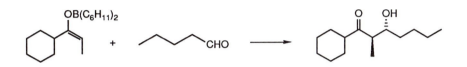

d.   In this directed aldol reaction, LDA deprotonates the ester at the α-position. The resulting enolate ion adds to the ketone carbonyl group, and workup with aqueous acid yields the β-hydroxy ester. The 3° alcohol may undergo elimination if the acid is too strong, in which case the product will be the β,γ-unsaturated ester because no proton is alpha to the ester carbonyl group.

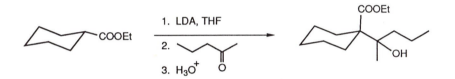

e.   An aldehyde undergoes self-condensation to form a β-hydroxy aldehyde. Elimination does not occur because there are no protons α to the aldehyde carbonyl group.

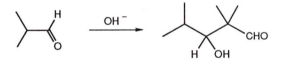

f.   In this directed aldol reaction, LDA reacts to form an enolate ion at the less hindered methyl group of the ketone (the kinetic enolate ion). The enolate ion adds to the aldehyde carbonyl group, and acid workup promotes elimination of water to form a conjugated π system.

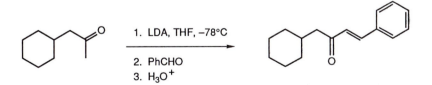

21.26. In the Henry reaction, an amine reacts with a nitro compound to form a carbanion. This nucleophile adds to the aldehyde carbonyl group. Elimination of water forms the α,β-unsaturated nitro compound.

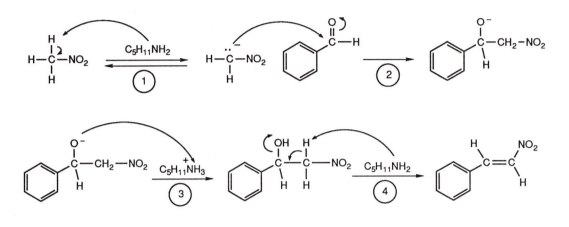

21.27. In the Knoevenagel reaction, an amine reacts with a β-keto ester to form a carbanion. This nucleophile adds to the aldehyde carbonyl group. Elimination of water forms the α,β-unsaturated keto ester.

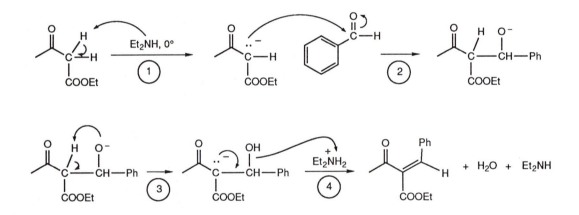

21.28. From the information presented in Chapters 20 and 21, we can draw a structure for each of the intermediates in the glycolysis scheme presented at the bottom of text page 1091. These structures are shown below along with important intermediates involved in each transformation.

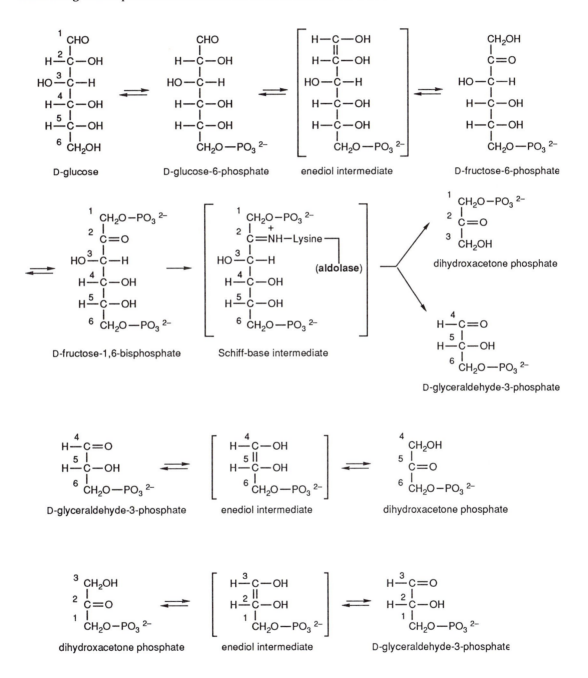

21.28. (continued)

a.   According to the general scheme outlined on page 539, C1 of D-glucose becomes C1 of dihydroxy-acetone phosphate, which becomes C3 of D-glyceraldehyde-3-phosphate.

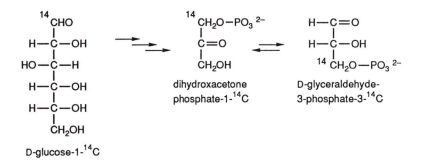

b.   According to the general scheme outlined on page 539, C2 of D-glucose becomes C2 of dihydroxy-acetone phosphate, which is also C2 of D-glyceraldehyde-3-phosphate.

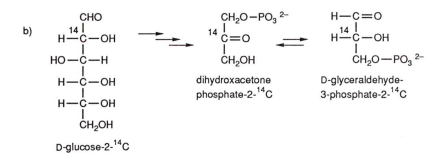

c.   According to the general scheme outlined on page 539, the oxygen atom at C2 of D-glucose becomes the carbonyl oxygen atom of D-fructose. The aldolase-catalyzed cleavage of D-fructose-1,6-bisphosphate occurs by formation of a Schiff-base between the fructose carbonyl group and a lysine residue of the enzyme. The oxygen atom is lost as $H_2^{18}O$ at this step, so it does not appear in either of the three-carbon sugars.

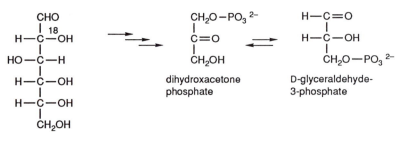

21.29. If the enzyme that interconverts D-glyceraldehyde-3-phosphate and dihydroxyacetone phosphate is missing, then the labeled carbon atom in the compounds of parts (a) and (b) in exercise 21.28 will be found only in the dihydroxyacetone phosphate product.   The carbon atoms in D-glyceraldehyde-3-phosphate originally derive from C4, C5, and C6 of D-glucose. The answer to part (c) in exercise 21.28 is the same: the oxygen atom is lost during Schiff-base formation between D-fructose-1,6-bisphosphate and the lysine residue of aldolase.

21.30. As shown in the solution to exercise 21.28, we first assign structures for each intermediate in the general scheme of gluconeogenesis:

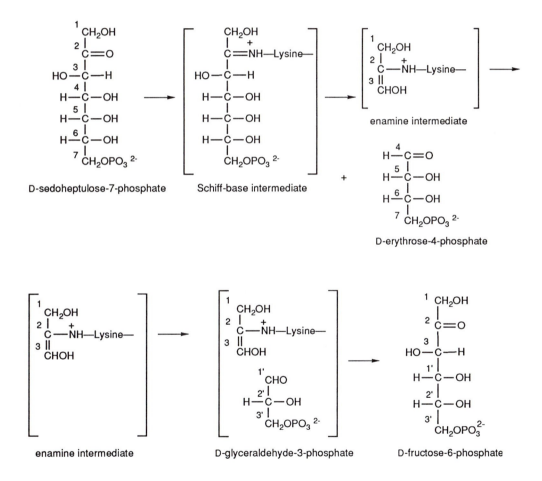

21.30. (continued)

According to the general scheme for gluconeogenesis shown on page 541, C1, C2, and C3 of D-sedoheptulose-7-phosphate become C1, C2, and C3, respectively of D-fructose-6-phosphate.

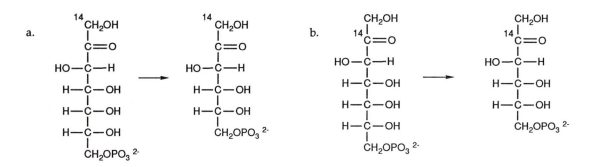

c. According to the general scheme for gluconeogenesis shown on page 541, C4, C5, and C6 of D-fructose-6-phosphate are derived from C1, C2, and C3, respectively, of D-glyceraldehyde-3-phosphate.

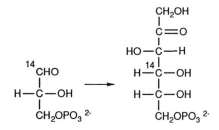

21.31. The use of FDP aldolase as a catalyst in directed aldol reactions that are performed in the laboratory is discussed on text page 1056. The aldehyde carbon atom becomes attached to C3 of DHAP, and the new chiral center (circled in the structures below) has the OH group oriented to the right when the product is drawn as a Fischer projection.

a.

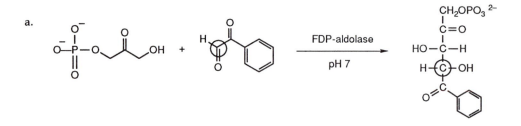

21.31.  (continued)

b.

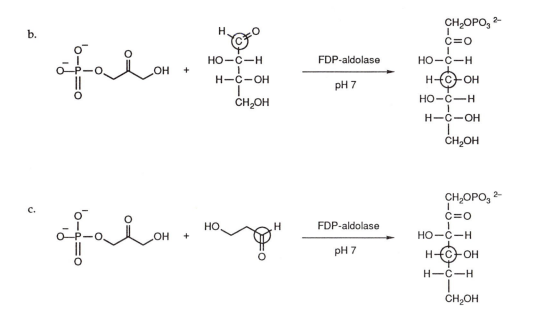

c.

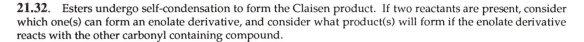

21.32. Esters undergo self-condensation to form the Claisen product. If two reactants are present, consider which one(s) can form an enolate derivative, and consider what product(s) will form if the enolate derivative reacts with the other carbonyl containing compound.

a. This is a "standard" Claisen condensation—the enolate ion produced by deprotonation undergoes addition–elimination in its reaction with another equivalent of ester. Acid workup is required to obtain the neutral product.

CH_3CH_2COOEt $\xrightarrow[\text{2. } H_3O^+]{\text{1. } OEt^-, \text{ EtOH}}$

b. In this crossed Claisen reaction, the enolate ion generated by deprotonation of the ester undergoes addition–elimination in its reaction with the acid chloride. Chloride ion is a weak base, so aqueous acid in the workup step is not required.

$CH_3COOt\text{-Bu}$ $\xrightarrow[\text{2. } PhCH_2COCl]{\text{1. LDA, THF, -78°}}$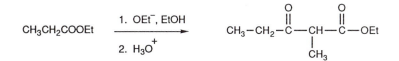

21.32.   (continued)

c.   This is a standard Dieckmann condensation, the ring-forming variation of the Claisen condensation. The chiral centers are not affected by the reaction.

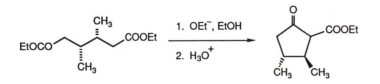

d.   This is a crossed Claisen reaction between an enolizable ketone and a non-enolizable ester. The product is a keto aldehyde, which normally exists in its hydroxmethylene form.

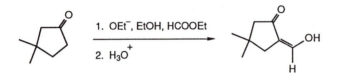

21.33.   In fatty acid metabolism, every other carbon atom appears as the carbonyl carbon atom in acetyl-CoA, starting with the carboxyl carbonyl group of the original acid. The carbon atoms that do not have a dot in the scheme below appear at the methyl carbon atom position in acetyl-CoA molecules that are formed.

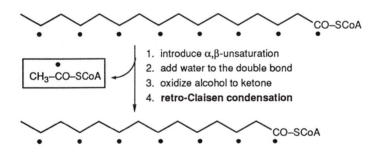

Therefore, you only have to identify whether an odd or even numbered carbon atom is labeled to figure out at which position of acetyl-CoA the labeled carbon atom appear.

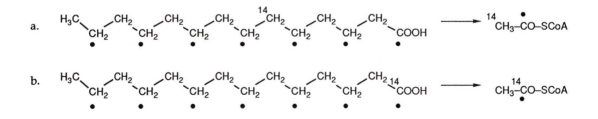

21.34. In acetoacetate biosynthesis, three equivalents of acetyl-CoA combine ultimately to make the product, acetoacetate. The carbonyl groups of acetoacetate derive from the carbonyl group of acetyl-CoA.

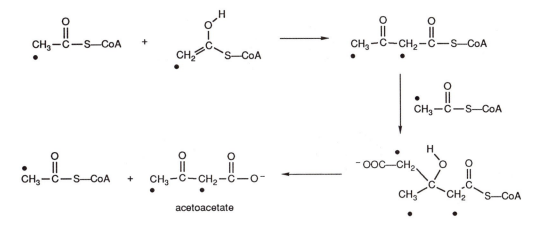

The synthesis of acetoacetate requires more than one equivalent of acetyl-CoA, so any molecule of product is labeled at two carbon atoms.

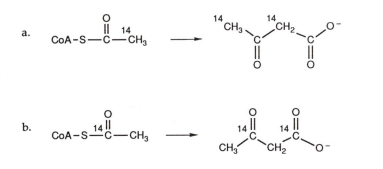

21.35. In the Thorpe reaction, a carbanion is formed by deprotonation at the position alpha to the nitrile group. This nucleophile adds to the carbon-nitrogen triple bond of the other nitrile group in this example. Several proton transfer steps occur eventually to form the enamine. In the second stage of the reaction, the enamine is hydrolyzed to the ketone, and the nitrile group is hydrolyzed to form the carboxylic acid [step (9)—all of the steps of this transformation are not shown, but see text page 953 for the mechanism.] Finally, the keto ester undergoes decarboxylation in step (10), and tautomerism produces the ketone product.

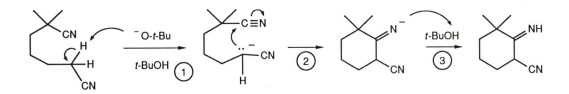

21.35. (continued)

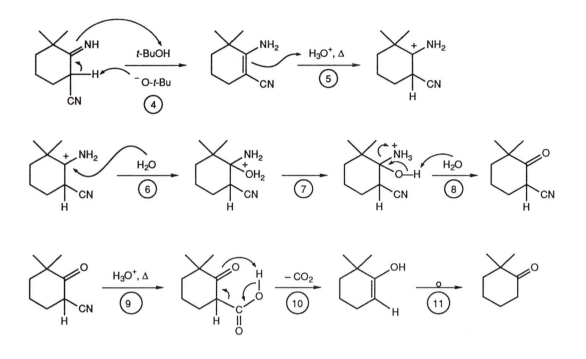

21.36. The Perkin reaction begins with formation of the enolate ion of acetic anhydride. In the first step with the substrate, this enolate ion adds to the carbonyl group of the aldehyde. In the next step, an acetyl group is transferred to the oxyanion nucleophile via a tetrahedral intermediate. The acetyl migration generates a good leaving group beta to the carboxylic acid group, and deprotonation in step (4) displaces acetate ion to form the double bond. In the workup, the carboxylate salt is protonated to form the carboxylic acid product.

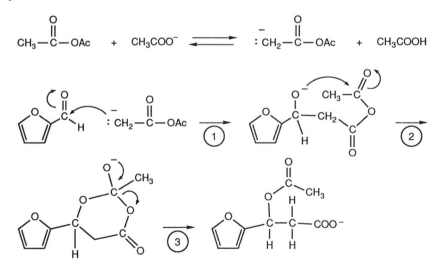

21.36. (continued)

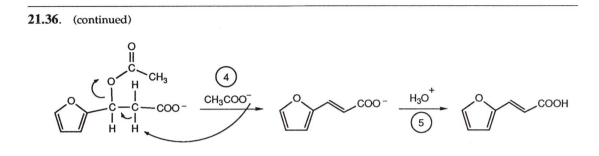

21.37. The Darzens reaction begins with formation of the enolate ion of ethyl chloroacetate. This step is facilitated by the acidity of the methylene protons of the α-halo ester that results from the presence of two strongly electron-withdrawing groups. The anion formed by deprotonation adds to the aldehyde carbonyl group, and the resulting alkoxide ion undergoes an intramolecular S$_N$2 reaction that generates the epoxide ring.

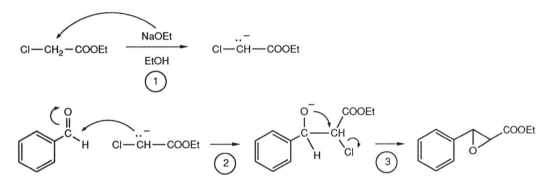

21.38. IR spectra reveal which functional groups are present in a molecule. To deduce what differences are be observed in the IR spectra of reactants and products of a specific reaction, identify the functional group in the starting material and in the product and decide what absorption bands will be observed for each.

a.   A diester and a β-keto ester each have two carbonyl groups, but the ester groups are equivalent in the starting material, so only a single absorption is observed for the carbonyl stretching vibrations. The keto ester product is likely to display either two bands or a broad absorption band that results from overlapping absorptions.

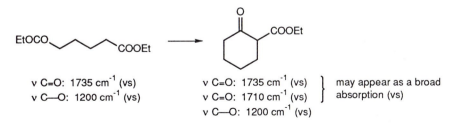

ν C=O: 1735 cm^{-1} (vs)   ν C=O: 1735 cm^{-1} (vs) ⎫ may appear as a broad
ν C—O: 1200 cm^{-1} (vs)   ν C=O: 1710 cm^{-1} (vs) ⎬ absorption (vs)
                                     ν C—O: 1200 cm^{-1} (vs) ⎭

21.38. (continued)

b.   Both starting material and product are aldehydes, but the product has a C=O stretching vibration at lower frequency because the carbonyl group is conjugated with the alkene double bond. The C=C stretching vibration may also be observable (unless overshadowed by the very strong carbonyl stretching vibration.) The alkene should also have strong absorptions in the out-of-plane-bend region.

v C=O: 1720 cm^{-1} (vs)

v C=O: 1690 cm^{-1} (vs)
v C=C: 1650 cm^{-1} (m-s)
v C=C (out-of-plane):
        900 and 1000 cm^{-1} (s)

c.   An ester and a β-keto ester each have carbonyl groups, but the ester group has only a single absorption in the carbonyl stretch region. The keto ester product is likely to display two bands or a broad absorption band that results from overlapping absorptions.

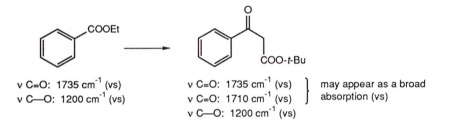

v C=O: 1735 cm^{-1} (vs)
v C—O: 1200 cm^{-1} (vs)

v C=O: 1735 cm^{-1} (vs)
v C=O: 1710 cm^{-1} (vs)   } may appear as a broad
v C—O: 1200 cm^{-1} (vs)   absorption (vs)

21.39.   When performing a reaction, you can predict which resonances will be observed in the proton NMR spectra of the starting material and product by using data from the tables and figures in Chapter 15. For the resonances given below, the first value is the chemical shift relative to the signal for TMS, the number with the capitalized H is the integrated intensity (2H means two protons, etc.), and the lower case letter is the spin-spin coupling pattern (s = singlet, d = doublet, t = triplet, q = quartet, m = multiplet, br s = broad singlet). The *J* value for the spin-spin coupling is the final parameter given.

a.

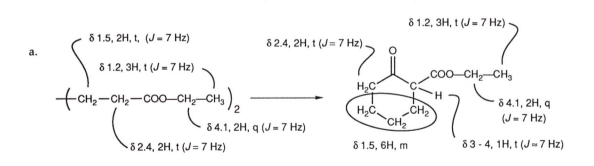

21.39.   (continued)

b.

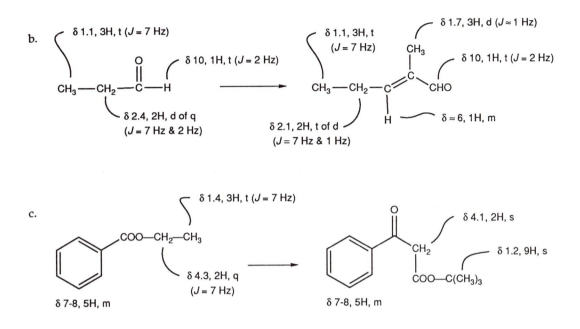

c.

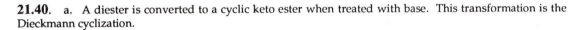

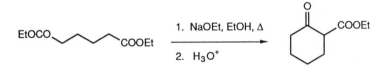

21.40.   a.   A diester is converted to a cyclic keto ester when treated with base. This transformation is the Dieckmann cyclization.

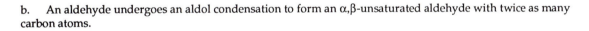

b.   An aldehyde undergoes an aldol condensation to form an α,β-unsaturated aldehyde with twice as many carbon atoms.

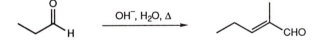

c.   An ester is converted to a keto ester via the Claisen condensation. In this instance, a crossed Claisen condensation is required.

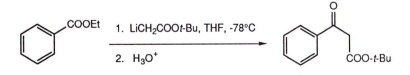

21.41. The instructions indicate that each compound has a carbonyl group, so we know that the compound is either an aldehyde or ketone because the formula has only a single oxygen atom in each case.

a. For compound **41A**, the proton NMR spectrum assignments are tabulated below. From the formula (one oxygen atom), we know that the compound is an aldehyde (proton observed in the NMR spectrum). One additional site of unsaturation is present and the NMR spectrum has an alkene proton resonance.

δ	integr	splitting	J (Hz)	assignment
9.4	1H	s	-	–CHO
6.5	1H	t	7	CH of a double bond, coupled with a CH$_2$ group
2.4	2H	qn	7	CH$_2$ attached to an alkene C atom; coupled with CH and CH$_3$ groups
1.7	3H	s	–	CH$_3$ attached to an alkene C atom
1.1	3H	t	7	CH$_3$; coupled with a CH$_2$ group

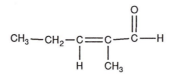

b. For compound **41B**, the proton NMR spectrum assignments are tabulated below. From the formula (one oxygen atom), we know that the compound is a ketone (no aldehyde proton in the NMR spectrum and no N or Cl). Two additional sites of unsaturation are present: a double bond (deduced from observing alkene proton resonances in the NMR spectrum), and a ring (most likely).

δ	integr	splitting	J (Hz)	assignment
6.7	1H	d	10	CH of a *cis* double bond
5.8	1H	d	10	CH of a *cis* double bond
2.5	2H	t	7	CH$_2$ adjacent to a carbonyl group; coupled with a CH$_2$ group.
1.9	2H	t	7	CH$_2$; coupled with a CH$_2$ group
1.2	6H	s	–	two CH$_3$ groups attached to a C atom

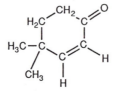

21.42. A Dieckmann reaction is used to prepare a cyclic β–keto ester from the appropriate diester.

a.   The desired product is a cyclic β–keto ester, so a Dieckmann reaction yields the product directly. The starting diester has more than six carbon atoms, so the diacid (six carbon atoms) is first converted to its diester derivative.

retrosynthesis:

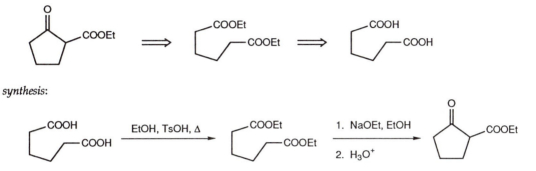

synthesis:

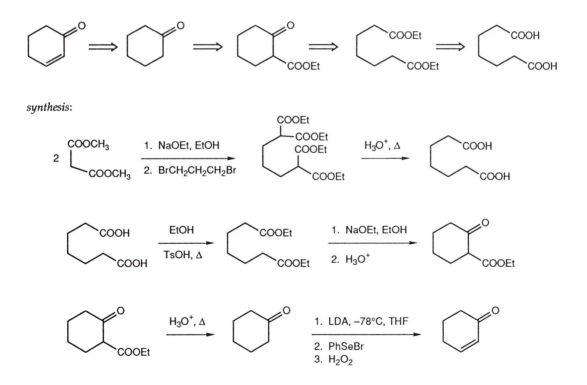

b.   The product is a ketone, so in planning a retrosynthesis that will make use of a Dieckmann cyclization, an ester group is attached to the carbon atom alpha to the ketone carbonyl group. The cyclic β–keto ester is prepared from a seven-carbon diacid, which in turn is made by alkylating diethyl malonate, followed by hydrolysis and decarboxylation.

retrosynthesis:

21.43. a. A chiral benzylic alcohol is made by asymmetric reduction of an alkyl aryl ketone (text page 844). The β-keto ester is made by a crossed Claisen condensation between an ester enolate and an acid chloride.

retrosynthesis:

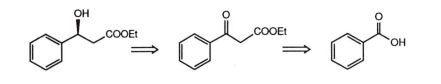

synthesis:

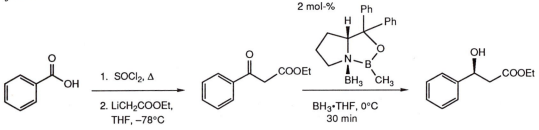

b. A chiral diol is made by asymmetric dihydoxylation of an alkene. The exocyclic double bond is formed by Wittig olefination of the cyclic ketone. The ketone is made by Dieckmann cyclization.

retrosynthesis:

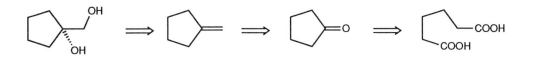

synthesis:

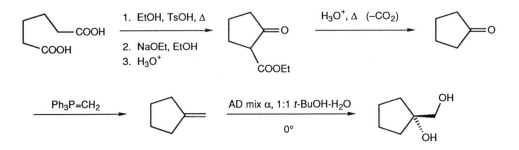

NUCLEOPHILIC ADDITION TO α,β–UNSATURATED CARBONYL COMPOUNDS

22.1. The most general way to prepare an α,β-unsaturated carbonyl compound from its saturated analog is to make an enolate derivative and use it to attach a PhSe group alpha to the carbonyl group. Oxidation of the Se atom with hydrogen peroxide and gentle warming leads to elimination of PhSeOH and formation of the double bond.

When preparing an α,β-unsaturated carbonyl compound in which the carbon skeleton must be enlarged, an aldol or crossed-aldol reaction is used to attach the additional carbon atoms and introduce unsaturation in conjugation with the carbonyl group.

a.

$$\text{1. LDA, THF, } -78°C$$
$$\text{2. PhSeCl}$$
$$\text{3. H}_2\text{O}_2, \Delta$$

b.

CH₃—C(=O)—O-*t*-Bu

$$\text{1. LDA, THF, } -78°C$$
$$\text{2. CH}_3\text{CHO}$$
$$\text{3. H}_3\text{O}^+$$

O-*t*-Bu

22.2. A 2° amine reacts with formaldehyde by nucleophilic addition of the amine nitrogen atom to the carbonyl group, followed by transfer of a proton from nitrogen to oxygen.

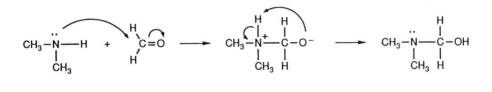

553

22.3. The Mannich reaction takes place in two stages. The mechanism of the first stage was shown in the solution to exercise 22.2 in which hydroxymethyldimethylamine is formed.

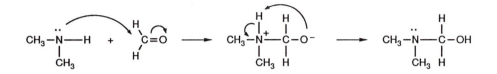

Then, an iminium ion is generated by the following two steps:

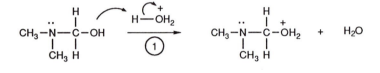

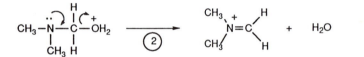

Once the required iminium ion has been made, the enol form of the ketone reacts by addition to the carbon-nitrogen double bond.

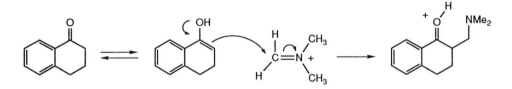

Transfer of a proton from oxygen to nitrogen generates the protonated Mannich base.

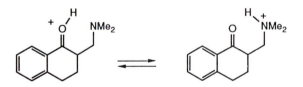

22.4.   Hydrolysis of an enol ether derivative of a β-diketone occurs by addition of water to the double bond. This addition step is catalyzed by acid, which serves to activate the carbonyl group and make the β-position more electrophilic. Water intercepts the carbocation formed in step (1), after which a proton is transferred from one oxygen atom to the other in step (3). The carbonyl group is regenerated to produce the enol form of the 1,3-diketone.

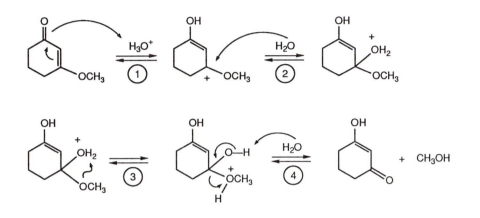

22.5.   Regeneration of citrate from aconitate proceeds by conjugate addition of water to the alkene double bond. The process starts with addition of hydroxide ion, generated from water by reaction with a base in the enzyme active site.

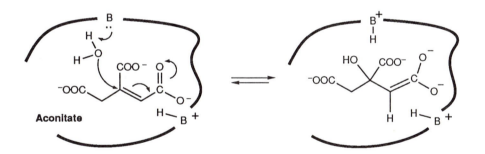

Regeneration of the carboxylate carbonyl group occurs by protonation at the α-position, forming citrate.

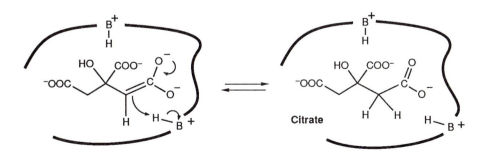

22.6. Conjugate addition of cyanide ion that makes use of diethylaluminum cyanide (text page 1108) occurs by complexation of the diethylaluminum group to the carbonyl oxygen atom. This binding activates the conjugated π system for conjugate addition, and cyanide ion subsequently adds to the β-carbon atom to form the diethylaluminum enolate, compound **A**. In steps (2) through (4), the diethylaluminum fragment is liberated from the oxygen atom by hydrolysis.

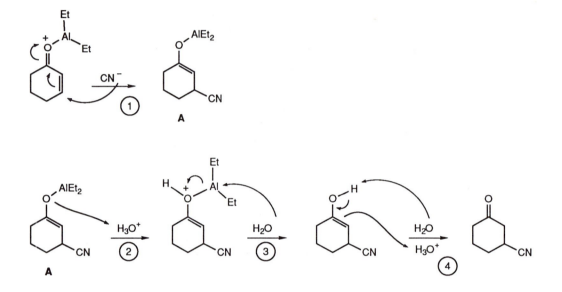

22.7. a. An amino group attached to the carbon atom β to a carbonyl group requires conjugate addition of an amine to the corresponding α,β–unsaturated compound, which is made according to the scheme shown in the solution to exercise 22.1.

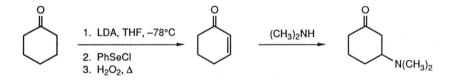

b. An aminomethyl group is the reduced equivalent of a cyano group. Its placement β to a carbonyl group makes use of a conjugate addition of cyanide ion to the corresponding α,β–unsaturated compound.

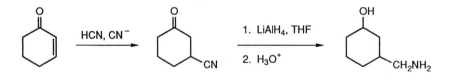

22.7.   (continued)

c. The dialkylaminomethyl group attached at a position α to a carbonyl group calls for the use of the Mannich reaction, which takes place according to the scheme shown in the solution to exercise 22.3.

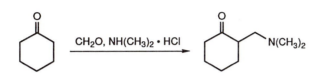

22.8.   The catalyzed addition of a thiol to an α,β–unsaturated ester takes place in three steps. The base generates a small amount of the thiolate ion in step (1). That nucleophile subsequently adds to the terminus of the conjugated system in step (2). The enolate ion so formed deprotonates another equivalent of thiol in step (3) to regenerate thiolate ion (for further reaction) and to form the product.

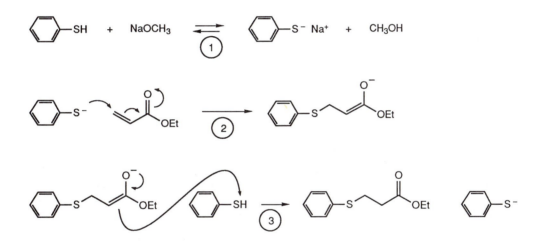

22.9.   The structural similarity of the two epoxide products suggests that they are made from a common precursor, dienol **X**.   One double bond is an allylic alcohol fragment, so it is epoxidized with use of a metal oxo complex and *t*-butyl hydroperoxide (text page 449).   The other double bond is epoxidized preferentially with a peracid because it is trisubstituted (text page 449).   Neither double bond is conjugated with a carbonyl group, so hydrogen peroxide and base is not a suitable reagent combination for forming either epoxide ring.

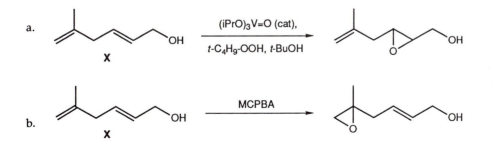

22.10. The Mannich base derivative of a phenol normally reacts via a quinone methide intermediate. Cyanide ion acts initially as a base in step (1), producing the quinone methide. The HCN and hydroxide ion that are produced react to regenerate cyanide ion, which subsequently undergoes conjugate addition in step (3) to form the nitrile. Hydrolysis of the nitrile in step (4) yields the carboxylic acid according to the process described on text page .953.

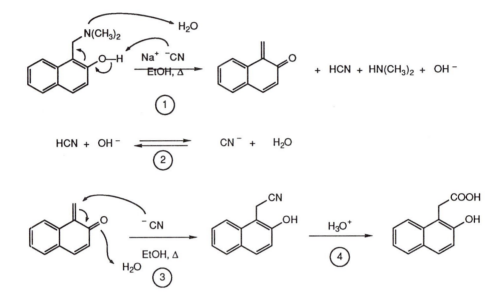

22.11. The mechanism for conjugate addition of glutathione (RSH in the schemes below) to each oxidized derivative of morphine involves two steps: 1. reaction of the thiol sulfur atom at the terminus of the unsaturated system, and 2. proton transfer (shown as an intramolecular process below, but more likely intermolecular).

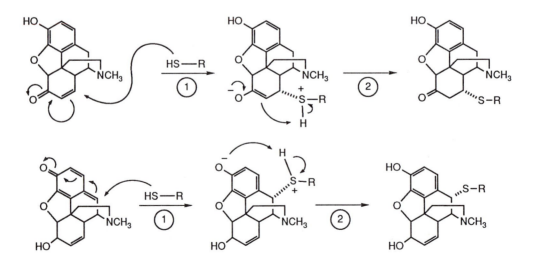

22.12.   In planning a synthesis that makes use of the Michael reaction, consider what stabilized enolate ion is needed to attach the required fragment β to the carbonyl (or nitrile) group.

a.   Addition of ethyl cyanoacetate ion is followed by workup with aqueous acid. Hydrolysis of the ester under mild conditions followed by decarboxylation forms the desired product.

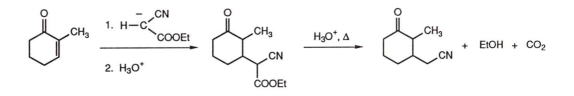

b.   The Michael acceptor is made by Friedel-Crafts acylation of benzene, followed by the standard series of reactions needed to introduce unsaturation (see the solution to exercise 22.1). The other fragment that is needed is made by alkylating ethyl cyanoacetate with benzyl bromide. Addition of the enolate ion of the cyano ester to the unsaturated ketone yields the product after workup with aqueous acid.

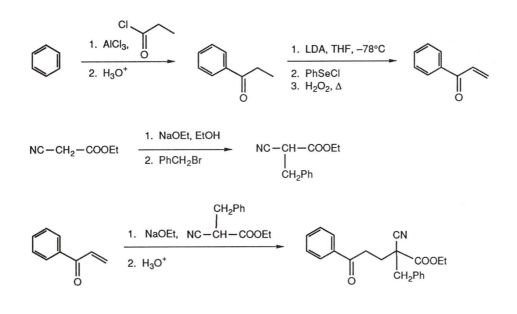

22.13.   The amino acids without a nucleophilic group in their side chain are those in which the R group is a hydrocarbon:

alanine   glycine   isoleucine   leucine   phenylalanine   proline   valine

These amino acids do not initiate polymerization of Super Glue, but the others listed in Table 25.1 (text page 1265) will.

22.14. An enolate ion undergoes conjugate addition with an α,β–unsaturated ester or nitrile. Subsequent steps occur by the normal reactions expected for the functional groups that are present.

a. After the Michael reaction, the ester groups are hydrolyzed, and the resulting diacid undergoes decarboxylation.

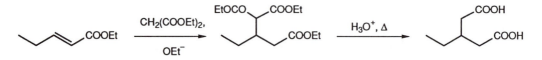

b. After the Michael reaction, the functional groups that are present are reduced by the hydride reagent (text pages 961 and 965).

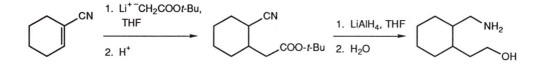

22.15. Interconversion of the two enolate ions shown on text page 1122 occurs by acid-base reactions in which the solvent functions as the proton donor. These processes are equilibria, but the equations have been written separately to show how the electrons move during the process in each direction.

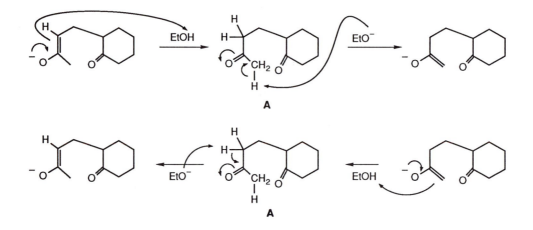

22.16. An E1cb reaction occurs via formation of an enolate derivative. When the carbonyl group is regenerated, hydroxide ion is expelled from the β–carbon atom to form the unsaturated ketone.

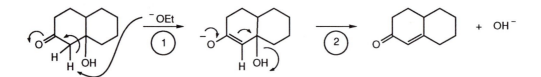

22.17. In planning a synthesis that makes use of conjugate addition of a hydrocarbon group from an organocopper reagent, identify the fragment attached β to the carbonyl group.

a. A Robinson annulation is used to construct the bicyclic ring system. Conjugate addition of a vinyl group makes use of the higher order cuprate reagent.

retrosynthesis:

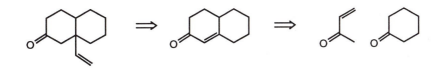

synthesis:

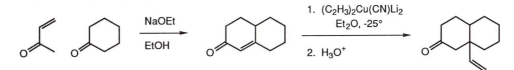

b. Conjugate addition of the propyl group makes use of the higher order cuprate reagent.

synthesis:

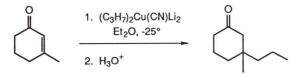

c. Conjugate addition of the C$_5$H$_9$ group makes use of the higher order cuprate reagent.

synthesis:

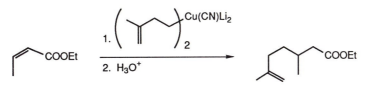

22.18. In planning a synthesis that makes use of conjugate addition of a hydrocarbon group from an organocopper reagent, identify the fragment attached β to the carbonyl group.   If another alkyl group is attached on the same side of the carbonyl group at the α position, then use a tandem addition/alkylation procedure.

a.   Conjugate addition of the ethyl group from the organocuprate reagent is followed by trapping the enolate ion with methyl iodide.

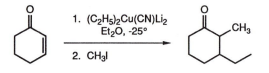

b.   Conjugate addition of the ethyl group from the organocuprate reagent is followed by trapping of the enolate ion with a PhSe group. Oxidation causes elimination and regeneration of the double bond. The methyl group is subsequently added by a second conjugate addition reaction.

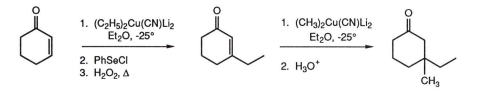

c.   3-Ethyl-2-cyclohexenone is prepared as shown in part b, directly above. Conjugate addition of the methyl group from the organocuprate reagent is followed by trapping the enolate ion with acetaldehyde (a crossed aldol reaction). Workup with acid leads to dehydration.

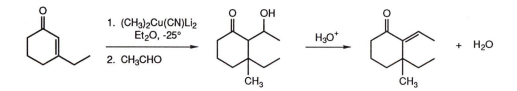

22.19. When 5-fluoro-2'-dexoyuridate reacts with *Thymidylate synthase*, conjugate addition of the thiolate ion to the heterocyclic ring in step (2) generates the enolate derivative. Regeneration of the carbonyl group leads to reaction with the methylene group of N^5,N^{10}-methylene-THF.

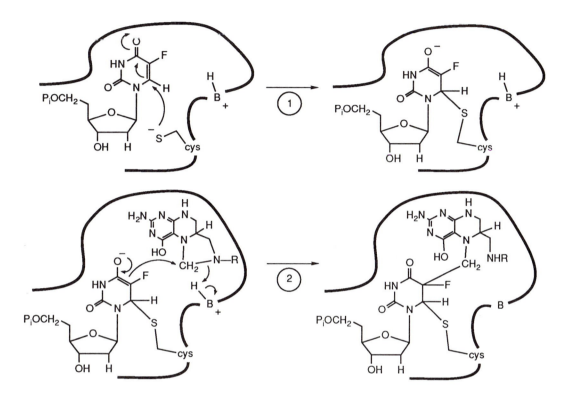

22.20. An allylic alcohol is routinely made by reducing the carbonyl group of an enone. An enone is made by the route shown in the solution to exercise 22.18b, which consists of conjugate addition of an alkyl group from an organocuprate, trapping the enolate ion with phenylselenyl chloride, and oxidation/elimination of PhSeOH.

a.   After formation of the carbon skeleton, the carbonyl group is reduced with use of $NaBH_4$ and $CeCl_3$.

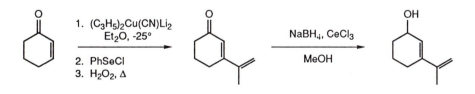

22.20. (continued)

b. Before the double bond of the enone is introduced, alkylation at the alpha position is accomplished by deprotonation with LDA followed by treatment with 1-bromobutane. In the last stage, the carbonyl group is reduced with use of NaBH$_4$ and CeCl$_3$.

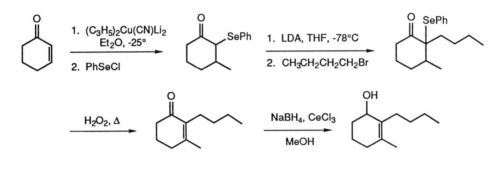

22.21. Dissolving metal reduction of a conjugated ketone leads to the formation of the specific enolate ion, which is alkylated with the appropriate alkyl halide.

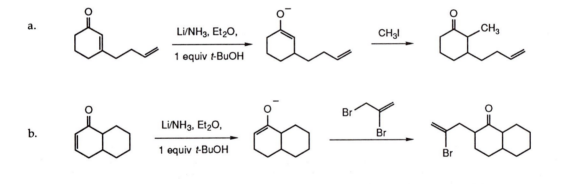

22.22. Interpret the name according to the directions given in Chapter 2.

a. methyl 2-phenylselenylbutanoate

b. (S)-2-isopropylidene-3-methylcyclopentanone

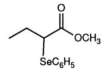

22.22.   (continued)

c.   2,3-epoxycyclohexanone

d. 1-carboxymethyl-2-naphthol

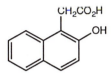

22.23.   Follow the procedure given in the solution to exercise 7.28.

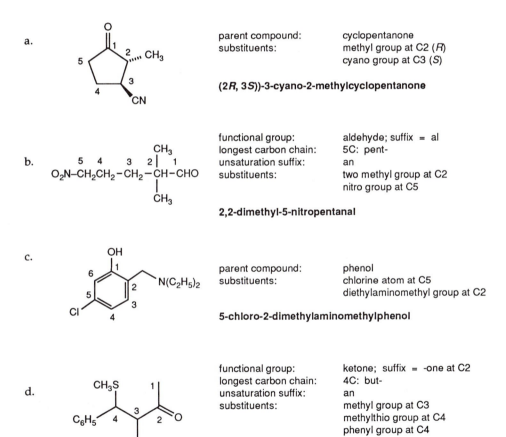

a.

parent compound:	cyclopentanone
substituents:	methyl group at C2 (*R*)
	cyano group at C3 (*S*)

(2*R*, 3*S*))-3-cyano-2-methylcyclopentanone

b.

functional group:	aldehyde; suffix = al
longest carbon chain:	5C: pent-
unsaturation suffix:	an
substituents:	two methyl group at C2
	nitro group at C5

2,2-dimethyl-5-nitropentanal

c.

parent compound:	phenol
substituents:	chlorine atom at C5
	diethylaminomethyl group at C2

5-chloro-2-dimethylaminomethylphenol

d.

functional group:	ketone; suffix = -one at C2
longest carbon chain:	4C: but-
unsaturation suffix:	an
substituents:	methyl group at C3
	methylthio group at C4
	phenyl group at C4

3-methyl-4-methylthio-4-phenyl-2-butanone

22.24.   Conjugate addition of water to the β-chloro-α,β-unsaturated ester shown in this exercise creates a situation in which two good leaving groups are attached to the β-carbon atom. When the ester carbonyl group is regenerated in step (2), expulsion of chloride ion produces an enol which tautomerizes to form the β-keto ester. The remainder of the mechanism comprises hydrolysis of the ester followed by decarboxylation, yielding the ketone. See the solution to exercise 20.22 for the steps involved in decarboxylation.

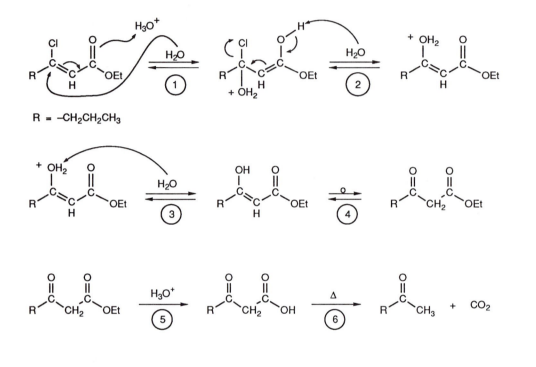

R = –CH₂CH₂CH₃

R = $-CH_2CH_2CH_3$

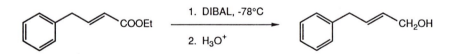

22.25.

a.   DIBAL is used to reduce an α,β-unsaturated ester to the corresponding allylic alcohol (text page 1134).

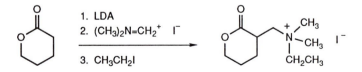

b.   LDA converts a lactone to its lithium enolate derivative. Addition to the preformed dimethyliminium salt forms the Mannich base (text page 1101), which can be alkylated at nitrogen by treatment with the ethyl iodide.

22.25.   (continued)

c.   HCN undergoes conjugate addition to an α,β-unsaturated ester (text page 1108).

d.   The combination of hydrogen peroxide and base does not react with isolated double bonds or carbonyl groups.

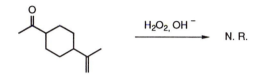

e.   The carbanion derived from ethyl acetoacetate undergoes conjugate addition to an unsaturated nitrile.

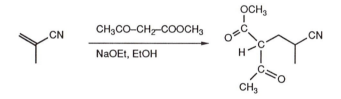

f.   Lithium dimethylcuprate promotes conjugate addition of a methyl group to an α,β-unsaturated ketone. The enolate ion formed by this 1,4-addition is trapped with allyl bromide to yield the ketone in which the two alkyl fragments are attached on the same side of the carbonyl group (text page 1125).

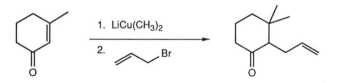

g.   Sodium borohydride in the presence of CeCl₃, reduces only the carbonyl portion of an α,β-unsaturated ketone (text page 1133).

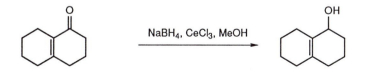

22.25. (continued)

h. A 2° amine undergoes conjugate addition to an α,β-unsaturated ketone (text page 1109).

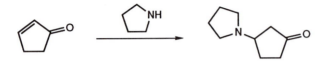

22.26. In performing the reaction in which our chemist friend adds water and base to the 2-isopropylidenecyclohexane, conjugate addition of hydroxide ion occurs as expected, forming compound **A**. This substance is a β-hydroxy ketone, so it undergoes a retroaldol reaction under the basic conditions being used (text page 1037). The retroaldol reaction generates acetone and cyclohexanone (compound **B**) as products.

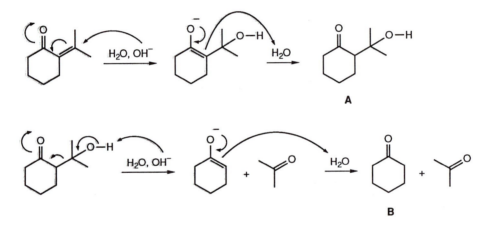

22.27. Diethyl malonate cannot be alkylated with *tert*-butyl bromide because the latter undergoes elimination under the moderately basic reaction conditions. Diethyl malonate undergoes the Knoevenagel reaction with acetone, producing the unsaturated isopropylidene diester (text page 1091). The desired diethyl *tert*-butylmalonate is made by conjugate addition of an alkyl group. Even though lithium dimethylcuprate can be used in this second step (as shown below), the simple Grignard reagent, CH3MgI, actually suffices to carry out the conjugate addition. The unsaturated diester is obviously a potent reactant toward conjugate addition, so the presence of copper is not required.

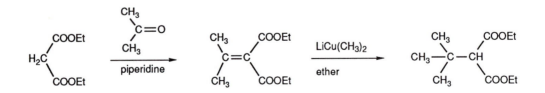

22.28.

1. **Citrate → Aconitate:** In this first step of the Krebs cycle, dehydration of citrate occurs via an E1cb mechanism. The enolate derivative is formed by deprotonation by a base, B, and hydroxide ion is expelled in (1b) when the carbonyl group is regenerated.

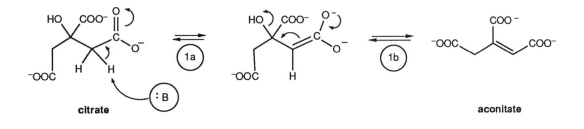

2. **Aconitate → Isocitrate:** Hydration occurs in (2) by the reverse of the dehydration mechanism.

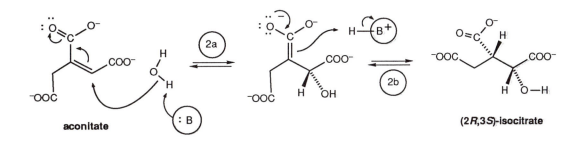

3. **Isocitrate → Ketoglutarate:** Oxidation of the alcohol to the ketone proceeds with involvement of NAD⁺ as a coenzyme in step (3a). Spontaneous decarboxylation occurs in step (3b) because the intermediate is a β-keto acid. The enol tautomerizes in step (3c) to yield α-ketoglutarate.

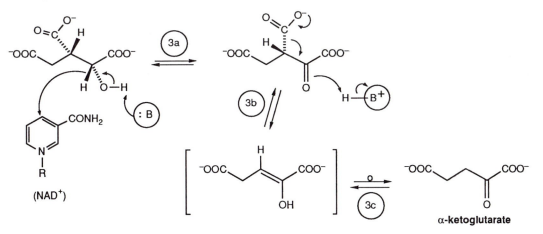

22.28. (continued)

4. Ketoglutarate → Succinate → Fumarate → (S)-Malate: After succinate is formed by another decarboxylation step (discussed later in the text), dehydrogenation produces fumarate. This compound undergoes hydration in step (4) by the same mechanism shown in step (2).

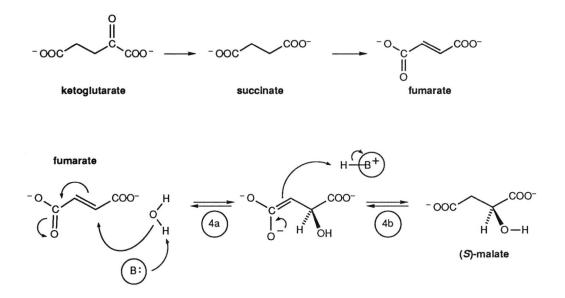

5. (S)-Malate → Oxaloacetate: Oxidation of the alcohol, (S)-malate, occurs in step (5).

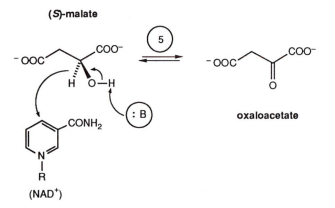

22.28. (continued)

6.   **Oxaloacetate → Citrate:** In step (6), a crossed Claisen condensation takes place between oxaloacetate and acetyl-coenzyme A. The resulting thioester undergoes hydrolysis via a tetrahedral intermediate to form citrate.

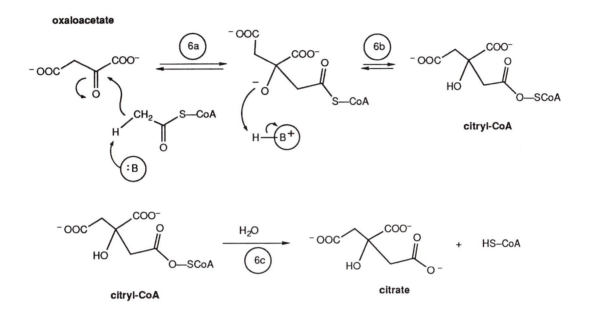

22.29. The identical carboxymethyl groups in citrate have priorities 3 and 4 in the Cahn-Ingold-Prelog system. If one of the carbon atoms of the group coming out of the plane of the page (in the orientation drawn) is labeled with ^{14}C, the (R) isomer of citrate is formed. Therefore, the carboxymethyl group in front is pro-(R) and the one behind the page plane is pro-(S).

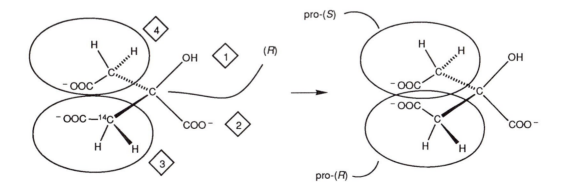

22.29. (continued)

For the hydrogen atoms of each carboxymethyl group, replacement of one of the hydrogen atoms with deuterium makes that substituent the #3 priority group in assigning absolute configurations. The remaining hydrogen atom has the #4 priority. The absolute configuration of each methylene carbon atom is thereby assigned after each hydrogen atom in turn is replaced by deuterium. The assignments are given below.

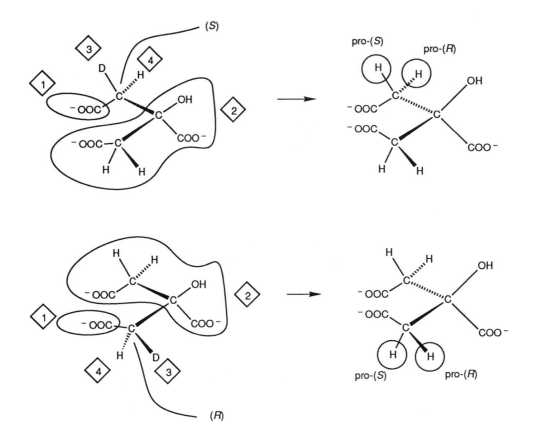

22.30.

The circled hydrogen atom in the structure at the right is removed when citrate is converted to aconitate and isocitrate. This provides a reference point to follow the fate of the labeled atom in the three cases given in this exercise. Only in reaction (c) is the label lost, when citrate is dehydrated to produce aconitate.

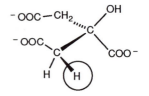

22.30.   (continued)

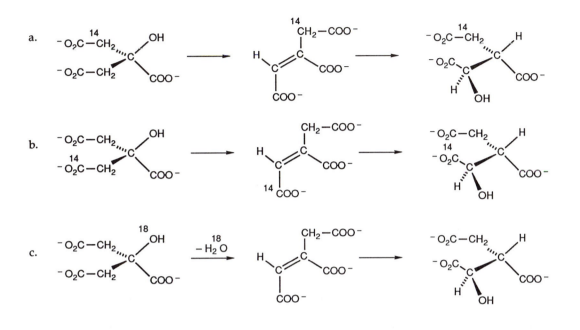

22.31.   The scheme shown in the solution to exercise 22.30 is used to follow the labeled atom(s) in the second and third steps of the following sequence:

From the mechanism (Claisen condensation) for the reaction between oxaloacetate and acetyl-coenzyme A that produces citrate, we can identify the origin of each carbon atom in citrate:

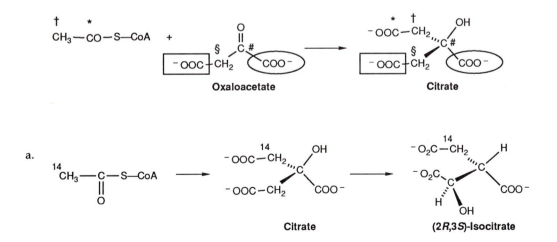

22.31. (continued)

b.

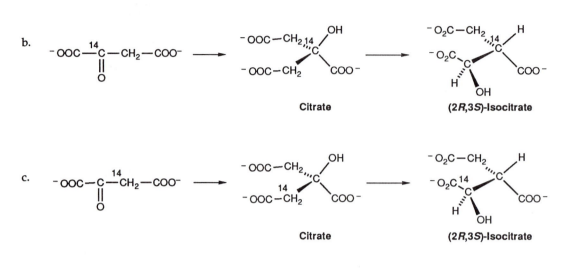

Citrate   **(2R,3S)-Isocitrate**

c.

Citrate   **(2R,3S)-Isocitrate**

22.32. Michael addition involves the reaction between an active methylene compound and an α,β-unsaturated carbonyl or nitrile. The disconnection in a retrosynthesis that will employ the Michael reaction in the forward direction breaks the bond between carbon atoms that are β and γ to the carbonyl group.

a.   *retrosynthesis:*

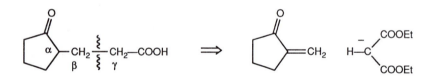

synthesis: After the Michael addition is complete, hydrolysis with concomitant decarboxylation yields the carboxylic acid.

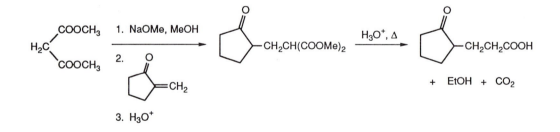

22.32.   (continued)

b.   *retrosynthesis*:

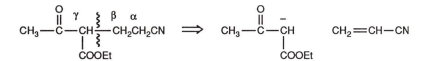

synthesis:

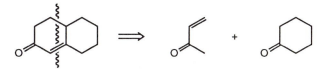

22.33.   Retrosynthesis based on the use of the Robinson annulation with methyl vinyl ketone breaks the bonds that remove the four-carbon ketone fragment from the product.   If the six-membered ring has other functional groups, they are included as part of the cyclohexanone starting material (see part b).

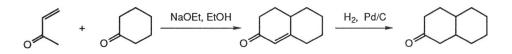

a.   Reduction of the α,β-unsaturated ketone produces the saturated ketone.

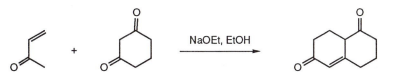

b.   This Robinson annulation uses 1,3-cyclohexanedione and methyl vinyl ketone to make the α,β-unsaturated ketone.

22.33. (continued)

c.   Conjugate addition of a methyl group from the Gilman reagent occurs on the under side of the fused ring system.

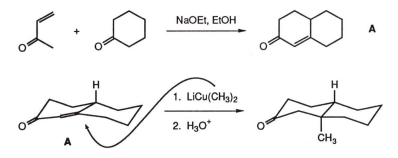

d.   Conjugate addition of a methyl group from the Gilman reagent occurs on the under side of the fused ring system. The intermediate enolate ion formed from conjugate addition is treated with methyl iodide instead of acid, as in part c, which produces the dimethyl compound.

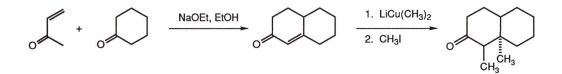

22.34.   In this transformation, which is called the Hillman-Baylis reaction, methoxide ion undergoes 1,4-addition to the unsaturated ester, forming an enolate ion. This nucleophile adds to the aldehyde carbonyl group in step (2). Upon workup with acid, a molecule of water is lost after protonation of the alcohol OH group. Dehydration forms the unsaturated ester.

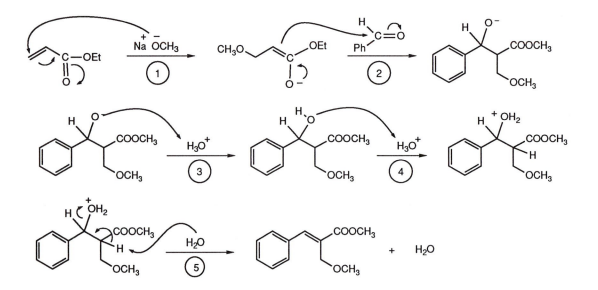

22.35. Follow the procedure given in the solution to exercise 19.28.

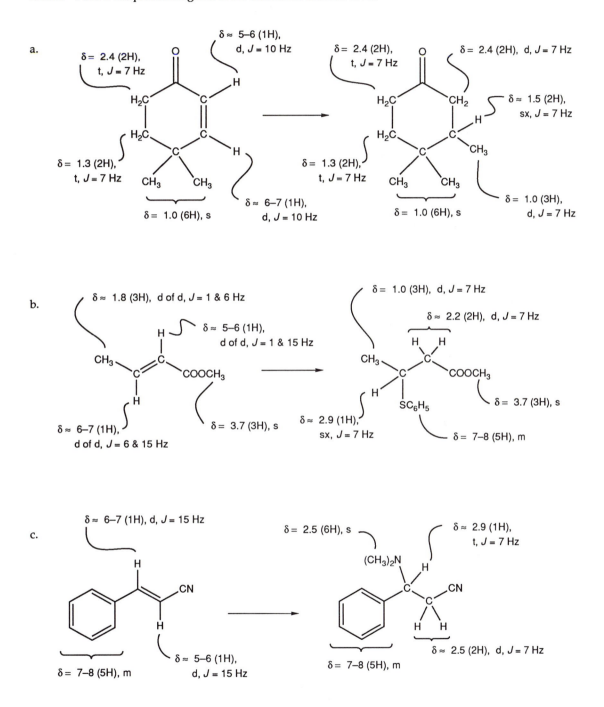

22.36. Predict what resonances you expect to see by using the tables and figures in Chapter 15. The broad band decoupled spectrum consists only of singlets, so it is only a matter of finding chemical shift values for the different carbon atom resonances.

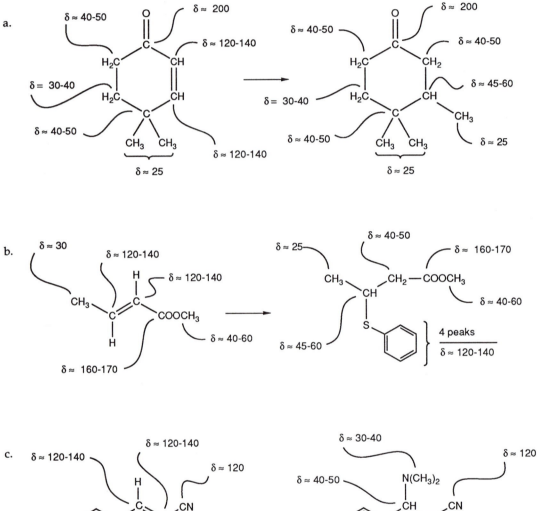

a.

b.

c.

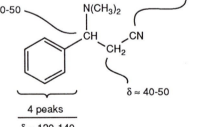

22.37. **a.** Conjugate addition of an alkyl group to an α,β-unsaturated ketone makes use of an organo-cuprate.

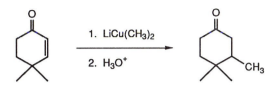

b. Conjugate addition of a thiol to an α,β-unsaturated ester is accomplished by using the corresponding thiolate ion as a catalyst.

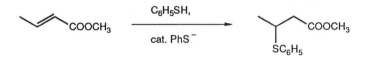

c. An amine readily undergoes conjugate addition to an α,β-unsaturated nitrile.

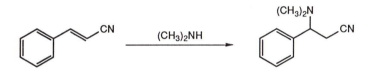

22.38. **a.** For compound **38A**, proton NMR spectrum assignments are tabulated below. We are told that the compound has a carbonyl group, and from the formula, we deduce the compound is a ketone. We also calculate that one additional site of unsaturation is present. From the alkene proton resonances in the NMR spectrum, we conclude that the compound has a carbon-carbon double bond. The coupling patterns among the alkene protons is complicated because their chemical shift values are so similar. In such cases, the *n+1* rule does not apply (the spectrum is no longer "first order"), and very complex patterns are observed. The fact that there are three alkene protons, however, tells us that the compound has a monosubstituted double bond; the compound is methyl vinyl ketone.

δ	integr	splitting	J (Hz)	assignment
6.6	2H	m	–	CH$_2$ of a double bond
5.9	1H	d	12	CH of a double bond
2.3	3H	s	–	CH$_3$ attached to a carbonyl group

22.38. (continued)

b. For compound **38B**, proton NMR spectrum assignments are tabulated below. We are told that the compound has a carbonyl group, and from the formula (two oxygen atoms), we conclude that the compound is either an acid or ester. We also calculate that one additional site of unsaturation is present. From the alkene proton resonances in the NMR spectrum, we conclude that the compound has a carbon-carbon double bond.

δ	integr	splitting	J (Hz)	assignment
12.2	1H	br s		–COOH
5.7	1H	septet	1	CH of an alkene coupled with two CH$_3$ groups
2.18	3H	doublet	1	CH$_3$ attached to an alkene coupled with a CH group
1.93	3H	doublet	1	CH$_3$ attached to an alkene coupled with a CH group

---

22.39. **a.** The product has a seven membered ring, so we start with cycloheptene. The methyl group and C=C bond can be introduced by the procedure shown in the solution to exercise 22.20a.

retrosynthesis:

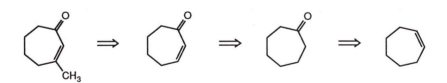

synthesis: Hydration of cycloheptene followed by oxidation yields cycloheptanone. Introduction of the PhSe group α to the carbonyl group followed by oxidative elimination generates the conjugated system. 1,4-Addition of a methyl group followed by trapping of the enolate ion with a PhSe group and oxidative elimination yields the desired product.

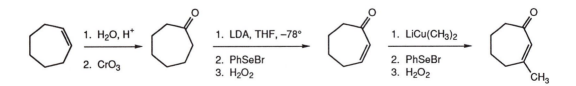

22.39. (continued)

b.   The bicyclic ring system is constructed by application of the Robinson annulation.

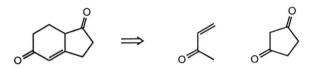

c.   The product has a seven membered ring, so we start with cycloheptene. An exocyclic double bond is made by elimination from a Mannich base.

retrosynthesis:

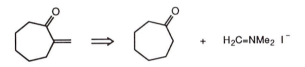

synthesis: Cycloheptene is converted to the ketone as shown in part (a). Formation of the enolate ion followed by trapping with the preformed methylene dimethylammonium salt yields the Mannich adduct. Methylation at nitrogen followed by heating causes elimination of trimethylamine.

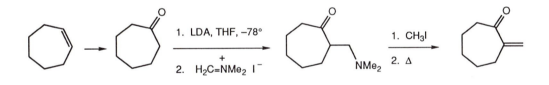

22.40.   First, summarize the reactions described in the accompanying paragraph.

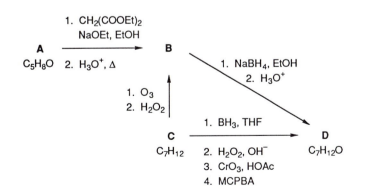

22.40. (continued)

The fact that compound **A** undergoes a Michael reaction suggests that it is an α,β-unsaturated ketone or aldehyde (only one oxygen atom is present). Addition of diethyl malonate followed by hydrolysis (and decarboxylation) adds a hydrogen atom and the –CH₂COOH group, which means that compound **B** has the formula $C_7H_{12}O_3$.

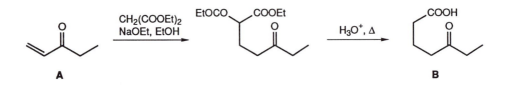

Compound **C** has the same number of carbon and hydrogen atoms as compound **B**, so ozonolysis must be cleaving a ring to form a ketone and carboxylic acid (an aldehyde cannot be made under conditions that employ oxidative workup). If compound **B** has ketone and acid groups, compound **A** is a ketone. It cannot be a methyl ketone, however, because the proton NMR spectrum of compound **B** has no feature that is a singlet, besides the resonance for the carboxylic acid proton.

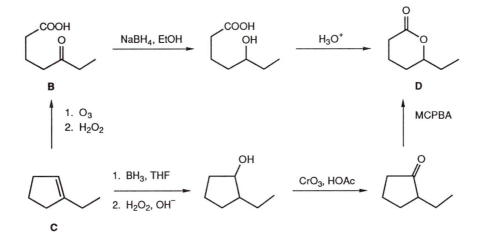

22.41. a. Chirality is introduced with use of an oxaboraoxazolidine. The carbon skeleton is made by Friedel-Crafts acylation using glutaric anhydride.

retrosynthesis:

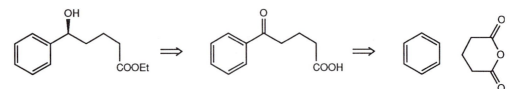

22.41.   (continued)

synthesis:

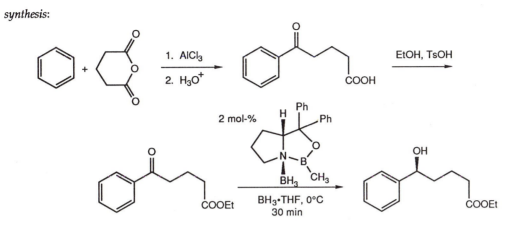

b.   Chirality is introduced via asymmetric hydroboration.

retrosynthesis:

synthesis: The side chain is added by a Grignard reaction with ethylene oxide, and the OH group is protected as its silyl ether. Asymmetric hydroboration forms the 2° alcohol.

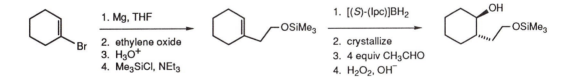

After deprotecting the 1° alcohol, it is converted to its mesylate derivative. It is possible to make a mesylate of a 1° alcohol in the presence of a 2° or 3° alcohol because the former is less hindered. The temperature may have to be lowered somewhat to increase selectivity. Generating the alkoxide ion from the 2° OH group leads to cyclization via an intramolecular S_N2 process.

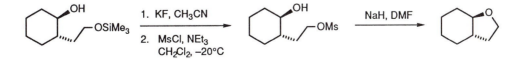

22.42.   Conjugate addition of a radical is similar to nucleophilic conjugate addition except that the electrons move individually rather than in pairs. Initiation takes place as you have seen before (text page 494), by reaction of the radical derived from AIBN with tributyltin hydride.

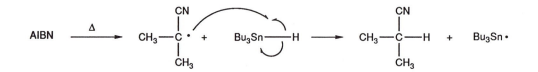

In the first step for the organic substrate, the bromine atom is abstracted by the tributyltin radical.

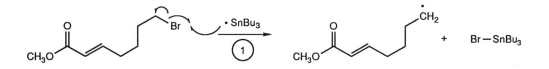

Addition of the radical to the double bond occurs in step (2) and involves the carbonyl π electrons, which is what makes this process a conjugate addition. The resulting oxygen-centered radical reacts with tributyltin hydride in step (3), forming the enol derivative of the ester and regenerating the tributyltin radical, which reacts with more bromoalkyl substrate. The enol finally undergoes tautomerism in step (4) to form the product, the cyclized ester.

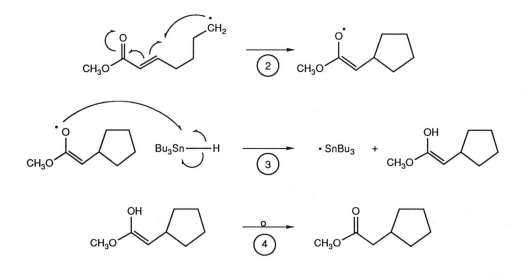

22.43. The carbonyl double bond of an aldehyde is susceptible to addition by radicals because it is unhindered. The mechanism is like that shown in the solution to exercise 22.42. The tributyltin radical adds to the ketone carbonyl group, forming a radical that adds to the unsaturated ester, as described in the solution to exercise 22.42.

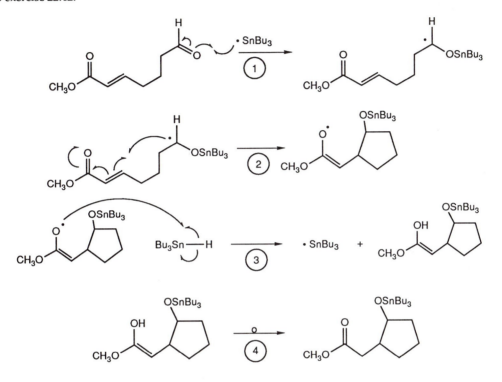

Once the ester is formed, hydrolysis of the O–Sn bond occurs as the alkoxy oxygen atom adds to the carbonyl group via formation of a tetrahedral intermediate. This addition step leads to lactone formation when the carbonyl group is reestablished.

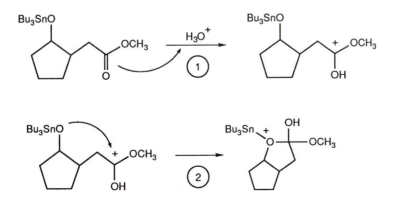

22.43. (continued)

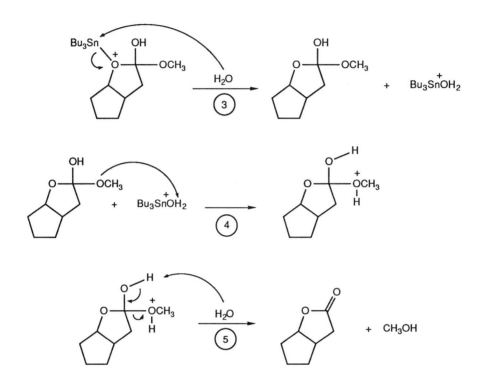

THE CHEMISTRY OF AMINES AND OTHER NITROGEN–CONTAINING COMPOUNDS

23.1. *p*-Methoxyaniline has five important resonance structures:

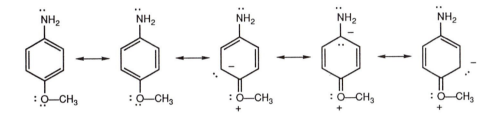

The structure that places a negative charge on the carbon atom bearing the amino group is expected to facilitate protonation on the nitrogen atom because the positive charge in this protonated form places opposite charges on adjacent atoms.

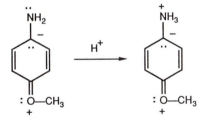

23.2. The conjugate base derivative of succinimide is like the anion derived from an active methylene compound (text page 1011). The charge is spread over the nitrogen and both oxygen atoms.

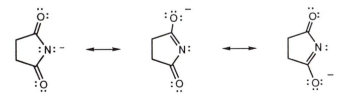

23.3. The shape of an azaalkane is the same as its carbocyclic analog, so envelope and chair forms are most likely for five- and six-membered rings, respectively. For an *N*-methyl derivative, the methyl group is equatorial. If another substituent is attached to the ring, it will be in the equatorial position, too.

23.4. The reaction between an amine and a dihaloalkane proceeds via two S$_N$2 reactions. Normally, a base is added to remove HX that is formed.

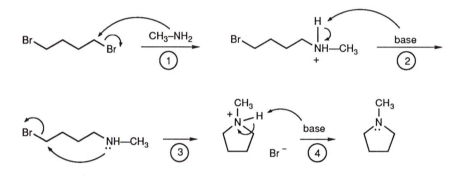

23.5. An amine with a chloroethyl group is unstable because the nitrogen atom displaces chloride ion to form an aziridinium ion. When hydroxide ion is present, ring opening occurs to form the hydroxyethylamine analog.

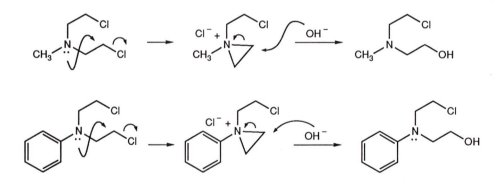

The aromatic compound reacts more slowly than its aliphatic analog because the nitrogen atom is less basic (hence less nucleophilic) as a result of electron delocalization with the π system of the ring.

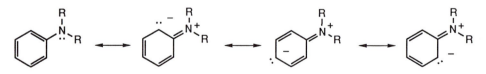

23.6. Formation of the seven-membered ring heterocycle shown in this exercise begins with deprotonation of the thiol group. The thiolate ion subsequently reacts with aziridine, opening the three-membered ring. The 1° amino group unmasked during steps (1) – (3) reacts with the ester group to form a tetrahedral intermediate that subsequently collapses to form the lactam ring.

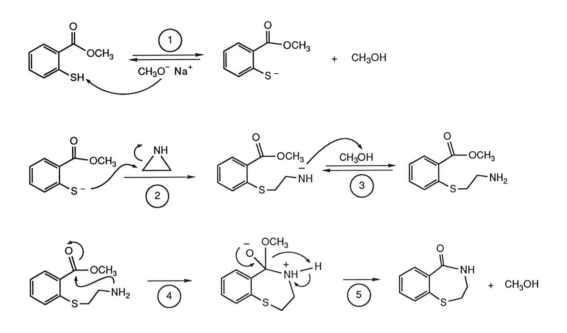

23.7. Dimerization of α-amino esters to form a diketopiperazine involves two separate amide-forming reactions between amine and ester functional groups.

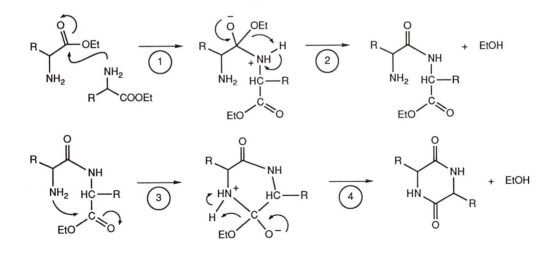

23.8. Kevlar is the polyamide formed by combining the aromatic diacid with diaminobenzene.

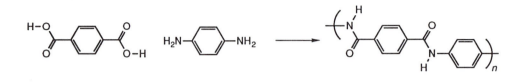

23.9. Hydrazine is a difunctional molecule that has two nucleophilic nitrogen atoms. It reacts by nucleophilic addition to one carbonyl group of an *N*-alkylphthalimide, forming a tetrahedral intermediate. Regeneration of the carbon-oxygen double bond displaces the amido group. The other nitrogen atom of the hydrazine molecule then adds to the other carbonyl group, repeating the same steps. An amine and the hydrazide derivative of phthalic acid are formed as products of this transformation.

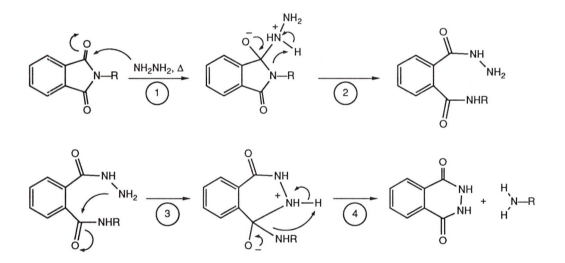

23.10. A primary amine is made from a primary alkyl bromide via nucleophilic substitution with potassium phthalimide followed by hydrazinolysis (see the solution to exercise 23.9).

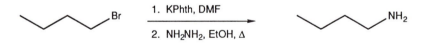

23.10.   (continued)

A secondary amine is made from a primary amine via its benzamide derivative. Formation of the anion of
N-butylbenzamide followed by alkylation yields the 3° amide. Hydrolysis of the benzoyl group from the
nitrogen atom produces the 2° amine.

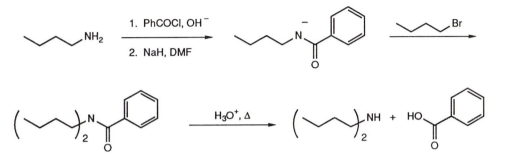

A tertiary amine is best made by reducing a tertiary amide with use of lithium aluminum hydride. The 3°
amide is made from reaction between an acid chloride and the corresponding secondary amine, the synthesis
of which is shown directly above. The acid chloride is made by converting 1-bromobutane to 1-butanol
followed by oxidation and reaction with thionyl chloride.

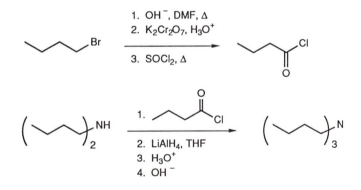

23.11.   **a.**   An iodine atom is most often incorporated into aromatic compounds via a diazonium salt,
available ultimately from a nitro compound. The hydroxy group is an *ortho, para* director, so in this synthesis,
it must be made from a functional group that is a *meta* director. The Baeyer-Villiger reaction of the
corresponding ketone is the way to accomplish this transformation.

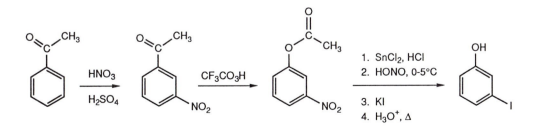

23.11. (continued)

b. A cyano group in introduced via a diazonium salt. *p*-Bromoaniline is made from the corresponding nitro compound, which is accessible by electrophilic aromatic substitution.

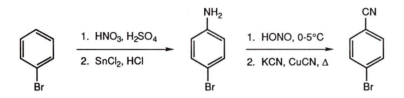

c. Synthesis of an *ortho*-disubstituted compound normally requires that the *para* position be blocked. A nitro group is a good blocking substituent because it can be reduced to an amino group and eliminated via a diazonium intermediate upon reaction with hypophosphorus acid.

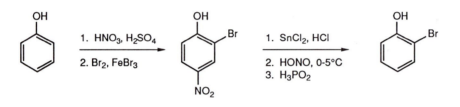

23.12. Alizarin Yellow R is made by diazotizing *p*-nitroaniline and coupling that intermediate with the phenol component, salicylic acid.

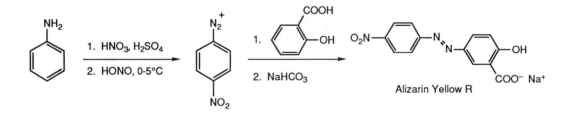

23.13. The Cope elimination takes place via concerted movement of electrons starting at the negatively-charged oxygen atom. This transformation yields the alkene and a molecule of dimethylhydroxylamine.

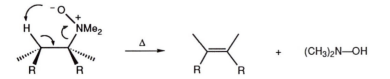

23.14. Substitution of a halogen atom *ortho* to a nitro group occurs by addition/elimination. The first step of this mechanism involves reaction of the thiolate ion at the carbon atom bearing the chlorine atom. This step generates a complex in which the nitro group accepts electrons from the nucleophile. Regenerating the nitrogen-oxygen double bond in step (2) displaces chloride ion to form the product.

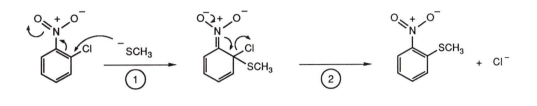

23.15. Formation of the DNP derivative of an amino acid occurs in the same fashion as shown in the solution to exercise 23.14. The first step of this mechanism involves reaction of the amine at the carbon atom that bears the fluorine atom. This generates a complex in which the nitro group accepts the electrons from the nucleophile. Regenerating the nitrogen-oxygen double bond leads to displacement of fluoride ion, which deprotonates the amino group. The HF that is formed reacts with added base.

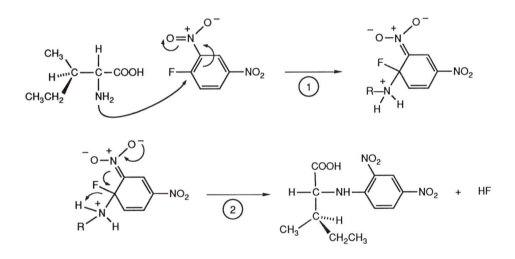

23.16. Reaction between the triflate derivative of the illustrated *o*-silyl phenol and fluoride ion generates the corresponding benzyne intermediate. The "extra" π bond of the benzyne is a good dienophile and adds across the "diene" portion of furan.

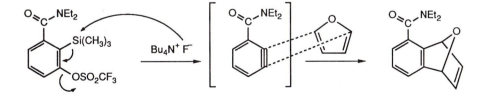

23.17. Substituting alkoxide ion for hydroxide ion does not change the early steps in the mechanism of the Hofmann reaction. In methanol solution, the alcohol intercepts the isocyanate product in step (6), which leads to formation of the urethane.

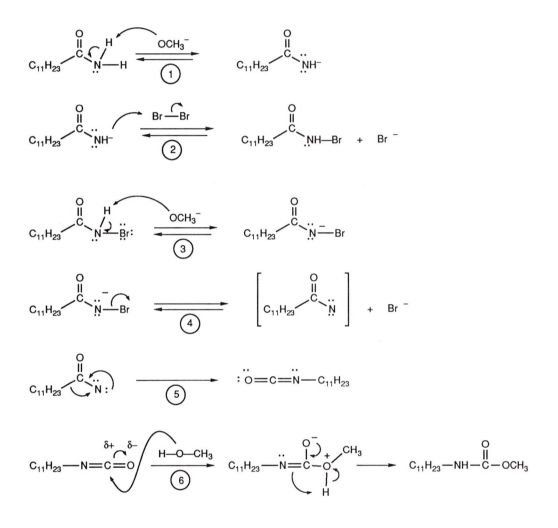

23.18. An acyl azide is prepared by treating an acid chloride with azide ion. This addition step forms a tetrahedral intermediate, which collapses and to displace chloride ion.

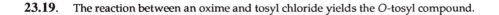

23.19. The reaction between an oxime and tosyl chloride yields the *O*-tosyl compound.

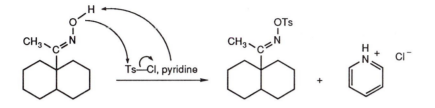

One of the alkyl groups subsequently migrates from carbon to nitrogen, displacing the tosylate group. Water intercepts the carbocation to form the amide after acid-base reactions.

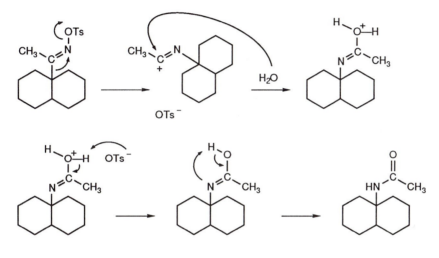

23.20. The mechanism for the conversion of cyclohexanone oxime to caprolactam follows the same steps as illustrated on text pages 1191-1192.

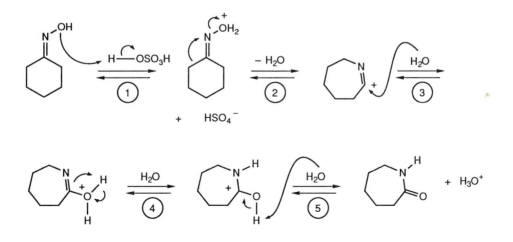

23.21. Follow the procedure given in the solution to exercise 7.28.

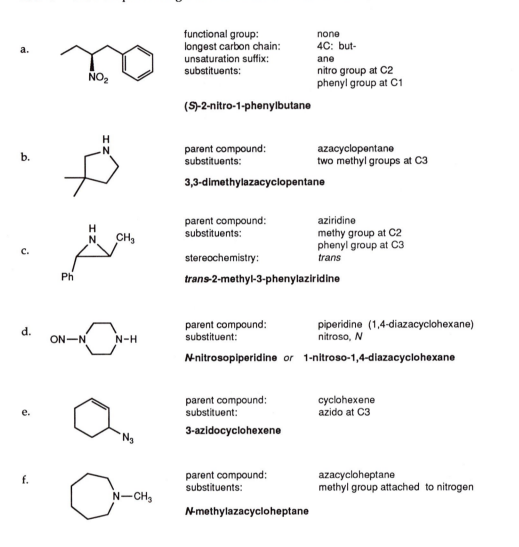

a.

functional group:	none
longest carbon chain:	4C: but-
unsaturation suffix:	ane
substituents:	nitro group at C2
	phenyl group at C1

(*S*)-2-nitro-1-phenylbutane

b.

| parent compound: | azacyclopentane |
| substituents: | two methyl groups at C3 |

3,3-dimethylazacyclopentane

c.

parent compound:	aziridine
substituents:	methy group at C2
	phenyl group at C3
stereochemistry:	*trans*

***trans*-2-methyl-3-phenylaziridine**

d.

| parent compound: | piperidine (1,4-diazacyclohexane) |
| substituent: | nitroso, *N* |

***N*-nitrosopiperidine** *or* **1-nitroso-1,4-diazacyclohexane**

e.

| parent compound: | cyclohexene |
| substituent: | azido at C3 |

3-azidocyclohexene

f.

| parent compound: | azacycloheptane |
| substituents: | methyl group attached to nitrogen |

***N*-methylazacycloheptane**

23.22. Interpret the name according to the directions given in Chapter 2.

a. *p*-chlorophenyldiazonium chloride

b. *N*-ethyl-*N*-nitroso-3-bromoaniline

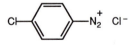

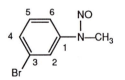

23.22.   (continued)

c.   *trans*-2,5-dimethylazacyclopentane

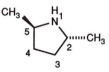

d.   *cis*-2,3-aziridinedicarboxylic acid

e.   2-azido-5-methylphenol

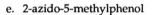

23.23.   Examples of reactions summarized in Table 23.1 (text page 1152).

Alkylation of an amide at nitrogen via its anion:

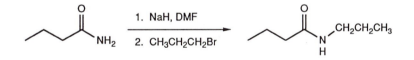

Halogenation of an amide at nitrogen (Hofmann rearrangement):

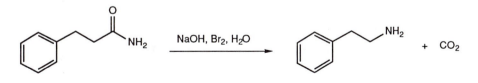

Alkylation of an amine at nitrogen:

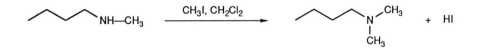

23.23. (continued)

Addition/elimination of an amine with aldehydes and ketones:

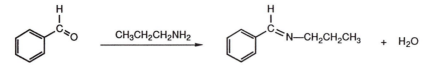

Addition/elimination of an amine with carboxylic acid derivatives:

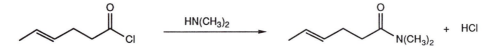

Michael addition of an amine:

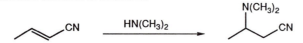

Esterification of carboxylic acids with use of diazomethane:

Alkylation at the α-position of a ketone via its enamine:

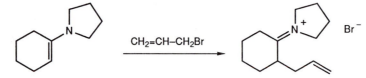

Alkylation at the nitrogen atom of an imide (Gabriel synthesis):

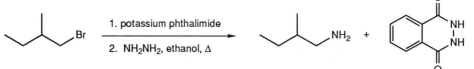

1. potassium phthalimide

2. NH₂NH₂, ethanol, Δ

Alkylation at the α-position of an aldehyde via its imine carbanion derivative:

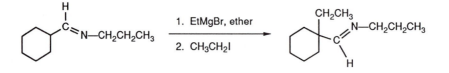

23.23.   (continued)

Michael addition of an α-carbanion of a nitro compound:

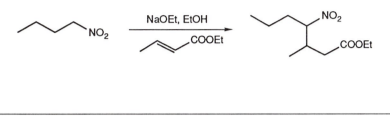

23.24.

a.   The basicity of a cyclic amine increases with decreasing ring size because ring strain is relieved upon protonation. In the free base form, any contribution of sp^2 hybridization of nitrogen increases ring strain. In the protonated form, the nitrogen atom has only sp^3-hybridization.

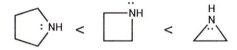

b.   The basicity of a 1° amine increases when an alkyl group is attached to nitrogen because the electron donating ability of a methyl group makes the nitrogen atom more electron rich. With the addition of a second methyl group, steric effects inhibit formation of the tetrahedral geometry required in the protonated form, so the basicity of a 3° amine is lower than its 2° analog. A 3° amine is still more basic than a 1° amine because of electron donation from the additional substituents.

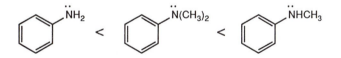

c.   An amide is not basic at all. An sp^3-hybridized nitrogen atom is more basic than the corresponding sp^2-hybridized nitrogen atom

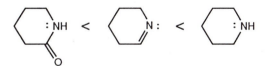

23.25. A base reacts with an acid by removing a proton from the latter, generating the conjugate base and conjugate acid of the reactants. The equilibrium lies to the side that has the weaker acid (text page 225). Specific pK_a values for some of the amine salts can be found on text pages 1154-1156.

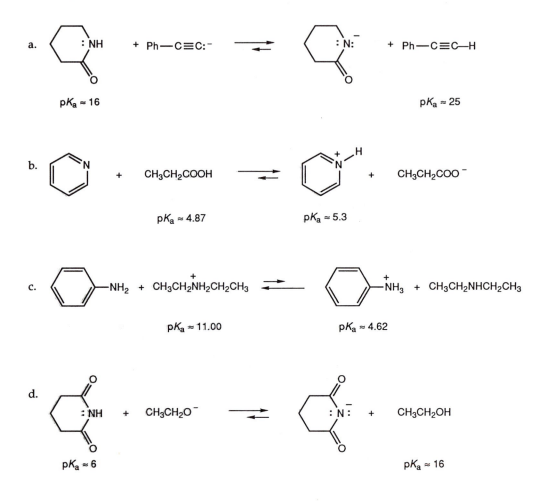

23.26. Nitrogen containing compounds have a chiral nitrogen atom only when the compound is a quaternary ammonium salt, when it is "locked" into rings (usually bicyclic), or it is an aziridine. An unsymmetric oxime exists as *syn* and *anti* isomers.

a. There are four isomers (two at the stereogenic carbon atom and two at the oxime double bond). The enantiomeric pairs are indicated. All other pairs are diastereomers.

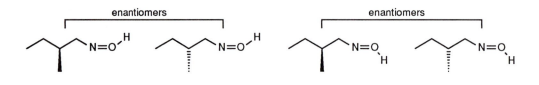

23.26.   (continued)

b. There are two isomers of this compound because the carbon atom bearing the phenyl group is stereogenic. The compounds are enantiomers. The nitrogen atom is not a stereogenic center because it undergoes pyramidal inversion.

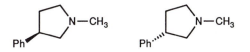

c. There are four isomers (two at the stereogenic carbon atom and two at the stereogenic nitrogen atom.) The enantiomeric pairs are indicated. All other pairs are diastereomers.

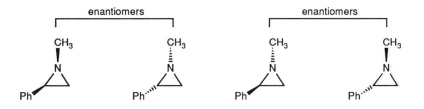

d. There are only two isomers because the nitrogen atom is stereogenic. The compounds are enantiomers.

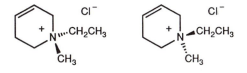

23.27.

a.   An amine reacts with a carboxylic acid anhydride to form an amide.

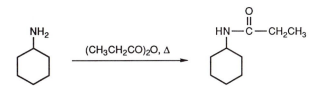

23.27. (continued)

b. A Grignard reagent reacts with a 3° dimethylamide to form a tetrahedral intermediate. Workup with aqueous acid converts the anion to a *geminal* amino alcohol, which eliminates a molecule of dimethylamine to form a ketone.

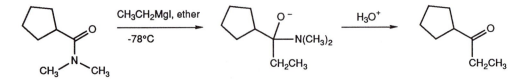

c. An aromatic amine reacts with HONO to form a diazonium salt. Hydroxide ion and water convert the diazonium intermediate to the corresponding phenol.

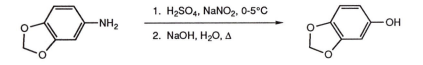

d. A 1° alcohol reacts with methanesulfonyl chloride to form a mesylate derivative. Nucleophilic substitution with azide ion yields the alkyl azide. Reduction of the –N$_3$ group of an alkyl azide with hydrogen and a metal catalyst produces the amine.

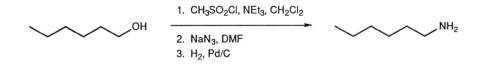

e. The diazonium derivative of an amine is reduced by hypophosphorous acid, which replaces the original amino group with a hydrogen atom.

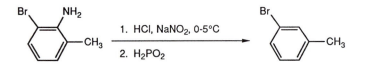

f. A primary amide reacts with bromine and alkoxide ion in alcohol to produce the rearranged product, a urethane (see the solution to exercise 23.17).

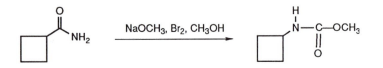

23.27.   (continued)

g. A ketone reacts with hydroxylamine to form an oxime derivative. Concentrated sulfuric acid initiates a Beckmann rearrangement. The aryl group migrates in this case because the oxime adopts a geometry such that its OH group points away from the methyl group (the group *anti* to the OH group is the one that migrates during the Beckmann rearrangement).

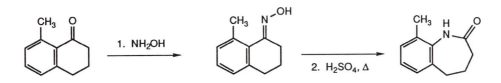

h. An aromatic nitro compound does not undergo diazotization.

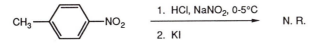

i. An amine undergoes Michael addition with an α,β-unsaturated ester.

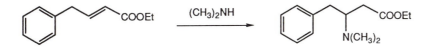

j. A halogen atom *ortho* or *para* to a nitro group is substituted by a nucleophile.

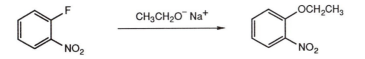

23.28.   The presence of the two dimethylamino groups proximal to each other focusses the two unshared pairs of electrons at a single point. The proton is bound at that point, symmetrically hydrogen-bonded to the two nitrogen atoms. The monosubstituted analog has only one pair of electrons that can react with a proton, so binding is not as strong.

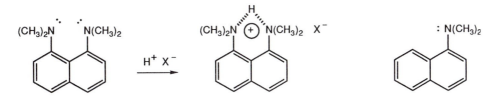

23.29. Three common routes are used to make a 1° amine:

- Substitution of a leaving group by a nitrogen-containing nucleophile   RX or ROMs
- Reduction by LiAlH₄ of a nitrile, an amide, or an oxime   RCN, RCONH₂, or RCHNOH
- Hofmann rearrangement of a 1° amide with one additional carbon atom   RCH₂CONH₂

The three compound types shown directly above provide key synthetic intermediates in the schemes of preparative routes to 1-aminohexane.

a. 1-hexene

b. hexanal

c. hexanoic acid

d. hexanenitrile

e. 1-hexanol

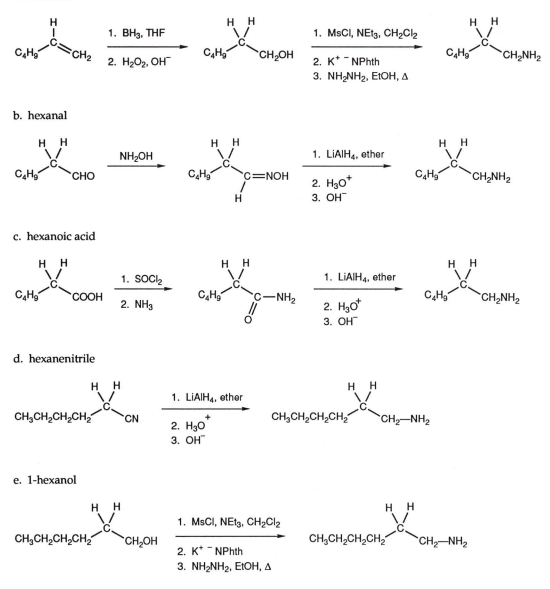

23.29. (continued)

f. 1-bromohexane

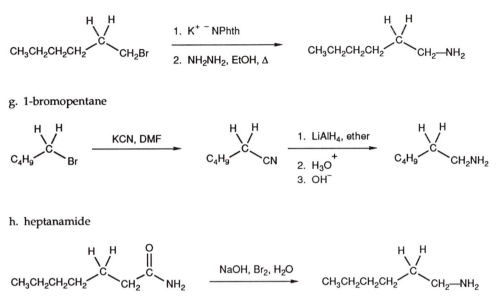

g. 1-bromopentane

h. heptanamide

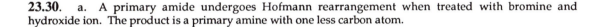

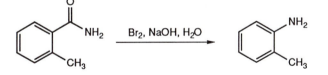

23.30. a. A primary amide undergoes Hofmann rearrangement when treated with bromine and hydroxide ion. The product is a primary amine with one less carbon atom.

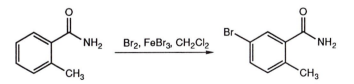

b. Bromine in the presence of a Lewis acid is used for electrophilic bromination if the substrate has an aromatic ring. The methyl group is an *o,p*-director, and the amide is a *m* director. The major product therefore has a bromine atom *para* to the methyl group.

23.30.   (continued)

c. An amide that has a hydrogen atom attached to nitrogen can be alkylated after generating the conjugate base (the amide ion) and treating that species with an alkyl halide.

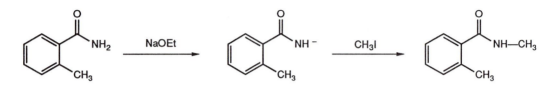

d. Amides are very weak nucleophiles, so *o*-methylbenzamide does not react readily with acetic anhydride.

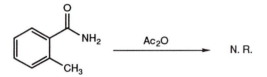

e. With hot, aqueous acid, an amide undergoes hydrolysis to form a carboxylic acid.

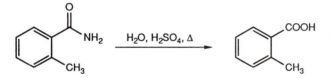

f. A primary amide is reduced by several hydride reagents to form a primary amine. Lithium aluminum hydride is the most commonly used of these reagents.

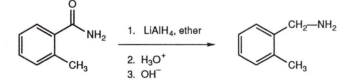

g. A primary amide undergoes dehydration to form the corresponding nitrile.

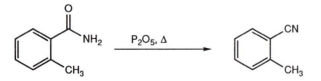

23.31. Phthalic anhydride is used to protect a primary amine by formation of a phthalimide derivative, which begins by addition of the amino group to one of the carbonyl groups of the anhydride. After the tetrahedral intermediate collapses to form the amido acid in step (3), ring closure leads to elimination of water and formation of the imide.

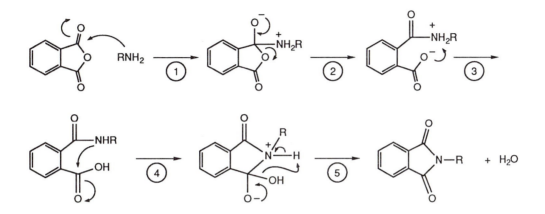

Deprotection of the derivatized amine is a hydrolysis reaction. Hydrolysis of an imide is like that for any acid derivative. Hydroxide ion adds to a carbonyl group to form the tetrahedral intermediate. The intermediate subsequently collapses to regenerate the carbonyl group and eliminate the heteroatom-containing fragment, in this case the amide group. The second stage of this reaction constitutes hydrolysis of an amide [steps (3) and (4), below].

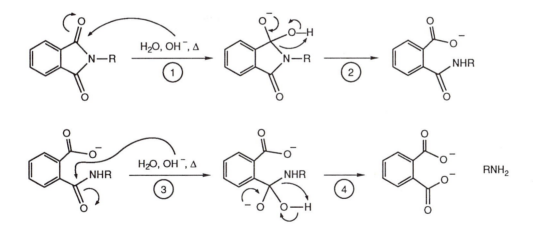

23.32. A tertiary amine is made by alkylating a primary amine or by reduction of a tertiary amide with use of LiAlH$_4$. With the constraint that we use compounds with five or fewer carbon atoms, as well as the fact that a neopentyl groups has five carbon atoms, carboxylic acid derivatives or an aldehyde are reasonable starting materials.

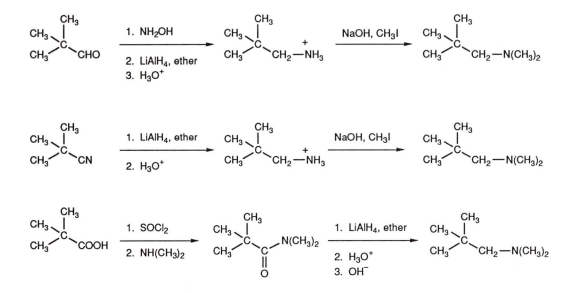

23.33. a. For compound **33A** the proton NMR spectrum assignments are tabulated below. From the formula (one nitrogen atom, no oxygen atom), we know that the compound is probably an amine.

δ	integr	splitting	J (Hz)	assignment
7.2	2H	d	10	CH of *para*-substituted benzene ring
6.5	2H	d	10	CH of *para*-substituted benzene ring
3.7	1H	br s	–	NH
2.8	3H	s	–	CH$_3$ attached to a nitrogen atom

23.33. (continued)

b.. For compound **33B** the proton NMR spectrum assignments are tabulated below. Integration of the signals in the aliphatic proton region suggests that two equivalent ethyl groups are present. If these are attached to nitrogen, then subtracting six carbon and four hydrogen atoms for a disubstituted benzene ring leaves only one hydrogen atom and one oxygen atom, which together constitute an OH group.

δ	integr	splitting	J (Hz)	assignment
7.05	1H	t	-	benzene ring, *m* or *o* substituted
6.2	3H	m	–	benzene ring, *m* or *o* substituted
5.8	1H	br s	–	OH group
3.3	4H	q	7	CH$_2$ attached to a nitrogen atom, coupled with a methyl group
1.1	6H	t	7	CH$_3$ coupled with a methylene group

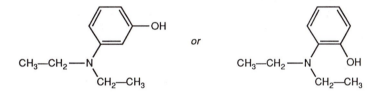

or

23.34. The rearrangement reaction known as the Demyanov ring expansion starts with formation of the aliphatic diazonium compound. The N$_2$ group is a good leaving group, so the electrons in a carbon-carbon bond displace it during rearrangement. The carbocation that forms is just a protonated carbonyl group, so deprotonation generates the product. The starting amino alcohol is prepared by reducing the cyanohydrin that is made from the ketone.

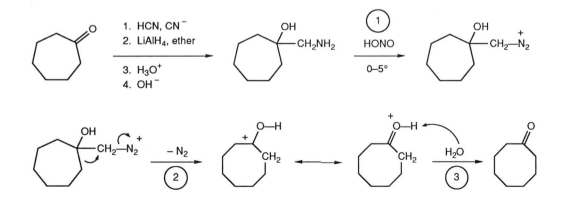

23.35. The carbon atom of diazomethane is nucleophilic and adds to a carbon-oxygen double bond, forming the same type of intermediate shown in the solution to exercise 23.24. Rearrangement yields the ketone product after deprotonation in step (3).

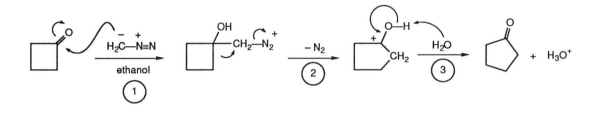

23.36. A scheme summarizing the given reactions is as follows:

$$\underset{C_5H_{11}NO}{A} \xrightarrow{Br_2,\ NaOH} B \xrightarrow{(CH_3)_2C(O_2)} C$$

The IR spectrum for **A** indicates that the compound has a carbonyl group ($\nu CO = 1650\ cm^{-1}$). The fact that the carbonyl stretch appears at a low frequency as a "doublet" suggests that this compound is a primary amide, which is confirmed by the presence of absorptions for N–H stretching vibrations at 3200 and 3400 cm^{-1}. The reaction conditions for converting **A** to **B** (bromine and hydroxide ion) are those needed for a Hofmann rearrangement, which requires that **A** is a primary amide. Assuming that **B** is an amine, its formula must be $C_4H_{11}N$. Dimethyldioxirane converts a primary amine to the corresponding nitro compound. The IR spectrum confirms that **C** is a nitro compound ($\nu NO = 1530$ and 1380 cm^{-1}). The NMR spectrum of **C** shows only a singlet. Compound **C**, which must have the formula $C_4H_9NO_2$, is therefore 2-methyl-2-nitropropane.

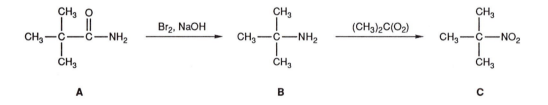

23.37. Anthranilic acid undergoes diazotization in the usual way. Chloride ion, functioning as a base, deprotonates the carboxylic acid group, which initiates electron movement that creates the benzyne triple bond by displacing molecular nitrogen as a leaving group.

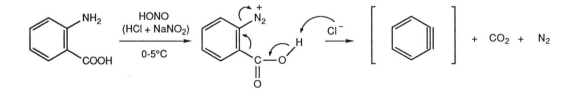

23.38.   Follow the procedure outlined in the solution to exercise 19.27.

a.   The starting material has a very strong absorption band for the carbon-oxygen double bond stretching vibration. Both reactant and product have a doublet associated with N–H stretching vibrations.

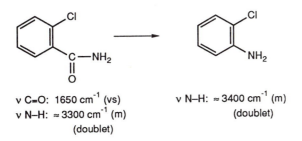

ν C=O: 1650 cm⁻¹ (vs)   ν N–H: ≈ 3400 cm⁻¹ (m)
ν N–H: ≈ 3300 cm⁻¹ (m)   (doublet)
    (doublet)

The starting material has ν C=O: 1650 cm^{-1} (vs), ν N–H: $\approx$ 3300 cm^{-1} (m) (doublet). The product has ν N–H: $\approx$ 3400 cm^{-1} (m) (doublet).

b.   The starting material has a very strong absorption band for the carbon-oxygen double bond stretching vibration. The product has an absorption band at lower frequency for the carbon-nitrogen double bond stretching vibration. The product also has N–H and O–H groups, and these stretching vibrations should produce absorption bands at about 3400 cm^{-1}.

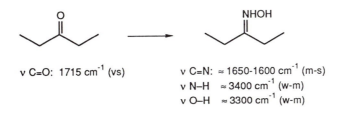

ν C=O: 1715 cm⁻¹ (vs)   ν C=N: ≈ 1650-1600 cm⁻¹ (m-s)
                      ν N–H ≈ 3400 cm⁻¹ (w-m)
                      ν O–H ≈ 3300 cm⁻¹ (w-m)

c.   The starting material has two very strong absorption bands assigned to the nitrogen-oxygen stretching vibrations of the nitro group. The aniline derivative has two N–H stretching vibrations around 3400 cm⁻¹, and the phenol has a single broad absorption at about 3400 cm⁻¹.

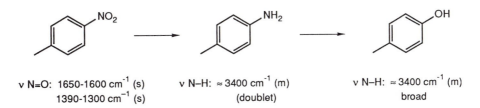

ν N=O: 1650-1600 cm⁻¹ (s)   ν N–H: ≈ 3400 cm⁻¹ (m)   ν N–H: ≈ 3400 cm⁻¹ (m)
        1390-1300 cm⁻¹ (s)   (doublet)   broad

23.39. Follow the procedure outlined in the solution to exercise 19.28.

a. There are two differences in the proton NMR spectra between reactant and product. The aromatic proton resonances will be farther downfield when the electron withdrawing amide group is attached to the ring. The electron donating amino group tends to shift the resonances upfield slightly from the "base value" of δ 7.25 observed for benzene itself. Amide proton resonances are likely farther downfield than those observed for the amine protons of the product. Both amide and amine protons produce resonances that can appear over a large range of chemical shift values. Furthermore, resonances for protons attached to nitrogen are often broadened.

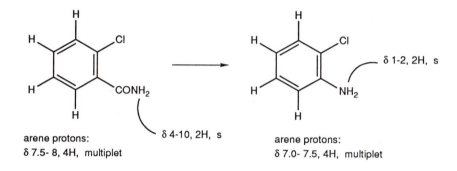

b. The spectrum of 3-pentanone is simple, comprising a triplet and a quartet. For the oxime derivative, additional resonances for the NH and OH protons will appear, and the two ethyl groups are made inequivalent—one is *syn* and the other *anti* to the OH group of the oxime. This difference may be small, so the two sets of signals may overlap.

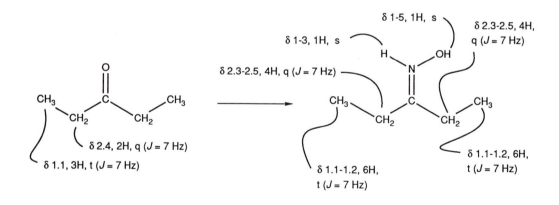

23.39. (continued)

c. The aromatic proton resonances will be farther downfield when the electron withdrawing nitro group is attached to the ring rather than the electron donating amino or hydroxy groups, so the aromatic region for the three compounds will appear slightly different. All three compounds should have a pair of doublets associated with the two types of protons in the *para*-substituted benzene ring. The signal for the methyl group may vary some also because of the electron withdrawing or donating influence of the other substituent. The amine and hydroxyl groups will have resonances not observed for *p*-nitrotoluene.

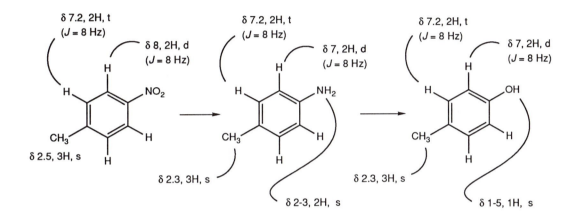

23.40. a. The Hofmann rearrangement converts a 1° amide to the corresponding primary amine.

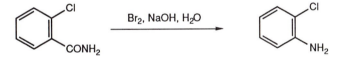

b. Hydroxylamine reacts with a ketone to form the oxime derivative.

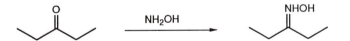

c. An aromatic nitro group is reduced to the corresponding aniline derivative with several reagent combinations, of which tin(II) chloride and HCl is common. Diazotization of the amino group followed by hydrolysis yields the phenol.

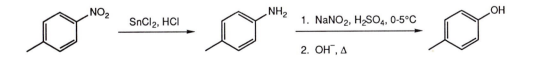

23.41. **a.** Construction of a six-membered ring aza compound from a five-membered ring ketone suggests that a Beckmann rearrangement be used as the key step to enlarge the ring.

retrosynthesis:

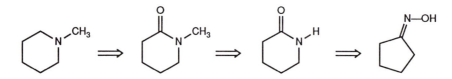

synthesis: After Beckmann rearrangement, the amide nitrogen atom is methylated, and the amide is reduced to the amine with use of LiAlH₄.

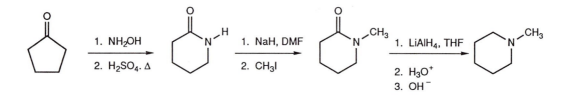

b. An amine is made from a ketone via the imine. The required starting ketone is the product of a Robinson annulation between cyclopentanone and methyl vinyl ketone.

retrosynthesis:

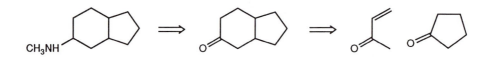

synthesis:

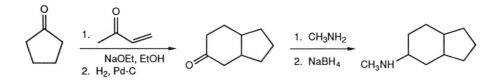

23.41. (continued)

c.   The amine can be made from the corresponding alcohol via substitution. The alcohol is available from the ketone via Horner-Emmons olefination followed by reduction.

retrosynthesis:

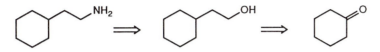

synthesis: Horner-Emmons olefination is followed by reduction with use of LiAlH₄. Mesylation is followed by substitution with azide ion, and the azidoalkane is reduced to form the 1° amine.

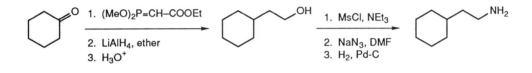

23.42.   a.   The amine is made from the corresponding alcohol via nucleophilic substitution. The alcohol is available via enantioselective hydroboration.

retrosynthesis:

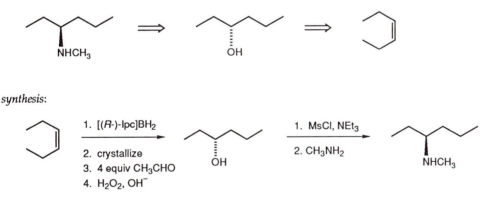

synthesis:

b.   The amine is made from the corresponding nitrile, available from the chiral diol via substitution.

retrosynthesis:

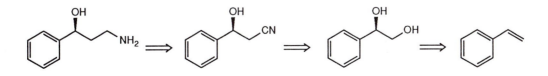

23.42. (continued)

synthesis: The carbon skeleton is available by a variety of strategies; the one shown uses organocuprate coupling with an aryl halide. Asymmetric dihydroxylation establishes the chiral benzylic alcohol group. The mesylate of a primary alcohol can be made in preference to the mesylate of a secondary alcohol. Cyanide ion displaces the leaving group to form the cyano alcohol. Reduction of the nitrile group leads to formation of the amino alcohol.

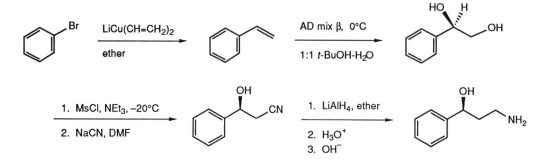

c. The alcohol is made from the corresponding ketone via enantioselective reduction. The fluoro group is introduced via the diazonium derivative.

retrosynthesis:

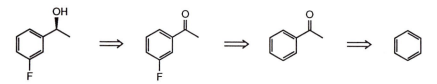

synthesis: Nitration of acetophenone gives the *meta* isomer. Reduction of the nitro group followed by diazotization and replacement by fluorine follows the standard protocol shown on text page 1175. Reduction of the ketone group is accomplished with use of the oxazaborolidine reagent (text page 844).

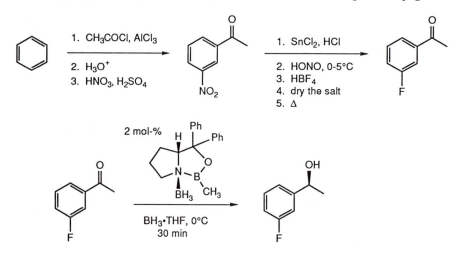

Solutions to Exercises for CHAPTER **24**

THE CHEMISTRY OF
HETEROCYCLIC COMPOUNDS

24.1. Non-aromatic oxygen- and sulfur-containing heterocycles react as any aliphatic ethers or sulfides do. Ethers are susceptible to cleavage under acidic conditions, but they are inert to basic and nucleophilic reagents. Sulfides undergo reaction with Raney nickel with removal of the sulfur atom as hydrogen sulfide.

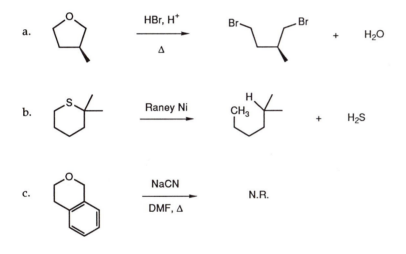

24.2. Pyridine derivatives undergo electrophilic substitution reactions slowly, but an activating group attached to the ring makes the reactivity toward substitution more like that of benzene. An activating group directs substitution to positions *ortho* and *para* to itself.

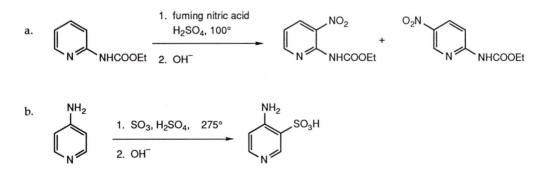

617

24.3. Pyridine has approximately the same reactivity as nitrobenzene in the sense that the ring is electron deficient compared with benzene itself. Recall from text page 1182 that a halogen atom *ortho* or *para* to a nitro group is readily substituted by a nucleophile.

Resonance structures of 4-chloropyridine show that C4 carries a positive charge, which makes it reactive toward nucleophiles.

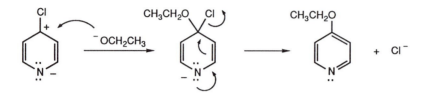

24.4. The positive charge on nitrogen in pyridine-*N*-oxide makes the adjacent position susceptible to reaction with nucleophiles, so a Grignard reagent adds to this activated double bond, and each atom retains an octet of electrons. When the reaction product of step (1) is treated with aqueous acid during workup in step (2), an *N*-hydroxy compound is formed. If this OH group is protonated again, as shown in step (3), then elimination of water takes place to regenerate the double bond and reestablish the aromatic pyridine ring.

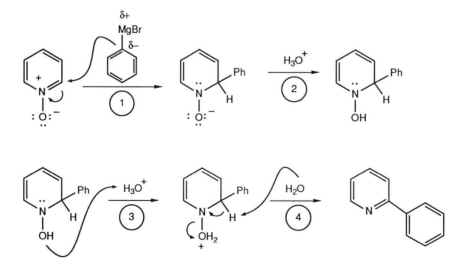

24.5. 2-Vinylpyridine reacts readily with nucleophiles, and Michael addition occurs as it does for an α,β-unsaturated ketone. The stabilized carbanion adds to the double bond that is conjugated with the C=N bond. Rearomatization occurs in step (2) with proton transfer from the solvent, regenerating the pyridine ring.

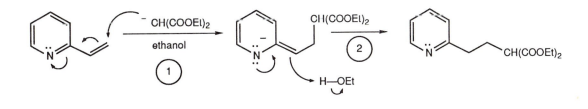

24.6. This reaction, shown on text page 1218, is a crossed aldol reaction. A carbanion is formed by deprotonation of the methyl group of 2-methylpyridine-N-oxide, and this nucleophile adds to the carbonyl group of the aldehyde. Proton transfer occurs in step (3), and a molecule of water is lost by an E1cb pathway to form the unsaturated product.

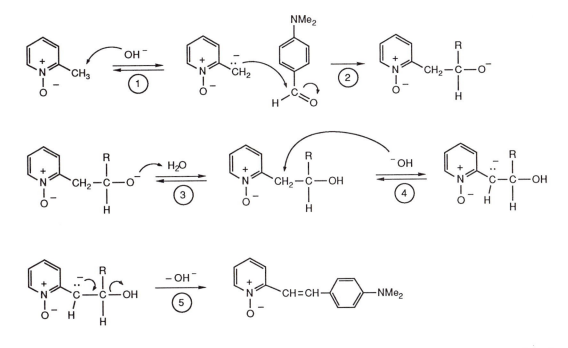

Resonance forms that account for the relatively high acidity of the methyl group in 2-methylpyridine-N-oxide are shown below:

24.7. The reaction shown in this exercise is a variant of the crossed Claisen condensation. The heterocycle is deprotonated by the strongly basic amide ion, and the resulting carbanion adds to the carbonyl group of the ester. Regeneration of the carbonyl group displaces ethoxide ion. In step (4), deprotonation occurs. Aqueous workup constitutes the fifth step, which regenerates the neutral product.

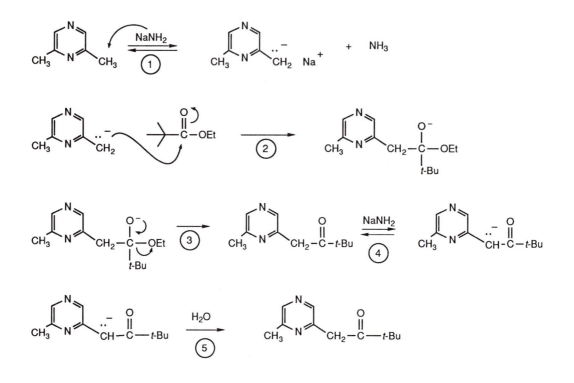

24.8. a. The combination of bromine and a silver(I) salt in sulfuric acid generates Br⁺, which participates in electrophilic substitution with quinoline. The non-heterocyclic ring is more reactive, so substitution occurs there.

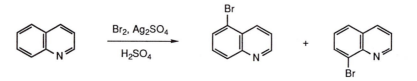

b. A chlorine atom adjacent to nitrogen in an aromatic heterocycle is readily substituted by a nucleophile. In this reaction, the ethoxy group replaces chlorine .

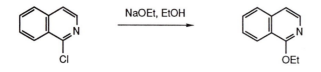

24.9. The Friedlander synthesis of the tetracyclic product shown in this exercise starts with an acid-catalyzed crossed aldol reaction between the enol form of the ketone and the aldehyde. This portion of the reaction creates an α,β-unsaturated ketone.

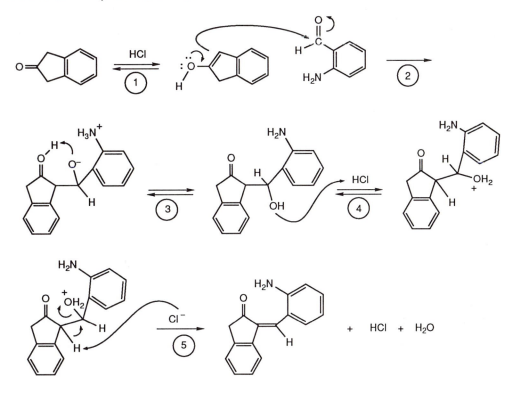

In the next stage of this transformation, the amine reacts with the ketone to form the imine bond, which creates heterocycle.

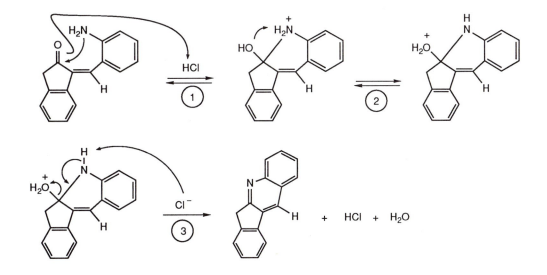

24.10. The five main steps in the pyrrole synthesis are given on text page 1225; the mechanism shown below has two extra steps just to show electron movement in more detail. The first step is acid-catalyzed addition of the amine to one of the carbonyl groups of the diketone. In step (2), proton transfer creates a good leaving group and in step (3), the imine bond is formed. Step (4) is addition of the imine nitrogen atom to the other carbonyl group. Step (5) creates another good leaving group, which leads to formation of a second double bond by elimination of water in step (6). Deprotonation in the last step produces the *N*-alkylpyrrole.

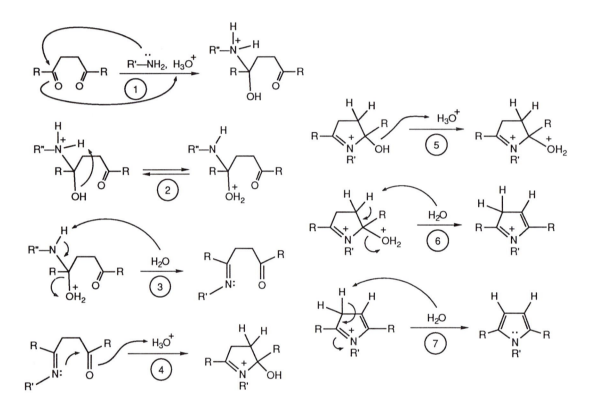

24.11. Resonance forms that we drawn for furan and thiophene are like those shown for pyrrole on text page 1224, which delocalizes an electron pair onto every atom of the ring.

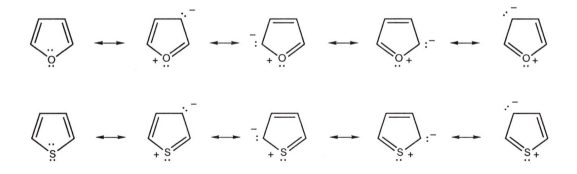

24.12. The final step in porphyrin synthesis is the cyclization of the linear tetracyclic material. The OH group is protonated in step 91), and a molecule of water dissociates in step (2) to produce a carbocation. Cyclization occurs in step (3), and deprotonation in step (4) completes the transformation.

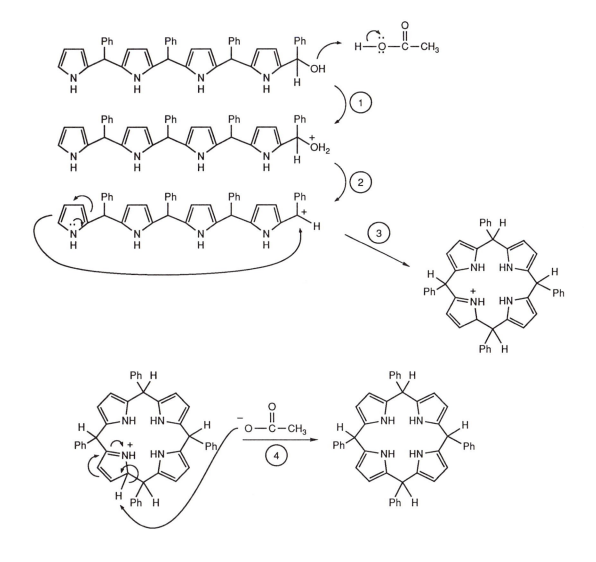

24.13. a. The first step in this sequence is Friedel-Crafts acylation, and substitution takes place at the 2-position. The second step is Wolff-Kishner reduction, which converts the acyl group to the hydrocarbon substituent.

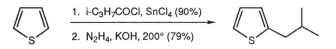

1. i-C$_3$H$_7$COCl, SnCl$_4$ (90%)

2. N$_2$H$_4$, KOH, 200° (79%)

24.13. (continued)

b. The Mannich reaction adds a dimethylaminomethyl group to the 5-position of 2-methylfuran.

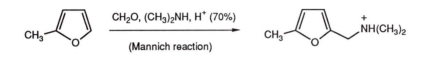

24.14. Recall that electrophilic aromatic substitution involves two steps (cf. text page 525). In the biosynthesis of tryptophan, the electrophilic carbon atom of the side chain precursor is interception by the π electrons of the heterocycle. Deprotonation of the cationic intermediate regenerates the aromatic π system.

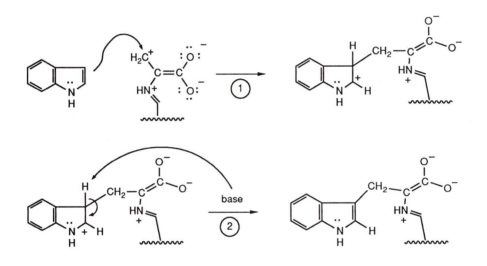

24.15. LCAO representations of the heterocycles oxazole, thiazole, and pyrazole are like the one shown for imidazole on text page 1235. In thiazole and oxazole, one unshared electron pair on S or O is part of the π system. The other electron pair on O or S is perpendicular to the π bonds. For pyrazole, the orbital representation of each atom is the same as the corresponding atom in imidazole; the difference is that the nitrogen atoms are adjacent in pyrazole.

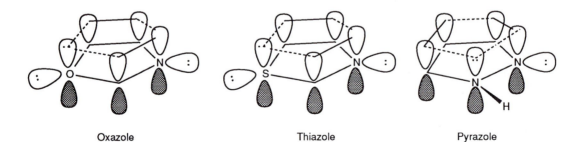

| Oxazole | Thiazole | Pyrazole |

24.16. In the reaction scheme for the synthesis of the given dimethylthiazole, the first step creates the C—S bond. The amine reacts with the ketone by nucleophilic addition to the carbon-oxygen double bond. An acid–base reaction in step (2) transfers a proton from the nitrogen to the oxygen atom, then electrons move from nitrogen toward sulfur, concomitant with the positive charge moving from S to N. This movement creates a protonated nitrogen atom, and the OH group serves as a base, forming water, which functions as a leaving group in step (5). Chloride ion, which was displaced in the first step, reacts as a base to regenerate the aromatic π system of the heterocycle.

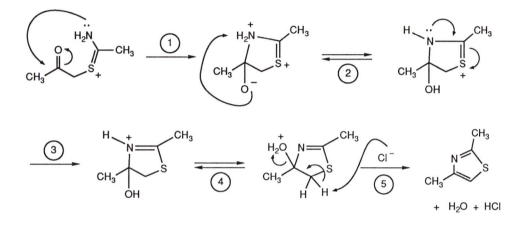

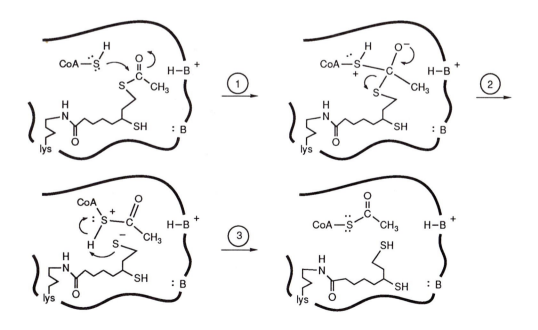

24.17. Like any transesterification procedure, the nucleophile (the sulfur atom of acetyl-CoA) adds to the carbonyl group of the thioester, forming a tetrahedral intermediate. Regeneration of the C=O bond in step (2) displaces thiolate ion, which subsequently removes the proton in step (3) from the positively charged sulfur atom of the acetyl-CoA molecule.

24.18. The mechanism for the conversion of α–ketoglutarate to succinyl-CoA is the same as that shown for decarboxylation and thioesterification of pyruvate. In fact, we can reproduce the figures for that scheme, substituting $^-OOCCH_2CH_2-$ for the CH_3- group of pyruvate. In the schemes below, R is used in place of the carboxyethyl group. After step (6), transesterification occurs by the mechanism shown in the solution to exercise 24.17.

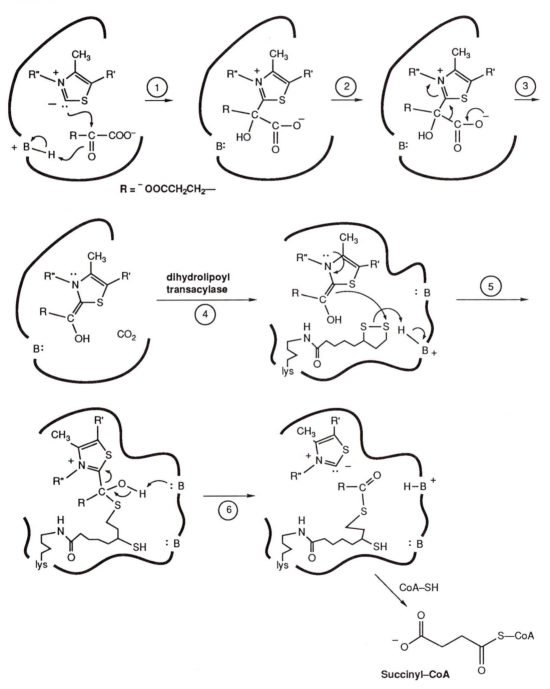

$R = \, ^-OOCCH_2CH_2-$

24.19. The mechanism of the benzoin condensation begins with formation of the thiazolium carbanion, which adds in step (2) to the carbon-oxygen double bond of benzaldehyde. Intermediate **A** is an unusual species because the benzylic hydrogen atom (circled) is relatively acidic so is removed in step (3) by the neighboring anionic oxygen atom. The resulting carbanion adds to the carbon-oxygen atom of another molecule of benzaldehyde in step (4), and regeneration of the carbonyl group occurs by an intramolecular acid-base reaction that displaces the thiazolium anion as a leaving group.

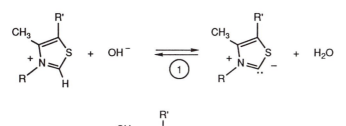

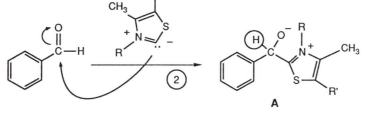

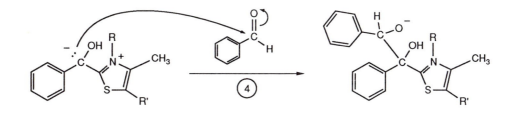

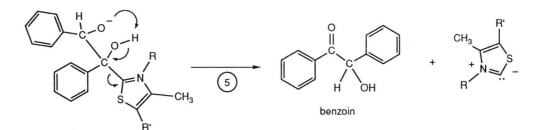

24.20. The mechanism of the cyanide-catalyzed benzoin condensation begins with formation of the cyanohydrin anion, then follows the same series of step that were shown in the solution to exercise 24.19. Cyanide, like the thiazolium ion, has the feature that it makes the benzylic hydrogen atom of intermediate **B** relatively acidic.

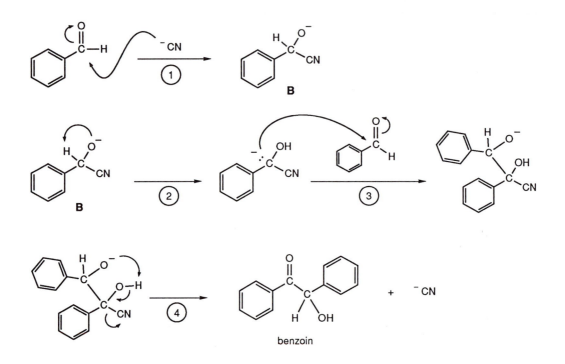

24.21. Interpret the name according to the directions given in Chapter 2.

a. 2-chloro-5-methylpyridine

b. (*S*)-3-methyltetrahydrothiophene

c. 3-thiacyclobutane-1-one

24.21. (continued)

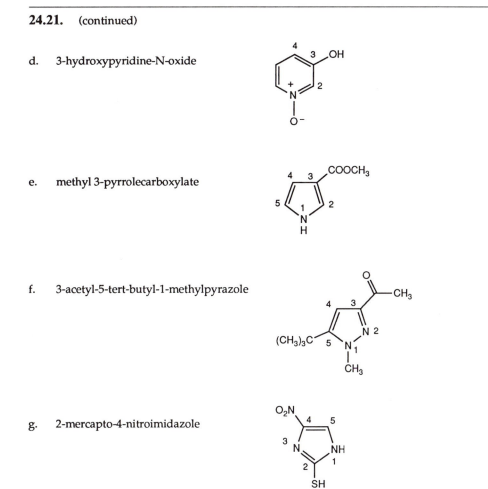

d.   3-hydroxypyridine-N-oxide

e.   methyl 3-pyrrolecarboxylate

f.   3-acetyl-5-tert-butyl-1-methylpyrazole

g.   2-mercapto-4-nitroimidazole

24.22.   First, decide what the parent heterocycle is. Then number the atoms of the ring (the heteroatom is normally assigned the #1 position). Indicate substituents with the appropriate number.

a.   **2-Amino-6-chloropyridine**

b.   **2-Methoxyoxazole**

24.22. (continued)

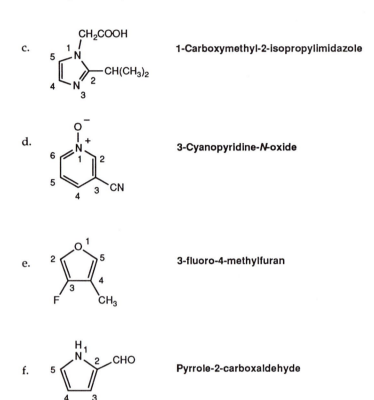

c. 1-Carboxymethyl-2-isopropylimidazole

d. 3-Cyanopyridine-*N*-oxide

e. 3-fluoro-4-methylfuran

f. Pyrrole-2-carboxaldehyde

24.23. The nitrogen heterocycles present in these drug molecules are identified after removing all substituents.

Imidazole

miconazole
antifungal

Pyridine

propiram
analgesic

Piperidine

Pyrazole

Isolan®
insecticide

Pyrimidine

metachloridine
antimalarial

24.24.   2-Pyridone is a lactam, which has a carbonyl group (hence the –one suffix).

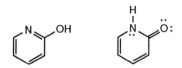

When 2-hydroxypyridine is deprotonated, the negative charge is delocalized over several atoms, most notably the nitrogen and oxygen atoms.

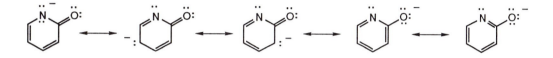

When the anion of 2-pyridone reacts with methyl iodide, alkylation takes place either at oxygen or nitrogen.

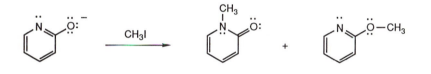

24.25.   Resonance structures of pyridine-*N*-oxide place a negative charge at C2 and C4, which explains why electrophiles react at those positions.

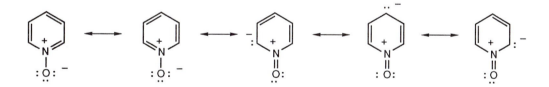

When we draw resonance structures for the intermediates formed after reaction with an electrophile at either the 2-, 3-, or 4-position, we see that only when reaction occurs at C2 and C4 are there contributors that have only a single charge. As to why reaction occurs at C4 rather than C2, steric effects between the incoming electrophile and the oxygen atom attached to nitrogen probably play a role.

reaction at C2

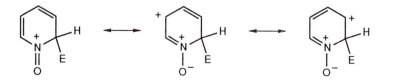

24.25.   (continued)

reaction at C3

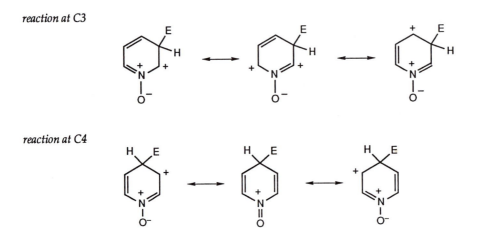

reaction at C4

24.26.   Pyridine–N–oxide is alkylated at oxygen by an S_N2 process. The nucleophile, cyanide ion in this exercise, subsequently adds to the π bond(s) at C2 or C4, and electrons flow toward the positively charged nitrogen atom, the electrophilic center. The *N*-methoxy group undergoes β–elimination in step (3) to regenerate the aromatic π system of pyridine.

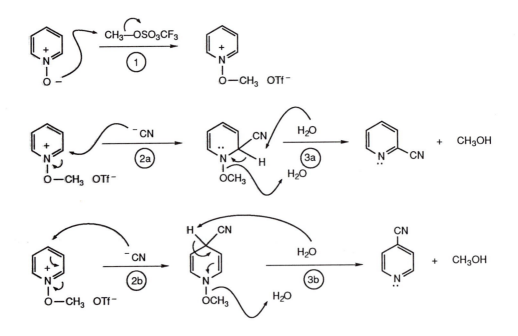

24.27. To assign chemical shifts to the resonances of the aromatic protons in these heterocycles, consider resonance forms as well as splitting patterns. A proton attached to a carbon atom with a positive charge in a resonance form tends to be deshielded, and its NMR signal appears farther downfield. A proton attached to a carbon atom with a negative charge in a resonance form tends to be more shielded, so its peaks appears farther upfield. Splitting patterns reflect the proximity of different protons to each other, although the magnitudes of coupling constants are often small in heterocycles. The integrated intensity is also used to make assignments if the molecule is symmetric.

a. **Pyridine**

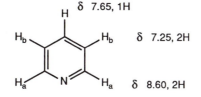

b. *N*-methylimidazole

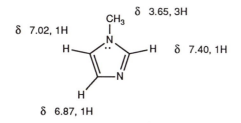

c. *N*-methylpyrrole

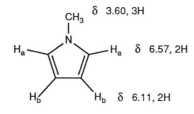

24.28. 3-Methylthiophene undergoes electrophilic substitution reactions at C2 or C5. The methyl group hinders C2 somewhat, so reactions occur preferentially at C5. Hydrogenation reactions are normally poisoned by the presence of a sulfur atom in organic compounds, and aromatic rings are not susceptible to catalytic hydrogenation.

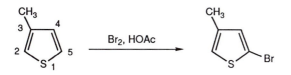

24.28. (continued)

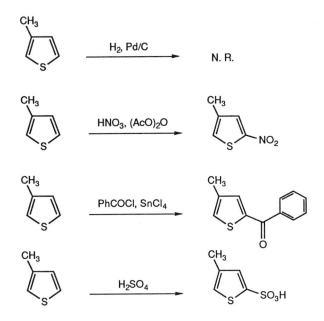

24.29. The mechanism for the synthesis of 3,5-dimethylisoxazole is similar to that for the synthesis of 3,5-dimethylpyrazole, which is shown in the solution to exercise 18.27. In the first stages of the reaction, the nitrogen atom of hydroxylamine adds to one of the ketone carbonyl groups, forming an oxime. Then the hydroxy portion adds to the other carbonyl group. Under acidic conditions, two elimination steps generate the double bonds of the heterocycle. The last step is an acid-base reaction that forms the free base.

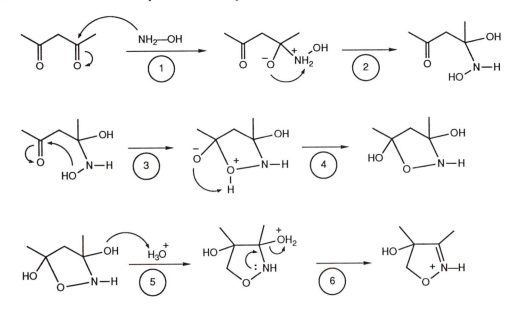

24.29. (continued)

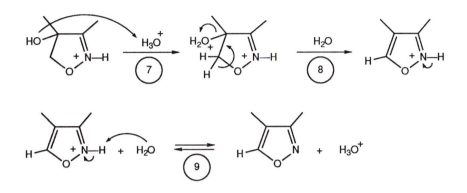

24.30. a. The 3-position of the indole ring is the most activated, so the electrophile, Br⁺, substitutes there.

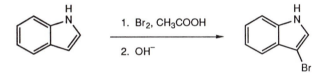

b. Oxazole is deprotonated at the carbon atom between the nitrogen and oxygen atoms. The resulting carbanion adds to the carbonyl group of pentanal, yielding a secondary alcohol after aqueous acid workup.

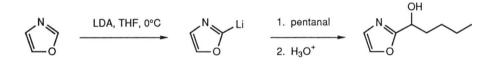

c. The 3-position of the pyridine ring is least activated, but this position is activated by the methoxy group. Bromination therefore occurs at C3.

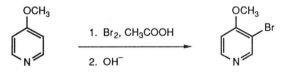

24.30.   (continued)

d. An alkyllithium reagent adds to the carbon-nitrogen double bond of quinoline. In refluxing toluene, elimination of LiH occurs to restore the aromatic system. Acid workup destroys the LiH produced, and aqueous base neutralizes the acid used in step 2.

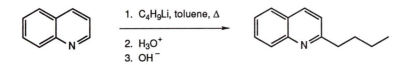

e. The protons of a methyl group adjacent to the nitrogen atom in an aromatic heterocycle are slightly acidic, so a carbanion derivative is formed in step 1. Addition of this carbanion to the aldehyde carbonyl group followed by hydrolysis yields the benzylic alcohol as the product. Under acidic conditions, elimination occurs to produce the alkene. The aqueous base in the last step neutralizes the acid used in step 3.

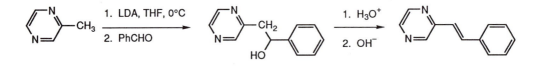

f. The nitrogen atom of a pyridine derivative is nucleophilic, so it reacts with an alkyl halide by an S_N2 pathway, displacing the halogen atom.

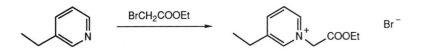

g. The carbon-nitrogen double bond of the isoquinoline ring is susceptible to addition by an organometallic reagent. Quenching the adduct with water yields the reduced, alkylated product.

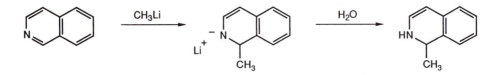

24.30.   (continued)

h. An imidazole ring is alkylated at one of its nitrogen atoms when its conjugate base is treated with an alkyl halide. Subsequent treatment with LDA and an alkyl halide leads to alkylation at the 2-position. Hydrolysis of the methoxymethyl group (this is equivalent to an acetal functional group) yields the product, 2-methylimidazole. The aqueous base in the last step neutralizes the acid used in step 5.

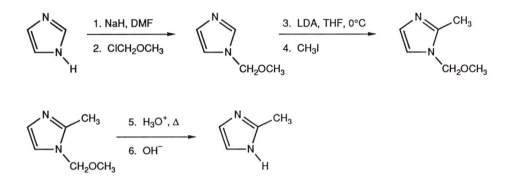

24.31.   If we look at the structures of the compounds involved in the carbon-carbon bond forming processes of the pentose phosphate pathway, we see that carbon atoms C3, C4, and C5 of D-xylulose-5-phosphate are found in D-glyceraldehyde-3-phosphate, and C1 and C2 are found in S7P after reaction of the enol intermediate with D-ribose-5-phosphate. Carbon atoms 3 through 7 of S7P derive from carbon atoms 1 through 5, respectively, of D-ribose-5-phosphate

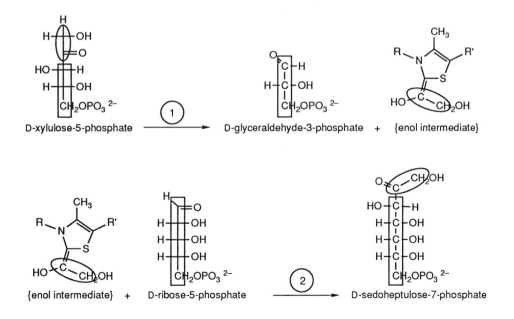

24.31.   (continued)

a. D-Xylulose-5-phosphate has the labeled atom; C1 of D-xylulose-5-phosphate becomes C1 of S7P.

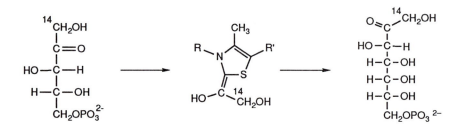

b. D-Ribose-5-phosphate has the labeled atom; C1 of D-ribose-5-phosphate becomes C3 of S7P.

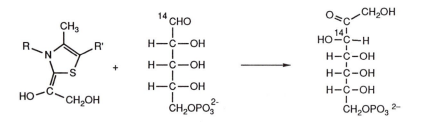

c. D-Xylulose-5-phosphate has the labeled atom; C3 of D-xylulose-5-phosphate becomes C1 of D-glyceraldehyde-3-phosphate.

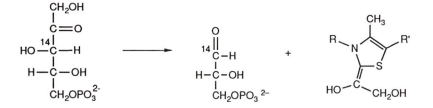

d. D-Ribose-5-phosphate has the labeled atom; C3 of D-ribose-5-phosphate becomes C5 of S7P.

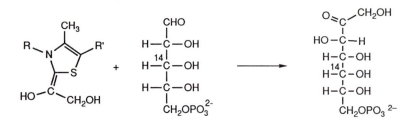

24.32.   Formation of an oxazole ring follows a mechanism that is similar to that involved in the formation of thiazole, as shown in the solution to exercise 24.16. The oxygen atom of formamide displaces chloride ion in an S_N2 process, then the amino group condenses with the ketone carbonyl group to form the cyclic precursor. Dehydration produces the oxazole.

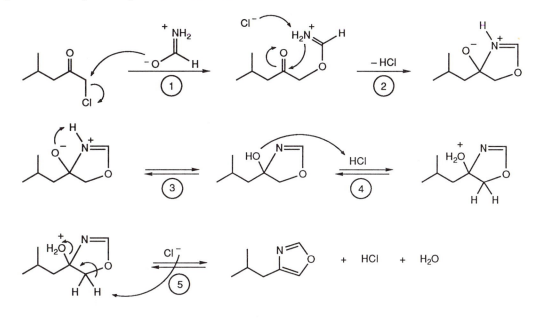

24.33.   The benzimidazole ring system uses the same orbitals as those used to construct an imidazole ring. The benzene portion has additional p orbitals that overlap with the π bonds of the heterocyclic ring.

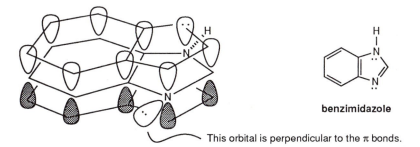

benzimidazole

This orbital is perpendicular to the π bonds.

b. In acid solution, the nitrogen atom with the unshared electron pair reacts with a proton. This step is reversible, but the other proton can be removed by a molecule of water, too, forming the tautomer.

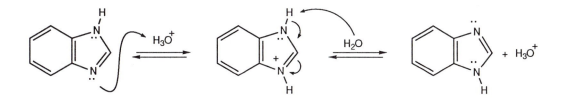

24.33.   (continued)

c.   A strong base produces the anion of benzimidazole. Reaction with an alkyl halide yields the alkylated product by an S$_N$2 pathway.

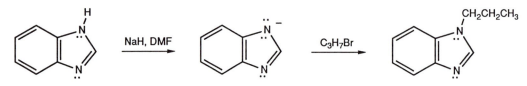

d.   In the first step of benzimidazole synthesis, an amino group of *o*-phenylenediamine adds to the carbonyl group of the carboxylic acid. Formation of a hydrogen bond between the carboxylic acid group and the other amino group of *o*-phenylenediamine circumvents the acid-base reaction that normally occurs between a carboxylic acid and an amine. In step (2), a proton is transferred to form a *gem*-diol.

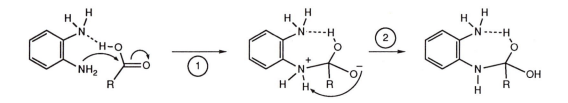

In the next step, one of the OH groups of the *gem*-diol is protonated to form a good leaving group, and this activation provides a pathway to regenerate the carbon-oxygen double bond in step (4).

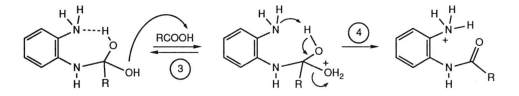

The carbonyl group then removes a proton from the ammonium ion, and the resulting amino group adds to the carbon-oxygen double bond.

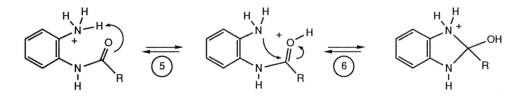

24.33.   (continued)

Proton transfer in step (7) creates another good leaving group. Elimination of water forms the heterocycle.

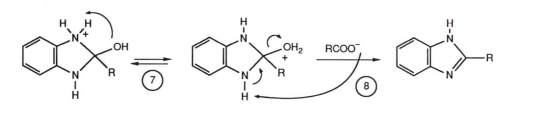

24.34.   Follow the procedure outlined in the solution to exercise 19.27.

a. All three compounds have an *ortho*-disubstituted benzene ring, so the carbon-carbon and carbon-hydrogen stretching and bending vibrations for all of the compounds should be similar. The starting compound, indene, has an additional double bond, for which the C=C stretching mode should be apparent, although it may be somewhat obscured by the vibrations of the benzene ring, which occur in the same region of the spectrum. The dialdehyde will have two very strong C=O stretching vibrations (their absorptions may be observed as a very strong but broadened peak because they are close together in frequency). The aldehyde C–H stretching vibrations should also be observable. When the dialdehyde is converted to the amine, the bands assigned to the C=O stretching vibrations will disappear, and a weak to medium band for the N–H stretching vibration will be observed.

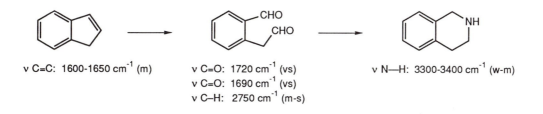

ν C=C: 1600-1650 cm^{-1} (m)

ν C=O: 1720 cm^{-1} (vs)
ν C=O: 1690 cm^{-1} (vs)
ν C–H: 2750 cm^{-1} (m-s)

ν N—H: 3300-3400 cm^{-1} (w-m)

b. All four compounds have an *ortho*-disubstituted benzene ring, so the carbon-carbon and carbon-hydrogen stretching and bending vibrations for all of the compounds should be similar. The first compound is the same as in part a. The ketone has a very strong C=O stretching vibration. When the ketone is converted to the amide, the band assigned to the C=O stretching vibration shifts to lower frequency (a second band for the N–H bending vibration will probably appear), and a band for the N–H stretching vibration will appear at about 3300 cm^{-1}. When the amide is converted to the amine, the band for the carbonyl stretching vibration disappears, but the band for the N–H stretching mode will be still be observed. Its frequency will probably shift slightly to a higher value, and its intensity will likely decrease some.

 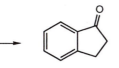 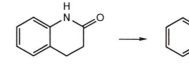

ν C=C: 1600-1650 cm^{-1} (m)   ν C=O: 1690 cm^{-1} (vs)

ν C=O: 1680 cm^{-1} (vs)
ν N–H: 3300 cm^{-1} (m-s)

ν N—H: 3300 cm^{-1} (m)

24.35. Follow the procedure outlined in the solution to exercise 19.28.

a. In this sequence of reactions, the most notable changes occur in the non-aromatic portions of the spectra. In the first step, the alkene resonances between δ 5 and 6 disappear, and new resonances at δ 10 (the aldehyde protons) appear. In the second reaction, the aldehyde resonances disappear, and resonances for the new aliphatic hydrocarbon and the NH groups are observed.

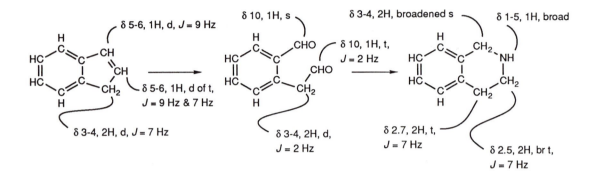

b. In this sequence, the most notable changes occur in the non-aromatic portions of the spectra. In the first step, the alkene resonances between δ 5 and 6 disappear, and a new resonance at δ 2.5 is observed for the methylene group adjacent to the ketone carbonyl group. In the second reaction, a new resonance appears for the amide NH proton. In the third reaction, the new methylene group contributes peaks in the aliphatic region.

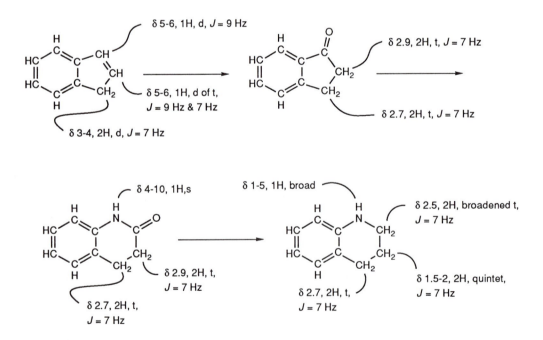

24.36. a. Ozonolysis of the double bond of indene, followed by reductive workup, yields the dialdehyde, which is reduced to the corresponding diol with sodium borohydride. Formation of the dimesylate derivative followed by two successive S$_N$2 reactions with the nitrogen atom of an ammonia molecule affords the six-membered ring 2° amine.

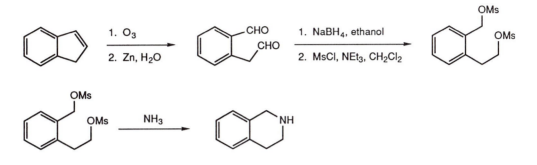

b. Hydration of the double bond of indene occurs by Markovnikov addition, and the benzylic alcohol so formed is oxidized to the ketone. Beckmann rearrangement of the oxime derivative of the ketone leads to ring expansion and formation of the amide. Reduction of the amide forms the corresponding amine.

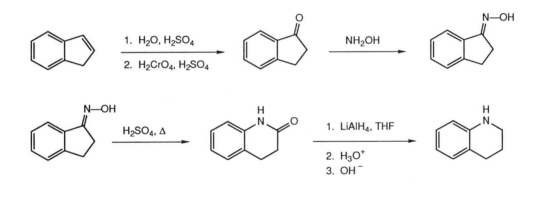

24.37.

a. This compound is made by Michael addition of the heterocycle to an α,β-unsaturated ester.

b. The product is made by a substitution reaction between the ion derived from piperidine and 2-chloropyridine.

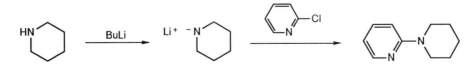

24.37. (continued)

c. A logical disconnection in a retrosynthetic analysis is at the double bond, which is made by a Horner-Emmons reaction.

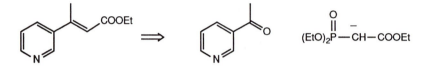

The needed ketone is made from the six-carbon nicotinic acid via reaction between a cuprate reagent and the corresponding acid chloride.

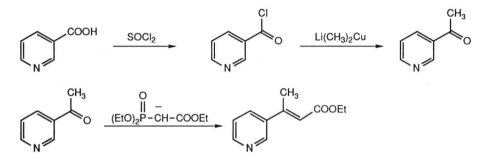

24.38. The NMR spectrum of **A** reveals the presence of an aldehyde group and 3 protons in the aromatic region. This compound cannot be a benzene derivative because it contains only five carbon atoms, so it is most likely an aromatic heterocycle. The formula suggests that compound **A** is a furan, so two possible structures are **I** and **II**, below.

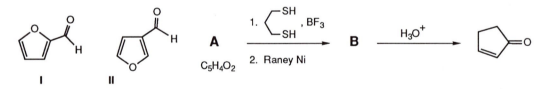

If structure **I** is correct, then the first series of reactions produces 2-methylfuran via the thioacetal. Hydrolysis produces the keto aldehyde, then a crossed aldol reaction (under acidic conditions) leads to formation of the observed cyclopent-2-ene-1-one.

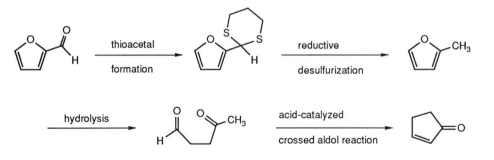

24.39. In the reaction between pyrazole and formaldehyde, nucleophilic addition of the nitrogen atom of pyrazole to the carbonyl group is followed by proton transfer to form the hydroxymethyl derivative.

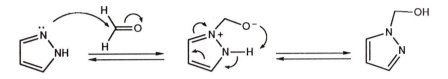

When 1-hydroxymethylpyrazole reacts with an amine, an equilibrium is established that regenerates formaldehyde and pyrazole. Formaldehyde reacts with the amine to form an imine; then pyrazole adds to the imine carbon-nitrogen double bond to produce the aminomethyl derivative.

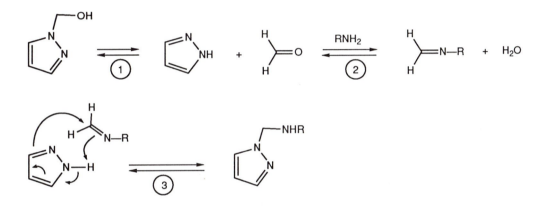

24.40. The aldol reaction between two allysine chains follows the same mechanism as any aldol condensation: the enolate derivative of one aldehyde chain adds to the carbonyl group of a second chain.

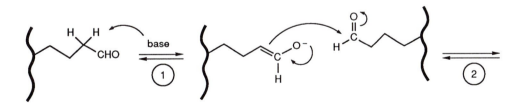

24.40. (continued)

After protonation to form the hydroxy compound, an E1cb pathway leads to formation of the unsaturated aldehyde.

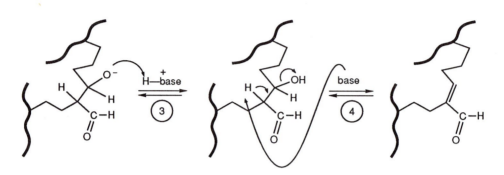

In the second stage of the crosslinking process, the histidine ring adds in a Michael fashion to the unsaturated aldehyde.

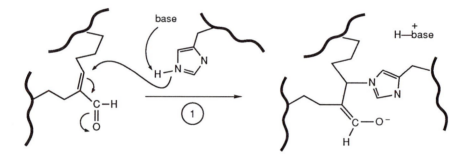

Proton transfer accompanies regeneration of the carbonyl group, and the three chains are now linked.

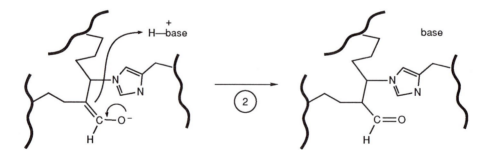

AMINO ACIDS, PEPTIDES, AND PROTEINS

25.1. Isoleucine and threonine each have two stereogenic centers. Assignments of absolute configuration of each stereogenic carbon atom are shown below.

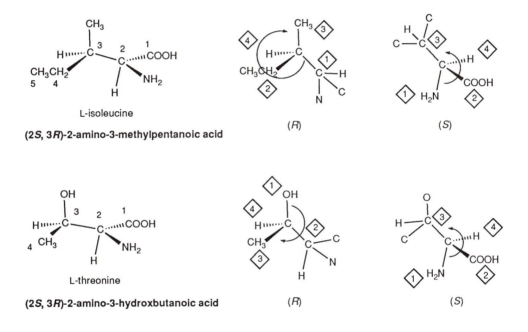

25.2. For an acid-base reaction, the percentage of the conjugate base, A⁻, of an acid, HA, is calculated with use of the equation shown in the box below, which is derived from equation 25.1.

$$pH = pK_a + \log \frac{x}{1-x} \longrightarrow \boxed{\frac{x}{1-x} = 10^{(pH-pK_a)}}$$

x = percentage of the conjugate base of the acidic group HA

If the pH and pK_a differ by ≥ 2, then a calculation shows that either the acid or conjugate base form of a compound constitutes >99% of the mixture, so we will assume that the appropriate species is "100%".

25.2. (continued)

a. Histidine at pH 7. The pK_a values for the three ionizable groups are given with the structure. Notice that each group is shown in its acidic form. The numeric difference between pH and pK_a values for the α-carboxy and α-amino groups is greater than 2.0, so the acid group is in the $-COO^-$ form (pH > pK_a) and the amino group is in the $-NH_3^+$ form (pH < pK_a). Calculating the percentage of the protonated side chain, we find that the imidazole ring exists in the deprotonated form to the extent of 91%.

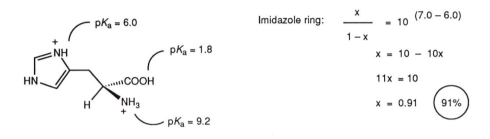

Imidazole ring:
$$\frac{x}{1-x} = 10^{(7.0-6.0)}$$

$$x = 10 - 10x$$

$$11x = 10$$

$$x = 0.91 \quad \boxed{91\%}$$

b. Aspartic acid at pH 5. The pK_a values for the three ionizable groups are given, with the structure. The difference between pH and pK_a values for the α-carboxy and α-amino groups is greater than 2.0, so the acid group is in the $-COO^-$ form (pH > pK_a) and the amino group is in the $-NH_3^+$ form (pH < pK_a). Calculating the percentage of the protonated side chain, we find that the carboxy group exists in the deprotonated form to the extent of 93%.

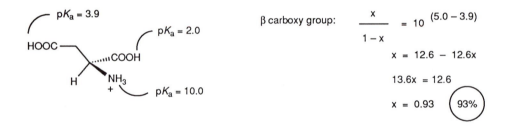

β carboxy group:
$$\frac{x}{1-x} = 10^{(5.0-3.9)}$$

$$x = 12.6 - 12.6x$$

$$13.6x = 12.6$$

$$x = 0.93 \quad \boxed{93\%}$$

c. Tyrosine at pH 8. The pK_a values for the three ionizable groups are given, with the structure. The difference between pH and pK_a values for the α-carboxy group and for the phenol OH group is greater than 2.0, so the acid group is in the $-COO^-$ form (pH > pK_a) and the phenol group is protonated ($-OH$). Calculating the percentage of the α-amino group, we calculate that it exists in the deprotonated form to the extent of 7.4% .

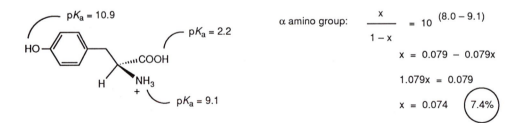

α amino group:
$$\frac{x}{1-x} = 10^{(8.0-9.1)}$$

$$x = 0.079 - 0.079x$$

$$1.079x = 0.079$$

$$x = 0.074 \quad \boxed{7.4\%}$$

25.3. The electron pair on the nitrogen atom is part of the 10 π electrons that make the indole ring aromatic This electron pair, therefore, is not able to engage in hydrogen bonding as an acceptor, which would disrupt the aromaticity of the heterocycle.

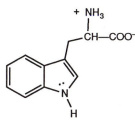

25.4. Classification of amino acids by type reflects the nature of the groups that are present. An OH group makes a side chain polar, and an amino group is basic. An aldehyde group is slightly polar, but most of the side chain of allysine is hydrocarbon in nature, so allysine is likely nonpolar.

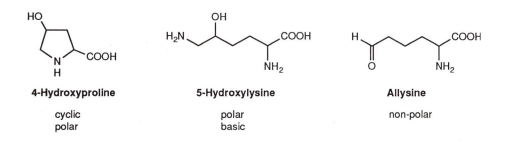

4-Hydroxyproline	**5-Hydroxylysine**	**Allysine**
cyclic	polar	non-polar
polar	basic	

25.5. The aldehyde needed to make an amino acid by a Strecker synthesis is determined by considering first what nitrile is needed. The carbon atom the cyano group is the aldehyde carbon atom in a retrosynthesis. The aldehyde needed in part b may be difficult to work with because it is likely to be very water soluble. It may also undergo self-condensation because both carbonyl and alcohol groups are present. The aldehyde needed to make the other two amino acids are expected to be stable and to present no problem.

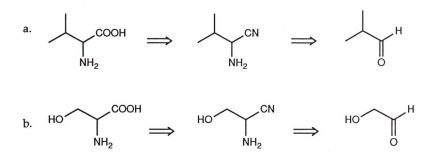

25.5.   (continued)

c.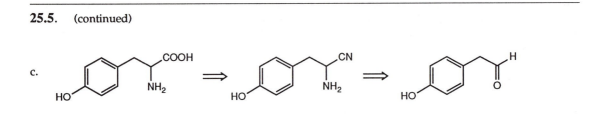

25.6.   Each of the amino acids shown in this exercise is made from the corresponding α-bromo acid by reaction with potassium phthalimide followed by hydrazinolysis of the phthalimido group (text page 1273).

a.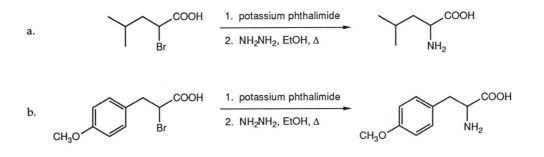

b.

25.7.   *tert*-Butyl and benzyl esters are most often prepared by reaction of the corresponding acid chloride with *tert*-butyl and benzyl alcohol, respectively. This route avoids the use of strong acid that can generate a carbocation intermediate from the alcohol.

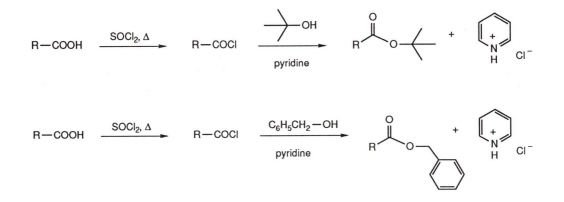

25.8. A *tert*-butyl ester is deprotected (converted to the carboxylic acid) by reaction with a strong acid such as trifluoroacetic acid. Protonation of the ester carbonyl group in step (1) yields a cation that dissociates in step (2) to form the carboxylic acid and *tert*-butyl carbocation. The *tert*-butyl carbocation is deprotonated to form isobutylene, which also regenerates the acid catalyst.

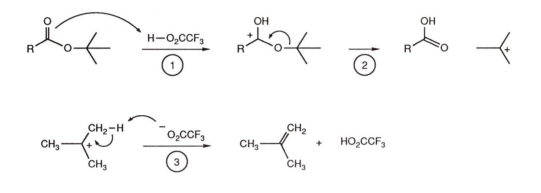

25.9. Each of the amino acids shown in this exercise is made by the sequence outlined below. The starting material in each case is made from the corresponding acid chloride and the chiral oxazolidone.

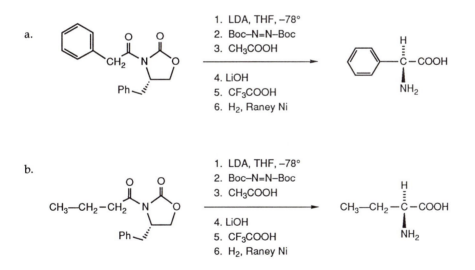

a.

1. LDA, THF, −78°
2. Boc−N=N−Boc
3. CH₃COOH

4. LiOH
5. CF₃COOH
6. H₂, Raney Ni

b.

1. LDA, THF, −78°
2. Boc−N=N−Boc
3. CH₃COOH

4. LiOH
5. CF₃COOH
6. H₂, Raney Ni

25.10. Each of these amino acids shown in this exercise is made by alkylating the chiral glycine enolate described on text page 1281. The order of alkylation in (b) and (c) inverts the configuration of the α carbon atom.

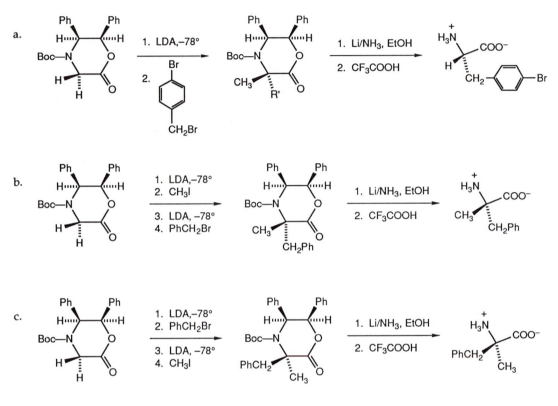

25.11. Asymmetric hydrogenation employing the (*S,S*)-DuPhos catalyst leads to formation of the product with the (*S*) configuration at the α carbon atom of an enamide or vinyl ester.

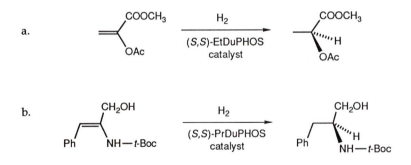

25.12. The synthesis of the tripeptide shown in this exercise is carried out according to the iterative procedure described in the text. First, tyrosine is protected as its methyl ester, and the amino group as its Boc amide. Then, the phenol OH group is acetylated by treating it with acetic anhydride, and the amino group is deprotected by reaction with trifluoroacetic acid. If the amino group is not protected before reaction with acetic anhydride, then acetylation will occur at the amino group, too. Phenylalanine and isoleucine are simply converted to their Boc derivatives before peptide bond formation is carried out.

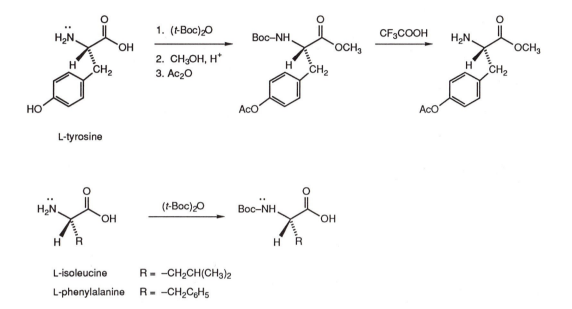

In the first stage of peptide synthesis, isoleucine and tyrosine are coupled with use of DCC.

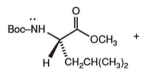

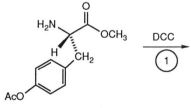

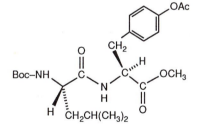

N-Boc-L-isoleucine   *O*-acetyl-L-tyrosine methyl ester   *N*-Boc-ile–(AcO)tyr methyl ester

25.12. (continued)

The amino terminus is deprotected in step (2) by reaction with CF_3COOH.

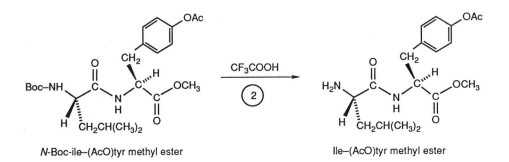

N-Boc-ile–(AcO)tyr methyl ester   Ile–(AcO)tyr methyl ester

Next, phenylalanine is coupled to the new *N*-terminus of the dipeptide, again with use of DCC. This step gives the tripeptide in its protected form.

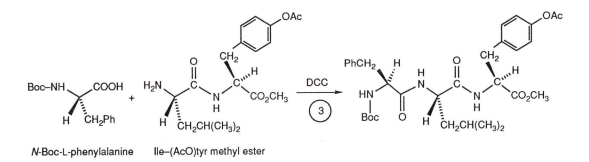

N-Boc-L-phenylalanine   Ile–(AcO)tyr methyl ester

If the free tripeptide is required, the Boc group is removed by treatment with CF_3COOH, and the methyl ester and acetate group are hydrolyzed by treatment with weak aqueous base.

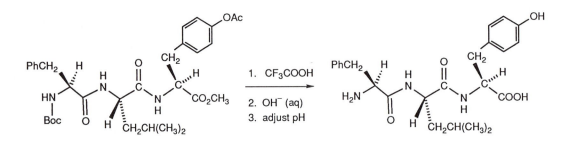

25.13. Rearrangement during the Edman reaction to generate the *N* phenylthiohydantoin product requires that the ring is opened to form a thiocarbonyl group. This process takes place like any ester (or lactone) hydrolysis reaction, which proceeds via protonation of the carbonyl group and formation of a geminal dihydroxy species.

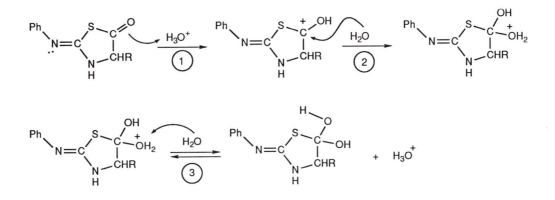

Ring opening subsequently occurs with regeneration of the carbonyl group of the carboxylic acid, which also generates the thioamide group.

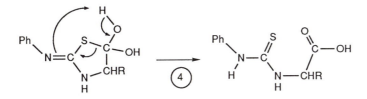

Ring formation occurs next by addition of the nitrogen atom of the thioamide group to the carbonyl group of the carboxylic acid. Proton transfer followed by loss of water from the *gem*-diol yields the product.

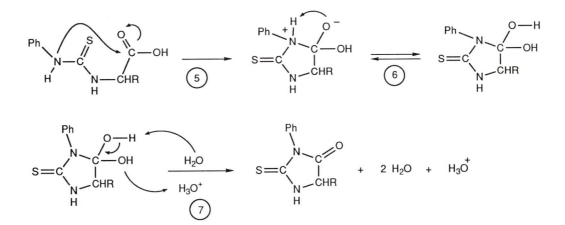

25.14. In the Edman degradation, the structure of each amino acid that is sequentially cleaved is identified by looking at the R– group of the heterocycle produced during each round. The R group is the same as the side chain of the amino acid that was cleaved. The assignments are made as shown below:

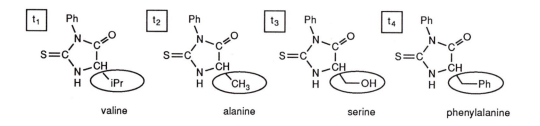

The starting polypeptide therefore has the sequence: H_2N—val—ala—ser—phe— • • • •

25.15.   a. The mechanism for hydrolysis of an imine was presented on text pages 874-875. Water adds to the C=N bond in step (1), then an acid-base reaction occurs in step (2) to transfer a proton from oxygen to nitrogen.

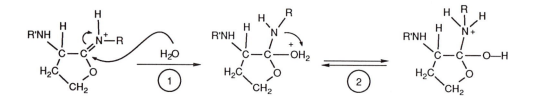

The third step regenerates the carbonyl group, which displaces the amine. In this step, the amino group is the new *N*-terminus of a polypeptide that was originally on the *C*-terminus side of the methionine residue.

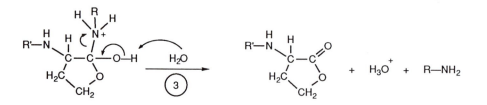

25.15. (continued)

b. The mechanism for the acid-catalyzed hydrolysis of a lactone is the same as that for hydrolysis of an ester, which was presented in the solution to exercise 19.12. Water adds to the carbonyl group in step (1), and after two acid-base reactions, the carbonyl group is regenerated in step (4), which leads to opening of the ring.

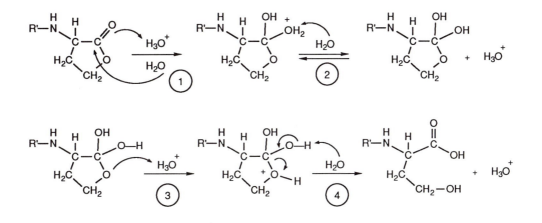

25.16. Given the structures of polypeptide fragments that have been formed by cleaving a protein, we deduce the sequence of the polypeptide by aligning identical portions. Recall that CNBr cleaves on the carboxy side of a methionine residue and produces a homoserine residue (hse).

$$H_2N-glu-ser-ile-COOH$$

$$H_2N-ala-gly-pro-arg-glu-ser-ile-COOH$$

$$H_2N-ala-val-phe-ala-met-ala-gly-pro-arg-COOH$$

$$H_2N-val-ala-ile-gly-lys-ala-val-phe-ala-hse-COOH$$

$$H_2N-val-ala-ile-gly-lys-COOH$$

$$H_2N-val-ala-ile-gly-lys-ala-val-phe-ala-met-ala-gly-pro-arg-glu-ser-ile-COOH$$

25.17. In an antiparallel β-sheet, the hydrogen bonds between strands take place as shown below.

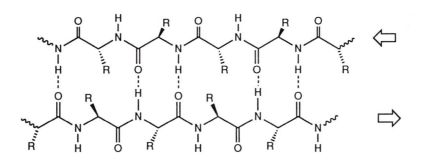

25.18. Several of the amino acids have the general structure shown, in which a group "X" is separated from the *alpha* carbon atom by a single methylene group. The identity of the substituent X is shown below.

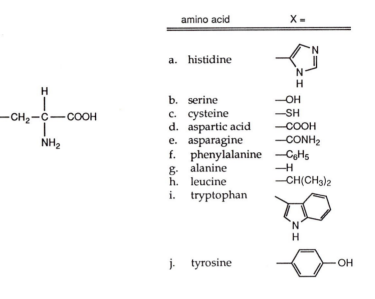

	amino acid	X =
a.	histidine	(imidazole ring)
b.	serine	—OH
c.	cysteine	—SH
d.	aspartic acid	—COOH
e.	asparagine	—CONH₂
f.	phenylalanine	—C₆H₅
g.	alanine	—H
h.	leucine	—CH(CH₃)₂
i.	tryptophan	(indole ring)
j.	tyrosine	(phenol ring)

25.19. The structure of an amino acid at a particular pH value reflects the predominant form of each ionizable group. Use the procedure outlined in the solution to exercise 25.2 to calculate whether each group (—COOH, —NH₂, —side chain) is in its protonated (or deprotonated) form to the extent of >50%. If the percentage is greater than half, render the group in that predominate form.

a. Tyrosine. At pH 1, all the groups are protonated. At pH 7, the zwitterion form predominates. At pH 11, the side chain phenol group is deprotonated, and the amino group is in its free-base form.

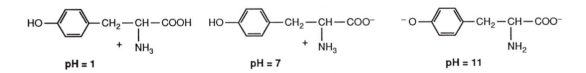

b. Histidine. At pH 1, all the groups are protonated, which means that the imidazole group bears a positive charge. At pH 7, the zwitterion form predominates, and the imidazole group is neutral (at pH 6 it exists protonated to the extent of 50%). At pH 11, the amino group is in its free-base form.

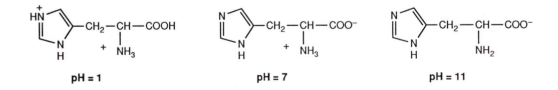

25.19. (continued)

c. **Serine.** At pH 1, all the groups are protonated. At pH 7, the zwitterion form predominates. At pH 11, the amino group is in its free-base form.

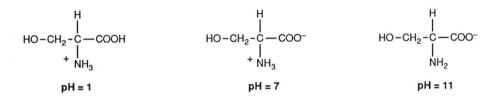

d. **Glutamic acid.** At pH 1, all the groups are protonated. At pH 7, the zwitterion form predominates, and the γ-carboxy group is in its deprotonated form. At pH 11, the amino group is in its free-base form.

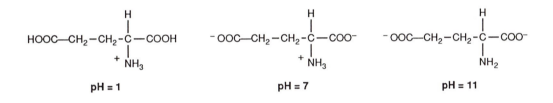

25.20. The peptide is drawn in the sequence given, then the ionization state of each group is evaluated by the procedure outlined in the solution to exercise 25.2. At pH 7, an acid group exists in its deprotonated form, an amino group exists in its protonated form, and a side chain groups exists according to the specific pK_a value of its ionizable group, if one is present.

a. **ala-glu-val:** the terminal amino group will be protonated and the carboxy terminus will be deprotonated. The side chain of glu is deprotonated.

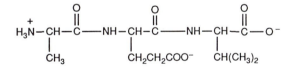

b. **phe-tyr-lys:** the lysine side chain is protonated; the other two side chains are neutral.

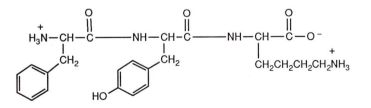

25.20. (continued)

c. **leu-his–asn–ser:** no side chain exists in its protonated form at pH 7.

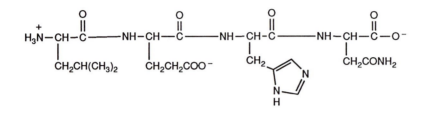

25.21. a. The valine residue at the *N*-terminus reacts with DNFB at its amino group. Hydrolysis cleaves the amide groups of the polypeptide, liberating the individual amino acids.

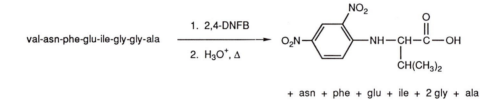

+ asn + phe + glu + ile + 2 gly + ala

b. The alanine residue at the *N*-terminus reacts with DNFB. Reaction of the hexapeptide with cyanogen bromide cleaves the peptide on the carboxyl side of the methionine residue, creating the homoserine lactone.

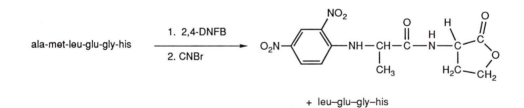

+ leu–glu–gly–his

25.22. When the pentapeptide H_2N—ile-ala-phe-lys-ser—COOH is treated with 2,4-DNFB, reaction occurs between any free amino group and the reagent. The structure of the starting peptide is shown below with its reactive amino groups circled.

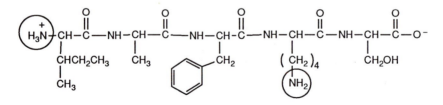

25.22. (continued)

The structure of the 2,4-DNFB derivative of the pentapeptide is illustrated below:

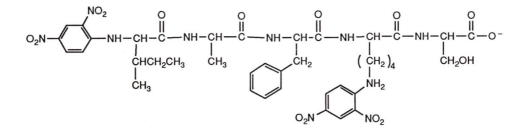

Hydrolysis of the 2,4-DNFB derivative of the pentapeptide produces the following five products:

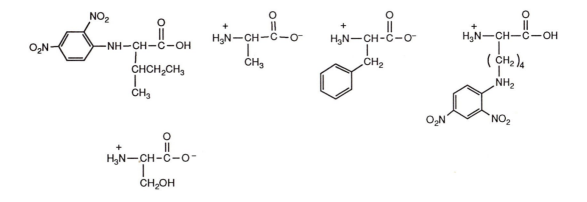

The amino acid that was at the *N*-terminus is the only one that does not exist as a zwitterion. Lysine reacts with DNFB at its side-chain amino group, which will have different properties that the derivative with the DNFB group attached at the α-amino group.

25.23.

a. Acetic anhydride acetylates an amino group, forming the corresponding acetamide derivative.

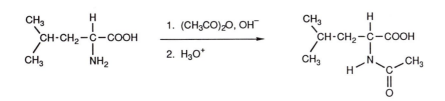

25.23. (continued)

b. The Strecker synthesis provides a way to make racemic amino acids. The carbon atom of the aldehyde group in the starting material becomes the alpha carbon atom of the amino acid.

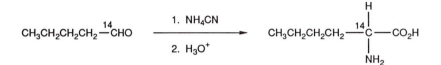

c. The (*S*,*S*)-DuPHOS catalyst forms an (*S*)-amino acid from the corresponding enamide.

d. The first step in this reaction sequence yields the tosylate derivative of the 1° alcohol. DBU is a base that promotes elimination (text page 336), leading to formation of the unsaturated amino acid in its protected form.

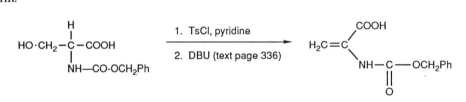

e. NaH is a strong base, so it reacts to deprotonate the phenol OH group. The phenolic anion is a nucleophile, and it reacts with bromoethane via an S_N2 pathway, forming the *O*-ethyl derivative of tyrosine.

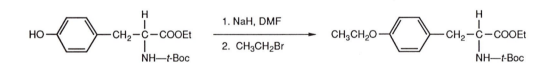

f. The first step is the given reaction sequence produces an (*S*)-amino acid in its protected form. The second step is a Suzuki reaction (text page 604), which couples the phenyl group of the boronate compound to the aryl ring of the amino acid, replacing the bromine atom with a phenyl group.

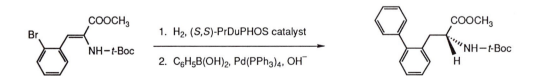

25.23. (continued)

g. An amino acid ester reacts the same way with phenylisothiocyanate that any amine does. The product is the *N*-phenylhydantoin.

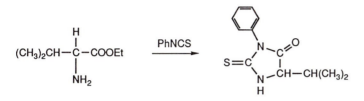

g. Alkylation of the chiral heterocycle occurs as its enolate derivative from below the plane of the ring, as illustrated on text page 1281. Treating that product with lithium in liquid ammonia cleaves the diphenylethane group, forming the protected chiral amino acid.

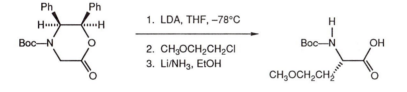

25.24. Any of the methods described in chapter 25 can be used to prepare the illustrated amino acid because the functional groups that appears in their side chains are inert to most reagents. Shown below are routes that make use of the chiral glycine enolate.

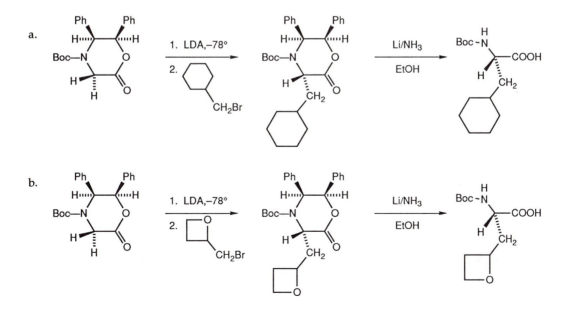

25.25. The first steps needed in the preparation of ala-gly-ile are those to prepare protected forms of the amino acids. Isoleucine, which is at the C-terminus, is converted to its methyl ester; the other two are converted to their Boc derivatives.

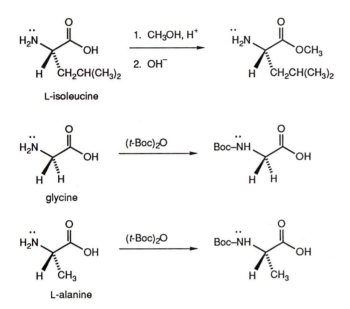

In the first step of peptide synthesis, glycine and isoleucine are coupled with use of DCC.

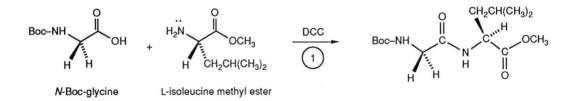

Then, the amino terminus is deprotected by reaction with trifluoroacetic acid.

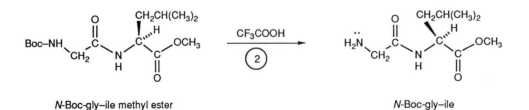

25.25. (continued)

Finally, alanine is coupled to the new *N*-terminus with use of DCC.

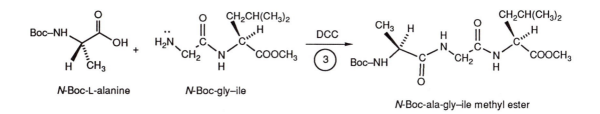

If the free tripeptide is required, the Boc group is removed by treatment with CF_3COOH and the methyl ester is hydrolyzed with weak aqueous base.

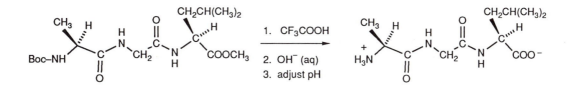

25.26. Follow the procedure outlined in the solution to exercise 25.16. Aligning the fragments reveals the sequence of the 12-residue polypeptide.

```
                 ala-ile-tyr-his-val

                     tyr-his-val

ala-phe-ser          ile-tyr-his

ala-phe          ala-ile-tyr-his

        ser-ser-ala-ile              phe-gly

      phe-ser-ser-ala-ile            his-phe-gly

ala-phe-ser                     val–his-phe-gly

ala-phe-ser-ser
```

ala-phe-ser-ser-ala-ile-tyr-his-val-his-phe-gly

25.27. The sequence of the dipeptide aspartame has aspartic acid at the *N*-terminus and phenylalanine at the *C*-terminus. The carboxy group of the phenylalanine residue is present as its methyl ester. At pH 7, the amino group is in its conjugate acid form, and the carboxy group in the side chain of aspartic acid is in its conjugate base form. There are no other acidic or basic groups in the molecule.

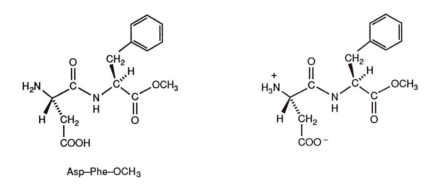

Asp–Phe–OCH₃

25.28. If the isolated chains of insulin are treated with trypsin, the A chain is not cleaved because it contains no lys or arg residues. The B chain is cleaved in two places, on the carboxyl side of lys and arg.

A chain:

$\overset{+}{H_3N}$—gly–ile–val–glu–gln–cys–cys–ala–ser–val–cys–ser–leu–tyr–gln–leu–glu–asn–tyr–cys–asn—COO⁻

B chain:

—cys–gly–ser–his–leu–val–glu–ala–leu–tyr–leu–val–cys—

$\overset{+}{H_3N}$—phe–val–asn–gln–his–leu⤳ ⤳gly–glu–glu–arg–gly–phe–phe–tyr–thr–pro–lys–ala—COO⁻

↑ ↑
T T

25.29. The polypeptide shown in this exercise has a very potent nucleophile—the side chain thiol group of the cysteine residue. In the first step, aziridine serves as a base, deprotonating the SH group. The thiolate ion then reacts with the protonated aziridine in step (2), opening the ring to make the *S*-(2-aminoethyl) derivative.

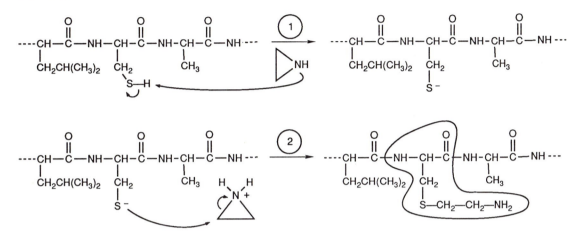

The structure of this derivatized peptide (circled above) is similar to that of an analogous peptide that has a lysine residue in the same position (circled below). The length of a lysine side chain is approximately the same as that of the derivatized cysteine residue (the only difference is the radius of a methylene carbon atom versus that of a sulfur atom). Because trypsin cleaves at the carboxy side of a lysine residue, it reacts with the derivatized polypeptide shown above.

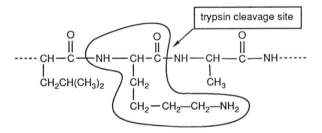

25.30. Follow the procedure outlined in the solution to exercise 25.16. Aligning the sequences of the shorter polypeptides and placing phenylalanine at the *N*-terminus, we write the sequence of the 18-mer as follows:

phe-ser-ala-lys

phe-ser-ala-lys-ala-glu-asn-glu-phe-trp-ser-hse

phe-ser-ala-lys-ala-glu-asn-glu-phe-trp-ser-met-gly-arg

ala-glu-asn-glu-phe-trp-ser-met-gly-arg-asn-ala-val-ile

gly-arg-asn-ala-val-ile

asn-ala-val-ile

H_2N—phe-ser-ala-lys-ala-glu-asn-glu-phe-trp-ser-met-gly-arg-asn-ala-val-ile—COOH

25.31. γ-Carboxyglutamic acid is like any compound that has two carboxylic acid groups attached to a single carbon atom—it is inherently unstable. A carboxylic acid group β to another carbonyl group is prone to undergo decarboxylation (text page 1016). During amino acid analysis of proteins, hydrolysis of γ-carboxyglutamic acid under acidic conditions is accompanied by decarboxylation.

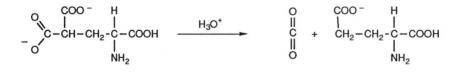

25.32. In the synthesis of γ-carboxyglutamic acid that starts with the optically active iodide analog derived from serine, malonate ion causes elimination, which forms the dehydro amino acid. Then, malonate ion undergoes Michael addition to generate the protected form of γ-carboxyglutamic acid. Elimination is facilitated by the presence of the ester group, which makes the proton on its alpha carbon atom more acidic than the corresponding proton of an alkyl iodide. The presence of the ester group likewise makes addition to the π bond possible, which makes the overall carbon-carbon bond forming reaction successful. Racemization occurs, however, because the dehydro amino acid is achiral.

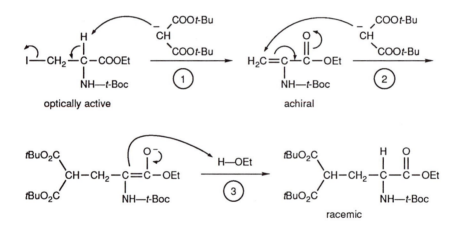

25.33. At pH 7, amino groups at the ε position of the side chain of lysine exist in their protonated form. These ammonium ions disrupt the hydrogen bonds that stabilize an α helix. At pH 10, the amino groups are converted to their free-base form, and hydrogen bonds reform to create the stable α helical structure.

NUCLEIC ACIDS AND
MOLECULAR RECOGNITION

26.1. The names of the given bases, nucleosides, and nucleotides are interpreted by comparing them to structures shown in Tables 26.1 and 26.2.

8-mercaptoadenine

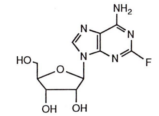

2-fluoroadenosine

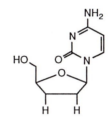

2',3'-dideoxycytidine

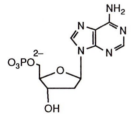

2'-deoxyadenosine-5-phosphate

5-trifluoromethyluracil

26.2. LCAO representations of pyrimidine and purine are drawn according to the procedure outlined in the solution to exercise 24.15. Both unshared electron pairs in pyrimidine are in the plane of the ring.

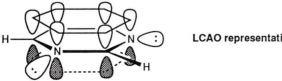

LCAO representation for pyrimidine

26.2.   (continued)

The LCAO representation of purine shows that three of the four pairs of electrons are coplanar with the sigma bonds. Only the unshared electron pair in the 5-membered ring of purine is part of the π system.

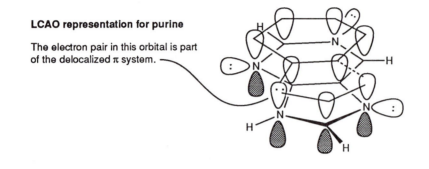

LCAO representation for purine

The electron pair in this orbital is part of the delocalized π system.

26.3.   Tautomeric forms of the given heterocycles are generated by isomerizing each amide group, moving the hydrogen atom from nitrogen to oxygen.

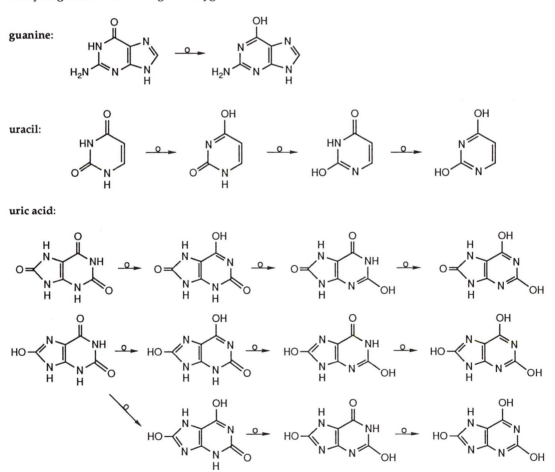

guanine:

uracil:

uric acid:

26.4. The synthesis of a pyrophosphate ester requires two steps. The alcohol first reacts at the middle phosphorus atom of the triphosphate group of ATP, creating a trigonal-bipyramidal, anionic intermediate. The trigonal-bipyramidal intermediate collapses in step (2) to form the pyrophosphate derivative and AMP.

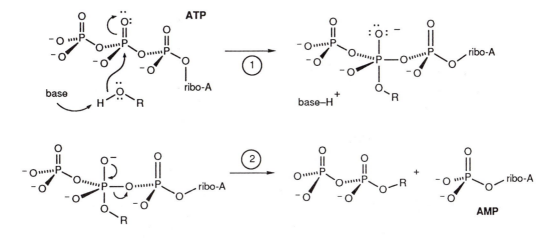

26.5. The structure of the given tetranucleotide is drawn so that the letter at the left end of the sequence corresponds to the 5′ end of the molecule, which has a phosphate group attached. The letter T tells you that this molecule is a sequence of DNA, so each sugar ring has an O atom attached at every carbon except C2. A phosphate group connects each nucleotide 3′-OH group with the 5′-OH group of its neighbor. The 3′-OH group of the nucleotide at the right end of the sequence has no phosphate group attached.

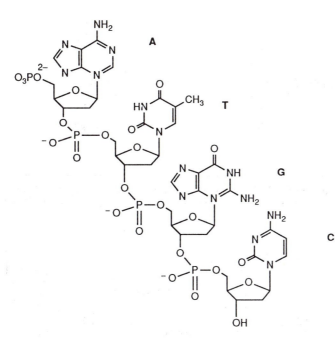

26.6. A DMTr group is removed by treating the derivatized alcohol with trifluoroacetic acid, which initiates an S_N1 reaction by protonating the alcohol oxygen atom. The alcohol molecule dissociates and the dimethoxytrityl carbocation is intercepted by trifluoroacetate ion.

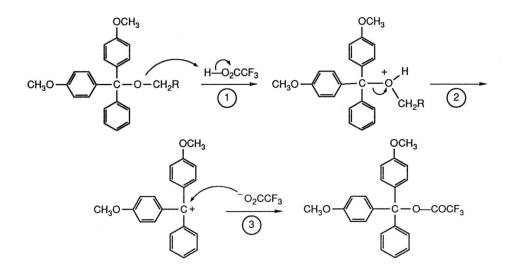

The dimethoxytrityl group forms a particularly stable carbocation because the positive charge is delocalized among the atoms in three rings, two of which have a methoxy group. There are many resonance forms (at least 17 good ones!) that can be drawn, but three particularly important ones are shown below:

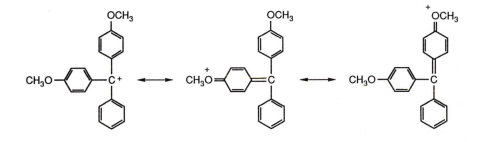

26.7. The synthesis of the tetranucleotide ACTT follows the steps illustrated on text pages 1332-1334. First, the nucleotide that appears at the 3' end of the sequence is attached to the solid support, and the 5'-OH group is deprotected.

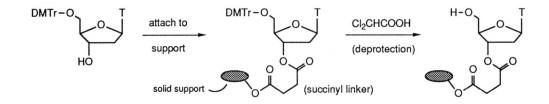

26.7. (continued)

Then, the second nucleotide is coupled, and the protecting group at the 5' end is removed. The other two nucleotides are attached by the same methodology, and the final product is deprotected.

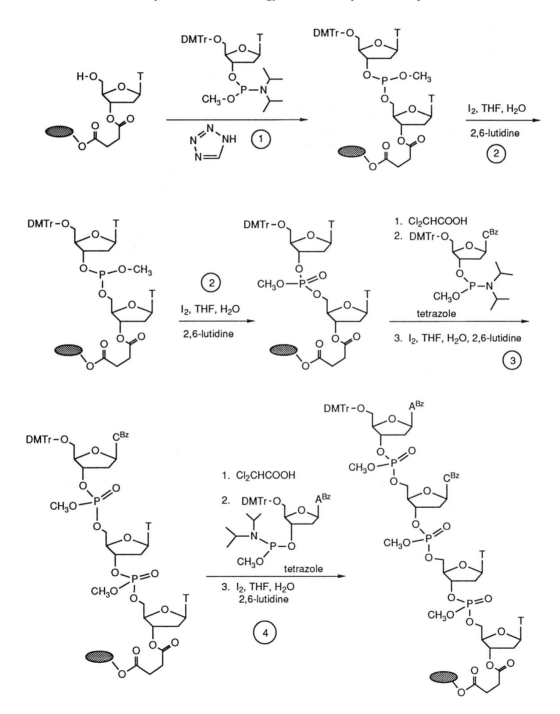

26.7.   (continued)

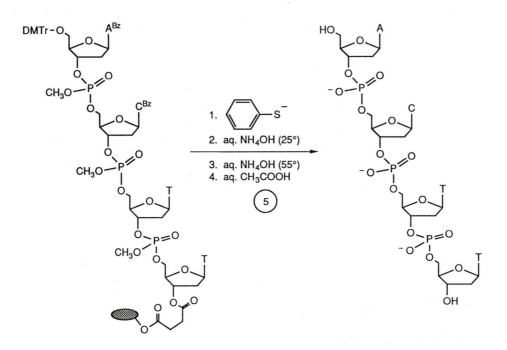

26.8.   The synthesis of dibenzo-18-crown-6 is done in a stepwise fashion, but simply mixing the catechol dianion with the ditosylate derivative of bis(2-hydroxyethyl)ether produces some of the desired compound, in addition to oligomers. Using the dipotassium salt of the catechol assists the synthesis of the crown ether product because of a template effect.

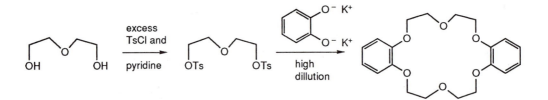

26.9. The synthesis of the illustrated pyridino-crown proceeds in two stages. First, diol **A** is prepared by alkylating bis(2-hydroxyethyl) ether. Then, lutidine is brominated by a radical process at its two benzylic positions with use of NBS. Dibromolutidine reacts with the dianion of **A** under high dilution conditions and with the template effect of potassium ion to form the crown ether.

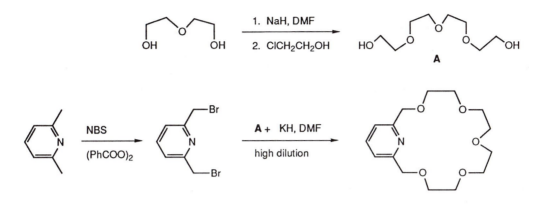

Binding strength is maximized when three hydrogen bonds form between the ammonium salt and the crown ether, so the binding constant is expected to decrease in the following order for the alkylated ammonium salts:

three H-bonds formed   two H-bonds formed   one H-bond formed

26.10. The synthesis of cyclotriveratrylene occurs via three consecutive Friedel-Crafts alkylation reactions between an arene and carbocation intermediates. The transformation begins with protonation of the benzylic alcohol, which forms a carbocation that is trapped by reaction with a second ring.

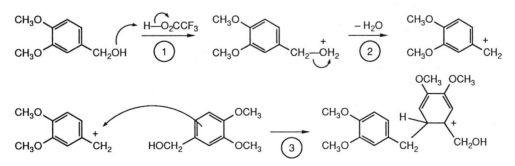

26.10. (continued)

After rearomatization in step (4), a second carbocation is formed, which is intercepted by a third ring in step (7).

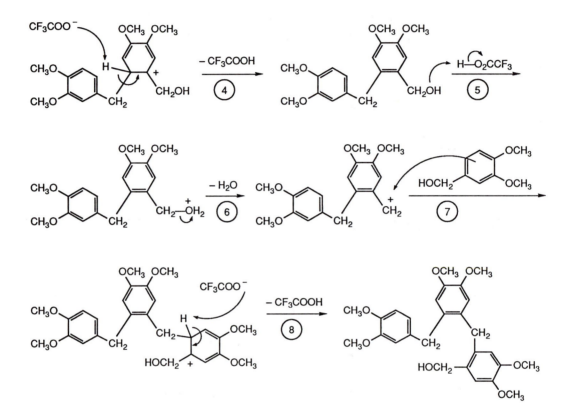

Finally, a third carbocation is formed, which is trapped in an intramolecular fashion in step (11) to form the tricyclic product.

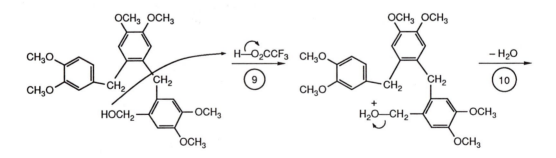

26.10.   (continued)

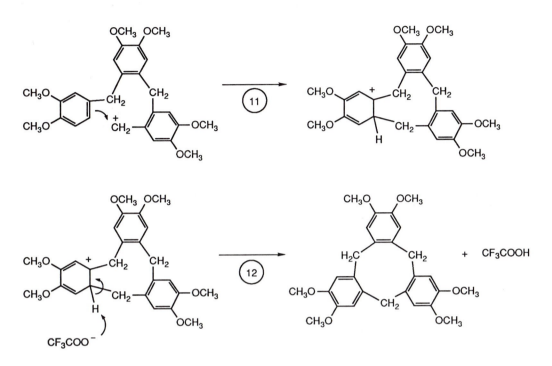

26.11. Preparation of a urea derivative from an amine and an isocyanate takes place in two stages. First, the nitrogen atom of the amine adds to the carbon-oxygen double bond of the isocyanate to form a zwitterion intermediate. In the second step, an acid–base reaction occurs as the carbonyl group is regenerated.

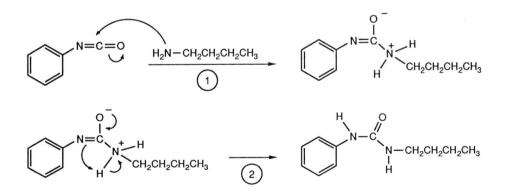

26.12. An alkylphosphate ion forms strong hydrogen bonds with urea. Both hydrogen bonding and electrostatic attraction are important in the interaction between an alkylphosphate ion and a guanidinium ion.

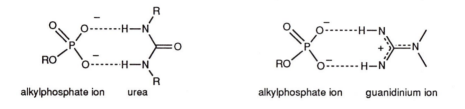

26.13. Receptor **A** binds to both the *syn* and *anti* electron pairs on each oxygen atom of the acetate ion. Dimethylurea forms hydrogen bonds only to two unshared electron pairs of the acetate ion;

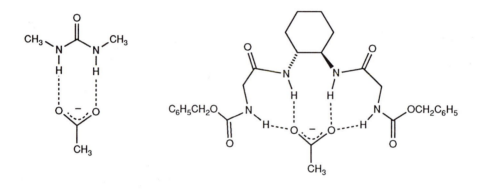

26.14. Pyridine and pyrazine, respectively, form one and two hydrogen bonds to convergent carboxylic acid groups of receptor **3**.

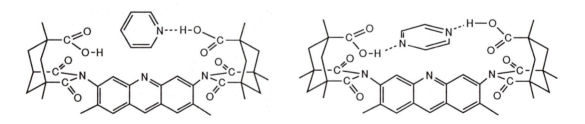

26.14.   (continued)

Purine forms three hydrogen bonds to the carboxylic acid groups of receptor **3**, or it forms hydrogen bonds to two carboxylic acid groups and one hydrogen bond to the central nitrogen atom.

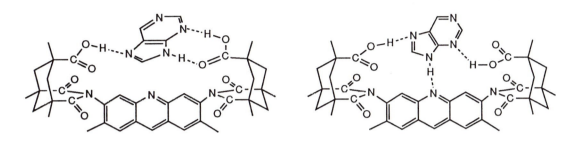

26.15.   The hydrogen bonding pattern between the receptor illustrated in this exercise and thymine is apparent if you look at the groups involved in hydrogen bonding: the receptor has two N–H groups with a N atom (acceptor) in between; thymine has one donor (N–H) flanked by two acceptor groups (the carbonyl oxygen atoms).

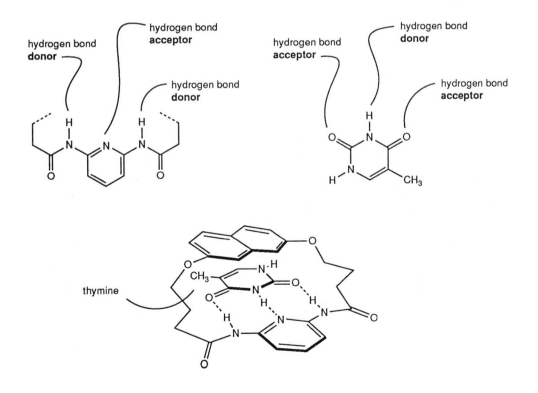

26.15. (continued)

Synthesis of the receptor is accomplished by reaction between 2,6-diaminopyridine and δ-butyrolactone. This transformation forms the two amide bonds and leaves alcohol OH groups at the ends of the chains. These OH groups are converted to good leaving groups (the bis-tosylate derivative), which are displaced by the nucleophilic dianion derived from 2,7-dihydroxynaphthalene.

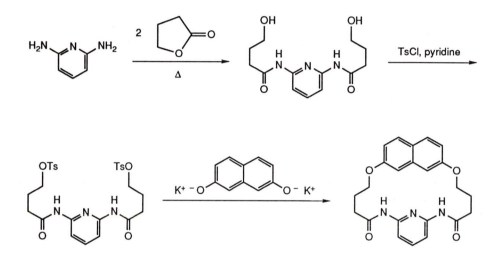

26.16. In the first step of self-replication, each reactant is bound by formation of complementary hydrogen bonds with the appropriate group of the template molecule.

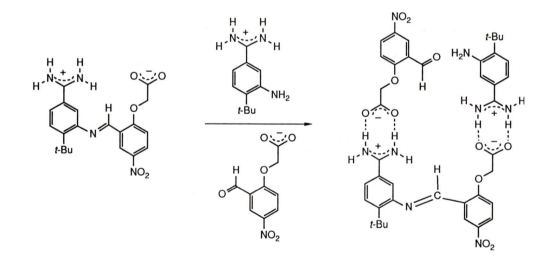

26.16. (continued)

In the next step, the amino group of one reactant adds to the aldehyde carbonyl group of the other. By the usual mechanism (see text pages 870-872), the imine bond is formed. This process results in formation of another molecule of template. Each template molecule binds the reactants, leading to replication of more template.

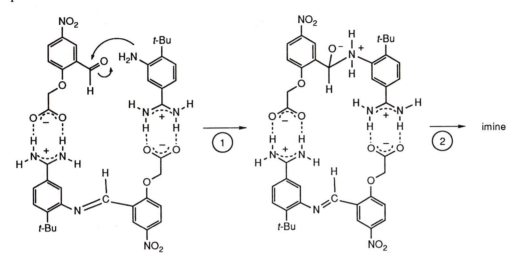

26.17. The names of the given bases, nucleosides, and nucleotides are interpreted by comparing them to structures shown in Tables 26.1 and 26.2 on text pages 1320 and 1321, respectively. Nomenclature of crown ethers is shown on text page 1346 and for ureas on text page 1360 (cf. exercise 26.11).

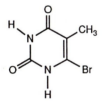

6-bromothymine

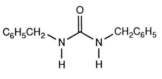

N,N'-dibenzylurea

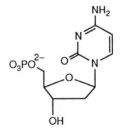

2'-deoxycytidine-5'-monophosphate

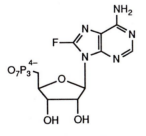

8-fluoroadenosine-5'-triphosphate

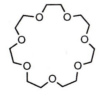

21-crown-7

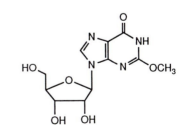

2-methoxyinosine

26.18. The names of the given bases, nucleosides, and nucleotides are interpreted by comparing them to structures shown in Tables 26.1 and 26.2 on text pages 1320 and 1321, respectively. Note use of the prefix *"epi"* in part (e), which indicates that the normal configuration of the 2' position is inverted.

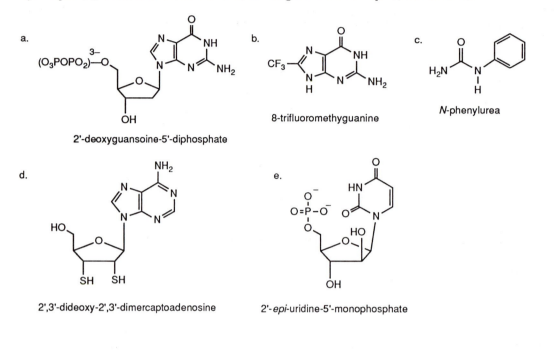

a.

2'-deoxyguansoine-5'-diphosphate

b.

8-trifluoromethyguanine

c.

N-phenylurea

d.

2',3'-dideoxy-2',3'-dimercaptoadenosine

e.

2'-*epi*-uridine-5'-monophosphate

26.19. Follow the procedure outlined in the solution to exercise 26.5.

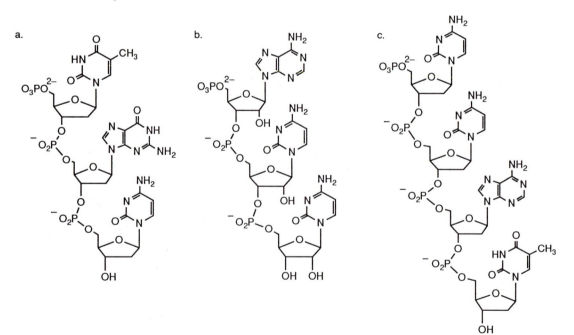

a.

b.

c.

26.20. The synthesis of the trinucleotide follows the steps illustrated on text pages 1332-1334.

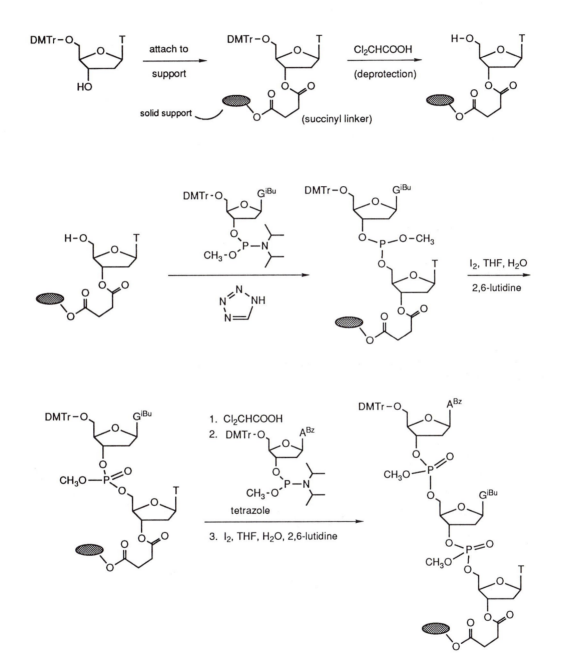

26.20. (continued)

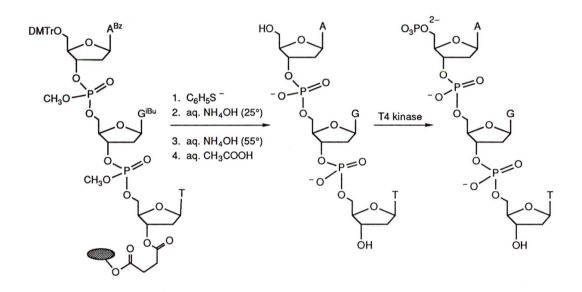

26.21. Inosine has both hydrogen bond donor and acceptor groups. These groups match complimentary substructures in A, C, or U.

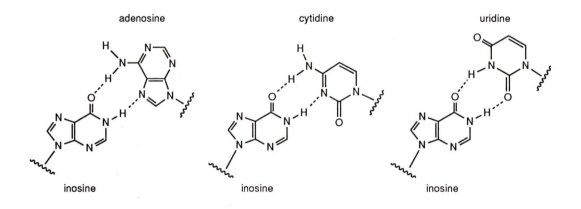

26.22. Pseudouridine forms hydrogen bonds with adenosine in the same way that uridine does. Attaching the heterocycle to the ribose ring through a carbon atom creates two lactam functional groups, so two different orientations are possible for a strand of RNA [indicated by (R), below] that contains pseudouridine.

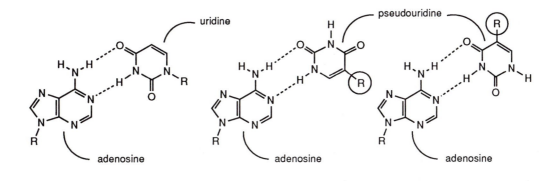

26.23. Because AZT has no OH group attached at C3 of the deoxyribose ring, no nucleophile is present to react with the 5'-triphosphate group of a growing polynucleotide chain.

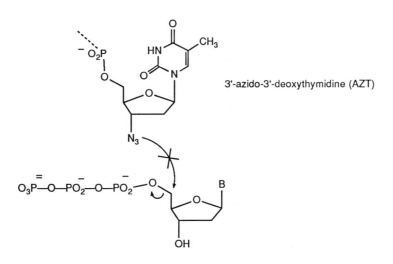

3'-azido-3'-deoxythymidine (AZT)

26.24. 5-Bromouracil has a structure similar to that of thymine, except that a bromine atom is present in place of the methyl group in T. The nitrogen atom attached to the anomeric carbon atom of the sugar ring requires that the adjacent oxygen-containing functional group exists in the carbonyl form. The NH and its neighboring carbonyl group (circled, below) defines the portion of the heterocycle that tautomerizes. The structures of the lactam and lactim form are shown below:

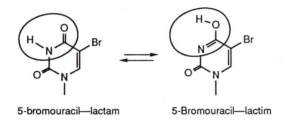

5-bromouracil—lactam   5-Bromouracil—lactim

If we place the two forms of 5-bromouracil in equivalent positions relative to the bases of A and G, we see that the lactam tautomer forms a hydrogen bond with A in the same manner that T does. The lactim tautomer forms three hydrogen bonds with G.

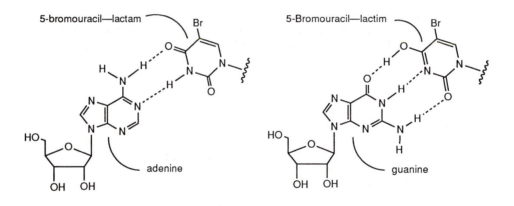

26.25. 7,9-Dimethylguanine exists as a zwitterion in its neutral form. Protonation occurs at the negatively charged oxygen atom, and tautomerism generates the carbonyl group, which is the preferred structural form of the protonated material.

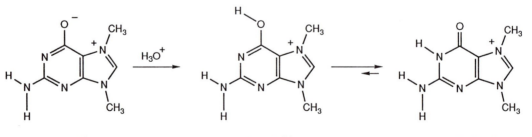

neutral form   protonated form   protonated tautomer

26.25.   (continued)

Hydrogen bonding between the neutral and tautomeric protonated forms of 7,9-Dimethylguanine occurs as follows:

protonated tautomer

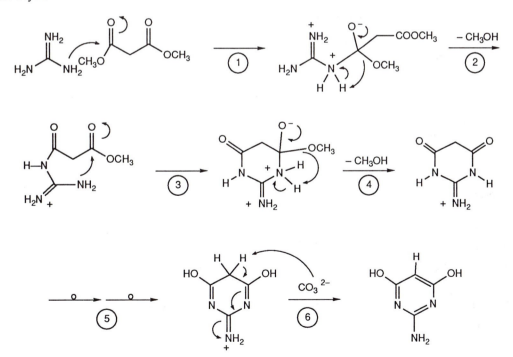

neutral form

26.26.   The condensation reaction between guanidinium carbonate and dimethyl malonate occurs via two addition–elimination sequences, in which the amino groups of the guanidinium ion react with the ester groups of dimethyl malonate. The initially formed cyclic product tautomerizes twice in step (5), and carbonate ion acts as a base to remove the proton from the tetrahedral carbon atom to form the aromatic heterocycle.

26.27.   Synthetic routes to prepare crown ethers take advantage of procedures involving S_N2 reactions between nucleophilic oxyanions (alkoxide or phenoxide ions) and bissulfonate ester derivatives of diols.

a. *retrosynthesis*:

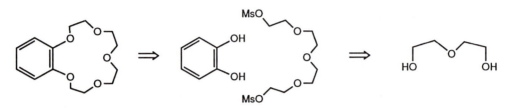

synthesis: Bis(2-hydroxyethyl)ether is alkylated at each end with use of a protected form of chloroethanol. After deprotection, the diol is converted to its dimesylate derivative, which is treated with the dianion of catechol. Sodium ion provides the needed template to ensure formation of the macrocycle.

b. *retrosynthesis*:

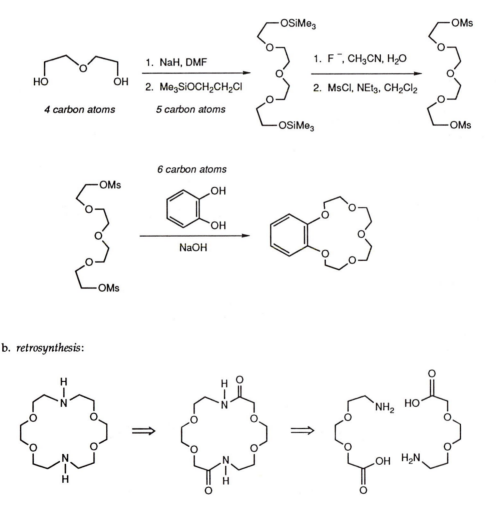

26.27.   (continued)

synthesis: DCC is used to couple two equivalents of the amino acid revealed by retrosynthetic analysis. Under high dilution conditions, this dimerization reaction produces the macrocycle. Reduction of the amide groups to form amino groups is accomplished with use of lithium aluminum hydride.

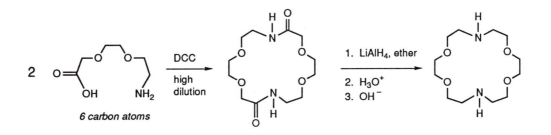

c. *retrosynthesis*:

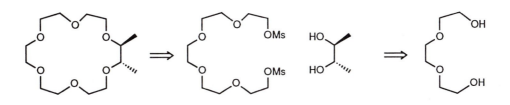

synthesis: The chiral diol is made by catalytic asymmetric dihydroxylation of *trans*-2-butene (text page 784), then the remainder of the carbon skeleton is constructed as in part (a) starting with bis(2-hydroxy-1-ethoxy)ethane. Coupling these two pieces under high dilution conditions makes use of the template effect of potassium ion to form this chiral 18-crown-6 analog.

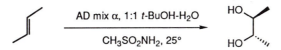

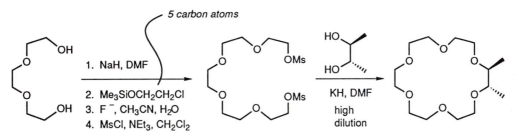

26.28. To prepare guanidine derivatives bearing an alkyl group at N^9, the nitrogen atom of the starting heterocycle is first deprotonated, and the resulting anion is used as a nucleophile in a reaction with a bromoalkane. This step alkylates the nitrogen atom.

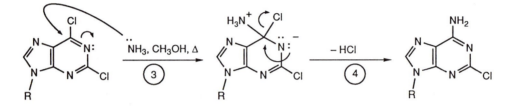

In the next stage, ammonia adds to the heterocyclic ring, which rearomatizes, displacing chloride ion.

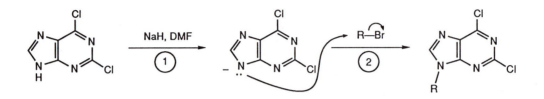

The next stage is diazotization (not every step of electron movement is shown here; see text page 1171 for additional details), which generates the diazo group.

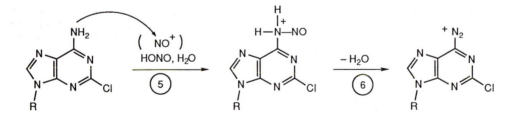

Then a molecule of water undergoes addition/elimination to the C=N bond. These two steps lead to replacement of dinitrogen by a hydroxyl group.

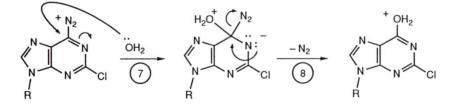

26.28. (continued)

Tautomerism of the lactim generates the lactam form of the heterocycle.

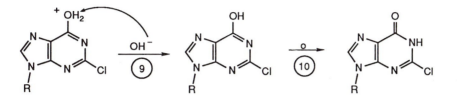

Finally, ammonia adds to the C=N bond of the heterocyclic ring a second time. Rearomatization occurs with concomitant displacement of chloride ion.

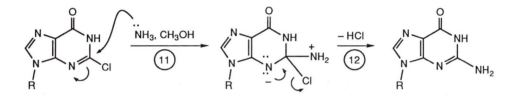

26.29. The three zwitterionic forms of receptor **A** are formed by transferring a proton from the phenolic OH group to one of the nitrogen atoms.

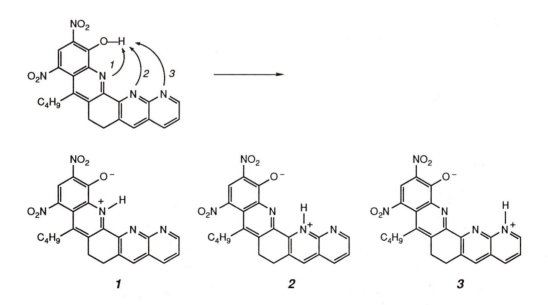

26.29. (continued)

Zwitterionic form 3, which has the proton attached to nitrogen farthest from the phenolic oxygen atom, forms a complex with creatinine that is stabilized by three hydrogen bonds.

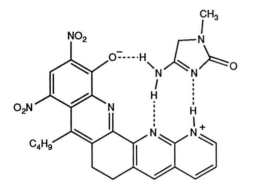

26.30. Both receptors shown in this exercise form the same number and types of hydrogen bonds, but receptor **Y** also makes use of π stacking between the naphthalene and pyrimidine rings in the recognition process. This stacking interaction is strengthened by attraction between pyrimidine's electron deficient π system and the electron rich π network of dihydroxynaphthalene.

X · pyrimidine

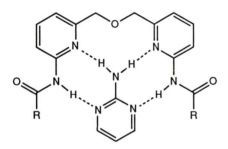

Y · pyrimidine

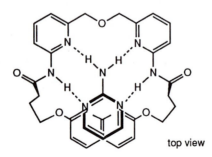

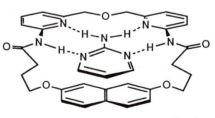

top view   side view

26.31. The synthesis of receptor **X** makes use of amide formation between amino groups of the heterocycle and the carbonyl group derived of an acid chloride.

retrosynthesis:

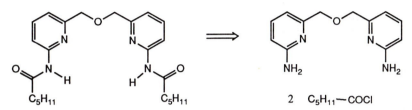

synthesis: 6-Chloro-2-methylpyridine is brominated at the benzylic carbon atom with use of NBS. Hydrolysis of the bromide forms the corresponding alcohol.

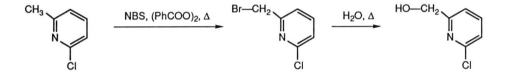

Williamson ether synthesis utilizes the reaction between the salt of the benzylic alcohol and the benzylic bromide analog, which produces the dipyridine compound.

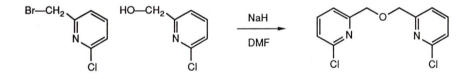

The chlorine atoms are replaced by reaction with NH_2^- to form the diamine.

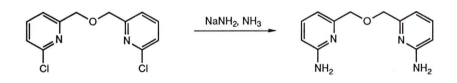

Finally, the acid chloride reacts with the diamine to form **X**.

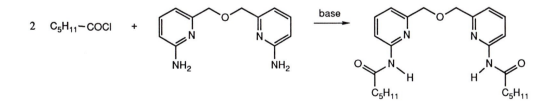

26.32. Both receptors shown in this exercise bind dihydrogen phosphate ion by taking advantage of the fact that the OH groups of $H_2PO_4^-$ act as hydrogen bond donors to urea carbonyl groups, as shown below on the left.

The NH portions of the urea groups also act as hydrogen bond donors, but the geometry required for this other mode of binding means that the three acceptor groups have to be in the same plane, so one oxygen atom of $H_2PO_4^-$ has to form two hydrogen bonds, as shown at the right, below.

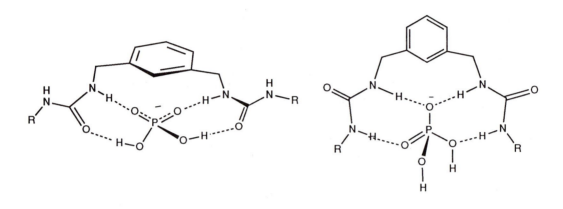

26.33. Derivatives of urea are commonly made by the reaction between an amine and an isocyanate. The receptor shown in this exercise is made by performing a free radical bromination of *m*-xylene, which produces the α,α′-dibromo derivative. Reaction of this dibromide with azide ion, followed by reduction, yields the diamine. The diamine reacts with two equivalents of butyl isocyanate to form the corresponding urea.

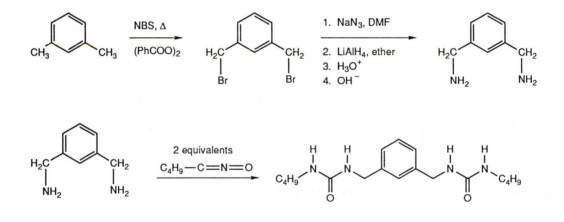

NAMING ORGANIC COMPOUNDS

Follow the procedure outlined in the examples shown in Appendix A.

a.

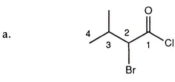

functional group:	acid chloride; suffix = -oyl chloride
longest carbon chain:	4C: but-
unsaturation suffix:	an
substituents:	bromine atom at C2
	methyl group at C3

2-bromo-3-methylbutanoyl chloride

b.

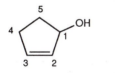

functional group:	alcohol; suffix = -ol at C1
ring size:	5C: cyclopent-
unsaturation suffix:	an, at C2
substituents:	none

2-cyclopentene-1-ol

c.

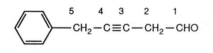

functional group:	aldehyde; suffix = -al
longest carbon chain:	5C: pent-
unsaturation suffix:	yne, at C3
substituent:	phenyl group at C5

5-phenyl-3-pentynal

d.

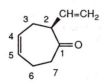

functional group:	ketone; suffix = -one
ring size:	7C: cyclohept-
unsaturation suffix:	ene, at C4
substituent:	vinyl group at C2
stereochemistry: (stereogenic carbon atom)	(*S*)

(*S*)-2-vinyl-4-heptenone

(continued)

e.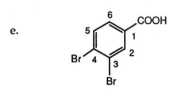

parent compound: benzoic acid
substituents: two bromine atoms, one at C3
    and one at C4

3,4-dibromobenzoic acid

f.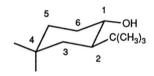

functional group: alcohol; suffix = -ol at C1
ring size: 6C: cyclohex-
unsaturation suffix: ane
substituents: *t*-butyl group at C2
    two methyl groups at C4
stereochemistry: *t*-butyl is *trans* to the OH group
  (substituent orientation)

***trans*-2-*tert*-butyl-4,4-dimethylcyclohexanol**

g.

functional group: carboxylic acid; suffix = -oic acid
longest carbon chain: 5C: pent-
unsaturation suffix: ene, at C3
substituent: methyl group at C3
stereochemistry: (*E*)
  (double bond geometry)

(*E*)-3-methyl-3-pentenoic acid

h.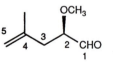

functional group: ketone; suffix = -one
ring size: 6C: cyclohex-
unsaturation suffix: ane
substituent: bromine atom at C2
stereochemistry: (*S*)
  (stereogenic carbon atom)

(*S*)-2-bromocyclohexanone

i.

functional group: aldehyde; suffix = -al
longest carbon chain: 5C: pent-
unsaturation suffix: ene, at C4
substituents: methoxy group at C2
    methyl group at C4
stereochemistry: (*R*)
  (stereogenic carbon atom)

(*R*)-2-methoxy-4-methyl-4-pentenal

(continued)

j.

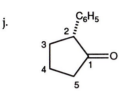

functional group:	ketone; suffix = -one
ring size:	5C: cyclopent-
unsaturation suffix:	ane
substituent:	phenyl group at C2
stereochemistry:	(*R*)
(stereogenic carbon atom)	

(*R*)-2-phenylcyclopentanone

k.

functional group:	none
longest carbon chain:	5C: pent-
unsaturation suffix:	ene, at C2
substituents:	methyl groups at C2, C3, and C4

2,3,4-trimethyl-2-pentene

l.

parent compound:	benzoic acid
substituents:	allyl group at C6
	hydroxy group at C2

6-allyl-2-hydroxybenzoic acid

m.

functional group:	alcohol; suffix = -ol at C1
longest carbon chain:	5C: pent-
unsaturation suffix:	ene, at C2
substituent:	methyl group at C3
stereochemistry:	(*Z*)
(double bond geometry)	

(*Z*)-3-methyl-2-pentene-1-ol

n.

functional group:	none
ring size:	4C: cyclobut
unsaturation suffix:	ene, at C1
substituent:	*N*-benzylamino, at C3

3-(*N*-benzylamino)cyclobutene

(continued)

o.

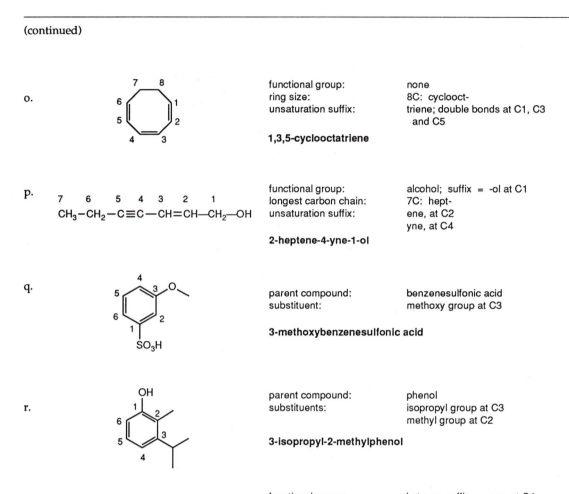

functional group: none
ring size: 8C: cyclooct-
unsaturation suffix: triene; double bonds at C1, C3
and C5

1,3,5-cyclooctatriene

p.

7   6   5   4   3   2   1
$CH_3-CH_2-C\equiv C-CH=CH-CH_2-OH$

functional group: alcohol; suffix = -ol at C1
longest carbon chain: 7C: hept-
unsaturation suffix: ene, at C2
yne, at C4

2-heptene-4-yne-1-ol

q.

parent compound: benzenesulfonic acid
substituent: methoxy group at C3

3-methoxybenzenesulfonic acid

r.

parent compound: phenol
substituents: isopropyl group at C3
methyl group at C2

3-isopropyl-2-methylphenol

s.

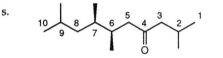

functional group: ketone; suffix = -one at C4
longest carbon chain: 10C: dec-
unsaturation suffix: ane
substituents: methyl groups at C2, C6, C7,
and C9
stereochemistry: (*R*) at C6, (*R*) at C7
(stereogenic carbon atoms)

(6*R*, 7*R*)-2,6,7,9-tetramethyl-4-decanone